T0399631

Environmental Engineering

Environmental Engineering

Fundamentals and Applications

Subhash Verma
Varinder S. Kanwar
Siby John

CRC Press is an imprint of the
Taylor & Francis Group, an **informa** business

First edition published 2022
by CRC Press
6000 Broken Sound Parkway NW, Suite 300, Boca Raton, FL 33487-2742

and by CRC Press
2 Park Square, Milton Park, Abingdon, Oxon, OX14 4RN

First edition published by CRC Press 2022

CRC Press is an imprint of Taylor & Francis Group, LLC

Library of Congress Cataloging-in-Publication Data
Names: Verma, Subhash (Professor), author. | S. Kanwar, Varinder, author. | John, Siby, author.
Title: Environmental engineering: fundamentals and applications/Subhash Verma, Varinder S. Kanwar, Siby John.
Description: First edition. | Boca Raton : CRC Press, 2022. | Includes bibliographical references and index. |
Summary: "Presenting an in-depth coverage, this textbook brings together and integrates key topics including water resources, wastewater, air, and solid waste in a single volume. The textbook introduces a unique approach that emphases on the water and wastewater treatments with its distribution system and engineering. It begins by discussing the public health and sanitation, then covers the wastewater collection system and design, wastewater characteristics, natural purification water, different wastewater treatments, industrial and rural wastewater. Finally, the emerging technologies in the reuse/recycle of waste and processes to conserve the environmental resources are discussed. The text will be useful for senior undergraduate and graduate students in the fields of civil and environmental engineering. Pedagogical features including solved problems, exercises and multiple-choice questions are interspersed throughout the book for better understanding"– Provided by publisher.
Identifiers: LCCN 2021040180 (print) | LCCN 2021040181 (ebook) |
ISBN 9780367750503 (hardback) | ISBN 9781003231264 (ebook)
Subjects: LCSH: Environmental engineering.
Classification: LCC TA170 .V47 2022 (print) | LCC TA170 (ebook) | DDC 628–dc23
LC record available at https://lccn.loc.gov/2021040180
LC ebook record available at https://lccn.loc.gov/2021040181

ISBN: 9780367750503 (hbk)
ISBN: 9781032138824 (pbk)
ISBN: 9781003231264 (ebk)

DOI: 10.1201/9781003231264

Typeset in Times
by Newgen Publishing UK

Contents

List of Figures

List of Tables

Preface

One of the major realizations of the 20th century is that natural resources do not last forever. Since the concept of sustainable development came into existence, the public and its leaders have been unable to ignore the catastrophic effects of depleting natural resources, festering waste dumps and subsequent pollution on the planet. This has led to the development of an interdisciplinary academic practice of environmental engineering.

Environmental Engineering covers a range of topics where engineering can be applied to conserve natural resources and preserve or restore environmental quality. It offers basic concepts and applications in the multidisciplinary subject of Environmental Engineering and Technology and is primarily designed for students in the civil, water resources and environmental engineering fields. Being multidisciplinary, a basic knowledge in chemistry, hydraulics and biology is assumed.

In an urban cycle, water is treated and distributed to meet various demands. The wastewater or sewage generated by homes and businesses is collected and treated before returning to the natural environment. Topics on water in this book follow the urban water cycle and fall under two main categories: water treatment and distribution; and wastewater collection, treatment and disposal. Other main subject areas included are air and noise pollution, and solid and hazardous waste management.

There is ample material in this book for a typical one-semester course. Traditionally, one-semester courses are offered in Water Supply Engineering, Wastewater Engineering, Air Pollution and Noise Control, and Solid and Hazardous Solid Management. Most of the chapters in this book are independent of each other – this provides flexibility to choose combinations of topics that suit the requirements of a given course.

The qualities that distinguish this book are its clear, easy-to-read style and use of unit cancellation in solving problems. Examples and diagrams are used throughout to illustrate and clarify important topics. Numerous practice problems with answers and discussion topics follow each chapter.

This introductory textbook addresses a wide range of environmental topics that are covered in greater detail and depth in more narrowly specialized and advanced texts. They are presented here in a form and level that is more accessible to students studying the subject for the first time. Every effort has been made to maintain a balance between the theoretical and the practical when covering each topic. This book is equally useful for practising engineers who want to review the basics when looking for information on environmental topics. A plethora of example problems have been chosen to illustrate basic concepts and the application of concepts in the design and solution of real-world problems.

We sincerely hope this book will enhance student learning and motivate and prepare readers to study environmental engineering and technology at a higher level. We hope it will provide a great resource for students to refresh themselves with the basics when taking graduate courses in the environmental field or preparing for professional examinations.

Environmental Engineering is a unique book that covers all of the emerging topics in environmental engineering, both in developed and developing nations. The authors sincerely hope it will be a valuable resource for the students, academicians, researchers, scientists, NGOs and anyone else working in the fields of environmental technology and engineering.

About the Authors

Prof Subhash Verma

Subhash Verma retired as a Professor from the Sault College of Applied Arts and Technology in Ontario, Canada, where he headed a programme in Water Resources Engineering Technology. During his teaching career at the college, Subhash taught courses including Hydraulics, Hydrology, and Water and Wastewater Engineering, and was responsible for developing the lab manuals for those courses. As part of the distance education programme in Environmental Engineering Technology, Subhash also prepared course manuals and study guides for said courses.

He led a team to develop a training programme in Sampling and Monitoring for the Ministry of the Environment in Ontario, and developed and delivered a training programme to water and wastewater plant operators to prepare them for writing licence examinations. Before leaving for Canada, Subhash was Assistant Professor at the College of Agricultural Engineering at Punjab Agricultural University, Ludhiana.

His current pastime is developing and teaching online courses related to water and wastewater technology. To date, he has developed 10 courses for the OntarioLearn programme. Subhash is also in the process of authoring three titles – *Applied Hydraulics*, *Engineering Hydrology*, and *Groundwater and Wells* – for undergraduate engineering programmes.

Prof (Dr) Varinder S. Kanwar

Dr Varinder S. Kanwar is a Professor of Civil Engineering and Vice Chancellor at Chitkara University, Himachal Pradesh. He has a master's degree in Structural Engineering and a PhD in Civil Engineering from Thapar University, Patiala. He also has a postgraduate diploma in Rural Development.

Dr Kanwar has more than 24 years of research, teaching and administrative experience. He is an active member of various professional societies, including ASCE, IEEE, IRC, the Indian Concrete Institute, the Punjab Science Congress and the Institution of Engineers (India), where he is a fellow.

His main research areas include the health monitoring of structures and alternate construction materials, for which he has obtained research funding from various government funding agencies. He has authored five books – *Water Supply Engineering*, *Health Monitoring of Structures*, *Modern Temples of Resurgent India: Engineer's Pilgrimage to Bhakra, Beas and Ranjit Sagar Dams*, *Sustainable Civil Engineering Practices*, and *Characteristics of Asphalt Modified with Industrial Waste Sludge* – published 40 journal articles and 17 research papers in conference proceedings; and edited 10 conference proceedings. He has filed nine patents in India.

Dr Kanwar is actively involved in joint research activities carried out by Chitkara University, Glasgow Caledonian University, ESTP Paris, Edith Cowan University and Federation University Australia. He is the recipient of several national and international academic awards.

Prof (Dr) Siby John

Dr Siby John is a Professor of Civil Engineering at Punjab Engineering College (Deemed To Be University), Chandigarh, where he is also Deputy Director. He received his doctorate from IIT, Kanpur.

Dr John has more than 30 years of academic and research experience and has authored more than 70 peer-reviewed articles. He has contributed more than 150 papers in conferences and seminars, and written four books in environmental engineering. He is a member of numerous major professional societies in civil and environmental engineering, and the recipient of several national and international academic awards and scholarships.

1 Introduction

Environmental engineering as an undergraduate course, in general, deals with water and wastewater, air and noise pollution, and solid and industrial wastes. The study of the environment and how it interfaces with society and technology is a frontier area in today's engineering education. This book covers all the main topics in environmental engineering. The first section deals with water supply and treatment, followed by a section on wastewater collection and treatment. The third section covers solid waste and environmental pollution, including air and noise. Different sections can be selected to meet the requirements of a given course.

1.1 WATER

Water is essential for life and no life is possible without water. Water is one of the most abundant and essential constituents of all living things. Almost 70% of human body weight is due to water present in the tissues. Besides being essential for life, water is used for many other purposes: for drinking and personal needs, in agriculture and industry, and for transportation and recreation. In India, about 70% of the total water available is used in agriculture, 20–22% in industry, and only 8% for domestic use.

1.1.1 Historical Perspective

Historically, many civilizations developed around water bodies that could support their agriculture, transportation and water for domestic use. Based upon what we know, the human search for water must have begun in prehistoric times. Though it is only speculation, it seems some individuals may have conveyed water through trenches dug into the earth. Thousands of years probably passed before our recent ancestors learnt to build cities with piped water supplies.

The earliest recorded knowledge of water treatment is in Sanskrit medical lore and inscriptions on walls in Egypt. Sanskrit writings in India that date back to 2000 BC describe water purification by boiling it in copper vessels, exposing it to sunlight, filtering it through charcoal, then cooling it in earthen pots. There is mention of digging wells in *Rig Veda* as far back as 4000 BC.

The first engineering report on water supply and treatment was written in 97 AD by S. J. Frontinus, a water commissioner from Rome, who produced two books on the subject. In the 17th century, the English philosopher Francis Bacon wrote of experiments in purifying water by filtration, boiling, distillation and clarification by coagulation. The first known illustrated description of sand filters was published in 1685 by Luc Porzio, an Italian physician. A comprehensive article on the water supply of Venice was published in 1863 in the *Practical Mechanic's Journal*.

Henry Darcy patented filters in France and England in 1856. He came up with Darcy's law for groundwater flow and a design for filters. Filtered water was piped to homes for the first time in 1807 in Glasgow, Scotland. In around 1890, rapid sand filters were developed, and coagulation was later introduced to improve their performance. Techniques for water clarification and its filtration were improved and modified thereafter, followed by the introduction of chlorination for disinfection in 1905. Thanks to the technical and scientific developments of the late 20th and early 21st centuries, engineers can meet the quantity and quality challenges of supplying water.

The drains and sewers of Nippur and Rome are among some of the great structures built to carry away run-off from storms and flush streets. The need for the regular cleaning of Rome and

DOI: 10.1201/9781003231264-1

the flushing of its sewers was well recognized by commissioner Frontinus. However, since then, no marked progress in sewage was made for some time. The history of the improvement of sanitation in London probably gives a clearer picture of what happened in the middle of the 19th century. In 1855, British parliament passed an act and subsequently the Commission of Sewers was formed to take responsibility for establishing an adequate sewerage system.

Edwin Chadwick (1800–1890), an administrator in the UK, introduced the idea of sanitation. Chadwick called for street and house cleaning by means of a supply of water and improved sewage collection. He specifically outlined the provision of services and advice from civil engineers and not physicians. Chadwick proposed the following engineering solutions:

1. A supply of drinking water to every house.
2. The collection of wastewater via a network of pipes.
3. The spread of wastewater into agricultural land away from the city.

1.1.2 Hydrologic Cycle

About 75% of the Earth's surface is covered with water, and because of this it is often called the water planet. Most of the water (about 97%) is in the seas and other saline water bodies (salt water). This water is too salty to be used for drinking and irrigation. Thus, only a tiny fraction (about 3%) of Earth's water can be called fresh water. Of this, a little more than 2% is locked up in glaciers, soil or atmospheric moisture, or is buried so deep that its extraction becomes cost-prohibitive. Therefore, for daily use and use in industry and agriculture, humans depend on less than 1% of the fresh water readily available in rivers, lakes, streams and groundwater.

Water in nature is in a constant state of motion, as depicted in the hydrologic cycle shown in **Figure 1.1**. The water cycle is driven by energy from the sun. The major components of the cycle are the evaporation of water from sea or land, its precipitation as rain or snow over the oceans and land, and its return from the land to the sea via streams and rivers. With slight variations, the water content of the sea remains constant.

As urban development expands, so does the need for municipal water systems. Thus, in urban areas, water also goes through a cycle to serve the needs of municipal establishments, known as the urban water cycle. Water is pumped from rivers or groundwater, treated, and distributed via pipe networks. After use, water is then discharged into sanitary sewers and treated before being discharged into receiving water bodies. All these systems, including water treatment, water distribution, wastewater collection and wastewater treatment, are engineered systems. Basic knowledge of these systems is essential for workers in civil, environmental and public health engineering. In addition, environmental engineers deal with solid waste and air pollution.

1.2 WASTEWATER TREATMENT

Wastewater is water carrying waste. In the past, when populations were more scattered and lifestyles were simple, natural purification was good enough to treat waste discharges. As our lifestyles changed and areas became more urbanized, it became beyond the capacity of natural systems to purify this waste. Therefore, to protect our water resources and natural environment, it is pertinent that man-made or engineered systems are in place.

In a developing country such as India, most of the urban areas inhabited by slums are plagued by acute problems related to the indiscriminate disposal of sewage. Even though a substantial amount of money is allocated to wastewater treatment, disposal is still in a dismal state. According to the Central Pollution Control Board, the estimated deficit in wastewater treatment in India is about 72%. It is not uncommon to find that a large portion of resources is spent by local governments manning sewerage systems for their operation and maintenance (O&M). Despite this, there has been

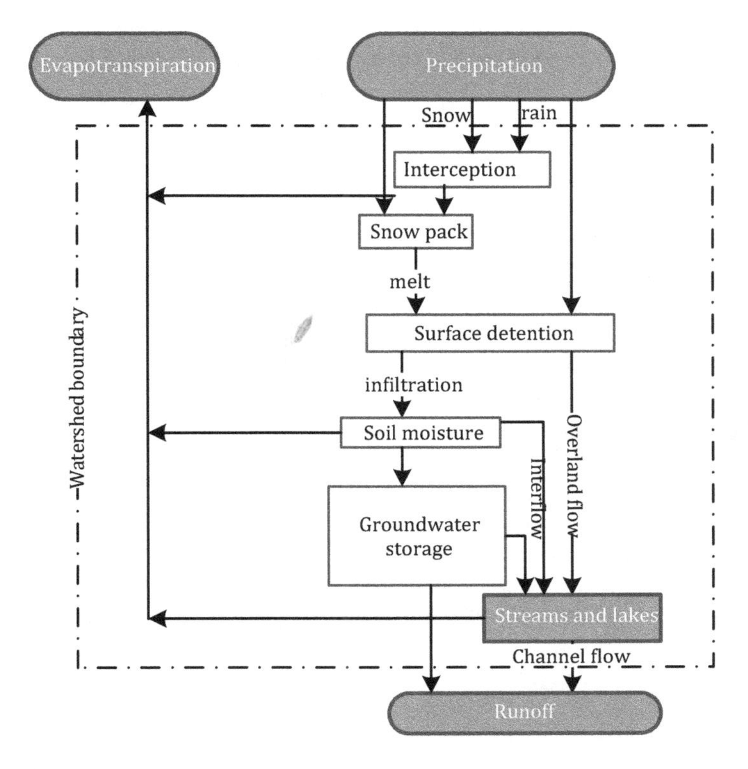

FIGURE 1.1 Hydrologic Cycle

a decline in the standard of services with respect to the collection, transportation, treatment and safe disposal of treated sewage, as well as measures for ensuring the safeguard of public health and the environment.

Sewerage and sewage treatment are a part of public health and sanitation and, according to the Indian Constitution, fall within the purview of the State List. Since this is non-exclusive and essential, the responsibility for providing these services lies within the public domain. In developed countries, water and wastewater treatment fall under the jurisdiction of municipal governments. This activity, being of a local nature, is entrusted to the urban local bodies (ULBs), which undertake the task of sewerage and sewage treatment service delivery with their own staff, equipment and funds. In a few cases, part of said work is contracted out to private enterprises. Cities and towns that have sewerage and sewage treatment facilities are unable to keep up with the increased burden of providing such facilities efficiently to the desired level. The leading cause of water pollution is the unintended disposal of untreated, partly treated and non-point sources of sewage, and its effect on human health and the environment is of great importance.

1.2.1 Need for Wastewater Treatment

Sanitation can be perceived as the conditions and processes relating to people's health, especially the systems that supply water and deal with human waste. Such a task would logically cover other matters such as solid wastes, industrial and other special/hazardous wastes, and storm water drainage. However, the most potent of these pollutants is sewage.

When untreated sewage accumulates and is allowed to become septic, the decomposition of its organic matter leads to nuisance conditions, including the production of malodorous gases. In addition, untreated sewage contains numerous pathogens that dwell in the human intestine tract.

Sewage also contains nutrients that can stimulate the growth of aquatic plants and may contain toxic compounds or compounds that are potentially mutagenic or carcinogenic. For these reasons, the immediate and nuisance-free removal of sewage from its sources of generation, followed by its treatment, reuse or disposal into the environment in an eco-friendly manner, is necessary to protect public health and the environment.

1.2.2 Wastewater Treatment Technologies

In the development of wastewater treatment technologies, the basic principles of engineering, such as hydraulics, are applied to solve issues associated with collection and conveyance. The basic principles of microbiology, chemistry and engineering are applied to the treatment and environmental issues in the disposal and reuse of treated sewage. The main objective is the protection of public health with due regard to environmental, economic, social and political concerns. In order to protect public health and the environment, it is necessary to have knowledge of:

- Constituents of concern, especially pollutants, in wastewater.
- The effects of these pollutants when sewage is returned to the natural environment.
- The transformation and long-term fate of these contaminants in treatment processes.
- Treatment methods and technology available.
- Methods for the beneficial use or disposal of solids generated by the treatment systems.

1.2.3 Microbiological Contaminants

Untreated or raw wastewater can contain any number of harmful microorganisms that can pose a serious health risk to the public. Bacteria are living organisms present in wastewater and may or may not be harmful. Harmful organisms are called pathogens. Viruses are non-living DNA chains. If a certain trigger is present, then the viruses come alive and can cause diseases. Untreated wastewater can also contain worms such as roundworms, tapeworms, hookworms or whipworms that can cause severe abdominal pain and illness. **Table 1.1** indicates the types of organisms and diseases associated with them.

Bacterial contamination of a drinking water source can lead to that source becoming unsafe for consumption without treatment. Bacterial contamination of surface water sources can also lead to unsafe drinking water conditions, as well as beach closures or conditions deemed unsafe for recreational uses.

1.2.4 Nutrient Contaminants

Untreated wastewater can contain high levels of nutrients, including nitrogen compounds and phosphorous. Nitrogen species that can present a public health or environmental risk include ammonia,

TABLE 1.1
Diseases and Causative Organisms

Disease	Causative Organism	Type of Organism
Typhoid	Salmonella typhosi	Pathogenic bacteria
Bacillary Dysentery	Shigella dysenteriae, Escherichia coli (E. coli)	Pathogenic bacteria
Hepatitis A, B, C		Virus
Amoebic Dysentery	Entamoeba histolytica	Parasite
Cryptosporidiosis	Cryptosporidium	Parasite

which can be toxic to fish and other aquatic life. High levels of nitrate, another nitrogen species present in untreated wastewater, can cause a condition called *methemoglobinemia*, which can be very dangerous to infants. This condition is also called blue baby syndrome. The high loading of nutrients encourages aquatic growth and cause eutrophication. Excess phosphorous can also lead to blue-green algae outbreaks, some of which are toxic to humans.

1.2.5 Hazardous Contaminants

Depending on the source, untreated wastewater can contain high levels of various chemicals, some of which may be hazardous. Hazardous chemicals can present serious public health and environmental risks if the collection system is not carefully installed, operated and maintained. Some communities with high-risk industries will initiate sewer bylaws limiting the amount, concentrations or discharges of certain chemicals that can be discharged to the public wastewater collection system.

1.2.6 Effluent Disposal

In the past, the disposal of wastewater in most municipalities and communities was carried out by the easiest method possible, without much regard to unpleasant conditions produced at the place of disposal. Irrigation was probably the first method of wastewater disposal, although dilution was the earliest method adopted by most municipalities. With increased industrial and urban development, effluent disposal and its effects on the environment now require special consideration. In many locations where the available supply of fresh water has become inadequate to meet water needs, it is clear that the once-used water collected from communities and municipalities must be viewed not as a waste to be disposed of but as a resource.

1.2.7 Role of the Engineer

Practising wastewater engineers are involved in the conception, planning, evaluation, design, construction, operation and maintenance of the systems needed to meet wastewater management objectives. Knowledge of the methods used to determine wastewater flow rates and characteristics is essential for understanding all aspects of wastewater engineering. The engineer must also study the subjects of source control, collection, transmission and pumping if truly integrated wastewater systems are to be designed. The issues of treatment, disposal of effluent and the possible reuse of wastewater are also equally important. Old ideas are being re-evaluated and new concepts are being formulated. To play an active role in the development of this field, the engineer must know the fundamentals.

1.2.8 Wastewater Effluent Standards

Water pollution arising from the discharge of wastewater into watercourses has led to the establishment of standards for wastewater discharges. The United States Environmental Protection Agency (EPA) and the Central Pollution Control Board in India are examples of agencies that prescribe such standards. In India, general effluent discharge standards are prescribed in the Environment (Protection) Rules, 1986.

1.3 GLOBAL ISSUE

The enormous demands being placed on water supply and wastewater facilities have necessitated the development of broader concepts, management scenarios, and improvement and innovations in environmental technology. As mentioned earlier, in the urban water cycle, water is taken out of a water source and then treated and distributed. Used water, otherwise known as wastewater or sewage, is collected and processed to remove contaminants to an acceptable level. This forms part of a complex natural system comprising physical and chemical as well as biological processes.

While the standards of water quality have become more stringent, the quality of raw water has become poorer due to increased demand and anthropogenic activities that limit the efficacy of natural purification. Toxic and hazardous materials have entered the water system, which has heightened the worry over environmental issues. The world population is seeing alarming growth, making environmental issues ever more critical. Climate change has added another dimension to environmental issues and problems. Therefore, the use, management and control of environmental resources must be sustainable.

1.4 A LOOK TO THE FUTURE

Besides being technically skilled, the engineers and scientists of tomorrow need to see beyond the political, legal and economic constraints of the problems they will face. These experts must set forth and assess viable alternatives and come up with realistic solutions. The future of this "one world" rests upon the decisions of environmental experts and the actions that come out of such decisions. These decisions will shape the lives of future generations.

As the quality of fresh available water declines, the demand for better management of water resources increases. With recent advancements in water treatment technology, it is possible to remove most undesirable contaminants and make water suitable for drinking (potable). However, the cost of producing drinking water is increasing significantly, and is beyond the capabilities of third-world countries. Therefore, it is better to manage our water resources and prevent the water from being polluted in the first place. In developing countries, the concept of source water protection is applied on a watershed basis. Similarly, better management strategies for other environmental vectors such as air, land, etc., should be prioritized over treatment efforts.

Discussion Questions

1. Write short notes on source water protection and water conservation.
2. Briefly describe the history of water and wastewater systems.
3. With the help of a flow chart, explain the various components of a water supply system.
4. World population is on the rise, but the availability of fresh water is declining. How will the water engineers of tomorrow meet this challenge?
5. Search the internet for countries that have experienced water shortages in the 21st century.
6. Make a flow chart of the urban water cycle, starting with the source of the water supply and ending with the discharge of wastewater effluent.
7. Environmental engineers and scientists have reported that global warming may be a real problem. What are the causes of global warming and how it might impact water supply?
8. In what sense will the jobs of future environmental engineers be more challenging?
9. Water is in a constant state of motion in nature. Explain.
10. Comment on the potential of wastewater reclamation as a water conservation strategy.
11. Give an account of the need for wastewater treatment.
12. Name the water quality parameters of concern from a public health point of view.
13. List the major microbiological contaminants and diseases caused by these parameters.
14. What basic knowledge is required for decision-making in wastewater treatment technology?

15. Describe the role of the engineer in wastewater management.
16. When and how was the discovery of activated sludge treatment made in the United Kingdom?
17. Search the internet to find effluent standards for the United States and India.
18. Explain why the secondary treatment of wastewater has become the norm?
19. Dilution is not a solution. Comment on this.
20. Explain the role of nutrients in wastewater treatment. What is lake eutrophication?

2 Sources of Water Supply

In order to design a water supply system for a town, the engineer must know the various sources of water in the vicinity of the town and their characteristics. As explained in Chapter 1, unless a steady source of water is available, the water supply system becomes unreliable.

The origin of all sources of water is precipitation. It can be collected as it falls as rain before it reaches the ground; as surface water when it flows over the ground surface; from lakes or ponds where it pools; as groundwater when it percolates into the ground and flows, or when it collects; or from the sea into which it finally ends up. The quality of the water varies according to the source as well as the media through which it flows. Water sources are either surface water in the form of lakes, reservoirs and rivers, or groundwater in the form of springs, infiltration galleries and wells.

2.1 ESSENTIALS OF A WATER SUPPLY SYSTEM

Figure 2.1 shows the flow diagram of a typical water supply system. The primary requirement of a water supply system is a reliable and safe source of water. After choosing such as water source, it is then essential to design and construct a water intake structure to collect and transmit water to the treatment plant.

At the treatment plant, water is treated either using chemicals (physiochemical) or by simple physical processes to improve its quality to meet consumer requirements. The treated water is stored in clear water reservoirs from where it is distributed to consumers.

2.2 SURFACE WATER

Surface water, as the name indicates, is exposed to the atmosphere and thus prone to pollution and contamination. However, for meeting higher demand, it is more reliable.

2.2.1 Lakes and Ponds

These refer to naturally formed large depressions in the ground filled with water. Lakes are usually found in mountainous regions where water from springs and streams flow in, whereas ponds are usually found on plains where water is collected from surface run-off. Water from these sources is more consistent in quality compared with water from flowing streams. Long periods of storage and detention leads to the sedimentation of suspended matter, bleaching of colour and the removal of bacteria. Self-purification, which is an inherent property of water, is usually less complete in smaller lakes than in larger ones.

The size of lakes and ponds varies greatly. For example, in Canada and the US, the Great Lakes are large bodies of fresh water that serve very large populations with their water supply. Lake Superior, the largest of the Great Lakes, has a detention time of more than 100 years. The turbidity and colouration of the water are almost negligible. In contrast, small lakes and ponds may be limited in capacity and only capable of serving smaller communities in hilly regions.

2.2.2 Rivers and Streams

Water from rivers and streams is generally dynamic and of variable quality, and less satisfactory than that of lakes and ponds. The quality of such water depends upon the character and area of

DOI: 10.1201/9781003231264-2

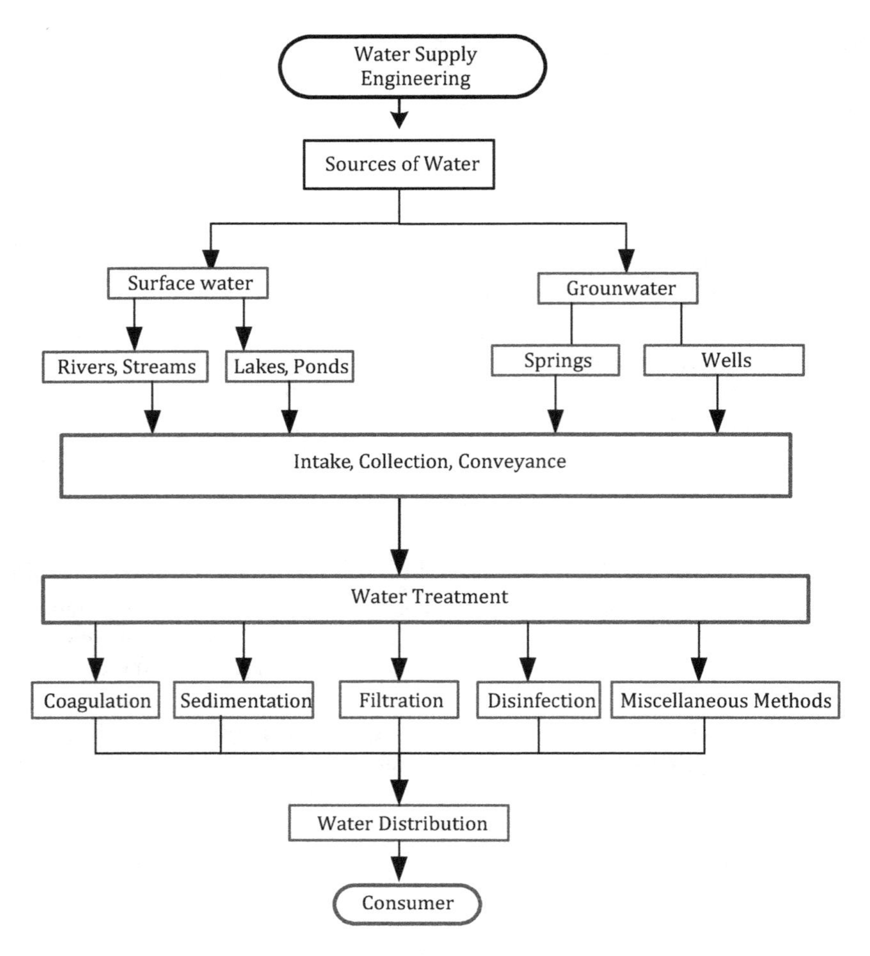

FIGURE 2.1 Flow Diagram of a Water Supply System

the watershed, its geology and topography, the extent and nature of urban development, seasonal variations and weather conditions. Streams from relatively sparsely inhabited watersheds carry suspended impurities from eroded catchments, organic debris and minerals. Substantial variations in the quality of the water may also occur between maximum and minimum flows. In populated regions, pollution from sewage and industrial waste will be direct. Natural and man-made pollution results in added colouration, turbidity, taste and odour, hardness, and bacteria and other microorganisms in such water supplies. The quantity of water available depends on the type of river, perennial or seasonal. Rivers are the most important sources of public water supply schemes.

2.2.3 Artificial Reservoirs

Artificial reservoirs are basins constructed in the valleys of streams. Reservoirs are formed by hydraulic structures thrown across river valleys, and the quality of water is more or less that of natural lakes and ponds. While the top layers of reservoir water are prone to developing algae, the bottom layers may be high in turbidity, carbon dioxide, iron, manganese, etc. The quantity of water available depends on the capacity of the reservoir. However, silting of the reservoir may reduce its capacity. Reservoirs are necessary for storing stream water during the monsoon season so that water supply works are not affected.

2.2.4 Sea Water

Although plentiful, it is difficult to economically extract potable water from sea water because it comprises about 3% salts, and is costly to treat before supply. There are various desalination processes available for the conversion of brackish water into drinking water. Desalting or de-mineralizing processes involve the separation of salt or water from saline water. However, this is a costly process and is only adopted in places where sea water is the only source available and potable water has to be obtained from it, such as in ships on the high seas, or places where an industry has to be set up and there is no other source of supply.

2.2.5 Wastewater Reclamation

Sewage or other community wastewater may be utilized for non-domestic purposes, such as water for cooling, flushing, lawns, parks, firefighting and for certain industrial purposes, after receiving necessary treatment to suit the nature of its use. Supply of this source to residences is prohibited because of possible cross-connection with the potable water supply system. In places like California, highly treated wastewater is supplied for use for flushing toilets and other similar purposes. This practice is becoming more common as water supply in many parts of the world is dwindling.

2.2.6 Stored Rainwater

At places where no water source is easily available, rainwater stored in cisterns or tanks can be used. The rainwater from roofs and paved pucca courtyards is collected in watertight tanks. Stored rainwater is of limited capacity and more susceptible to contamination.

2.2.7 Yield Assessment

Correct assessment of the capacity of a water source is necessary for deciding on its dependability for a water supply project. The capacity of flowing streams and natural lakes is decided by the area and nature of the catchment, the amount of rainfall and allied factors. The safe yield of a subsurface source is decided by the hydrological and hydrogeological features relevant to each case.

2.3 INTAKE WORKS

Intake works comprise certain works wherein a structure is constructed on a surface water source to impound water and arrangements are provided to withdraw water from this source and discharge it into an intake pipe. From this pipe, water is made to flow into a waterworks system. There are reservoir intakes, river intakes and canal intakes. River intakes are further classified into weir intakes, intake wells, pipe intakes and intake wells with approach channels.

2.3.1 Reservoir Intakes

This consists of a circular concrete well, the floor of which is well below the low level of the reservoir. A number of intake pipes with screens are installed at different depths to facilitate the drawing of clear water into the well. It is better to draw water from upper levels, as water at lower levels contains more silt and other impurities. Valves are provided for drawing off or shutting out water. The top of the dam features a gangway or bridge that connects the dam with the valve tower. This is located inside the river in order to get an adequate supply of water in all seasons. Water is drawn from the upstream side of the river where it is comparatively of better quality and not affected by pollutants discharged into the river from the city.

2.3.2 Twin Tower River Intake

A twin tower intake consists of a collector well and a jack well, both connected by an intake pipe.

Intake Well

This consists of a concrete/masonry well located inside a river. It is circular or, more preferably, oblong in shape. An intake well is located in the riverbed somewhat away from the riverbank, as shown in **Figure 2.2**.

Intake Pipe

The intake pipe connecting the intake well with the jack well is usually of a non-pressure type and is laid on a gentle slope. Flow velocity in an intake pipe should not exceed 1.2 m/s. The ends of the pipes are fixed with strainers. The water is drawn into a jack well and then pumped to the purification works. This is a cheap and simple arrangement. Water entering a jack well is pumped to the treatment works. Jack wells should be founded on hard strata.

2.3.3 Single Well Type River Intake

In alluvial rivers, water is usually ponded by constructing a weir across the river. This is done so that the river's water level in the dry season provides sufficient water. Intake water may be drawn from the weir through a channel into a sump well from which it can be pumped for supply. This arrangement will eliminate the need for the construction of a separate inlet well and inlet pipe, as required in the twin tower set-up shown in **Figure 2.3.**

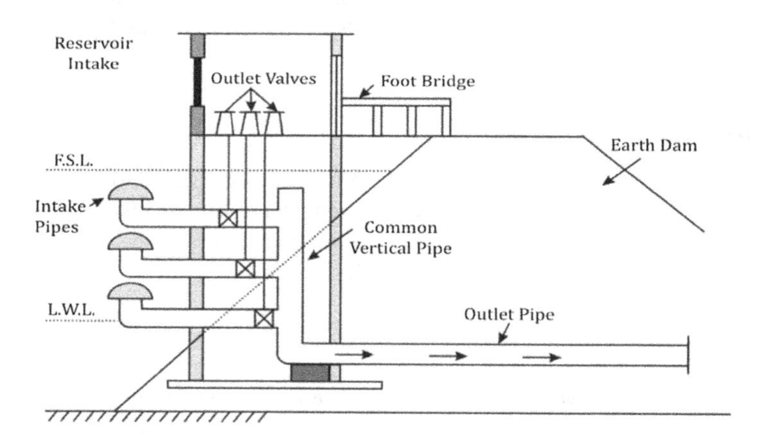

FIGURE 2.2 Reservoir Intake

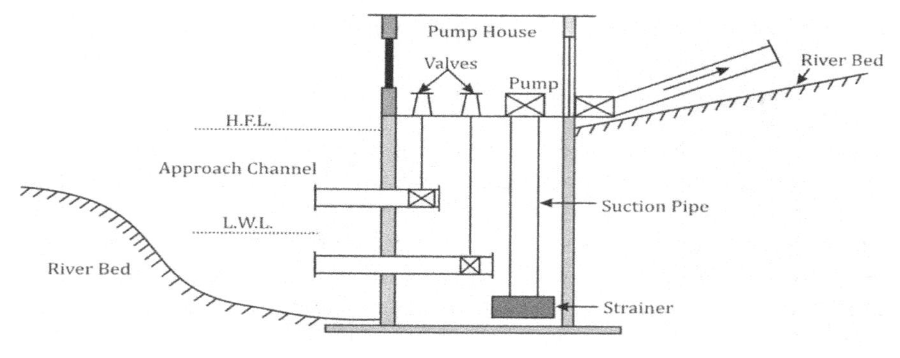

FIGURE 2.3 River Intake

Example Problem 2.1

Design a twin tower type river intake based on the following information.

R. L. of the riverbed	= 100.0 m
Lowest water level	= 102.5 m
Normal water level	= 116.0 m
Peak water level	= 121.0 m
Population served	= 65000 at 170 L/c.d
Maximum screen entrance velocity	= 0.15 m/s
Maximum intake pipe flow velocity	= 1.2 m/s

Assume water is pumped for 200 h/d and the screen occupies 30% of the area.

SOLUTION:

$$Average, Q = \frac{170L}{p.d} \times 65000\ p \times \frac{m^3}{1000L} \times \frac{h}{3600s} \times \frac{d}{20h} = 0.153 m^3/s$$

$$Q_{max} = 1.8 \times \frac{0.153 m^3}{s} = 0.276 m^3/s$$

$$Open\,area, A_o = \frac{Q}{v} = \frac{0.276 m^3}{s} \times \frac{s}{0.15 m} = 1.84 m^2$$

$$Total\ A_t = \frac{1.84 m^2}{0.70} = 2.63 m^2$$

A 1.0 m high screen would require a width of 2.63 m, say 2.7 m. This port can be provided at all three levels and can be fitted in an oblong well measuring 3.0 m × 2.0 m. This inlet well must be sunk into the riverbed in order to provide enough space for the accumulation of sand and silt. If the inlet well is sunk 3.0 m below the bed with a free board of 2.0 m above the peak water level, then the height of the inlet well would be 123 m - 97 m = 26 m.

$$Intake\ pipe, D_i = \sqrt{\left(\frac{4}{\pi} \times \frac{Q}{v}\right)} = \sqrt{\left(\frac{4}{\pi} \times \frac{0.276 m^3}{s} \times \frac{s}{1.2 m}\right)}$$

$$= 0.541 m = 600 mm\ say$$

$$Grade, S_i \text{ (Manning's Equation)} = \left(\frac{Q \times n}{0.311 D^{2.67}}\right)^2 = \left(\frac{0.276 \times 0.15}{0.311 \times 0.6^{2.67}}\right)^2$$

$$= 2.71 \times 10^{-3} \times 100\% = 0.30\%$$

2.3.4 Lake Intake

To obtain water from a lake, a submersible intake is mostly used. This intake is constructed in the bed of the lake below the low water level. It essentially consists of a pipe laid in the bed of the lake. The opening of the pipe is fitted with a bell mouth opening covered with a screen. The water enters

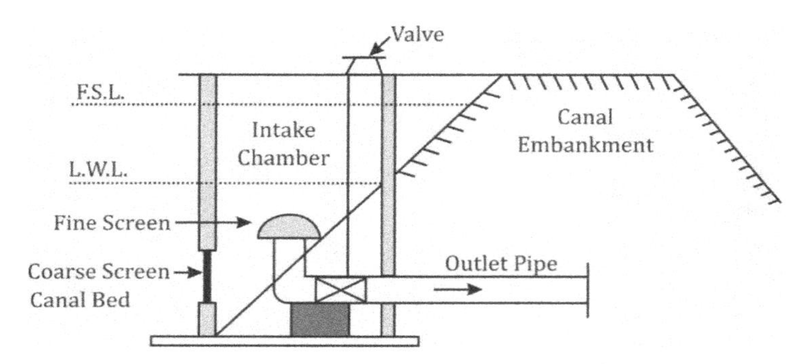

FIGURE 2.4 Canal Intake

the pipe through the bell mouth opening and flows under gravity to the sump well before being pumped to the treatment plants.

2.3.5 Canal Intake

This consists of a concrete well in a canal. An intake pipe laid in the canal bed leads into the well, as shown in **Figure 2.4**. As the full supply level (FSL) in the canal is fairly constant, inlets at different depths are not necessary. The inlet end of the pipe is provided with an enlarged bell mouth, to which is fixed a fine hemispherical screen that prevents floating material from entering the intake pipe. A coarse screen is also provided so that big floating particles are excluded. The water from the outlet of the intake pipe is led to a sump well or supply.

2.4 WATER TRANSMISSION

The collection of water is performed by conveyance or transportation from a water source to the treatment works. After treatment, water is conveyed to consumers via a water distribution system. Normally, the sources of supply are far away from towns and cities. Therefore, structures called conduits must be constructed. Conduits may flow under gravity or pressure. Pipes used for transmission are also called low-pressure pipes since they usually flow under low-pressure conditions.

2.4.1 Gravity Conduits

This is an arrangement in which water is made to flow by gravity with a free-water surface. These include aqueducts, canals, flumes and tunnels. A canal is an open excavation in natural ground. It may be lined or unlined. An aqueduct is a canal that is carried on a river. Normally, the drainage and canal intersect at right angles. A flume is an open channel of cement, concrete, wood or steel laid on natural ground and supported on trestles. A tunnel is a horizontal cutting or bore into a hill or mountain.

A gravity conduit is designed in such a way that it follows the natural hydraulic gradient. It has to follow long contours of the ground if hills, ridges, valleys, etc., are to be avoided. Otherwise, the construction of structures such as bridges, embankments, cuttings, tunnels, etc., are necessary. Open channels are also liable to loss of water by evaporation and percolation, and are susceptible to pollution from surface drainage. These have high maintenance costs.

2.4.2 Pressure Conduits

These are structures in which water flows under pressure. These consist of pipes made of iron, steel, reinforced cement concrete, etc. A pressure conduit can be constructed at any level below the hydraulic gradient. It can be located easily, following a direct and convenient alignment, maintaining low pressure, thereby achieving a lower cost in maintenance and construction. The type of conduit to be constructed depends on the topography of the area, the type of soil, the volume of water to be transported and the cost.

2.5 GROUNDWATER

Rainwater that has percolating into the ground and reached permeable layers (aquifers) in the saturation zone constitutes a groundwater source. As it seeps down through the ground, the water comes into contact with organic and inorganic substances and acquires chemical characteristics representative of the strata it passes through, such as hardness due to limestone, iron and manganese, fluoride, arsenic, etc. Percolation into the subsoil sometimes results in the filtering out of bacteria and other living organisms.

Generally, groundwater is clear and colourless, but is harder than the surface waters of the region. Groundwater is generally of uniform quality, although changes may occur because of waterlogging, overdraft from areas adjoining saline water sources and the recycling of water used in irrigation.

2.5.1 Water Wells

Water wells are structures made to tap groundwater from greater depths. In the past, wells used to be large in diameter and relatively shallow since modern methods of drilling did not exist. A larger size of open well allows groundwater to seep into it so it can act as small storage.

Modern wells are holes that are small in diameter and dug into the ground to tap groundwater. Tube wells feature a conduit section to accommodate a pump. At the lower end of the pipe is a strainer at the intake portion of the well.

2.5.2 Springs

Springs may be considered as outcrops of groundwater and often appear as small water holes at the foot of hills or along riverbanks. Springs may be either perennial or intermittent. A gravity spring may result from the outcropping of an impervious stratum underneath a water-bearing formation. It may also be due to overflow of the water table by a continuous rise in it.

An artesian spring is formed when two impervious strata sandwich an aquifer or a water-bearing stratum under pressure. Water flows through the weaker spots or faults or cracks in the rock. The yield of a gravity spring varies with the position of the water table and of the rainfall, whereas the yield of the artesian spring is more uniform. Their usefulness as sources of water supply depends on the discharge and its variability during the year. It is a common practice to construct collecting tanks at the point of springs to ensure uniform supply and protection from contamination. Spring waters from shallow strata are more likely to be affected by surface contamination than deep-seated waters. Protected spring water contains minerals that may supply essential nutrients for organisms (mineral water).

2.5.3 Infiltration Galleries

An infiltration gallery is another way of tapping groundwater. It is essentially a shallow tunnel (3–5 m deep) dug along the banks of the river through water-bearing strata. In the tunnel, pipes are laid under the ground to withdraw groundwater at moderate depths. The shape of the tunnel is generally circular, and it is covered by graded filter material to retain unwanted solids. This graded material is laid in layers around the perforated pipe collecting filtered water. The outer layer covering the pipe consists of relatively fine material, typically in the range of 3 mm to 10 mm in size. The inner layer is of coarser material, varying in size from 25 mm to 50 mm. This makes the gradation in the direction of flow from finer to coarser. The length of the gallery varies depending on water demand and permeability of the formation, and ranges from 10 m to 100 m.

2.5.4 Karez

A karez is an underground tunnel driven into the hillside to reach underground springs. The tunnel is graded to allow tapped water flow towards the collection point.

2.5.5 Collector Wells

A radial collector well is also called a horizontal well. It consists of a concrete cylinder, typically 5 m in diameter, which is sunk into the aquifer by excavating the earth material. When the desired level is reached, it is sealed at its bottom with a thick concrete plug. The concrete cylinder has precast ports at the bottom end. Through these ports, a number of radial collector pipes are jacked horizontally into the formation. Perforated pipes with proper screens are inserted in these horizontal pipes. Collector wells extract relatively large supplies of groundwater from valley fills and alluvial aquifers of high permeability. The entrance velocity of the water is low, thus the chances of premature plugging and scale deposition are minimized.

2.6 WATER QUANTITY

As the effect of rainfall is most direct on surface sources of water supply, the quantity of water available is abundant. However, since rainfall may not be uniformly spread throughout the year, especially in the case of tropical countries like India, considerable variations in the flow of surface waters are likely. Thus, the flow into streams or rivers may vary from a maximum during the rainy season, sufficient to result in floods, to a minimum during dry months, sufficient to cause long droughts. In the case of the impounding reservoirs, in addition to rainfall and runoff, the topography of the catchment area is important. It should be such as to drain off water from all remote points.

As regards the underground sources, the quantity of water available is usually less than that in the case of surface sources, the effect of rainfall now being mostly indirect. The quantity depends upon the available underground storage and the geological formations of the substrata, viz., permeable or non-permeable. In the case of shallow wells and springs, it is easier to get a water supply by tapping the upper water-bearing strata, but such storage may be temporary and fall off during the dry season, resulting in the failure of the source. The underground supplies drawn from greater depths, such as deep wells, are more constant in their yield and hence more reliable.

2.7 WATER QUALITY

Water impurities are normally of two types, suspended and dissolved. Surface waters are characterized by suspended impurities, whereas groundwaters are generally free from suspended matter but are likely to contain a large amount of dissolved impurities, which they gather during the course of their travel through underground strata comprising rocks and minerals. The suspended matter often contains pathogenic or disease-producing bacteria. As such, surface waters are not considered to be safe for water supply without necessary treatment. Groundwaters are comparatively safer and fit for use with or without minor treatment only. A comparison of surface water and groundwater supplies is shown in **Table 2.1**.

River water varies in quality. This variation is caused by a great difference in maximum and minimum flow. The maximum flow is caused by high floods, resulting in an increase in turbidity and bacteria due to the surface wash brought into the river. The minimum flow is due to the flow of

TABLE 2.1
Surface Water versus Groundwater Supplies

Feature	Surface Water	Groundwater
Quality	Low in minerals, soft water	High in minerals, hard water
Changes in quality	Can be low quality during lake turnovers and after storms	Little variation Unaffected by storms
Quantity	Large quantity	Low if well yield is low
Taste and odour	Problems from algae	From iron and H_2S
Contamination	High	Low
Costs to treat	Medium to high	Low to medium

groundwater into the river, resulting in a decrease in turbidity but an increase in dissolved impurities. River water is also usually found to be contaminated with sewage or industrial water from towns and cities. River water, therefore, must be thoroughly treated before being supplied for public use.

An impounding reservoir stores water by the construction of a dam across a natural watercourse. The storage provided may be as much as 60 days or less. This long storage enables the suspended matter to settle down and be removed. There is also a considerable reduction in the harmful bacteria and colour present. Long storage is, however, objectionable in one way in that it causes the growth of microscopic organisms in the water, impairing its general quality. While the top layers of the water are prone to developing algae, the bottom layers may be high in turbidity, carbon dioxide, iron, manganese and even hydrogen sulphide. Aeration and chlorination are thus normally required before the water is considered fit for supply.

The quality of groundwater is comparatively much better. This is because the water gets strained during its passage through the porous underground strata. The geological formations it comes into contact with also impart certain qualities, such as softness or hardness. In granite formations, groundwaters are soft, low in dissolved mineral and high in carbon dioxide, and are actively corrosive; while in limestone formations, they are very hard and tend to form deposits in pipes, but are relatively non-corrosive. The bacterial content of water from springs, infiltration galleries and deep wells is usually low due to the straining action involved. In general, groundwater is good in quality but may require some treatment to improve its chemical characteristics.

2.8 GROUNDWATER UNDER THE DIRECT INFLUENCE

Groundwater is clean because the water gets naturally filtered as it moves through subsurface layers. However, sometimes the filtration process is not complete as water takes shorter routes and is not filtered adequately, especially when surface water easily makes its way to shallow groundwater. When this happens, there is a concern that solids and microorganisms could get into groundwater. This results in a groundwater water supply that is under the direct influence of surface water (GUDI). In many developed countries, the minimum treatment necessary for GUDI is the same as that of surface water supply systems, namely chemical-assisted filtration. Here are some examples of GUDI systems.

- A drinking water system that gets its water from a well that is not drilled, or a well that does not have a watertight casing that extends to a depth of at least 6 m below the ground surface.
- A water system where the source of water is from an infiltration gallery.
- A shallow well within the vicinity of surface water (< 100 m). Open shallow wells in villages get contaminated during the rainy season.
- A drinking water system that has been contaminated by surface water.

2.9 CHOICE OF SOURCE OF WATER SUPPLY

Considerations in the selection of a particular source for supplying water are the quantity of water available, its quality and the cost of production. The quantity of water available from the source should be sufficient to cater to the needs of the town or city regarding domestic service, industrial demands, firefighting requirements and other public use. The quantity of water supplied should also include design requirements, which means the calculated quantity should be somewhat higher than the bare needs. The quality of water should be wholesome, safe and free from pollution of any kind. The health of the public should in no way be endangered due to epidemics associated with water-borne diseases.

The quantity and quality of water are prime considerations in the selection of any source of supply. Cost considerations regarding the development and operation of water supply are also significant. The cost of supply would depend on whether the system of supply is such that the water flows by gravity from the source or it has to be pumped first before supplying. The cost shall also depend upon the distance between the source of supply and the distribution system. A longer distance means greater cost of conduits and other appurtenances required. In short, the investment cost of a water supply must be reasonable compared to the number of people served and must bear a fair relation to the value of property served so that by equitable taxes and reasonable charges for water, the original cost of the system can be repaid at the end of the design period, which is usually 20 to 30 years.

Discussion Questions

1. What is an intake? What are the important considerations when selecting an intake?
2. Compare a reservoir intake versus a lake intake.
3. Describe two main types of river intake.
4. Compare groundwater supplies versus surface water supplies.
5. List and describe various sources of surface water supply.
6. What are the various methods of tapping groundwater?
7. What factors would you consider when choosing a source of water supply?
8. What are the reasons that surface water supplies are usually used for larger cities and towns?
9. Why do GUDI systems require rigorous treatment to make water potable? Under what conditions and situations should groundwater sources be treated as GUDI?
10. Search the internet and other sources to find the available sources of water supply in your region.

3 Water Wells

A well is a man-made structure. It is a vertical shaft or opening driven into an aquifer to tap groundwater. Based on the type and depth of aquifers tapped, there are shallow, deep, artesian, water table and infiltration wells. (**Figure 3.1**)

Shallow wells get their water supply from the subsoil water table. Deep wells derive water from more than one aquifer. Artesian wells are formed when a porous aquifer is enclosed between two impervious strata. Infiltration wells are shallow wells constructed on the banks of rivers or streams. Water infiltrates through the sides of these wells passing through layers of sand, undergoing natural filtration. The quantity and quality of water is limited in shallow wells, restricting its use, whereas deep wells offer more water of better quality. The quantity and quality of water obtained from infiltration wells depend on the sand medium in the substrata. Depending on the method of construction, wells are classified as dug, driven, bored and drilled.

Dug wells (also known as percolation wells) and driven wells are shallow wells that are usually confined to soft ground, sand and gravel. Dug wells are highly susceptible to contamination. The diameter of these wells ranges from 1 m to 4 m and has a depth of about 20 m. Tube wells are deep wells especially driven to obtain more yield. Due to the low yield of dug or driven wells, they are useful for a small locality, whereas tube wells can supply water to larger areas. Municipal wells are usually drilled.

In the Indian subcontinent, the wells used are typically open wells and tube wells. Open wells refer to dug wells that tap the upper or water table aquifer and thus have limited capacity. In the majority of cases, wells are not covered and hence are susceptible to pollution. They were common in the past and still serve as a source of water for smaller communities, usually in rural areas. With the advent of pumps, many open wells have been fitted with hand pumps or small powered pumps.

Recently, tube wells have been more commonly employed to withdraw groundwater for municipal supplies. Tube wells are created by lowering a pipe into a deep drilled hole so as to tap artesian aquifers for high yield and a reliable water supply. Tube wells are as deep as 300-400 m. Unfortunately, due to excessive draft, there has been a significant lowering of the water table in many areas. This problem has made pumping uneconomical, and in some cases, water quality has deteriorated to the extent that it is not suitable for drinking purposes. The main cause of this problem is that groundwater is extensively used for agricultural irrigation.

3.1 WELL DRILLING

Tube wells for municipal supply are usually drilled. The common methods of drilling are the cable tool, hydraulic rotary and reverse rotary methods.

3.1.1 Cable Tool Method

The cable tool method is also called percussion drilling since hard material is crushed under the heavy weight of crushing tools. This method, also called the standard method, is recommended for drilling into consolidated rock formations. A drilling assembly hoist, walking beam and power unit is mounted on a truck for portability. This is called a drilling rig. The string of percussion tools consists of a rope socket, a set of drilling jars, a drill stem and a drilling bit. Drilling is accomplished

DOI: 10.1201/9781003231264-3

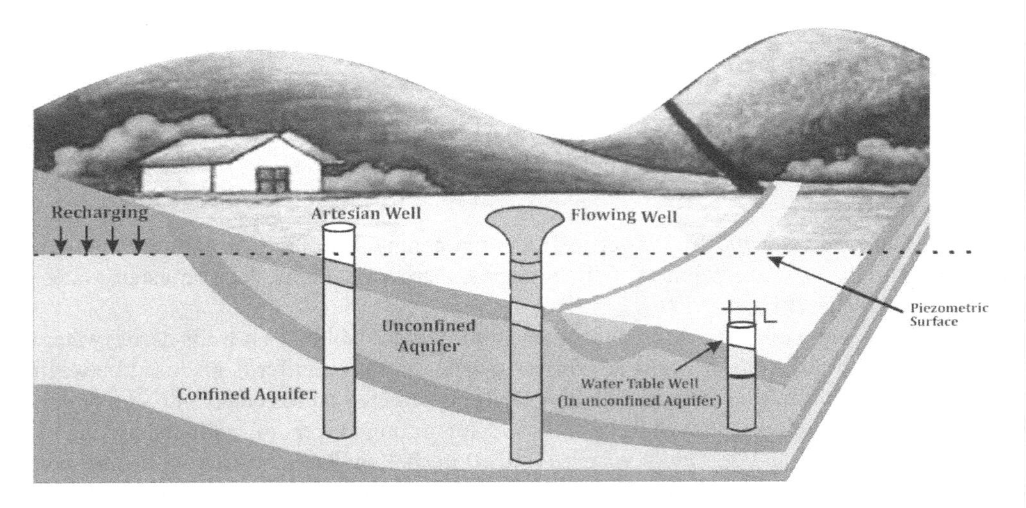

FIGURE 3.1 Types of Water Wells

by alternately lifting and dropping the drilling tools. The drilling line is continuously rotated to ensure the hole cut is round and drills straight. As drilling is continuous, the pipe casing and shoe are driven by drive clamps fastened to the drill stem. Once the bit has cut about 1–1.5 m in depth, depending on the type of rock, it is taken out and a sand pump, called a bailer, is lowered to remove the cut material. The bailer is fitted with a check valve at the bottom to hold the cuttings. During the downward stroke, a valve opens to let the cuttings move into the bailer.

As drilling proceeds, samples of cuttings from various depths are collected. Information on the depth versus material gradation is kept in a well log. This information is used to decide the well depth and design of the intake portion of the well. The percussion drilling method is limited to drilling holes of up to 300 mm in diameter, but as deep as 1600 m.

3.1.2 Hydraulic Rotary

Rotary methods are suitable for drilling larger holes in relatively soft formations. In the hydraulic rotary method, also called the direct rotary method, drilling is performed by a drilling bit attached to a string of hollow pipes. Unlike percussion drilling, cuttings are removed by circulating fluid of a specific consistency called drilling mud. In addition to removing cuttings, drilling fluid performs the following functions:

- It supports the wall of the borehole against sloughing.
- It seals the wall of the borehole against fluid loss.
- It suspends cuttings when circulation is stopped.
- It cools and cleans the drill bit.

The drill bit has hollow shanks and one or more centrally located orifices for jetting the mud in the bottom of the hole to carry the cuttings upward through the annular space. The drill rod, made of heavy pipe, carries the drill bit at one end and is screwed to a square section known as a kelly. The direct rotary drilling method offers the following advantages and disadvantages:

- It is relatively fast, especially in unconsolidated formations.
- The borehole is not cased.

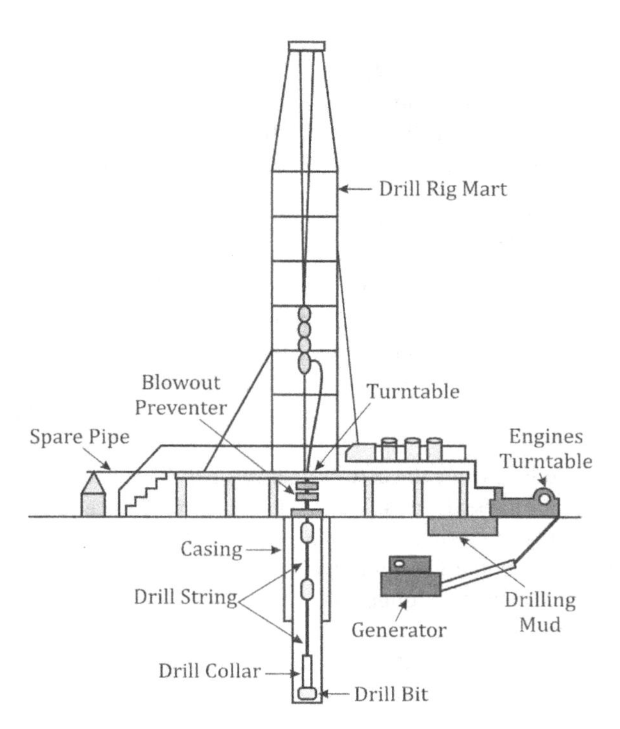

FIGURE 3.2 Hydraulic Rotary Drilling

- It is well suited to the construction of artificial packed wells.
- It is difficult to obtain representative samples.
- It needs large volumes of water and other additives.

3.1.3 Reverse Rotary

As the name indicates, the direction of fluid circulation is reversed in reverse rotary drilling. The drilling fluid is circulated down the annular space and backs up through the drill pipe. The procedure is essentially a suction dredging method, very suitable for drilling larger diameter holes in unconsolidated formations. Reverse rotary drilling has the following characteristics:

- It is more economical for drilling large diameter holes.
- It is more suitable for constructing artificial gravel packed wells.
- It does not need special materials such as bentonite to make drilling fluid.
- It requires a large water supply to make up for water loss.
- It is not practical for high water table conditions due to caving problems.

3.1.4 Air Rotary

Drilling fluid is lighter than water and can vary from air to foam to aerated mud. This is especially desirable when drilling in highly fractured formations. The air rotary method can also be combined with percussion drilling. Air percussion rotary drilling is more applicable in consolidated rock formations that require no casing.

3.1.5 Jet Drilling

In jet drilling, a high-velocity jet is used to cut the material. Cuttings are transported in the drilling fluid, which is pumped through the drill pipe as in the direct rotary method. Normally, a casing is driven as drilling proceeds. Jet drilling is highly successful in sandy formations.

3.2 WELL INTAKE PORTION

In screened wells, water from the aquifer enters the well through a screen or strainer with openings specially designed to suit the aquifer material. Only the parts of the well with selected water-bearing formations are screened. A screen essentially consists of a perforated or slotted pipe with a wire mesh wrapped around it and a small annular space between them. During well development, fine sand particles are removed such that the remaining material surrounding the screen is coarse and permeable.

In the majority of municipal wells, the screen is enveloped by a gravel pack of suitable material, and so these are called gravel packed wells. When there is a rock type aquifer or hard clay confining layer, the well is not screened and a cavity is created under the overlying clay layer. Cavity wells are of limited capacity and not suitable for municipal supplies. Proper design of a well intake section ensures high hydraulic efficiency and longer life of the well and provides sand-free performance. The gravel size and screen slot size are chosen to provide sand-free performance without causing excessive entrance velocities. The objective is to keep entrance velocities to less than 0.3 m/s in order to keep well losses in check and prevent the entry of sand and premature plugging.

3.2.1 Types of Well Screen

Depending on the type of slot, there are three main types of screens. However, there are a variety of well strainers made by different companies, including Cook, Tej, Brownlie, Ashford, Leggett and Phoenix.

Continuous Slot Screen

This kind of screen is more common in municipal wells. A continuous screen is made by winding a V-shaped wire of the desired size around a cylinder made of vertical bars. The spacing between consecutive wires is adjusted based on the size of particles around the intake portion. This type of screen provides more open area per unit length, and the V-shaped opening prevents sand from plugging up the screen.

Louvered Screen

In a metallic cylinder, holes are punched in the form of louvers. Care should be taken to prevent the metal from tearing. A louvered screen has a reduced open area but greater strength.

Perforated Screen

There are several types of perforated screens depending on how the slots or openings are made. The hydraulic efficiency of this type of screen is relatively poor. The design factors of this screen that influence well performance include its length and diameter, slot-size opening and resistance to corrosion and incrustation.

3.2.2 Screen Length

The length of the screen should be based on the grain size distribution, uniformity, depth of the aquifer and whether it is confined or unconfined. The following general criteria have been suggested for determining the screen length in various types of aquifers.

Homogeneous Artesian Aquifer

As a rule of thumb, screen length is chosen at 70–80% of the aquifer's thickness and centred vertically or divided into equal sections with spacer blanks. In the case of aquifers with a thickness of less than 6 m, it may be okay to choose the lower limit at 70% of the aquifer thickness. Experience has shown that 90% of the maximum specific capacity can be obtained by screening 75% of the aquifer thickness.

Non-Homogeneous Artesian Aquifer

The general rule of thumb is to screen the most permeable portion. The permeability of different layers can be estimated by visual inspection, checking the well log and more commonly doing sieve analysis.

Homogeneous Water Table Aquifer

In water table aquifers, during pumping, the aquifer is dewatered, so only the lower portion of the aquifer is screened. Mostly the bottom one-third of the aquifer is screened. Screening the bottom one-third will leave two-thirds of the aquifer thickness as available drawdown. In extreme cases, the bottom one half can be screened to achieve a higher specific capacity by reducing the convergence of flow, but this will decrease the available drawdown.

Non-Homogeneous Aquifer

The same rules apply as for a non-homogeneous artesian aquifer, except the screen is positioned in the lower part of the aquifer to achieve more available drawdown as the aquifer is actually being dewatered in this case.

3.2.3 Grain Size Analysis

Grain size distribution is required when selecting screen openings. It is assessed by passing a desegregated sample through a set of sieves. The cumulative weight of the particles retained on each sieve is plotted as a percent of the total sample weight versus grain size or sieve opening size. A typical grain size distribution curve is shown in **Figure 3.3**. Both the coarseness and the uniformity of the material can be studied from the grain size distribution curve.

On the particle distribution curve, effective size is defined as the particle size where 10% of the material is finer (passes through) and 90% is coarser (retained). The median size is the size where

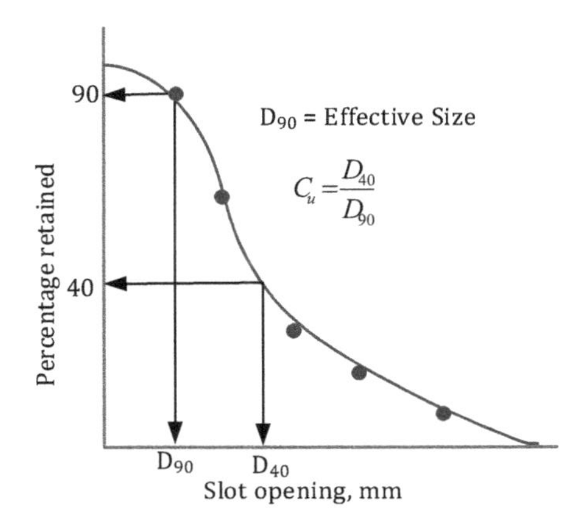

FIGURE 3.3 Distribution Curve

50% of the material is finer and 50% is coarser. For relatively uniform sand, where the major portion of the distribution curve is quite steep, the general slope of the distribution curve indicates the uniformity of the particles in the sample. Since the curve is not a straight line, the slope or steepness of the curve is defined by taking two points on the curve, D_{40} and D_{90}, and calling it the uniformity coefficient (C_u). A low coefficient value indicates steeper slopes and thus uniform sands, and a high coefficient value indicates non-uniform sands. Fine sands with a uniformity coefficient of less than 2 are considered relatively uniform.

3.3 NATURALLY DEVELOPED WELL

For a naturally developed well, the general rule is to select a slot opening that will retain 40-50% of the aquifer material. A larger slot (D_{40}) is chosen when the water is not corrosive and the bore samples are reliable. A smaller slot size (D_{50}) is recommended when the water is corrosive, samples are not certain and you plan on having a conservative design. Very large openings or slots (D_{30}) are suggested when the formation is non-uniform, as sand and gravel achieves a higher specific capacity and the water may be scale forming. Large openings require more development work to wash out fine material. Selecting screen openings of different sizes does not increase the cost, but it does increase well performance. In a stratified formation, two additional rules apply.

1. When there is fine material overlying coarser material, do not extend less than 1 m of a screen with slots designed for finer material.
2. In such cases, the screen size for coarser material should not be more than double the screen size used for finer material. This will provide a gradual transition.

Example Problem 3.1

An aquifer consists of two strata. The depths and grain size distributions are given below. Select the screen length and slot openings.

			Grain Size/Opening, mm			
Strata	**Depth, m**	**Thickness, m**	D_{30}	D_{40}	D_{50}	D_{70}
A	43–47	4.6	0.90	0.75	0.64	0.25
B	46–50	3.0	2.4	2.0	1.7	0.66
Full	43–50	7.6				

SOLUTION:

$$\text{Screen length} = 70\% \text{ of total thickness}$$

$$= 0.7 \times 7.6 = 5.3 = 5.0 \text{ m.}$$

As seen in **Figure 3.4**, aquifer strata B is more permeable, so the entire portion of this stratum should be screened. The screen opening is based on a slot size retaining 40% of aquifer material.

As shown in the table, D_{40} for A and B are 0.75 mm and 2.0 mm, respectively. However, the jump in slot size from 0.75 mm to 2.0 mm is too big. It is suggested that an intermediate size be selected for a small section extending into the coarser strata to provide a smooth transition.

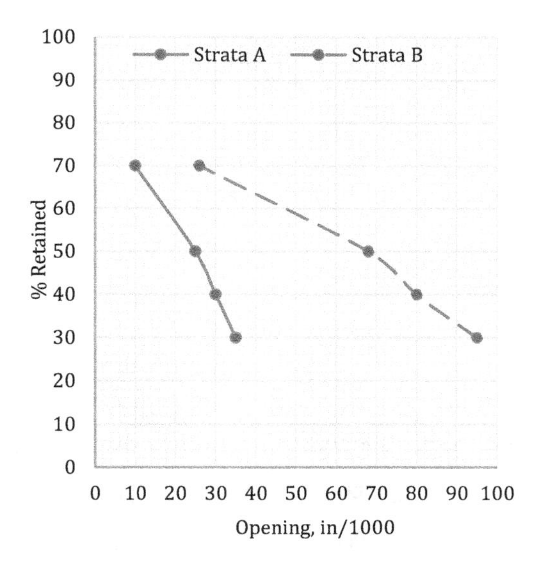

FIGURE 3.4 Plots (Ex. Prob 3.1)

Select a 0.5 metre length section with a slot size of 1.5 mm as the middle section. The selected slot sizes are tabulated below.

Strata	Depth, m	Length, m	Slot Size, mm
A	43–47	2.5	0.75
B	47–47.5	0.5	1.5
B	47.5–50	2.0	2.0

3.3.1 Screen Diameter

The basic principles of hydraulics dictate that the screen should have sufficient open area to maintain a desirable entrance velocity. The open area of the screen is determined by its length, diameter and slot openings. The length of the screen is primarily based on aquifer thickness while the slot openings are determined by the fineness of the formation material. Once two of these variables are fixed, diameter remains the only variable in the design of the intake section that can be changed to achieve an adequate open area. As discussed before, the diameter has no significant bearing on the well yield.

Screen Open Area

Screen open area should be such that the entrance velocity is less than 3.0 cm/s. This will ensure the good hydraulic efficiency of the well. At high entrance velocities greater than 3.0 cm/s, the problem of sand pumping and encrustation may also appear. Recent studies have shown that for screens with an open area greater than 3–5%, frictional head losses or well loss across the screen are at a velocity of less than 0.75 m/s at a minimum. The entrance velocity for a given pumping rate will depend upon the open area A_o. The open area of a given screen depends on the wire width (WW), slot size (SS) and length (L) of the screen and the diameter (D) of the screen. The open area can be expressed in terms of slot size and wire width as follows:

$$A_o = \frac{Q}{v_e} = \pi DL \times F_o = \pi DL \times \frac{SS}{(SS + WW)}$$

Q = pumping rate
v_e = entrance velocity
SS = slot size
WW = wire width
F_o = fraction open area

Example Problem 3.2

For the data given below, check if the screen has an adequate open area.

L = 10 m   D = 150 mm   Q = 60 L/s   WW = 4 mm   SS = 1.5 mm

SOLUTION:

$$F_o = \frac{SS}{(SS + WW)} = \frac{1.5\ mm}{(1.5 + 4.0)mm} = 0.27 = 27\%$$

$$v_e = \frac{Q}{\pi\, D\, F_o\, L} = \frac{0.06\ m^3}{s} \times \frac{1}{\pi \times 0.15\ m \times 0.27 \times 10\ m}$$

$$= 0.047 m/s = 4.7 cm/s$$

This velocity is greater than the recommended velocity of 3.0 cm/s, so the new diameter is:

$$D = \frac{Q}{\pi\, L\, F_o\, v_e} = \frac{0.06\ m^3}{s} \times \frac{s}{\pi \times 10\ m \times 0.27 \times 0.03\ m}$$

$$= 0.235\ m = \underline{250\ mm}$$

3.4 ARTIFICIAL GRAVEL PACK

A gravel pack is properly graded material that is placed in the annular space between the screen and the casing. The median size of gravel pack material is about 7–10 times larger than that of the finer strata of the aquifer. During well development, finer material is pumped out, leaving coarse material around the gravel pack. Thus, together the screen and gravel pack become the intake portion of the well.

The installation of an artificial gravel pack has become very common in water well construction. Unlike a naturally developed well where 50–60% of the finer material surrounding the screen is removed, an artificial-gravel packed well contains relatively coarse uniform graded material.

A gravel pack is almost essential for large capacity or municipal wells. In addition, the use of a gravel pack is strongly recommended when the aquifer material is uniform fine sand and the aquifer is relatively thick. It is advantageous to have a gravel pack in extensively stratified aquifers, as it removes the need to design and locate different screen sizes.

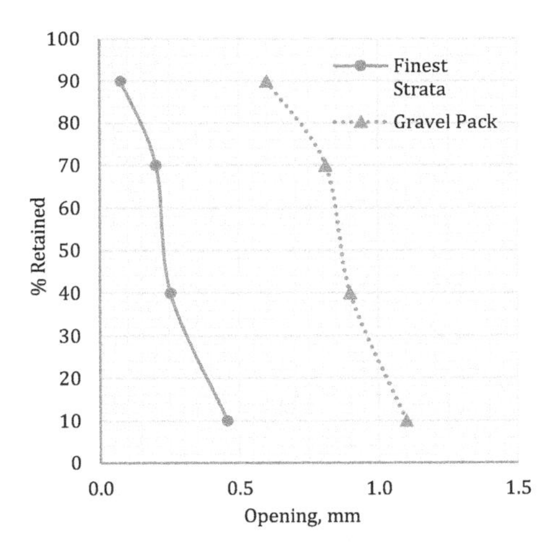

FIGURE 3.5 Design of Gravel Pack

3.4.1 Design of a Gravel Pack

The size and gradation of gravel pack material are based on the finest strata of the aquifer material. Here are the steps required for designing a gravel pack:

1. Construct sieve analysis and prepare particle size distribution curves. The finest sand strata will form the basis of the design, as shown in **Figure 3.5**.
2. Select D_{70} from step 1. Multiply this by a factor of four for fine uniform, five for coarse non-uniform or six for highly non-uniform material.
3. Draw a smooth curve through this point such that the uniformity coefficient is less than 2.5 and select a grade of gravel pack allowing for an 8–10% tolerance.

Example Problem 3.3

The grain size distribution curve for formation material is shown in **Figure 3.5**. Select the size of gravel pack material and screen openings.

Given: $D_{70} = 0.20$ mm
$D_{90} = 0.13$ mm
$D_{40} = 0.25$ mm

SOLUTION:

$$C_u = \frac{D_{40}}{D_{90}} = \frac{0.25\ mm}{0.13\ mm} = 2.0 = Relatively\ uniform$$

Based on an effective size of 0.13 mm, the aquifer material can be classified as fine uniform. A multiplier factor of 4 is chosen.

$$D_{70}(GP) = 4 \times 0.20\ mm = 0.80\ mm$$

Based on this point, a curve is drawn such that the uniformity coefficient is < 2. From GP curve:

$$C_u(GP) = \frac{D_{40}}{D_{90}} = \frac{0.90\ mm}{0.60\ mm} = 1.5$$

$$D_{90}\ (GP) = 0.70\ mm = 26\ slot$$

3.4.2 Gravel Pack Material

The material used for a gravel pack should be clean so that there is less loss of material during development and a shorter development time. In addition, the material should consist of well-rounded grains that are smooth and uniform. Preferably, the material should consist mostly of siliceous rather than calcareous particles so that there is no loss from dissolution. This is also important because acid treatment might be required in the well later. In theory, only a few mm of gravel pack material around the screen is required for retaining formation particles. However, it is impractical to place less than 50 mm. A minimum of 75 mm is considered practical. The upper limit of gravel thickness is 200 mm. A gravel pack that is too thick can make the development of the well more difficult and less efficient. A thicker envelope does not significantly increase the yield of the well.

3.5 WELL DEVELOPMENT

The final step of well construction is well development. Well development cleans out and removes fine particles around the intake portion of the well. The development of a well minimizes formation damage, improves hydraulic efficiency and stabilizes the aquifer or gravel material around the well to effect sand-free production.

In naturally developed wells, the finer material in the vicinity of the screen is removed to establish a zone of high permeability. In gravel packed wells, development eliminates the fines and bridges between gravel particles. The most commonly used methods of well development are described in the following sections.

Bailing

Bailing water from the well causes flow from the aquifer and hence the removal of fine material. This is a slow process. Due to the absence of reverse flow, this method is not very effective.

Mechanical Surging

This method is used mostly when the well is drilled by cable tool or rotary methods. The movement of a surge plunger or swab causes flow into and out of the screened area to break the bridges and wash out the fine material. The fine particles are removed by bailing or by fluid circulation. To avoid damaging the screen, surging should start slowly and the speed should be gradually increased. This method is effective in wells constructed with a thick gravel pack.

Air Surging

In air surging, compressed air (700–1000 kPa) is used to surge the formation and to lift the material brought into the well. To successfully develop a well, the submergence ratio (length of the airline divided by the total length) is important. The optimum working submergence should be about 60% for good results. This is an effective method of well development.

Jetting

Jetting moves fine particles and breaks down bridges and mud cake. In difficult formations or those drilled with clay-based mud, polyphosphate can be added to jetting water or directly to the well as a dispersing agent. The recommended dosage of sodium hexametaphosphate is 1.2 g/L. If organic mud has been used, a chlorine dosage of 0.4 g/L is suggested. A high velocity jetting method is suitable for the development of wells drilled into layered fine sands and clayey materials.

Backwashing

Backwashing by interrupted pumping is also called rawhiding. A surging effect or reverse of flow can be caused by alternatively lifting the water to the surface and then letting it run back into the well. Besides using an airlift pump, a turbine pump without a foot valve can be used for this purpose. The surging effect is less vigorous and not very uniform. Although it is very common in developing domestic wells, its overall effectiveness in high-capacity wells is limited. The effectiveness of this method can be enhanced using water under pressure or compressed air. Backwashing can be adapted for the development of wells drilled using rotary drilling and without excessively thick gravel packs.

Overpumping

Though not an effective method overpumping can be very effective when used in combination with other methods such as surging. It can be used to develop shallow wells fitted with agricultural strainers that may get damaged under high pressure. In wells drilled with clay-based muds, overpumping followed by surging and jetting has shown excellent results.

3.6 WELL HYDRAULICS

A water well is a hydraulic structure that, when properly designed and constructed, permits the economic withdrawal of water from an aquifer. For the proper design and operation of wells, it is important to understand some basic principles of well hydraulics. Darcy's law, combined with Laplace's equation, allows us to study the groundwater movement under a given set of conditions. In the following section, the derivation and application of equilibrium (steady state) well equations for artesian and water table aquifers are discussed. The application of non-equilibrium well equations pertaining to well testing and the determination of aquifer parameters are also discussed.

3.6.1 Steady Flow to an Artesian Well

Referring to **Figure 3.6** below, consider a cone of depression at a distance, r. The flow passing through the surface of this cylinder can be determined by applying Darcy's law. For steady state equilibrium conditions, this rate must be equal to the pumping rate, Q.

Under equilibrium conditions, the pumping rate must equal the water entering the cylinder defined by the outside boundary of the cone of depression.

$$Q = k(2\pi rb)\frac{dh}{dr}, \quad by\ integration \quad Q\int_{r_1}^{r_2}\frac{dr}{r} = 2\pi kb\int_{h_1}^{h_2} dh\ or$$

$$Q = \frac{2\ \pi kb(h_2 - h_1)}{\ln(r_2 / r_1)}$$

Subscripts 1 and 2 are the two observation wells to observe drawdowns. The equilibrium equation shows that h increases indefinitely with an increase in r and maximum $h = h_0$ at the radius of influence. Thus, from a theoretical aspect, steady radial flow in an extensive aquifer does not exist because the cone of depression must expand indefinitely with time.

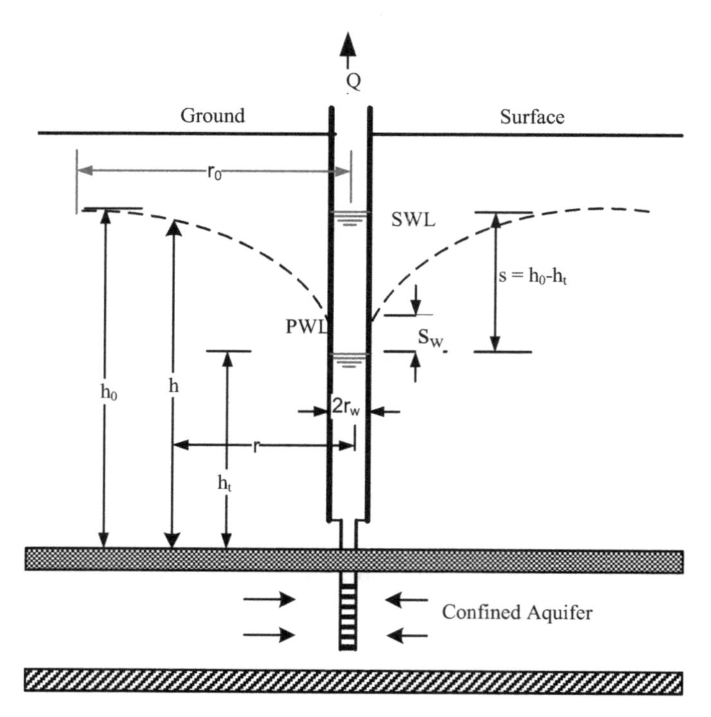

FIGURE 3.6 Pumping an Artesian well

Artesian Well Yield

When point 1 refers to the outside of a well and point 2 lies at the radius of influence, that is $r_1 = r_w$ and $r_2 = r_0$, and $(h_0 - h_w) = s_w$ = drawdown in the pumped well, the equilibrium well equation becomes:

$$\text{Well yield, } Q = \frac{2\ \pi kb\left(h_0 - h_w\right)}{\ln\left(r_0 / r_w\right)} = \frac{2\ \pi kbs_w}{\ln\left(r_0 / r_w\right)} = \frac{2\ \pi Ts_w}{\ln\left(r_0 / r_w\right)}$$

Specific Capacity

Well yield is more commonly expressed as the specific capacity of the well. The specific capacity of a well is the yield per unit drawdown in the pumped well. T = transmissivity

$$SC = \frac{Q}{s_w} = \frac{2\ \pi kb}{\ln\left(r_0 / r_w\right)} = \frac{2\ \pi T}{\ln\left(r_0 / r_w\right)}$$

It can be said that the yield of the well is directly proportional to the drawdown. This is not strictly true since some drawdown is caused due to well losses. In an ideal scenario, well losses are negligible. In real situations, the well losses will be greater than zero. In other words, the drawdown inside the pumped well will be greater than the theoretical drawdown given by the well equation. The difference in the actual and theoretical drawdown indicates the well losses. The well losses might increase over the years, thus causing a drop in well efficiency.

Well Efficiency

Well efficiency accounts for well losses as the water enters the well. In a perfect well, (ideal) losses would be zero or the well is 100% efficient. However, in the real world, there will always be some losses. Well drawdown and specific capacity as given by the well equation are theoretical and do not account for well losses. The equilibrium equation accounts for drawdown due to formation only. In real situations, the drawdown in the pumped well will be more and so well efficiency will be less than 100%.

$$\text{Well efficiency, } E_w = \frac{s_i}{s_a} = \frac{ideal}{actual} = \frac{\text{Formation loss}}{\text{Formation loss} + \text{Well loss}}$$

Example Problem 3.4

An artesian well is installed in a 14 m thick aquifer at a depth of 32 m from the ground surface. The diameter of the intake section is 0.60 m. The piezometric surface is at a depth of 9.0 m and the radius of the cone of depression is 350 m. Given that the permeability of the sandstone is 0.070 mm/s, what would be the well yield if the allowable drawdown in the pumped well is 13 m?

Given: $k = 0.070$ mm/s
$b = 14$ m
$h_w = 23\text{-}13 = 10$ m
$r_w = 0.30$ m
$s_w = 13$ m
$h_o = 32\text{-}9.0 = 23.0$ m
$r_o = 350$ m
$Q = ?$

SOLUTION:

$$Q = \frac{2\,\pi\,kbs_w}{\ln(r_0/r_w)} = 2\pi \times \frac{0.07\ mm}{s} \times 14\ m \times \frac{m}{1000\ mm} \times \frac{13\ m}{\ln(350/0.30)}$$

$$= \frac{0.01133\ m^3}{s} \times \frac{1000\ L}{m^3} = 11.33 = \underline{11\ L/s}$$

Estimating T or K

Applying the equilibrium equation, the transmissivity of an aquifer can be determined by observing the data at two points when the well installed in that aquifer is pumped.

$$Q = \frac{2\,\pi T(h_2 - h_1)}{\ln(r_2/r_1)} \quad or \quad \text{Transmissivity, } T = kb = \frac{\ln(r_2/r_1)\,Q}{2\,\pi(h_2 - h_1)}$$

All of the parameters on the right-hand side of this equation can be determined from a pumping test. The well is pumped at a constant rate Q to reach the equilibrium conditions. Drawdowns in two observation wells located at distances r_1 and r_2 from the pumped well are observed.

Radius of Influence

When drawdowns (s_1, s_2) in two observation wells 1 and 2 ($r_2 > r_1$) are known, the radius of influence can be estimated by employing steady state equations.

Example Problem 3.5

A 60 cm diameter well is constructed in a 15 m thick confined aquifer. The well is pumped at a rate of 13 L/s and equilibrium conditions are achieved after two days of pumping. The observed data is given in the following table. Calculate the permeability, specific capacity, radius of influence and maximum yield.

Parameter	Pumped Well	1	2	3
r, m	0.3 (r_w)	10	30	(r_0)
s, m	9.6	3.7	2.4	0
h, $h_0 - s$	?	16.3	17.6	20

Given: Q = 13 L/s
b = 15 m
$s_{w\text{ - actual}} = 9.6$ m
K, T, SC, s_w, E_w, s_0 = ?

SOLUTION:

$$T = \frac{\ln(r_2/r_1)\,Q}{2\,\pi(h_2 - h_1)} = \frac{\ln(30/10)}{2\pi} \times \frac{13\,L}{s} \times \frac{1}{(17.6-16.3)m}$$

$$= \frac{1.748L}{s.m} \times \frac{m^3}{1000L} \times \frac{3600\ s}{h} \times \frac{24}{d} = 151.0 = \underline{150\ m^2/d}$$

$$k = \frac{T}{b} = \frac{151.0\ m^2}{d} \times \frac{1}{15\ m} \times \frac{d}{24h} \times \frac{h}{3600\ s} \times \frac{1000\ mm}{m}$$

$$= 0.116 = \underline{0.12\ mm/s}$$

$$(s_w - s_1) = \ln\left(\frac{r_1}{r_w}\right)\frac{Q}{2\pi T} = \ln\left(\frac{10\ m}{0.3\ m}\right) \times \frac{13\ L}{s} \times \frac{s.m}{2\pi \times 1.748\ L}$$

$$s_w = s_1 + 4.15\ m = 3.7m + 4.15\ m = 7.85 = 7.9\ m$$

$$Specific\ capacity,\ Q_{sp} = \frac{Q}{s_w} = \frac{13\ L}{s} \times \frac{1}{9.6\ m} = 1.35 = 1.4\ \text{L/s}.m$$

$$Well\ efficiency,\ E_w = \frac{s_i}{s_a} = \frac{7.85\ m}{9.6\ m} \times 100\% = 81.7 = 82\%$$

$$Max.\ yield,\ Q_{max} = Q_{sp}s_{max} = \frac{1.36\ L}{s.m} \times 20\ m = 27.2 = \underline{27\ L/s}$$

$$r_0 = r_1\ e^{\frac{2\pi T s_1}{Q}} = 10\ m\ e^{2\pi \times \frac{1.748L}{s.m} \times \frac{s}{13L} \times 3.7\ m} = 227.8 = \underline{230\ m}$$

3.6.2 Unconfined Well Equation

The equilibrium equation for a water table well can be found by considering a cone of depression of radius r, thickness dr and height h such that the gradient at point r is dh/dr, as shown in **Figure 3.7**.

Under equilibrium conditions, the pumping rate must be equal to that of the water entering the cylinder defined by the outside boundary of the cone of depression.

$$Q = k(2\pi rh)\frac{dh}{dr} \qquad Integrating\quad Q\int_{r_1}^{r_2}\frac{dr}{r} = 2\pi k\int_{h_1}^{h_2} hdh$$

$$Q = \frac{\pi k\left(h_2^2 - h_1^2\right)}{\ln\left(r_2/r_1\right)}$$

Note, in the case of an unconfined aquifer, the thickness of the aquifer varies and the aquifer is actually getting dewatered. This leads us to conclude that by doubling the drawdown in the well, the yield will be less than doubled. Assuming average head = $(h_1 + h_2)/2$

$$Q = \frac{\pi k\left(h_2^2 - h_1^2\right)}{\ln\left(r_2/r_1\right)} = \frac{\pi k\left(h_2 - h_1\right)\left(h_2 + h_1\right)}{\ln\left(r_2/r_1\right)} = \frac{2\,\pi k\bar{h}\left(h_2 - h_1\right)}{\ln\left(r_2/r_1\right)}$$

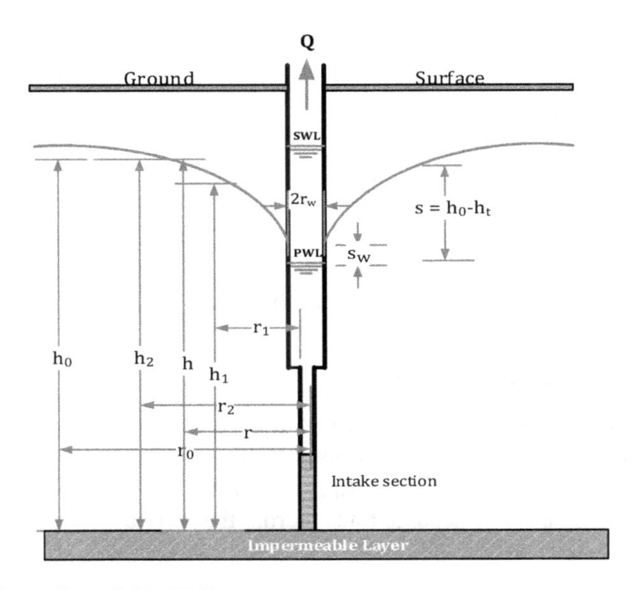

FIGURE 3.7 Pumping a Water Table Well

This is comparable to the equilibrium equation for an artesian well. Instead of using the thickness of the artesian aquifer, you can substitute in the average hydraulic head. This equation fails to accurately describe the drawdown curve near the well because of the large vertical flow.

Determining K

Observing h_1 and h_2 in two observation wells located at r_1 and r_2 from the pumped well allows us to determine the average hydraulic conductivity of the aquifer.

$$Q = \frac{\pi k\left(h_2^2 - h_1^2\right)}{\ln\left(r_2 / r_1\right)} \quad or \quad k = \frac{Q\ln\left(r_2 / r_1\right)}{\pi\left(h_2^2 - h_1^2\right)}$$

Example Problem 3.6

To determine the aquifer characteristics of an unconfined aquifer, a pumping test was conducted using a test well extending to the underlying impervious stratum at a depth of 20 m. During the test, the pumping rate was kept constant at 25 L/s. The observed data is shown in the table below. Find the coefficient of permeability.

Parameter	1	2
Distance r, m	20	120
Drawdown s, m	14.0	0.0
Head $h = h_o - s$, m	11.96	14.2

Given: Q = 25 L/s
$h_0 = 20 - 5 = 15$ m

SOLUTION:

$$k = \frac{Qln\left(r_2 / r_1\right)}{\pi\left(h_2^2 - h_1^2\right)}$$

$$= \frac{25\ L}{s} ln\left(\frac{120\ m}{20.\ m}\right) \times \frac{1}{\pi\left(14.2^2 - 11.96^2\right)m^2} \times \frac{m^3}{1000L} \times \frac{1000\ mm}{m}$$

$$= \frac{0.243\ mm}{s} \times \frac{3600\ s}{h} \times \frac{24\ h}{d} \times \frac{m}{1000\ mm} = 21.0 = \underline{21 m / d}$$

3.6.3 Flow into Infiltration Gallery

Consider an infiltration gallery of length L located at a distance R from the source. Let H_L be the height of the water level in the source, H_0 be the height of the water in the gallery and k be the hydraulic conductivity of the porous medium. Applying Darcy's law, at any point x from the gallery, under equilibrium conditions, the discharge rate entering the vertical section can be expressed as:

$$Q = -khL\frac{dh}{dx} \quad integerating,\ Q\int_0^R dx = -kL\int_{H_0}^{H_L} hdh \ \ or$$

$$Q = \frac{kL\left(H_L^{\,2} - H_0^{\,2}\right)}{2R}$$

If the flow enters from both sides, the discharge will be doubled.

Example Problem 3.7

An infiltration is situated at 12 m from the ditch feeding the gallery. The ditch penetrates the permeable zone fully and the water level is at a height of 6.5 m. What is the discharge rate into the gallery per unit length when the water level in the gallery is 2.0 m and the hydraulic conductivity is known to be 90 m/d?

Given: q = Q/L = ?
H_0 = 2.0 m
H_L = 6.5 m
k = 250 m/d
R = 12 m

SOLUTION:

$$\frac{Q}{L} = \frac{k\left(H_L^{\,2} - H_0^{\,2}\right)}{2R} = \frac{90\ m}{d} \times \left(6.5^2 - 2.0^2\right)m^2 \times \frac{1}{2\times 12\ m}$$

$$= 143.44 = \underline{140\ m^3/m.d}$$

3.6.4 Modified Non-Equilibrium Equation

Pumping conditions are usually unsteady. That is to say, well drawdown increases as pumping duration increases, though at a lesser rate. When pumping conditions are unsteady, aquifer parameters can be found using Cooper and Jacob modified non-equilibrium equations. Let us say an artesian aquifer with transmissivity, T, and coefficient of storage, S, is pumped at rate Q. When *r* is small and *t* is large, drawdown s at any radial distance r from the pumped well is given by the following equation:

$$Drawdown,\ s = \frac{Q}{4\pi T}\times \ln\left(\frac{2.25Tt}{r^2S}\right) = \frac{0.183Q}{T}\times \log\left(\frac{2.25Tt}{r^2S}\right)$$

Based on this modified equation, we can draw the following conclusions:

1. Drawdown increases in direct proportion with increasing pumping rate.
2. For the same pumping rate, drawdown will be higher in an aquifer with low T.
3. For a given pumping rate, drawdown per log cycle of time is constant. This allows for the prediction of drawdown at various continuous pumping times.
4. Drawdown decreases with logarithms of distance away from the pumped well.

Example Problem 3.8

The transmissivity and storage coefficient of an artesian aquifer is 200 m²/d and 5 × 10⁻⁴, respectively. Predict the drawdown just outside a pumped well of diameter 1.0 m after 1 day of pumping at 16 L/s.

Given: r = 0.5 m

T = 200 m²/d

S = 5 × 10⁻⁴

t = 1 d

Q = 16 L/s = 1382 m³/d

SOLUTION:

$$log\left(\frac{2.25Tt}{r^2S}\right)=log\left(2.25\times\frac{250m^2}{d}\times 1d\times\frac{1}{(0.5m)^2\times 5\times 10^{-4}}\right)=6.653$$

$$s=\frac{0.183Q}{T}log\left(\frac{2.25Tt}{r^2S}\right)=0.183\times\frac{d}{250\ m^2}\times\frac{1382\ m^3}{d}\times 6.653$$

$$=6.73=\underline{6.7\ m}$$

Well equations are derived based on assumptions like constant thickness, S and T, and laminar flow conditions, which in the real world are rarely found completely valid. However, these equations can yield reliable results provided good judgement is applied. If in a certain case there is a serious derivation from the above-mentioned assumptions, extra caution should be exercised in making decisions based on results obtained using equilibrium equations.

Time Drawdown Relationship

According to this equation, the plot of drawdown versus logarithm of time (s vs $log\ t$) will yield a straight line or a linear relationship. This is true for large values of t or small values of r when assumptions apply. The equation of the straight line represents the time drawdown relationship. Alternately, the data of s and $log\ t$ is regressed to yield the equation of a straight line.

$$\mathrm{s}=\mathrm{a}+\mathrm{b}\log\mathrm{t}$$

Constants a and b, respectively, are the intercept and slope of the straight line. The value of transmissivity and storativity may be found by fitting the equation and knowing the value of a and b. Writing the modified equilibrium equation for time t_1 and t_2 such that $s_2 > s_1$

$$s_2-s_1=\frac{0.183Q}{T}\log\left(\frac{t_2}{t_1}\right)$$

When $(t_2/t_1) = 10 = 1$ log cycle; $\log(t_2/t_1) = \log 10 = 1$; and $s_2 - s_1 = \Delta s$ = change in drawdown/log cycle of time. Δs represents the slope of the straight line. The value of the slope ($\Delta s = b$) is used to calculate transmissivity. By fitting the straight line to the time drawdown data, the slope of the line, Δs and time t_0 can be determined.

$$T=\frac{0.183Q}{\Delta s}\qquad S=\frac{2.25Tt_0}{r^2}\qquad s_2=s_1+\Delta s\log\left(t_2/t_1\right)$$

Example Problem 3.9

An artesian well is pumped at a constant rate of 1.4 m^3/min and the drawdown is observed in a test well located at a distance of 14 m. The time drawdown data observed over a period of 500 minutes indicates that the change in drawdown over a period of one log cycle is 5.3 m. Find the transmissivity of the aquifer using the modified non-equilibrium equation.

Given: Q = 1.4 m^3/min
r = 14 m
Δs = 5.3 m/log cycle
T = ?

SOLUTION:

$$T = \frac{0.183Q}{\Delta s} = \frac{0.183}{5.3\ m} \times \frac{1.4\ m^3}{min} \times \frac{1440\ min}{d} = 69.6 = \underline{70\ m^2\ /\ d}$$

Example Problem 3.10

A well penetrating a confined aquifer is pumped at a uniform rate of 2500 m^3/d. Drawdowns during the pumping period are measured in an observation well 60 m away and are as follows:

Time, t (min)	1	2	5	10	18	30	60	100	240
Drawdown, s (m)	0.2	0.3	0.45	0.57	0.67	0.76	0.9	0.96	1.12

Plot the time drawdown curve and fit the equation by regression analysis. Compute T and S using the Cooper-Jacob method. What will be the drawdown in the observation well after pumping the aquifer continuously for 20 h?

SOLUTION:

A plot of the data is shown in **Figure 3.8**.

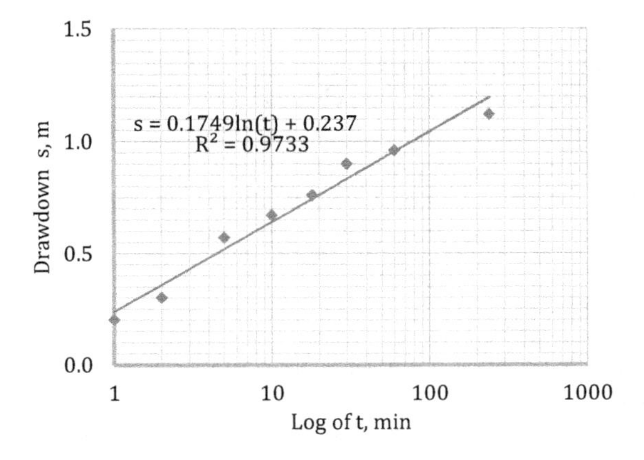

FIGURE 3.8 Plot of Time Drawdown Data

$$s = 0.2375 + 0.1749 ln\, t = 0.2375 + 0.4022 log\, t$$

$$\Delta s = \frac{slope}{log\, cycle} = 0.40\, m \quad and\, t_0 = e^{\left(-0.1749/0.2375\right)} = 0.26\, min$$

$$T = \frac{0.183Q}{\Delta s} = \frac{0.183}{0.4022\, m} \times \frac{2500\, m^3}{d} = 1137 = 1140\, m^2 / d$$

$$S = \frac{2.25Tt_0}{r^2} = 2.25 \times \frac{1137\, m^2}{d} \times \frac{0.33\, min}{(60\, m)^2} \times \frac{d}{1440\, min} = 1.42 \times 10^{-4} = \underline{1.4 \times 10^{-4}}$$

$$s = 0.2375 + 0.4022 log\left(20h \times \frac{60\, min}{h}\right) = 1.475 = \underline{1.5\, m}$$

Distance Drawdown Curve

Similar to the time drawdown relationship, a distance drawdown relationship can be developed and used to estimate aquifer parameters. After constantly pumping for a given period (t = constant), the values of drawdown (s) are observed at various distances (r varies) from the pumped well. The plot of s versus r is known as the distance drawdown curve. As indicated by the non-equilibrium equation, the plot of s versus log r should be linear. In other words, the plot of s versus r on semi-logarithmic paper should be a straight line. However, carefully observing the non-equilibrium equation reveals that the slope of the distance drawdown curve is opposite to that of the time drawdown. Therefore, the drawdown increases with pumping time but decreases as we move away from the well.

3.6.5 Recharge and Boundary Effect

If during pumping significant recharge is taking place, the slope of the time drawdown curve becomes flatter and so the transmissivity calculated from the flatter slope will be higher than the true value, and the storativity would be lower than the real value. If an impermeable boundary is encountered, it will make the time drawdown steeper and yield lower values of transmissivity.

3.7 FAILURE OF WELLS AND REMEDIATION

The clogging of wells by filling with sand or by corrosion or incrustation of the screen may reduce the yield greatly. Wells may be readily cleaned of sand by means of a sand pump or bucket, but if the strainers are corroded, they must be pulled out, cleaned, and renewed or replaced. If the clogging is due to fine sand collecting outside the tube, it may be removed to some extent by forcing water into the well under high pressure or by use of a hose or by other suitable means. Sometimes instead of the yield of a well becoming less through continued operation, it is actually increased owing probably to the gradual removal of the fine material immediately surrounding the well.

3.8 SANITARY PROTECTION

Water wells must be protected against pollution. Any activity or operation such as landfill that may impair groundwater must be prohibited within the well head area (50 m radius). Well casing should extend at least 0.4 m above the ground surface and be properly grouted. A sanitary seal should be provided to protect against the entry of contaminated water into the ground.

3.9 WELL ABANDONMENT

Once a well has stopped being useful, it must be abandoned properly. The well bore should first be filled with contaminated material and the top portion grouted with cement mixture. If this is not done properly, it can become a source of major pollution and a safety issue.

Discussion Questions

1. Discuss the various methods used for constructing municipal wells in terms of their suitability for drilling in different types of formation.
2. Compare open wells with tube wells.
3. Compare artesian wells and water table wells.
4. Describe the various methods of well development.
5. Derive an equation for steady radial flow into an artesian well.
6. Define and explain the following terms related to wells.
   a. Transmissivity and storage coefficient
   b. Specific capacity, well efficiency
   c. Shallow well, deep well
   d. Well rehabilitation, well development
7. Most municipal wells are gravel packed. Discuss.
8. What makes up the intake portion of a well? What are the steps required for designing the intake portion of a well?
9. Using the modified form of non-equilibrium equation, derive an expression for the transmissivity and coefficient of storage in terms of characteristics of time drawdown curve, $s = \frac{0.183Q}{T} log\left(\frac{2.25Tt}{r^2S}\right)$
10. Compare an infiltration gallery with a water table well both from a hydraulics and a construction point of view.
11. Write short notes on sanitary protection, well abandonment and well remediation.
12. From the modified form of non-equilibrium equation, prove that theoretically the slope of a distance drawdown curve is twice that of a time drawdown curve.
13. How might the chemistry of well water affect the design of a well intake section?
14. Search the internet to find out if there are any government regulations pertaining to groundwater wells (groundwater withdrawal, construction, design and abandonment). Search for the proper procedure for the abandonment of wells.

Practice Problems

1. An artesian aquifer has a transmissivity of 160 $m^3/m \cdot d$. The piezometric readings in two observation wells located at 10 m and 150 m are 12.45 m and 15.5 m, respectively. What is the steady pumping rate? (13 L/s)
2. Determine the permeability of a 25 m thick artesian aquifer being pumped at a constant rate of 125 L/s. At equilibrium, the drawdown in an observation well 25 m away is 3.5 m and the drawdown in the second observation well 200 m away is 0.25 m. (44 m/d)
3. A 30 cm well fully penetrates a 24 m thick confined aquifer. The well is pumped at the rate of 65 L/s. Two test wells are located at a distance of 15 m and 40 m, respectively. After reaching equilibrium conditions, the drawdowns in wells 1 and 2 were observed to be 3.1 m and 1.8 m, respectively. Estimate the transmissivity of the aquifer. Also,

find the theoretical drawdown in the pumped well and hence the theoretical specific capacity. If the actual drawdown in the well is observed to be 11.7 m, determine the efficiency of the well. (T= 1300 m/d, s = 11 m, SC = 6.2 L/s.m, Efficiency = 90%)

4. A 20 cm diameter well is pumping an unconfined aquifer with a coefficient of permeability of 40 m/d. After equilibrium conditions are achieved, the two observation wells located at distances of 10 m and 100 m indicate a water height of 7.3 m and 8.1 m, respectively. Determine the steady pumping rate. (7.8 L/s)
5. A well is pumped at a rate of 1.7 ML/d continuously for 30 days. The transmissivity and storativity of the aquifer, respectively, are 63 m^2/d and 6.4×10^{-4}. Applying the non-equilibrium equation, work out the drawdown just outside the 60 cm diameter main well. (40 m)
6. Predict the drawdown in a test well after 100 minutes of pumping, located at a distance of 60 m from an artesian well being pumped at a rate of 2500 m^3/d. The storativity and transmissivity of the aquifer are 2.0×10^{-4} and 1110 m^3/m·d, respectively. (1.0 m)
7. A municipal well is pumped at a constant rate of 20 L/s. For this well, the time drawdown equation is: $s = 0.5 + 3.9 \log t$. Calculate the transmissivity of the aquifer? (81 m^3/m.d)
8. A 75 cm municipal well is tapped into a confined aquifer. During a pumping test, the well is pumped at a constant rate of 55 L/s. The time drawdown data observed in a test well is regressed to yield the following data: r = 20 m, $s = 0.60 + 4.0 \log t$. where: s = m, t = min
   a. Predict the theoretical drawdown in the pumped well after 1 day of continuous pumping. (27 m)
   b. Determine the efficiency of the well if the actual drawdown in the pumped well is 32 m. (84%)
9. Water is proposed to be withdrawn from an infiltration gallery at a rate of 0.50 ML/d. The infiltration gallery is 80 m away from the feeding channel where the water table is 5.0 m. What must the length of the gallery be if the drawdown in the gallery should not exceed 3.0 m? Assume the hydraulic conductivity of the porous medium is 100 m/d. (38 m)
10. A 25 cm diameter water table well is pumped at a constant rate of 32 L/s. Before pumping started, the elevation of the static water level was recorded to be 22.0 m above the impervious layer. After achieving equilibrium conditions, drawdowns in the pumped well and observation well at 50 m respectively were 2.5 m and 0.45 m. What is the hydraulic conductivity of the aquifer? (63 m/d)
11. A well in a confined aquifer is pumped at a rate of 830 L/min for a period of 10 hours. Time drawdown data for an observation well located 25 m away was regressed to yield the following equation: $s = 1.61 + 1.89 \log t$ (s in m and t in h)
    Find T and S of the aquifer. (120 m^2/d, 2.4×10^{-3})
12. A fully penetrating artesian well is designed with a discharge rate of 50 L/s. The thickness of the aquifer is 22 m. Select the length and diameter of the screen assuming the available strainer has an open area of 15% and the entrance velocity is designed not to exceed 3.0 cm/s. (18 m, 200 mm, centred in the aquifer)
13. The following values are read from the grain size distribution curve of aquifer material.
    $D_{30} = 0.40$ mm $D_{40} = 0.35$ mm $D_{50} = 0.32$ mm $D_{90} = 0.11$ mm
    Do you think it is a wise decision to gravel pack this well? If yes, select the lower limit and upper limit of the median size of gravel pack material. Assume the median size of the gravel pack is 10–15 times that of the median size of the finest aquifer material. ($C_U = 3.2$, yes, 4.0 mm)

4 Water Demand

The proper design and execution of a water supply scheme requires an estimate of the total amount of water required for the community, called the demand. The demand is normally expressed as total annual demand in megalitres per annum (ML/a), average daily demand in megalitres per day (ML/d) or in litres per capita per day (L/c·d).

4.1 DESIGN PERIOD

The goal of designing a water supply scheme should be to meet the needs of the people served for a specified period in the future. The period of time for which the useful life of the project is defined is referred to as the design period. In a robust design, the lifespan of the individual components and the distribution system should exceed the design period. Also, the design of the scheme shall depend on the projected population to be served and their probable demand.

This warrants a study of the expected population increase in the community and an understanding of the trends of development in the region, including industrial and commercial growth. This period should not be too short, which demands immediate replacement, or too long, as the present population would require a huge investment for a long future. Water supply projects are normally planned to consider a design period of 30 years. The time lag between the design and completion of the project should also be accounted for in deciding the total design period, which should not exceed 2–5 years.

4.2 FORECASTING POPULATION

When designing a water supply scheme, the population of the town must be estimated by the end of the design period. The cost of any water supply scheme depends on the total population to be served and the total demand per capita per day. Any water supply scheme is usually designed in two stages: for an anticipated population after 25–30 years; and for a future population at the end of 50 or 60 years. Hence, it is important to know how to project the population at the end of a few decades based on the present trends in population growth.

There are several methods of predicting population. Most of these methods try to project the future population based on the present trend in population growth. Some of the most important methods are discussed in the following sections.

4.2.1 Arithmetical Increase

As the name indicates, population increase per decade, C, is assumed to be constant, thus indicating a linear relationship. In differential form, this is shown as $dP / dt = C$. By integrating the above formulation, the population at the end of n number of decades can be found.

$$P_2 = P_1 + C(t_2 - t_1) = P_1 + nC$$

This method is adopted for old and well-settled cities where the trend in population increase is more or less constant.

DOI: 10.1201/9781003231264-4

TABLE 4.1
Design Periods of Various Components

Component	Design Period
Storage reservoir/dams	50
Infiltration works	30
Pump house and civil works	30
Electric motors and pumps	15
Water treatment units	15
Pipe connections and small appurtenances	30
Raw and clear water conveyance mains	30
Overhead and ground-level service reservoirs	15
Distribution system	30

4.2.2 Geometrical Increase Method

In this method, the percentage increase in population from decade to decade is assumed to be constant. The population of two or more previous decades is considered, the percentage increase is calculated, and the mean percentage is adopted to calculate the population of future decades. If the present population is P and the average percentage growth is r, the population after n decades will be:

$$P_n = \left(1 + \frac{r}{100}\right)^n$$

4.2.3 Incremental Increase Method

In this scenario, the average increase per decade and the average incremental increase per decade are calculated and their sum is added to the present population. This process can be repeated for each successive decade until the population of the required decade is arrived at.

This method will yield results somewhere between the forecasts obtained using the arithmetic technique and the geometric method. This method yields better results since it assumes the growth rate varies from decade to decade.

Example Problem 4.1

In **Table 4.2** below, the population of a town is shown across six decades. Based on this data, forecast the future population for decades 2021 and 2031 using the three techniques discussed above.

SOLUTION:

In column 3 of **Table 4.2**, a change in population, ΔP, for every successive decade is calculated. At the end of the column, the cumulative and average values based on the data of six decades are shown.

Based on the worksheet, the average value of ΔP is calculated to be 30,200 per decade. In the next column, the change in population ΔP is expressed as a percentage. In column 5 of

TABLE 4.2
Excel Worksheet for Population Forecast

Year	Population ×1000	ΔP ×1000	ΔP %	Δ(ΔP) ×1000	Population Forecast × 1000		
					Arithmetic	Geometric	Incremental
1951	180						
1961	250	70	38.9				
1971	350	100	40.0	30.0			
1981	460	110	31.4	10.0			
1991	620	160	34.8	50.0			
2001	820	200	32.3	40.0			
2011	1100	280	34.1	80.0			
	Σ	920	211.5	210			
	Avg	153.3	30.2	42			
2021					1253	1432	1295
2031					1407	1865	1491

Table 4.2, the change of change of population or Δ(ΔP) is calculated. At the bottom of the column, the averages of these parameters are shown.

SAMPLE OF CALCULATION:

For the first two decades, $\Delta P = 250 - 180 = 70$
Based on the data of the first three decades, $\Delta(\Delta P) = 40 - 38.9 = 1.1$

Expected population in the decade 2021:

$$\text{Arithmetic}, P_{2021} = P_{2011} + \overline{\Delta P} = 1100 + 153.3 = 1253.3 = 1300$$

$$\text{Geometric}, P_{2021} = P_{2011}\left(1 + \frac{r}{100}\right)^n = 1100(1 + .3020)^1$$

$$= 1432.2 = 1400$$

$$\text{Incremental}, P_{2021} = P_{2011} + n\left(\overline{\Delta P} + \overline{\Delta(\Delta P)}\right) = 1100 + 1 \times (153.3 + 42)$$

$$= 1295 = 1300$$

As seen in **Table 4.2**, the arithmetical increase method gives a very low value, the geometric increase method gives a very high value, and the incremental increase method gives a reasonable value. Hence, the incremental increase method is generally adopted.

4.2.4 Decreasing Rate of Growth Method

In this method, contrary to the incremental increase method, it is assumed that the rate of percentage increase decreases and the average decrease in the rate of growth is calculated. Then the percentage growth rate is modified by deducing the decreasing rate of growth. This method is appropriate in cases where the rate of growth of the population shows a diminishing trend.

4.2.5 Graphical Extension Method

In this method, a graph is plotted of the population in thousands against time in decades. A smooth curve is extended to determine the future population. This gives approximate results.

4.2.6 Graphical Comparison Method

Here, for a town whose future population is to be predicted, other towns with similar population characteristics three or four decades in the past are considered. Their population curves are drawn and their curves are extended reasonably. With the help of these curves, the plotted data of the city in question can be extended so as to determine the population. In **Figure 4.2**, the population curve for city A was drawn up to 1991, when its population was 80,000. A population of 80,000 was reached by cities B, C, D, E in 1961, 1965, 1971 and 1978, respectively.

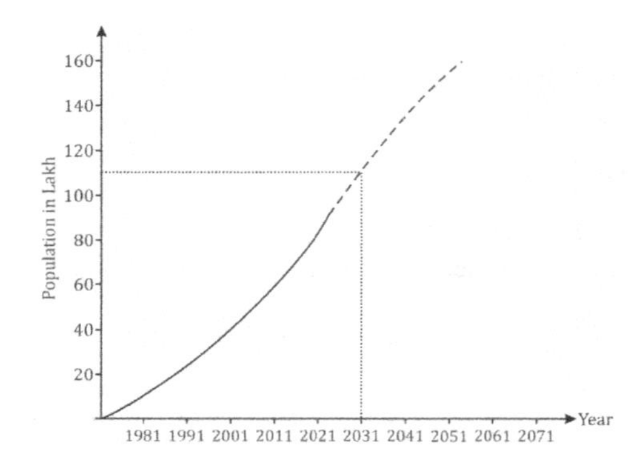

FIGURE 4.1 Graphical Extension Method

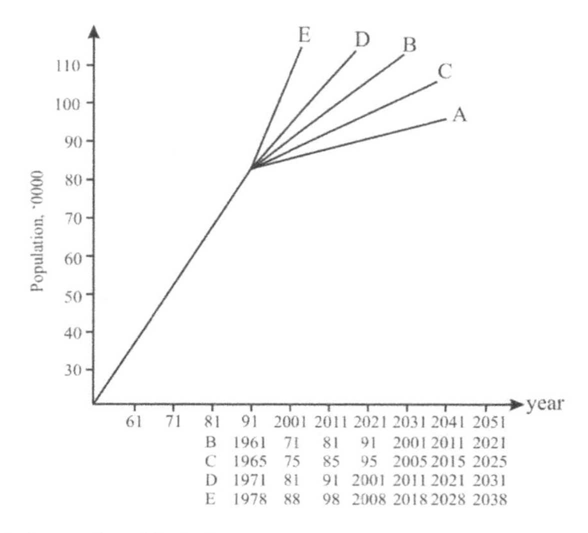

FIGURE 4.2 Graphical Comparison Method

The curves for B, C, D and E may now be drawn starting from the point corresponding to the year 1991 and population 80,000 of A. The curve of city A can now be continued considering the patterns of rates of growth of cities B, C, D and E. This method gives a fairly accurate forecast of the population. Hence, it is frequently adopted when data about the populations of similar cities are available.

4.2.7 The Zoning or Master Plan Method

The development of towns and cities cannot be done in a haphazard way, and master plans are prepared. The city is divided into various zones such as commercial centres, industrial areas, residential areas, schools, colleges, parks, etc. The future expansion of cities is strictly regulated by the various by-laws of corporations and other local bodies, according to the master plan.

Master plans are prepared for the development of cities for 25–30 years. The population densities for various zones of the towns to be developed are fixed. Thus, as the population of a particular zone is fixed, it is very easy to design its water supply scheme. The future development of a waterworks is also designed on the basis of the master plan. Common population densities for the preparation of master plans are presented in **Table 4.3**.

4.2.8 Logistic Curve Method

If the population of a town is plotted with respect to time, the curve so obtained under normal conditions shall be as shown in **Figure 4.3.**

The early growth of the city is shown by curves AB at an increasing rate of $dp/dt \propto P$. The growth rate between points C tand D follows the curve of $dp/dt = \text{constant}$. The transitional curve BD also passes through the point of inflexion C. Later on, the growth from D to N follows the decreasing rate, i.e. $dp/dt \propto (P_s - P)$, where P is the population of the town at point t from the origin A and p_s is the saturation value of the population. The s-shaped curve ABCDN is called the logistic curve.

The following mathematical analysis of the logistic curve can be used to calculate the population. The equation of the logistic curve is given by:

$$\ln\left(\frac{P_s - P}{P}\right) - \ln\left(\frac{P_s - P_0}{P_0}\right) = -kP_s t \text{ or } \ln\left(\frac{P_s - P}{P} \times \frac{P_0}{P_s - P_0}\right) = -kP_s t$$

$$\text{Substituting} \frac{P_s - P_0}{P_0} = m,\ kP_s = n$$

TABLE 4.3
Common Population Densities

Types of area	Persons/ha
Residential areas	10–40
(a) Large lots of single-family residences	40–90
(b) Small lots of single-family residences	90–250
(c) Multiple family quarters	250–3500
(d) Tenement houses	40–75
Commercial areas	10–40
Industrial areas	25–125
Parks, playgrounds, etc.	

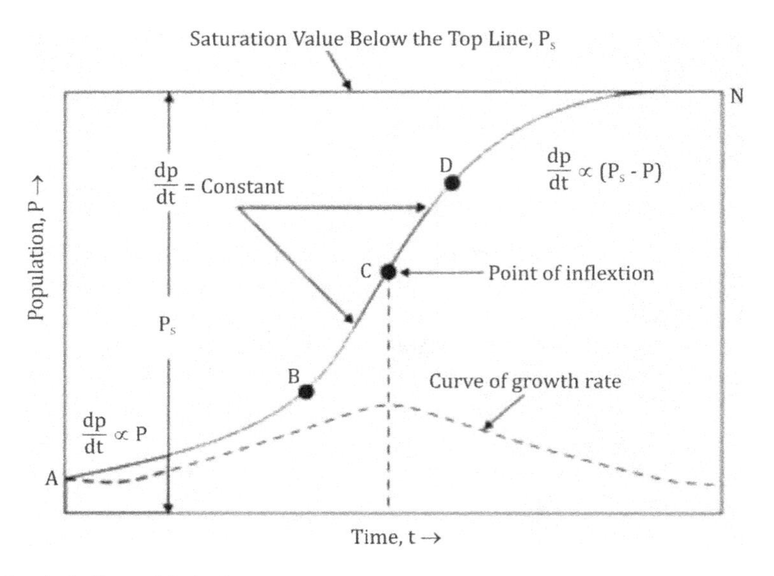

FIGURE 4.3 Logistic Curve Method

$$\frac{P_s}{P} = 1 + me^{-nt} \quad or \quad P = \frac{P_s}{1 + me^{-nt}}$$

P_o = the population of the town at the starting point
P_s = the saturation population
P = the population at time t from the origin
k = constant

If three pairs of the characteristics value $P_{0,}$ P_1 and P_2 at time t_0, t_1 and t_2, which are extending over the useful range of population, are so chosen that $t_0 = 0$, and $t_2 = 2\ t_1$, the saturation value P_s and constants m and n can be determined from the following equations:

$$Saturation\ population, P_s = \frac{2P_0\ P_1\ P_2 - P_1^2\left(P_0 + P_2\right)}{P_0\ P_2 - P_1^2}$$

$$m = \frac{P_s - P_0}{P_0}, \qquad n = \frac{1}{t_1} \ln\left[\frac{P_0\left(P_s - P_1\right)}{P_1\left(P_s - P_0\right)}\right]$$

4.2.9 The Apportionment Method

This is also known as the ratio method of forecasting future populations. In this method, the census population record is expressed as the percentage of the population of the whole country. The population of the city under consideration and the country's population for the past four to five decades are collected from the census department. The ratio of the town under consideration to the national population is calculated for these decades. A graph is then plotted between these ratios and the time. The extension of this graph will give the ratio corresponding to the future years for which the forecasting of the population is to be done. The ratio obtained is multiplied by the expected national population at the end of the designed period, for determining the expected national population of the town under reference. This method is suitable for towns and cities whose development is likely to take place according to the national growth.

4.3 ESTIMATING WATER DEMAND

Before designing a water supply scheme, it is necessary to evaluate water demand and availability.

4.3.1 Types of Demand

It is not possible to accurately assess the demand for water as it involves various factors that affect consumption. To arrive at a fair estimate, a few empirical formulae and rules of thumb are used. Demands can be classified as domestic, industrial and commercial, public use, firefighting, compensating various losses, and requirements for various buildings.

Domestic

This is the requirement of water for private buildings. It includes water required for drinking, cooking, bathing, maintaining lawns and gardens, domestic sanitation, washing, etc. The amount of water required by a person, called per capita consumption, varies over wide limits. As per Indian Standard IS:1172-1983, the normal value may be taken as 135 litres per day per person. However, 150 litres per day per person is recommended for metropolitan or megacities with a sewerage system. By comparison, in some of the most advanced countries like the US, it is more than 350 litres per day per person. The per capita consumption of 135 litres per day comprises the following, as shown in **Table 4.4**.

The total domestic demand for a town will be equal to the product of the total population of the town and the per capita consumption (domestic). This amounts to about 50% of the total consumption.

Industrial and Commercial

Water required by offices, factories, industries, hotels, etc., comes under this category. This accounts for 10% of total consumption.

Public Use

This is water required by public buildings such as hospitals, city halls and jails, and for washing streets, maintaining lawns and orchards. This accounts for 10% of the total demand.

TABLE 4.4
Consumption Rate

Purpose	L/c·d
Drinking	5
Bathing	55
Cooking	5
Washing of utensils	10
Washing of floors, etc.	10
Washing of clothes	20
Flushing of toilets	30
Total	135

TABLE 4.5
Water Consumption

Type of Structure	L/c.d
(a) Factories with bathrooms	45
(b) Factories without bathrooms	30
Hostels	135
Offices	45
Hospitals (including laundry, under 100 beds) per bed	340
Hospitals (including laundry, over 100 beds) per bed	450
Restaurants per seat	70
Schools: (a) day schools	45
(b) boarding schools	135
Cinemas, theatres, concert halls per seat	15
Medical quarters	135
Hostels per bed	180
Airports and seaports	70
Railway stations	70
i) Junction stations where mail/express stoppage is provided	45
ii) Terminal stations	45
iii) Intermediate stations	

Firefighting

This will be 5-10% of the total demand. Fire can be extinguished cheaply and quickly by water entailing heavy demands for brief periods. It depends upon the population, the number of buildings, and the planning and design of buildings. Satisfactory results can be achieved by applying the following empirical formula given by Kuichling:

$$\text{Kuichling formula}, Q = 3182\sqrt{P}$$

where Q = fire demand in L/min and P = population in thousands

$$\text{Freeman's Formula}, Q = 1136.5 \times \frac{P}{5} + 10$$

This gives high results. However, these formulae are not perfect and only give an idea of the demand.

4.4 TOTAL DEMAND

The total quantity of water required by a community is estimated using the following two pieces of information:

1. The anticipated population of a particular town for a given design period, say 30 years, worked out by any one of the methods of population forecasting.
2. The per capita demand, litres per person per day, which is the average amount of water required by a person every day, worked out by considering the water requirements for various purposes such as domestic use, commercial use and so on, divided by the total population and 365 days.

Based on these two data, the annual water requirement of the town can be worked out. For example, if the projected population of a town is 50 000 and the per capita demand is 200 L/c·d, the total amount of water required per year would be:

$$Q = \frac{200\,L}{p.d} \times 50000\,p \times \frac{ML}{10^6\,L} = 1.0\,ML\,/\,d \text{ (365 ML/a)}$$

4.4.1 Factors Affecting Per Capita Demand

The annual average demand for water is not the same for every country, city or town. In India, it varies from 100 to 300 litres per capita per day. Factors affecting water consumption are further discussed in the following sections.

Size of the Town

The bigger the town, the greater the per capita demand for water. This is because, in big cities, water requirements will be larger to cater for the maintenance of health and sanitation, and to meet the needs of industry and commerce.

Climatic Conditions

Water consumption in a hotter country will be more than in colder regions, as more water will be required for bathing, cleaning, air coolers, air conditioners, maintenance of gardens, lawns, etc.

Lifestyle

The quantity of water consumed varies between rich people and poor people. The highly affluent sections of society usually use more water than their poorer counterparts. The amount of water consumed usually depends upon the status of people.

Industrial and Commercial Activities

The industrial and commercial activities in a particular area increase the demand for water. Many industries require very large quantities of water. The demand due to industries and commerce depends on the number of industries and business activities and not on the population or size of the town.

System of Supply

In many cases, intermittent supply of water can result in reduced demand because there is a saving in water consumption due to a reduction in wastage and loss. But in some places, intermittent supply may produce increased demand due to the following reasons:

- Water stored by consumers during non-supply periods may be thrown away during the next supply period, resulting in losses and wastage of water.
- If taps are kept open during the non-supply period, and the consumer is not present when water is supplied, then a lot of water is wasted.

Quality of Water Supplied

Good quality water is welcomed by domestic consumers and factories, and demand for this can only increase. If the quality of water supplied is good, the consumer will never think of alternatives such as well water. This, in turn, results in an increase in demand for public water supply.

Pressure in the Distribution System

Higher pressures in the distribution system result in an increase in water consumption and thereby demand. If pressures are high, those living in the upper floors (or flats) can get water. Naturally, this results in increased demand. Further, higher distribution pressures produce higher losses in mains, taps, etc., which also increases demand. An increase in pressure from a head of 20 m to that of 30 m can produce an increase in loss of 30%.

Development of Sewerage Facilities

The conservancy system demands less water than the water carriage system. This is because, in the latter system, water is used liberally to flush toilets, urinals, etc.

Policy of Distribution

If the supply is unmetered and consumers are required to pay a fixed amount every month, they are more liable to waste water. But if the supply is metered, people are more cautious as they have to pay for the extra consumption.

Cost of Water

The cost of any material influences consumption and thereby demand. If the cost of water is made high, people will use it carefully and thus reduce the demand.

4.5 VARIATION IN DEMAND

The water demand discussed in the preceding sections is based on an annual average that is calculated based on annual water demand. The average demand represents the average daily demand over one year. There are peak periods as well as lean periods. There are wide variations in seasonal, daily and hourly water demands. Generally, demand is high during holidays and festivals; hot and dry days have more demand than wet and cold days; within a day, the demand in the mornings and evenings is higher. Accordingly, the variations in demand are broadly categorized in the following ways.

4.5.1 Seasonal Variation

The variation in demand from season to season is indicated in this category. Normally, demand peaks during summer. Fire breakouts generally occur more in summer, increasing demand. In winter or monsoon season, people generally consume less water.

4.5.2 Daily Variation

This depends on the variation in daily activity during the week. People draw out more water on Sundays and during festivals, thus increasing demand on these days.

4.5.3 Hourly Variation

This is very important as there is a wide range. The bulk of daily consumption is made during active household working hours, i.e. from 6 to 10 in the morning and 4 to 8 in the evening. During other hours, requirement is less or negligible. The maximum daily demand represents the amount of water required during the day with the maximum consumption in a year. Peak hourly demand represents the amount of water required during the hour with the maximum consumption in a given day.

Hence, in order to design a water supply project, one must consider demand during peak hours of the day and peak periods in a year. The design should ensure an adequate quantity of water to meet the peak demand. To cope with all the fluctuations in demand, the supply pipes, service reservoirs and distribution pipes must be properly sized. If water is supplied by pumping directly, then the pumps and distribution system must be designed to meet the peak demand. The effect of monthly variation influences the design of storage reservoirs, and the hourly variation influences the design of pumps and service reservoirs. As the population decreases, the fluctuation rate increases. Therefore, a careful study to understand these fluctuations must be made for each city from past water demand data. If such data is not available, the following equation can be used to estimate the maximum month, week, day and hour demand:

$$\textit{Percentage demand}, P = 180\,t^{-0.10}$$

where P = percent of annual average demand for time, t
t = time in days.

For example, for hourly demand, the multiplication factor is:

$$180 \times \left(\frac{1}{24}\right)^{0.10} = 247\% = 2.5\times$$

Also, based on experience, some empirical formulae are presented below to calculate the peak demand.

Maximum daily demand = 1.8 x Annual Average daily demand
Peak demand = 1.5 x Average hourly demand = 1.5 x Maximum daily demand
= 1.5 x (1.8 x Annual Average daily demand) = 2.7 x Annual Average daily demand
= 2.7 x Average hourly demand

Hourly demand does not require the unit /h, it can be any unit, e.g. ML/d or L/s, etc.

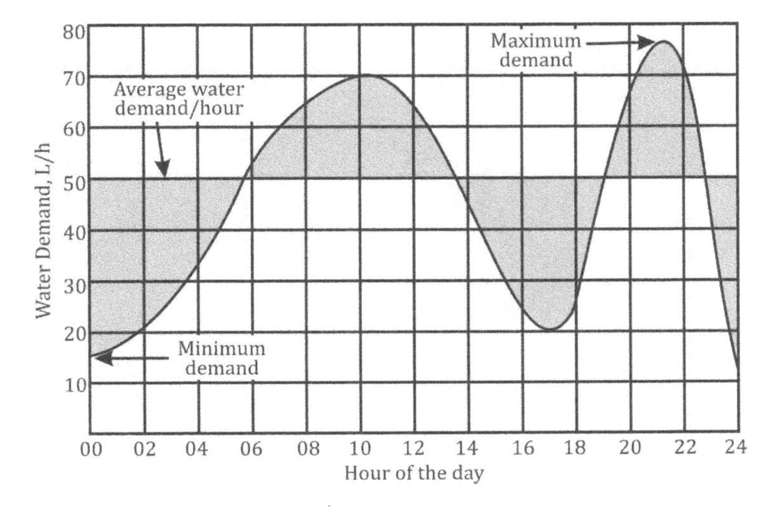

FIGURE 4.4 Hourly Variation (Diurnal Variation)

If the average demand for a town is 10 ML/d, the maximum daily demand is 15 ML/d and peak demand is 27 ML/d. **Figure 4.4** shows a typical hourly variation of water consumption. The maximum demand is observed to be at about 20:30 hours, and the minimum is at midnight.

Discussion Questions

1. Describe the common methods of population forecasting.
2. What factors determine per capita demand? Compare the water demand of developed versus developing countries like India.
3. Describe, with the help of a hydrograph, diurnal variation in water demand.
4. Explain various factors that affect population growth.
5. Discuss the logistics curve method for determining the future population of a city.
6. Derive a standard equation for determining future population.
7. Prepare a Microsoft Excel template to work out future population as shown in Example Problem 4.1.
8. Describe the incremental method for population forecasting. State its superiority over the arithmetic and geometric methods.
9. Explain the importance of variation in demand in the design of a water supply system.
10. What is peak demand, and how do you estimate it?

Practice Problems

1. The following data shows the variation in population of a town from 1941 to 1991.

Year	1941	1951	1961	1971	1981	1991
P, ×1000	72	85	110.5	144	184	221

Estimate the population in the year 2021 using:

a. The arithmetic method. (310,400)
b. The geometrical method. (435,000)
c. The incremental increase method. (346,400)

2. The following data shows the variation in population of a town from 1960 to 2000.

Year	1960	19760	1980	1990	2000	2010
P, ×1000	25	28	34	42	47	?

Estimate the population in the year 2010 using:

a. The arithmetic method. (57,000)
b. The geometrical method. (60,000)
c. The incremental increase method. (67,500)

3. Based on the data of three decades shown in the table below, estimate the population of the town in the year 2011.

Year	1981	1991	2001
P, ×1000	258	495	735

4. In two 20-year intervals, a city population grew from 30,000 to 172,000 to 292,000. Applying the logistic curve equation, determine the saturation population.
5. In two periods of 20 years, a city grew from 40,000 to 160,000 and then 280,000. Determine:
   a. The saturation population.
   b. The equation of the curve.
   c. The expected population for the next 15 years.

5 Water Quality and Treatment

The main purpose of water treatment is to provide a hygienic (potable) and palatable water supply. Few raw water sources can supply water that meets these objectives. Even groundwater, which is usually of good quality, requires at least disinfection to provide residual protection against possible contamination in the water distribution system. This is also called secondary disinfection. Groundwater containing excessive minerals needs additional treatment, including softening. Surface water should never be used for drinking without proper treatment. In most jurisdictions, water treatment plants with surface water as a source of supply must have chemically assisted filtration followed by disinfection as part of their process scheme.

5.1 TREATMENT PROCESSES

In river water supplies, the first step is usually preliminary treatment (**Table 5.1**). Screening, presedimentation and micro-straining are preliminary treatment processes intended to remove grit, algae and other small debris that can damage or clog plant equipment. Chemical pre-treatment conditions the water to further remove algae and other aquatic plants that cause taste and odour problems. The main treatment processes and their purpose are listed in **Table 5.2**.

5.2 SOURCE OF WATER SUPPLY

The primary water treatment methods used depend on the quality of raw water and the source of the water supply. Common water sources for municipal supplies are groundwater or surface water. Groundwater includes water wells, and surface water includes rivers, natural lakes and reservoirs.

Surface water usually requires a higher degree of treatment compared with groundwater. Surface waters are open to a variety of contaminants, from domestic and industrial wastes to the accidental spillage of harmful chemicals. River water is subject to extreme fluctuations and therefore is more difficult and costlier to treat. Nevertheless, surface waters are the most widely used water supplies because of their high yield and other advantages. The quality of water in a lake or a reservoir varies a great deal from season to season. To protect the water supply, the watershed feeding the river or lake must be properly managed. The same thing applies to the well fields and the recharge zone of groundwater aquifers. With good reason, the concept of watershed management in drinking water supplies is gaining popularity.

The scheme of water treatment processes chosen will depend upon the raw water characteristics. In the case of surface water systems, conventional treatment consists of coagulation, flocculation and sedimentation, followed by filtration and disinfection (**Figure 5.1**).

The simplest treatment system consists only of disinfection, as in the case of groundwater with a medium level of hardness. In some cases, fluoride is added to reduce the incidence of dental caries. Excess iron and manganese are removed from groundwater by adding chlorine or potassium permanganate, followed by filtration. Excessive hardness is commonly removed by precipitation softening, as shown in **Figure 5.2**. Aeration of groundwater is performed to strip out dissolved gases and to add oxygen.

Groundwater quality problems mostly concern a high level of hardness, and the presence of iron and manganese and dissolved gases. In general, groundwater is the preferred source of water supply from a quality and ease-of-treatment point of view. However, suitable aquifers with sufficient yields are not always available.

DOI: 10.1201/9781003231264-5

TABLE 5.1
Preliminary Water Treatment Processes

Preliminary	Purpose
Screening	Removes large debris that can foul or damage plant equipment.
Pre-treatment	Conditions the water for the eventual removal of algae and other aquatic nuisances that cause taste, odour and colour.
Pre-sedimentation	Removes gravel, sand, silt and other gritty material that can foul or damage plant equipment.
Micro-straining	Removes algae, aquatic plants and small debris.
Flow measurement	Measures the amount of water being treated.

TABLE 5.2
Main Treatment Processes

Main Processes	Purpose
Aeration	Removes odours and dissolved gases, adds aeration to improve taste.
Coagulation	Converts non-settleable particles to settleable particles.
Sedimentation	Removes settleable particles.
Softening	Removes hardness-causing chemicals.
Filtration	Removes finely divided particles, suspended flocs and most microorganisms.
Adsorption	Removes organics and colour.
Stabilization	Prevents scaling and corrosion.
Fluoridation	Adds fluoride to harden tooth enamel.
Disinfection	Kills disease-causing organisms.

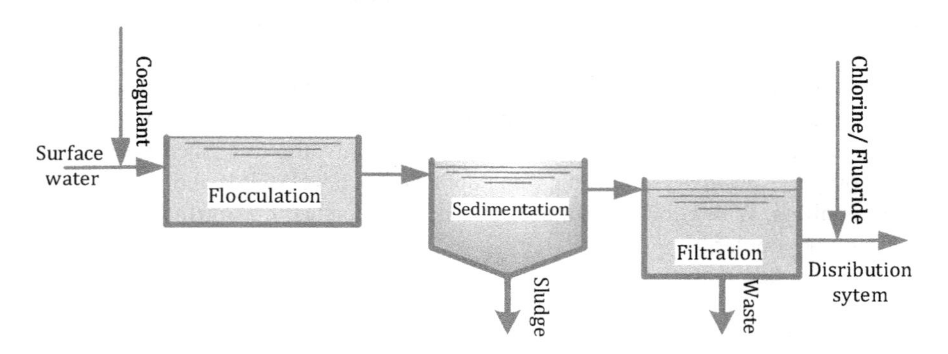

FIGURE 5.1 Process Schematic (Surface Water)

5.3 WATER QUALITY STANDARDS

Ideally, water delivered to the consumer should be clear, colourless, odourless and, more importantly, safe to drink (potable). It should not contain concentrations of chemicals that may be physiologically harmful, aesthetically objectionable or economically damaging.

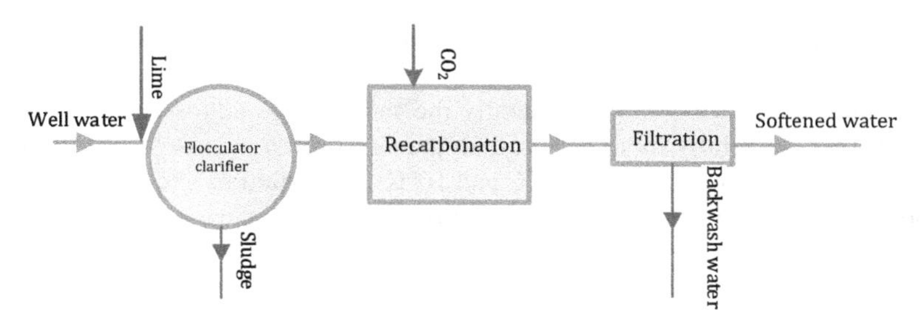

FIGURE 5.2 Process Schematic (Groundwater)

In most jurisdictions, the quality of drinking water is regulated by legislation. These standards pertain to the physical, chemical and bacteriological characteristics of the treated water. The World Health Organization provides guidelines for the minimum level of quality required in the public supply of water. However, the standards for quality control are evolved locally considering the limitations imposed by local factors. Normally, the check against transmitting diseases is comparatively more important than physical or chemical qualities, which are prescribed within a particular range.

In many jurisdictions, the monitoring and reporting of water quality is mandated by legislation. For example, in the United States and Canada, regular monitoring and data submission is required by law. In addition, municipal water systems are required by legislation to put a summary of their data on the web. As an example, in Ontario, Canada, drinking water quality is legislated under the Safe Drinking Water Act (SDWA). Under the SDWA, two regulations specifically relate to water quality. In Ontario, the drinking water quality standards are described in Regulation 169 and pertain to the physical, chemical and bacteriological characteristics of treated water. Regulation 170 details information related to sampling requirements, the reporting of adverse water quality and operational checks.

5.4 WATER ANALYSIS

Analysis of water quality supplements any information that exists on it. The parameters for such analysis depend on the specific use the water is to be put to. The treatment plant must be designed and operated based on the information on impurities present in the water. Therefore, analysis of water is essential for designing a water treatment system. Similarly, after the water is treated, it is analyzed again to ascertain that it has been purified and meets the prescribed quality standards. Since quality is a variable with respect to time and space, it is imperative to conduct periodic/routine tests to ascertain regular maintenance of the quality of water.

5.4.1 Physical Tests

Some of the most common tests falling into this category include temperature, turbidity, colour, taste and odour, and electrical conductivity, which is indicative of the total dissolved solids. Water is tested for turbidity to determine removal and adjust the operation of various processes such as chemical dosing and the backwashing of filters. In many developed countries, there is a legislative requirement that the turbidity of filtered water should be continuously monitored and must not exceed 1-NTU.

Temperature

From a treatment point of view, water temperature is not significant since it is not practical to change the temperature of water during treatment. However, the temperature of water does affect the rate of chemical reactions and particle settling velocity. For chlorine residual, disinfection is more powerful at high temperatures.

The temperature of surface water is generally the same as the atmospheric temperature, while that of groundwater is generally more or less than the atmospheric temperature. For public supply, the most desirable temperature is between 4°C and 10°C. Temperatures exceeding 25°C are undesirable, and above 35°C, water is unfit for public supply because of its lesser acceptability and the danger of it becoming more prone to bacterial contamination.

Turbidity

Turbidity is caused by the presence of suspended and colloidal matter in water. Turbid water will appear to be muddy or cloudy, and its appearance is not pleasing. Though turbidity is usually not harmful, water must be treated to remove turbidity more for aesthetic reasons. If not removed, turbidity can shelter pathogens, making disinfection inefficient.

Turbidity is expressed by a unit called a nephelometric turbidity unit (NTU). The turbidity produced by one part of finely divided silica in a million parts of distilled water is the standard. Turbidity exceeding 10 NTU is noticeable to the naked eye. The turbidity of filtered water should be less than 1.0 NTU. The instrument used to measure turbidity is called a turbidimeter and must be calibrated before use with certain standards.

Colour

The colour of water usually results from the presence of colloidal organic matter or mineral and dissolved organic and inorganic impurities. Its colour is usually due to dissolved substances in water that are hard to remove by simple sedimentation or filtration. Like turbidity, colour is an aesthetic parameter. Turbidity in water imparts some colour, hence this measurement is called apparent colour. The colour of water lacking turbidity is known as true colour. To read true colour, a water sample should first be filtered to remove turbidity.

The colour produced by one milligram of platinum in a litre of distilled water has been fixed as the unit of colour (CU). The permissible colour for domestic water is 20 mg/L on the Platinum-Cobalt Scale. Mostly, even the slightest colour in water is objectionable as it directly affects the palatability of water. For aesthetic purposes, the colour in water should be below 5 colour units (CU).

Taste and Odour

Taste and odour determine the palatability of drinking water. Taste and odour in water may be due to the presence of dead or living microorganisms; dissolved gases such as hydrogen sulphide, methane, carbon dioxide or oxygen combined with organic matter; and mineral substances such as sodium chloride, iron compounds, carbonates and sulphates. Tests are done by smell and taste because these are present in such small proportions that it is difficult to detect them by chemical analysis. The odour of water also changes with temperature. Water having a bad taste or odour is objectionable and should not be supplied to the public. The intensity of odour is measured in terms of threshold odour number (TON). A similar test to compute the flavour threshold number (FTN) is used to quantify taste. Experience has shown that most complaints received at water plants are related to taste and odour problems.

Electrical Conductivity

Electrical conductivity or the specific conductance of water is related to its dissolved salt content. It is a measure of the ability of the water to conduct an electric current. It is determined by means of a di-ionic electrode and is expressed in microsiemens per cm (μS/cm) at 25°C. The specific conductivity of water multiplied by a coefficient (ranging from 0.5 to 0.9; generally, 0.65) to directly obtain the total dissolved solids (TDS) content in mg/L.

5.4.2 Chemical Tests

Total Solids

Total solids include suspended, colloidal and dissolved solids. The quantity of suspended solids is determined by filtering the water sample through a fine filter pore size of 1 μm, then drying and weighing. The quantity of dissolved and colloidal solids is determined by evaporating the filtered water (obtained from the suspended solid test) and weighing the residue. The total solids in a water sample can be directly determined by evaporating the water and weighing the residue. As mentioned earlier, total solids can be estimated by measuring the electric conductivity of the water. The maximum acceptable limit of total solids in water is 500 mg/L.

Hardness

Hardness in water is caused by the presence of multivalent ions, primarily calcium and magnesium as their carbonates, sulphates, nitrates and chlorides. Hardness is exhibited by its difficulty to lather soap. Hard water scales water main, thus reducing its flow-carrying capacity. Hardness is usually expressed as an mg/L equivalent of calcium carbonate. It is generally determined by a titrimetric method using ethylenediaminetetraacetic acid (EDTA).

In general, under a normal range of pH values, water with a hardness of up to 75 mg/L is considered soft and above 200 mg/L is considered hard. Groundwater is generally harder than surface water. For boiler feed waters and laundry purposes, water must be soft with a hardness level below 75 mg/L or so. In the beverage industry, excessive hardness can cause taste problems. However, for drinking purposes, the prescribed limit for public supplies ranges between 75 mg/L and 150 mg/L.

pH

Depending on the nature of its dissolved salts and minerals, water from natural sources may be acidic or alkaline. Its acidity is determined on the pH scale, which is defined as the negative logarithm of hydrogen ion (H^+) concentration. This is expressed on a scale of 0 to 14, with 7 being the neutral value. A pH value of less than 7 is considered acidic and above 7 is considered alkaline.

The pH value of a water sample is determined using specially prepared electrodes with an instrument called a pH meter. It can also be measured using suitable indicators, but the results are not very accurate. For public water supplies, the pH value should be kept as close to 7 as possible, within a range of 6.5 to 8.5.

Alkalinity and Acidity

Alkalinity and acidity are the characteristics of natural water that provide its buffering capacity. Alkalinity, therefore, is a measure of the buffering capacity of water. The alkalinity of natural water is due primarily to the presence of bicarbonates and carbonates, although a small portion of alkalinity may also be due to silicate, borate, phosphate, hydroxide, etc. Apart from their mineral origin, these substances originate from the dissolution of CO_2, either from the atmosphere or from microbial decomposition.

Alkalinity plays an important role in the water treatment process, especially coagulation and water softening. For optimum coagulation, alkalinity must be added if natural alkalinity is insufficient to react with a coagulant like alum. After coagulation, residual alkalinity should not fall below 30 mg/L. The alkalinity of drinking water should not be more than 200 mg/L as equivalent $CaCO_3$. Low alkalinity water is aggressive and can cause corrosion. The minimum level of alkalinity is around 30 to 50 mg/L.

Chlorides

Natural water near mines and the sea is high in sodium chloride. Chlorides in water may also be present due to the mixing of saline water and sewage. Excess chlorides are dangerous and unfit for use.

Chloride concentration exceeding 250 mg/L causes a salty taste. The standards thus limit the concentration of chloride in water to 250 mg/L. Chloride concentration can be determined by titrating the water with silver nitrate and potassium chromate.

Residual Chlorine

Free chlorine is never found in natural water. It is present in treated water as a result of its disinfection with chlorine. The chlorine remains as residual in treated water for the sake of safety against pathogenic bacteria, known as secondary disinfection. The concentration of residual chlorine is determined by the starch iodide test or the orthotolidine test. Portable chlorine meters using diethylphenylenediamine (DPD) tablets are very common these days. The free chlorine residual in water should be between 0.5 mg/L to 0.2 mg/L so that it remains safe against pathogenic bacteria. The absence of chlorine residual or a level below 0.05 mg/L indicates adverse water quality. Such water may be unsafe to drink.

5.4.3 Metals and Other Chemicals

Iron and Manganese

These are generally found in groundwater. If iron in water exceeds 0.3 mg/L, it is not suitable for domestic, bleaching, dyeing and laundry purposes. The presence of iron and manganese in water gives it a brownish-red colour, causes the growth of microorganisms and corrodes water pipes. Iron and manganese also give it taste and odour. The quantity of iron and manganese in water is determined by colorimetric methods.

Fluorides

Fluorides in natural water are present either due to the leaching of rocks (geological origin) or the discharge of industrial wastes containing fluorides. Up to 1 mg/L fluorides in water are considered beneficial, especially for dental health. However, a concentration exceeding 1.5 mg/L causes fluorosis, a disease characterized by the mottling (excessive staining) of teeth and the disfigurement of bones. Water contains various types of minerals or metals such as iron, manganese, copper, lead, barium, cadmium, selenium, fluoride and arsenic. **Table 5.3** gives the maximum permissible level of these metals in water. Though initially present in trace amounts in natural water, their concentration increases due to the discharge of industrial wastes.

5.4.4 Nitrogen

The presence of nitrogen indicates the existence of organic matter. Nitrogen may be present in water as nitrites, nitrates, free ammonia and albuminoid nitrogen. Groundwater supplies, due to the

TABLE 5.3
Maximum Permissible Limit (MPL) of Chemicals

Chemical	MPL, mg/L	Chemical	MPL, mg/L
Magnesium	125.0	Cadmium	0.01
Zinc	15.0	Chromium	0.05
Silver	0.05	Sulphate	250
Arsenic	0.05	Phenolic substances	0.001
Lead	0.05 to 0.1	Cyanide	0.2
Fluoride	1.5	Iron	0.3
Copper	1.0 to 3.0	Manganese	0.05
Barium	1.0	Selenium	0.05

absence of oxygen, may contain free ammonia due to the reduction of nitrates in water. Albuminoids are normally derived from animal and plant life typical of an aquatic environment. Its presence indicates organic pollution.

The presence of nitrites in water due to partly oxidized organic matters is very dangerous. Therefore, in no case should nitrites be allowed in water. Nitrites are rapidly and easily converted to nitrates by full oxidation of organic matter. But in no case should the quantity of nitrate exceed 45 mg/L since this can cause blue baby syndrome (methemoglobinemia) in children.

The presence of free ammonia indicates recent pollution or decomposition, the presence of nitrites indicates partial decomposition, and the presence of nitrates indicates fully oxidized organic matter. The presence of free ammonia in water should not exceed 0.15 mg/L. Any sudden change in nitrate level indicates recent pollution of the water source. Nitrate content is measured by reduction to ammonia.

5.4.5 Dissolved Gases

The various gases that may get dissolved in water include nitrogen, oxygen, carbon dioxide, ethane and hydrogen sulphide. The nitrogen content is not important. A brief description of the other gases is presented in the following paragraphs.

Dissolved Oxygen

Dissolved oxygen in water, DO, indicates the amount of unstable organic matter present. The DO content of water is dependent on atmospheric pressure and water temperature. In a way, DO content affects the palatability of water: too much DO may lead to corrosion in water distribution pipes; and low DO content in raw water suggests the presence of biodegradable organic matter and is indicated by biochemical oxygen demand (BOD). Although BOD is of greater significance in determining the characteristics of wastewater, the BOD of raw water should not exceed 5.0 mg/L, and the BOD of treated water must be zero.

Carbon Dioxide

The carbon dioxide content of water may make water aggressive at low pH. Carbon dioxide and other undesirable gases such as hydrogen sulphide can be removed by stripping.

Hydrogen Sulphide

The presence of hydrogen sulphide, even in low concentrations, can lead to strong odours and can make water aesthetically unacceptable. As mentioned earlier, it is most commonly found in groundwater and can be removed by stripping or aeration.

5.4.6 Microbiological Tests

Microbiological composition is the most important aspect of drinking water quality because of the possible presence of disease-causing organisms or pathogens. Most microorganisms are harmless and play an important role in the natural purification of water. Such organisms are called non-pathogens and are beneficial to plants, animals and soil.

Pathogen organisms include bacteria, protozoa, viruses, cysts and parasitic worms. Typhoid fever, cholera, hepatitis, enteroviral disease, bacillary and amoebic dysenteries, giardiasis, cryptosporidiosis and many varieties of gastrointestinal diseases are caused by these microorganisms and can be transmitted by water.

Indicator Organisms

Disease-causing organisms are not easily identified, and the techniques for identification are complex and time-consuming. The most widely used method of testing the bacteriological quality of water involves testing for a single group of bacteria known as indicator organisms, principally the coliform group that includes total coliforms and faecal coliforms. Coliforms are non-pathogens and are always present if there is microbial contamination. Since coliforms exist in large numbers compared to pathogens and are sturdier, the absence of coliforms indicates water is hygienic and safe to drink.

The coliform group, whose name means rod-shaped, is made up of various types of bacteria. These are associated with the faeces of humans, warm-blooded animals and soil. The presence of faecal coliform bacteria in a water sample is better proof of sanitary pollution than the presence of total coliform bacteria because the faecal bacteria are restricted to the intestinal tract of mammals. Escherichia coli (E. coli) is the predominant member of the faecal coliform group and is present in large numbers but has a short life in the environment. Detection of E. coli in drinking water, therefore, is an indication of recent pollution.

The simplest and the most common method for the microbiological testing of coliforms is the membrane filter test. An aliquot of a sample is filtered through a specially designed sterile membrane. This membrane has pores that range in size from 5–10 nm on which bacteria will be retained if present. The filter is then rinsed with a sterile buffer solution, placed on a pad saturated with a suitable nutrient medium or broth, and incubated at an appropriate temperature. Bacteria, if present, will grow on the medium and form colonies, each colony representing one bacterium in the original sample.

When testing for total coliforms, m-Endo broth is used with incubation at 35°C for 20 to 22 hours. The colour and shine of the colony are characteristic of the type of bacteria. Coliform colonies are pink to dark red with a golden metallic sheen, often with a greenish tinge. Non-coliform bacteria lack this sheen. For a faecal coliform count, the medium used is m-FC broth, incubated at 44.5°C for 22 hours. Coliform colonies here are blue, and other bacteria that grow on this medium are grey to cream coloured.

Coliform groups of bacteria ferment lactose and form gas within a maximum period of 48 hours when incubated at 35°C. This characteristic is used to indicate the presence of coliform in a given water sample. This method is known as the multiple-tube fermentation technique. In this test, a broth containing lactose and other substances, which inhibit non-coliforms, is placed in a series of test tubes, five for each of the three dilutions. For the first dilution, 10 mL of the sample aliquot is used, 1 mL is used for the second set of five tubes and 0.1 mL is used for the last set. After incubation at 35°C for 24 hours, the formation of gas is noticed in the tubes. The presence of a gas is a positive presumptive test. If no gas is formed, the tubes are incubated for another 24 hours, thus making the total incubation period 48 hours. If no gas is found after 48 hours of incubation, the test for the presence of coliforms is indicated as negative.

The test tubes showing positive presumptive gas are further subjected to a confirmatory test, which will eliminate certain bacteria of non-sanitary importance. Based on the number of positive test tubes, statistical methods are used to determine the bacterial density called the most probable number (MPN). Based on the number of positive test results from each series of test tubes, the MPN can be read from tables.

Sample Collection for Microbiological Testing

Microbiological samples are collected to determine the safety of drinking water supplies. They are grab samples manually collected at designated locations in the water distribution system. Bottles used for collecting bacteriological samples should not be rinsed as they contain sodium thiosulphate to neutralize any chlorine residual.

- The sampling tap should be free from leaks, have no attachments, and provide a steady flow.
- Run the faucet for 2–5 minutes at a steady flow before collecting the sample.
- Opening the faucet to full flow generally is not desirable.
- Remove the cap from the bottle and hold it with the threads facing down during sample collection.
- Without touching the bottle, fill the sample, leaving some headspace.
- Do not rinse the bottle.
- Label the bottle and refrigerate it at 4°C.

The exception to the above procedure is sampling for lead and copper analysis, where the first draw is used to fill the bottle.

Preservation and Storage of Samples

Some samples require preservation to ensure the stability of the target compounds during transportation and storage, or to eliminate substances that may interfere with the analysis. In some cases, preservation of the sample is optional and, if selected, will allow for a longer storage period before analysis must be initiated. Other parameters require that analysis be conducted immediately following sample collection.

Storage time is defined as the time interval between the end of the sample collection period and the initiation of analysis. All samples should be stored for as short a time interval as possible and under conditions that minimize sample degradation. It may be possible to analyze a number of parameters from one container provided there is enough sample volume (i.e. sodium, ammonia, nitrate, nitrite, turbidity, alkalinity, pH, chloride, colour, sulphate). For required sample volume, one should check with the laboratory. As a rule of thumb, if not indicated by the laboratory, a minimum sample of 1-L should be collected.

Discussion Questions

1. Draw a schematic diagram of a water treatment plant that uses river water as the source. Discuss the purpose of each unit's operation or processes in the process schematic.
2. Draw a schematic diagram of a water treatment plant that uses groundwater as the source. Discuss the purpose of each unit's operation or processes shown in the process schematic.
3. What physical tests are conducted for the examination of drinking water supplies?
4. Describe the steps used to conduct a membrane filtration test.
5. What care should be exercised when collecting water samples for microbiological testing?
6. Compare the following terms:
   a. True colour, apparent colour.
   b. Potable, palatable.
   c. Hardness, alkalinity.
   d. Total coliforms, faecal coliforms.
7. List the water treatment processes in which alkalinity and pH play an important role.
8. Why is it important and necessary to preserve and store a sample?
9. What are indicator organisms and what role do they play in the bacteriological testing of water quality?
10. Discuss the various microbiological tests performed to determine the safety of a drinking water supply.

6 Coagulation and Flocculation

In the conventional treatment of potable water, two important treatment processes are coagulation and flocculation. The majority of the particles contributing to colour and turbidity are smaller than 1 μm. Colloidal particles contributing turbidity and colour are relatively very small particles, in the range 1–100 nm, and usually do not settle under normal conditions. The main purpose of these processes is to precondition the water for the removal of turbidity and colour by the following processes clarification and filtration. If the quality of raw water is reasonably good, clarification is eliminated, and coagulated water is directly fed to filters. For the same reason, this scheme of processes is called direct filtration.

6.1 COAGULATION

Coagulation is the chemical reaction that occurs when a coagulant (chemical) is rapidly mixed into water. The chemical reaction destabilizes the suspended particles and helps bring them together in a short time. Simply put, coagulated particles are insoluble and appear in the form of large, suspended particles. During flocculation, suspended particles are stimulated to come together and grow and make large floc. This makes it easier to remove the suspended materials from the water by the sedimentation and filtration processes. An important result of coagulation is the neutralization of negatively charged suspended particles in water.

Whereas coagulation is a chemical process, flocculation is primarily a physical process. It is important to note that coagulation takes place in less than 30 seconds. In fact, much of coagulation is completed in the first couple of seconds. Thus, it becomes necessary to quickly disperse the coagulant uniformly into water. This is accomplished by high-speed mixing, also called flash mixing. Coagulated water starts forming invisible microfloc. Microfloc still carries a positive charge from the coagulant. Flocculation must follow coagulation to allow the growth of microfloc to macrofloc.

Flocculation entails gentle or slow mixing for about 20–60 minutes. Whereas coagulation is basically a chemical process, flocculation is mainly a physical process. By agglomeration of microfloc, floc grows and becomes dense enough to be removed by settling and or filtration.

6.2 COAGULATING CHEMICALS

Coagulants refer to chemicals causing coagulation. When added to water, they form insoluble and gelatinous precipitates. These gelatinous precipitates absorb and entangle non-settleable solids during their formation and descend through the water.

6.2.1 Primary Coagulants

Primary coagulants are responsible for destabilizing the particles and making them clump together to form floc. The coagulants more commonly used in the water industry consist of positively charged ions. Trivalent cations, as in the case of aluminium and iron salts, are more effective since they carry triple the charge of monovalent cations. Coagulants with trivalent ions like aluminium and iron are 50–60 times more effective than bivalent ions like calcium. **Table 6.1** lists the chemical coagulants used in the treatment of water.

DOI: 10.1201/9781003231264-6

TABLE 6.1
Common Coagulants and Doses

Coagulant	Name	Formula	Dose, mg/L
Aluminium sulphate	Alum	$Al_2(SO_4)_3.18\ H_2O$	15–100
Copper sulphate		$CuSO_4$	5–20
Ferric sulphate	Ferrisul	$Fe_2(SO_4)_3$	10–50
Ferric chloride	Ferrichlor	$FeCl_3.6H_2O$	10–50
Ferrous sulphate	Copperas	$FeSO_4.7H_2O$	5–25
Sodium aluminates		$NaAlO_2$	5–50
Polymers		Various	0.1–1

Alum is the most common and universal coagulant. Alum is found effective in the pH range of 6.5 to 8.5. Alum is preferred over other coagulants since it also reduces taste and odour problems. It is cheap and makes tough floc more easily. One disadvantage of alum is that the sludge formed is difficult to dewater.

Hydrated ferrous sulphate, more commonly known as copperas, has too high solubility to act as a satisfactory coagulant at the usual pH range. Hence, it is first chlorinated to form trivalent ions and is known as chlorinated copperas.

$$6FeSO_4.7H_2O + 3Cl_2 = 2Fe_2\left(SO_4\right)_3 + 2FeCl_3 + 7H_2O$$

Chlorinated copperas is also effective in removing colour. Ferric sulphate and ferric chloride can also be used independently. Ferrous sulphate or copperas can react with natural alkalinity, but it is a slow reaction. To expedite the reaction, lime is added.

$$FeSO_4.7H_2O + \mathrm{Ca(OH)_2} = Fe\left(OH\right)_2 + CaSO_4 + 7H_2O$$

The ferrous oxide thus formed is efficient floc. However, it is oxidized to a ferric state by the dissolved oxygen, as shown below:

$$4Fe\left(OH\right)_2 + O_2 + 2H_2O = 4Fe\left(OH\right)_3$$

Floc thus formed is gelatinous in nature and heavier than alum floc. Since copperas is effective above a pH of 8.5, it is unsuitable for the treatment of soft coloured water. Iron coagulants make stronger floc and are effective over a wide range. However, iron coagulants leave the water more corrosive, and some iron residual can promote the growth of iron bacteria and cause staining.

Sodium aluminate, although an effective coagulant, is less commonly used due to its high cost compared to alum. Its reaction with natural alkalinity is as follows:

$$Na_3Al_2O_4 + 3Ca\left(HCO_3\right)_2 = CaAl_2O_4 \downarrow + Na_2CO_3 + 6CO_2 + H_2O$$

$$Na_3Al_2O_4 + CaSO_4 = CaAl_2O_4 \downarrow + Na_2SO_4$$

As shown above, sodium aluminate reacts with both temporary (carbonate) hardness and permanent (non-carbonate) hardness. It is effective in the pH range of 6 to 8.5.

6.2.2 Coagulant Aids

As their names indicate, coagulation and flocculation aids are chemicals or substances that, when added in small quantities, increase the effectiveness of the process. The main advantages of using aids are as follows:

- Assists in overcoming temperature drops that slow coagulation.
- Improves the quality of the filtered water.
- Reduces chemical costs.
- Increases plant capacity.
- Reduces the amount of sludge produced.

Activated Silica

The key chemical in activated silica is sodium silicate. Activated silica will improve coagulation by strengthening the floc and widening the pH range for effective coagulation. Improved colour removal and better floc formation in colder temperatures can also result. The main disadvantage is the precise control required in its preparation and feed rate.

Weighting Agents

As the name indicates, their primary role is to add weight to the floc to improve settleability. Bentonite clay is a very common weighting agent. Weighting agents are used to treat water high in colour, low in turbidity and low in mineral content. Typically, dosages of 10–50 mg/L are used. Keep in mind though this will add to the solids in the sludge.

Polymers/Polyelectrolyte

Polymers are long-chained synthetic organic compounds commonly referred to as polyelectrolyte. The molecular mass of polymers may vary from 10^2 to 10^7. Based on electrical charge, polymers are classified into three groups: anionic (negative charge), cationic (positive charge) and non-ionic (neutral or no charge).

Cationic polymers can be used as primary coagulants or coagulant aids. The primary advantage of polymers is in reducing the quantity of alum sludge produced. The sludge produced is also easily dewatered. Alum sludge is otherwise difficult to dewater.

The dosage of activated silica is usually less than 10% of the coagulant used. The normal dosage range of cationic and anionic polymers is 0.1 to 1.0 mg/L. Overdosing is not only wasteful, it can upset the process and allow the discharge of chemicals in the finished water. Overdosing can upset settling and filter operation.

TABLE 6.2
Recommended Doses of Coagulant Aids

Coagulant Aid	Typical Dose, mg/L
Activated silica	7–10% of the coagulant
Cationic polyelectrolyte	0.1–1
Anionic polyelectrolyte	0.1–1
Non-ionic polyelectrolyte	1–10
Bentonite (clay)	10–50

6.3 CHEMISTRY OF COAGULATION

Coagulants like alum react with the alkalinity of the water and form insoluble floc. Sufficient chemicals must be added to water to exceed the solubility limit of the metal hydroxide. The optimum pH range is between 5 and 8. There should be sufficient natural alkalinity present to react with the coagulating chemical and serve as a buffer. If the source (raw) water is low in alkalinity, the alkalinity may be increased with the addition of lime or soda ash.

6.3.1 Chemical Reactions

Coagulation reactions are quite complex. The chemical reactions presented below are hypothetical but are useful in estimating the quantities of reactants and products. The chemical reaction of alum with alkalinity is shown below:

$$Al(SO_4)_3.14.3H_2O + 3Ca(HCO_3)_2$$

$$= 2Al(OH)_3 \downarrow + 3CaSO_4 + 6CO_2 + 14.3H_2O$$

$$\frac{Alkalinity}{Alum} = \frac{3\ mol \times 100 g/mol}{1\ mol \times 600 g/mol} = 0.50$$

From the chemical reaction, it is evident that the alkalinity requirement to react with alum is half of the alum dose. If alkalinity is expressed as calcium bicarbonate, the alkalinity consumed will be 81% of the alum dose. When alum reacts with alkalinity, carbon dioxide is produced and sulphate ions are added to finished water. For each unit of alum reacted or consumed, natural alkalinity is reduced to half. Thus, the addition of alum results in reducing the pH, the alkalinity or both. In addition, some permanent hardness in the form of calcium sulphate is added and the production of carbon dioxide makes the water corrosive. Despite these drawbacks, alum is still the most commonly used coagulant.

Alum Floc (Sludge)

In the hypothetical coagulation equation, aluminium floc is written as Al $(OH)_3$. The quantity of sludge produced as $Al(OH)_3$ can be estimated as follows:

$$\frac{Alum\ floc}{Alum} = \frac{2\ mol \times 78 g/mol}{1\ mol \times 600 g/mol} = 0.26$$

Experiments have shown that the actual production of alum floc is twice as much. In addition to aluminium hydroxide, turbidity will be removed. The total solids produced in alum coagulation can be estimated by using the following empirical relationship.

$$= 0.44 \times Alum\ dose + 0.74 \times Turbidity\ removed\ in\ NTU$$

For a known concentration of dry solids produced as sludge, the volume of sludge can be estimated. Sludge with a solid content of 1% will contain 20 g of dry solids in every litre of sludge. In other words, 10 g of dry solids would make one litre of wet sludge. The consistency of alum sludge ranges from 1–3%. Alum sludge is difficult to dewater.

Example Problem 6.1

The raw water of a given river water system has an alkalinity of 35 mg/L as $CaCO_3$. The jar test indicated an optimum alum dosage of 30 mg/L to reduce the turbidity to less than 1 NTU. To assure complete precipitation, a minimum of 30 mg/L of residual alkalinity is recommended. Work out the dosage of alkalinity to be added to maintain the level of residual alkalinity. Find the dose of 85% hydrated lime in mg/L that will be needed to complete this reaction. What should be the setting on the lime feeder in kg/d when the flow is 10 ML/d? Work out the production of carbon dioxide.

Given: Raw Alk = 35 mg/L
Residual Alk = 30 mg/L
Alum dose = 30 mg/L
Alkalinity added = 10 mg/L = 0.1 mmol/L
Q = 10 ML/d

SOLUTION:

$$Alum\ dose = \frac{30\ mg}{L} \times \frac{mmole}{600\ mg} = 0.05\ mmole / L$$

$$Alkalinity\ consumed = 0.50 \times Alum = 0.5 \times 30\ mg / L$$

$$= 15.0 = 15\ mg / L$$

$$Added = Consumed + Residual - Natural$$

$$= (15 + 30 - 35)\frac{mg}{L} \times \frac{mmole}{100\ mg} \times \frac{74\ mg}{mmol\ Ca(OH)_2}$$

$$= 7.40 = \underline{7.4\ mg / L}$$

$$= \frac{7.4\ kg}{ML} \times \frac{10\ ML}{d} \times \frac{100\%}{85\%} = 87.05 = \underline{87\ kg / d}(lime)$$

$$Production\ of\ CO_2 = 6.0 \times \frac{0.05\ mmol}{L} \times \frac{44\ mg}{mmol} = 13.2 = \underline{13\ mg / L}$$

6.4 CHEMICAL FEEDING

Most of the processes in water treatment require chemical feeding. The correct dosing of chemicals is important to maintain the efficiency of the process and to avoid overdosing. Chemical dosage and feed rate are based on a mass balance equation (**Figure 6.1**)

Dilution formula is another expression for the mass balance of a chemical. Dilution formula is useful in performing the calculations for the volume needed to dilute a solution, making a solution of desired strength and volume, and determining the volume of solution to be fed in order to apply the desired dosage of a chemical. The following expressions are useful when doing calculations on dilutions or determining the feed pump rate of chemical solutions.

$$V_1 \times C_1 = V_2 \times C_2 \qquad Q_1 \times C_1 = Q_2 \times C_2$$

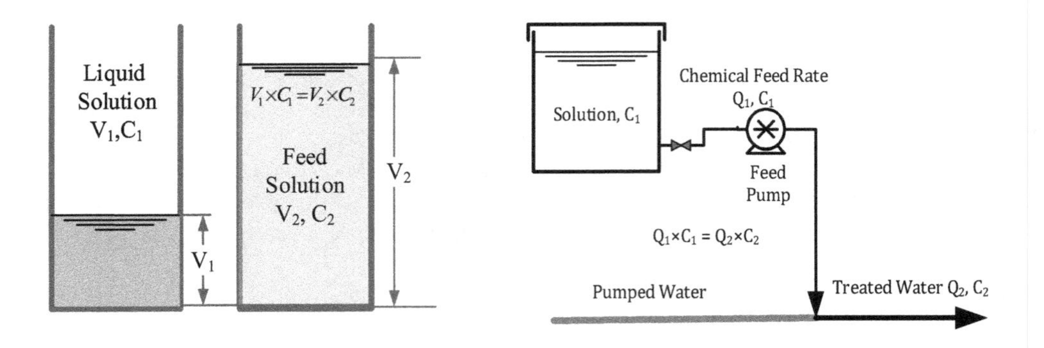

FIGURE 6.1 Liquid Chemical Feeding

When diluting, subscript 1 refers to the liquid chemical or the original solution, and subscript 2 refers to the desired solution. When dosing a given volume of water, subscript 1 refers to the liquid chemical or its solution, and subscript 2 indicates the water being dosed, as illustrated in **Figure 6.2.**

6.4.1 Density Considerations

Quite often, the strength of a solution is expressed as a percentage on a mass-to-mass basis. When the density of the liquid solution is different from that of water, density considerations must be made. Further, if the liquid chemical is not 100% pure or contains only a fraction of the chemical to be fed, the feed rate must be adjusted accordingly. When the density of the liquid chemical is different from that of water, a strength of 1% is not equal to 10 g/L. To do the correct conversion, you need to multiply by the density of the liquid chemical, ρ, as shown below:

$$C = C_{m/m} \times \rho = C_{m/m} \times SG \times \rho_w$$

6.4.2 Feed Pump Setting

Since water treatment involves the addition of chemicals, it is important to know how to set the chemical feed pump to apply a set dosage for treating a given flow rate. It is recommended to adjust the strength of the feed solution so that the feeder setting is about in the middle range (40–60%). This allows the dosage to go up and down and ensures the feed pump operates under optimal conditions.

Example Problem 6.2

The optimum alum dose as determined by jar testing is 10 mg/L. The chemical feed pump has a maximum capacity of 250 mL/min at a setting of 100% capacity. The liquid alum delivered to the plant is 48.5%, with an SG of 1.35.

a) What is the strength of the alum solution in g/mL?
b) Determine the setting on the liquid alum feeder for treating a flow of 12 ML/d.
c) How long would a container containing 1 kL of liquid alum last?
d) If you need to make 1-L of 1.0% alum solution for jar testing, what volume of liquid alum will you need?

TABLE 6.3
Data for Example Problem 6.2

Parameter	Liquid Alum = 1	Treated water = 2	Alum Solution
Concentration	48.5%	10 mg/L	1.0% =10 g/L
SG	1.35	1.0	1.0
Pump rate	?	12 ML/d	-
Volume	?	-	1.0 L

SOLUTION:

$$a)\ Liquid\ alum = \frac{48.5\%}{100\%} \times \frac{1.35\ g}{mL} = 0.6547 = 0.655 g/mL = \underline{655\ g/L}$$

$$b)\ Feed\ Rate,\ Q_1 = \frac{C_2 \times Q_2}{C_1}$$

$$= \frac{10\ mg}{L} \times \frac{L}{654.7g} \times \frac{g}{1000\ mg} \times \frac{12\ ML}{d} \times \frac{10^6 L}{ML}$$

$$= \frac{183.2\ L}{d} \times \frac{1000\ mL}{L} \times \frac{d}{1440\ min}$$

$$= 127.2 = \underline{130\ mL/min}$$

$$Percent\ Setting = \frac{130\ ml/min}{250\ ml/min} \times 100\% = 52.0 = \underline{52\%}$$

$$c)\ Usage = \frac{V}{Q} = \frac{1000\ L.d}{183.2\ L} = 7.86 = \underline{8.0\ d}$$

$$d)\ Volume\ of\ liquid\ alum\ V_1 = \frac{C_2}{C_1} \times V_2$$

$$V_1 = \frac{10\ g}{L} \times \frac{L}{650\ g} \times 1.0\ L \times \frac{1000\ mL}{L} = 15.38 = \underline{15\ mL}$$

6.5 FLOCCULATION PHENOMENON

Flocculation is the slow, gentle mixing of water to encourage tiny particles to collide and clump together to become settleable floc. Slow mixing is a physical process that helps to transform the microfloc formed during coagulation into macrofloc, which is easy to settle. However, as floc grows in size, it becomes more fragile, so it is important that the speed of the mixer is controlled. The idea is to get the floc particles to increase in size until they are heavy enough to settle rapidly. A typical floc size is 1–2 mm. However, floc may vary depending on the characteristics of raw water. If algae are present, they can also be trapped or caught up in the floc particles. When algae are present in large numbers, the flow will have a stringy appearance.

6.5.1 Mixers

Mixing is the most important part of the coagulation/flocculation process. Because the chemical reaction takes place in less than a minute, flash mixers are used in coagulation. The most commonly used mixers are mechanical mixers and in-line mixers or blenders. A minimum detention time of 30 seconds should be provided.

A different kind of mixing is needed for flocculation. The purpose is to move particles around so that they collide and clump together to make settleable floc. Because 95% of floc is water, it is fragile and has to be treated gently. High-speed flocculation can be damaging.

6.5.2 Completely Mixed Flow Reactor

In completely mixed flow reactors, the concentration of the reactant in the effluent is equal to that in the mixed liquid. Rapid flash mixers are completely mixed flow reactors. In completely mixed flow reactors, the detention time for the first and second-order reactions is given by the following equations:

$$C_t\left(Zero\ order\right) = C_0 - kt \quad or \quad t = \frac{1}{k}\left(C_t - C_0\right)$$

$$C_t\left(First\ order\right) = \frac{C_0}{1+kt} \quad or \quad t = \frac{1}{k}\left(\frac{C_0}{C_t} - 1\right)$$

C_0 = initial concentration
C_t = concentration after time t
k = reaction constant

6.5.3 Plug Flow Reactor

The concentration of the reactant decreases as the liquid moves along the direction of flow, remaining within the imaginary plug of water moving through the basin (**Figure 6.2**). Due to short-circuiting and intermixing, it may be difficult to achieve ideal plug flow. Flocculators in water treatment systems are designed as plug flow reactors.

$$C_t\left(Zero\ order\right) = C_t - kt \quad or \quad t = \frac{1}{k}\left(C_t - C_0\right)$$

$$C_t\left(First\ Order\right) = C_0 e^{-kt} \quad or \quad t = \frac{1}{k}\ln\left(\frac{C_0}{C_t}\right)$$

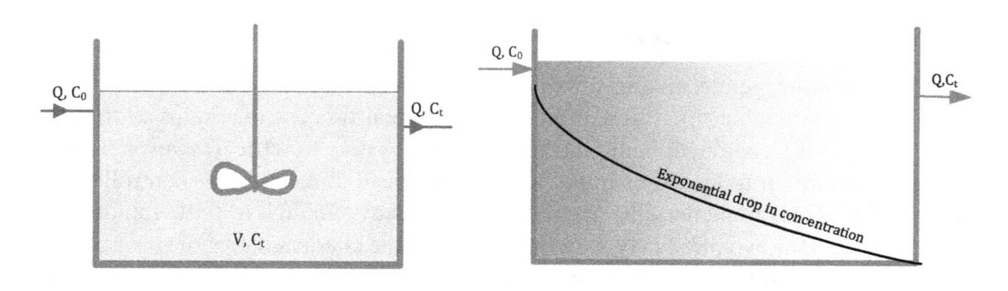

FIGURE 6.2 Types of Chemical Reactor

Example Problem 6.3

The flow rate to an ideal plug flow reactor is 6500 m^3/d, and the reaction is first order with a reaction rate constant of 900/d. Calculate the volume of the reactor for 95% removal.

Given: Q = 6500 m^3/d
k = 900/d
PR = 95%
V = ?

SOLUTION:

$$Percent\ removal,\ PR = \frac{(C_0 - C_t)}{C_0} \times 100\ \ or\ \frac{C_0}{C_t} = \left(1 - \frac{PR}{100\%}\right)^{-1}$$

$$= \frac{1}{0.05} = 20$$

$$Volume,\ V = Q \times t = Q \times \frac{1}{k}\ln\left(\frac{C_0}{C_t}\right) = \frac{6500\ m^3}{d} \times \frac{d}{900} \times \ln(20)$$

$$= 21.6 = \underline{22\ m^3}$$

Flocculation Tanks

Flocculation tanks for horizontal flocculators are usually rectangular; tanks for vertical flocculators are generally square. The depth of a flocculation tank is comparable to sedimentation tanks. The best flocculation results are obtained when the mixing intensity decreases in the direction of flow. As water flows through the tank, floc size grows, and reduced mixing intensity reduces the chances of floc break-up. The mixing intensity decreases in the direction of flow. Each successive compartment in the flocculation unit receives less mixing energy than the previous compartment. In the final stage, the peripheral speed of the pedals should be in the range of 25–35 cm/s. The minimum detention time recommended for flocculation ranges from 5–10 minutes for direct filtration systems and up to 30 minutes for conventional filtration. Minimum flow through velocity shall not be less than 2.5 or greater than 7.5 mm/s.

Chemical reactors in water treatment are designed as either completely mixed or plug flow basins. The desired intensity is achieved by varying the speed of the flocculators. A parameter called G-factor (velocity gradient), or simply G, is often used to express the intensity of mixing. Typical values for G in a flocculation tank decreases from a maximum of 75 to 30 /s, whereas in the case of rapid mixing, G values can be as high as 1000/s. Mathematically, the velocity gradient, G, can be expressed as follows:

$$G = \sqrt{\frac{P}{\mu V}} = \sqrt{\frac{N.m}{s} \times \frac{m^2}{N.s} \times \frac{1}{m^3}} = \frac{1}{s}\ \ or\ Gt = \sqrt{\frac{P}{\mu V}} \times \frac{V}{Q} = \frac{\sqrt{\frac{PV}{\mu}}}{Q}$$

P = power input in watts
μ = dynamic viscosity in Pa·s
V = volume of the tank in m^3

The power input is computed as a product of the drag force and relative velocity between paddles and water.

$$\boxed{Power\ input,\ P = Drag\ force \times Relative\ velocity}$$

$$\boxed{= \frac{1}{2} \times C_D \times A \times \rho \times v^3}$$

By knowing the drag coefficient, $C_{D,}$ relative velocity, v, and kinematic viscosity, ν, G can be found.

$$\boxed{G = \sqrt{\frac{C_D A v^3}{2V\nu}}}$$

A = area of cross section of paddles
V = volume of the tank
ν = kinetic viscosity
C_D = coefficient

Mechanically operated flocculation tanks seem to perform better than baffle type tanks.

Example Problem 6.4

A water plant is to process water at the rate of 30 ML/d. A square rapid-mix tank with vertical baffles and flat impeller blades will be used. The design detention time and velocity gradients are 30 s and 900 /s, respectively. Determine the dimensions of the mix tank and the power input required assuming an operating temperature of 20°C at which dynamic viscosity is 1.002×10^{-3} Pa.s.

Given: Q = 30 ML/d = 30 000 m^3/d
μ = 1.002×10^{-3} Pa.s
t_d = 30 s
G = 900/s

SOLUTION:

$$Volume\ of\ tank,\ V = Q \times t_d = \frac{30000\ m^3}{d} \times 30s \times \frac{d}{24 \times 3600s} = 10.417 = 10.4\ m^3$$

Assuming the depth of the tank is 5.5 m, each side of the square rapid mix tank, S, is:

$$S = \sqrt{\left(\frac{V}{d}\right)} = \sqrt{\frac{10.41\ m^3}{5.5\ m}} = 1.385 = 1.4\ m$$

$$Power\ input,\ P = G^2 \times V \times \mu$$

$$= \left(\frac{900}{s}\right)^2 \times 10.4\ m^3 \times 1.002 \times 10^{-3}\ Pa.s \times \frac{J}{m^3.\ Pa} \times \frac{W.s}{J}$$

$$= 8440.8\ W = 8.4\ kW$$

For the reliability of the operation, either provide multiple tanks or a spare electric motor.

Example Problem 6.5

A water plant has a daily flow rate of 50 ML/d. It has a flocculation basin measuring 22 m × 12m × 4.0 m. It is fitted with four horizontal shafts, each supporting four paddles measuring 13 m × 0.20 m centred 1.5 m from the shaft and rotated at 1.5 rpm. Assume the viscosity of the water at the prevailing temperature is 1.2×10^{-6} m²/s. Making the appropriate assumptions, find the velocity gradient and flocculation time.

Given: Q = 50 ML/d
Tank = 22 m × 12m × 4.0 m
Paddle = 13 m × 0.20 m, r = 1.5 m, N = 1.5 rpm
Assumed C_D = 1.9 and mean flow velocity as 30% of paddle peripheral velocity.

SOLUTION:

$$Peripheral\ v_P = 2\pi rn = \frac{2\pi \times 1.5\ m}{revolution} \times \frac{1.5\ revolutions}{min} \times \frac{min}{60\ s}$$

$$= 0.2356 = 0.24\ m/s$$

$$Differential\ v_r = 0.7\ v_P = 0.7 \times 0.2356\, m/s = 0.1649 = 0.16\ m/s$$

$$G = \sqrt{\frac{C_D A v_r^3}{2V\nu}} = \sqrt{\frac{1.9}{2} \times \frac{4 \times 4 \times 13m \times 0.2m}{22m \times 12m \times 4.0m} \times \left(\frac{0.16m}{s}\right)^3 \times \frac{s}{1.2 \times 10^{-6} m^2}}$$

$$= 11.30 = 11/s$$

$$Detention\ time,\ t_d = \frac{V}{Q} = \frac{22\ m \times 12\ m \times 4.0\ m.d}{50\ ML} \times \frac{ML}{1000\ m^3} \frac{1440\ min}{d}$$

$$= 30.4 = \underline{30\ min}$$

Example Problem 6.6

A four-arm paddle has a rotational speed of 4.0 rpm operating in a flocculation basin treating a flow of 6000 m³/d. Each arm has two paddles that are 3.0 m long and 10 cm wide. The distance from the shaft to the centre of the blade is 1.0 m for the inner blade and 1.5 m for the outer blade. The flocculation basin is 6.0 m long, 4.0 m wide and 4.5 m deep. Assume the kinematic viscosity of water, $\nu = 1.02 \times 10^{-6}$ m²/s, density of water, ρ = 998 kg/m³, drag coefficient C_D = 1.8 and relative velocity, $v_r = 0.75 v_p$. Determine how much power is dissipated into the water and the Gt value for the flocculation basin.

Given: Q = 6000 m³/d
Basin = 6.0 m × 4.0 m × 4.5 m = 108 m³
Paddle = 3.0 m × 0.10 m, r_1 = 1.0 m, r_2 = 1.5 m, N = 1.5 rpm
Assumed C_D = 1.8 and $v_r = 0.75 v_p$, ρ = 998 kg/m³

SOLUTION:

$$Inner\ paddle\ peripheral, v_P = 2\pi rn = \frac{2\pi \times 1.0\ m}{rev.} \times \frac{4.0\ rev.}{min} \times \frac{min}{60\ s}$$

$$= 0.418 = 0.42\ m/\mathrm{s}$$

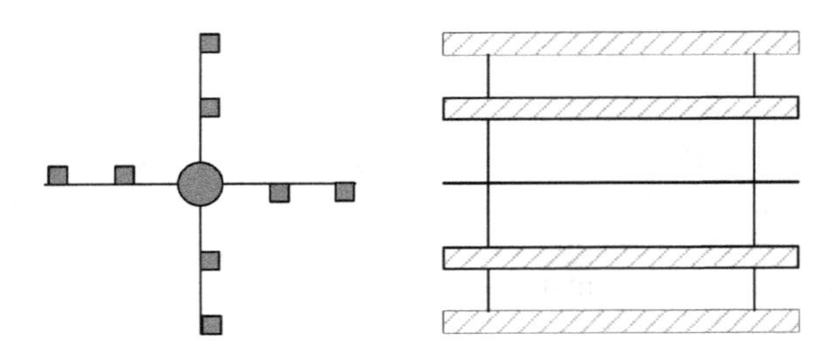

FIGURE 6.3 Flocculation Paddles

$$Power,\ P = \frac{C_D}{2} \times A \times \rho \times \left(v_r\right)^3$$

$$= \frac{1.8}{2} \times 4 \times 3.0\ m \times 0.1\ m \times \frac{998\ kg}{m^3}\left(0.75 \times \frac{0.418\ m}{s}\right)^3 \times \frac{W.s^3}{kg.m^2}$$

$$= 33.21 = 33\ W$$

$$v_P = 2\pi rn = \frac{2\pi \times 1.5\ m}{rev.} \times \frac{4.0\ rev.}{min} \times \frac{min}{60\ s} = 0.627 = 0.63\ m/s$$

$$P = \frac{1.8}{2} \times 4 \times 3.0\ m \times 0.1\ m \times \frac{998\ kg}{m^3}\left(0.75 \times \frac{0.627\ m}{s}\right)^3 \times \frac{W.s^3}{kg.m^2} = 112\ W$$

$$Total,\ P = 33\ W + 112W = 145\ W$$

$$G = \sqrt{\frac{P}{\mu V}} = \sqrt{145\ W \times \frac{m^2}{1.02 \times 10^{-3}\ N.s} \times \frac{1}{108\ m^3} \times \frac{N.m}{W.s}} = 36.28 = 36/s$$

$$t_d = \frac{V}{Q} = \frac{108\ m^3.d}{6000\ m^3} \times \frac{1440\ min}{d} = 25.9 = \underline{26\ min}$$

$$Gt = \frac{36.28}{s} \times 25.9\ min \times \frac{60\ s}{min} = 5.64 \times 10^4 = \underline{5.6 \times 10^4}$$

6.5.4 Factors Affecting Flocculation

The purpose of all flocculators is to provide gentle mixing that will produce a quick settling floc. The main factors affecting flocculation are as follows.

Degree of Mixing

If mixing is too gentle or too fast, it will prevent the formation of large floc. Very slow mixing fails to bring the suspended particles into contact with each other, while mixing too fast tears the floc particles. Mixing energy is reduced in the direction of flow to achieve better results.

Duration of Mixing

A minimum time of mixing is necessary for flocculation to be completed. In actual plant operations, depending on the temperature of the raw water, a period of 20–40 minutes is usually sufficient. To provide the required detention time, short-circuiting should be prevented. For this reason, at the entrance to the flocculator, the flow is directed downwards by placing a baffle.

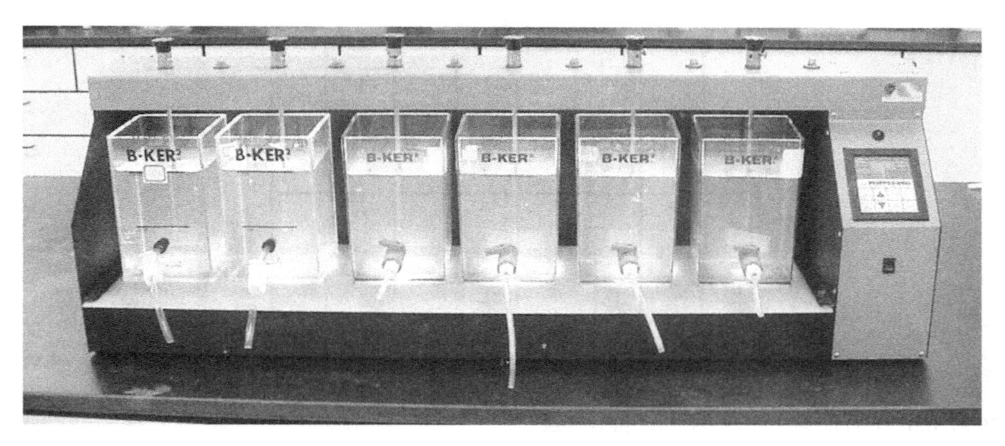

FIGURE 6.4 Jar Testing Apparatus

Number of Particles

Relatively clear water is harder to flocculate than turbid water containing a lot of suspended matter. A large number of particles allows for more collision, thus resulting in large-sized floc.

Degree of Coagulation

Coagulation destabilizes the particles causing turbidity, and flocculation clumps these particles together forming a settleable floc. Improper chemical dosage, change in the quality of raw water, and temperature can affect the coagulation and flocculation processes. Since there are so many factors affecting the coagulation/flocculation process, whenever there is a change in the type of coagulant or the raw water characteristics, a laboratory procedure called the jar test is performed.

6.6 JAR TESTING

Jar testing is basically a laboratory simulation of coagulation, flocculation and settling to determine the optimum dosage of coagulant and coagulation aids. The apparatus consists of a set of six paddle mixers driven by a variable speed motor (**Fig. 6.4**).

A glass beaker acting as a batch reactor is placed under each paddle mixer and a different dosage is added to each beaker. The set of dosages to be tried will depend upon the characteristics of the water. During jar testing, chemicals should be added in the same sequence as in plant operations. The working solutions for jar testing should preferably be made from the commercial chemical used by the plant. When adding a chemical, all jars should be dosed at the same time and paddle speed should be kept at more than 200 rpm to simulate rapid mixing.

Example Problem 6.7

Design the water depth for a mixing basin with baffles for hydraulic mixing to treat 45 ML/d. Assume the tank is partitioned into two sets of channels and the clear width is 7.5 m. Assume the spacing between the baffles to be 52.5 cm and the thickness of the wall to be 7.5 cm. Make appropriate assumptions for flow velocity and detention time.

Given: $B = 7.5$ m
$t_d = 30$ min
$v_H = 0.30$ m/s
flow width, $W = 0.525 \text{ m} - 0.075 \text{ m} = 0.45$ m
$Q = 45$ ML/d

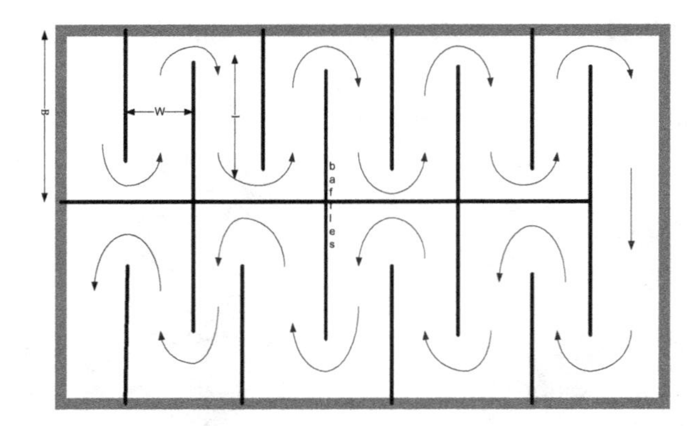

FIGURE 6.5 Flocculator Tank with Baffles

SOLUTION:

$$Capacity,\ V = Q \times t_d = \frac{45\ ML}{d} \times \frac{10000\ m^3}{ML} \times 30\ min \times \frac{d}{1440\ min}$$

$$= 937.5 = 938\ m^3$$

$$Length,\ l = v_H \times t_d = 30\ min \times \frac{0.30\ m}{s} \times \frac{60\ s}{min} = 540.0 = \underline{540\ m}$$

$$Area,\ A_X = \frac{V}{l} = \frac{937.5\ m^3}{540\ m} = 1.736\ m^2$$

$$Depth,\ d_w = \frac{A_X}{W} = \frac{1.736\ m^2}{0.45\ m} = 3.85 = 3.9\ \underline{m}$$

$$= 1.5 \times W = 1.5 \times 0.45\ mm = 0.67 = 0.70\ m$$

$$Channel,\ l_C = B - 2 \times \frac{1}{2} opening = 8.0\ m - 0.70\ m = 7.3$$

$$Number = \frac{l}{l_c} = 540\ m \times \frac{Channel}{7.3 m} = 74\, channels$$

$$= \frac{74\, channels}{2\, halfs} = 37\, channels\, /\, half$$

$$Tank.L = \# \times spacing = 37 \times 0.60\ m = 22.2 m = \underline{22\ m}$$

6.7 OPERATIONAL CONTROL TESTS

A brief discussion of tests used to optimize the coagulation and flocculation processes follows.

6.7.1 Acidity Tests

It is well established that coagulation/flocculation is very much affected by the chemical environment indicated by pH and the alkalinity of the water. If changes in pH are noticed due to natural changes in raw water or other treatment chemicals, the doses of coagulants need to be adjusted accordingly.

6.7.2 Turbidity Tests

The turbidity of raw water and settled water should be observed and compared with the jar test results to ascertain the actual removal after sedimentation. The turbidity removal should be pretty comparable. Any drastic difference from the jar testing should be investigated and corrected. For a well operating plant, the turbidity of the settled water should not be less than 10 NTU. Turbidity levels exceeding this level may prematurely seal the filter and shorten the filter runs.

6.7.3 Filterability Tests

Filterability tests are done using a pilot filter to determine how efficiently the coagulated water can be filtered. The pilot filter contains the same media as the actual filter and is equipped with an online turbidimeter. The amount of water passing through the filter before turbidity breakthrough can be used to judge the filter run under the same coagulant dosage. The water used in this test is the actual coagulated water from the plant.

6.7.4 Zeta Potential

Zeta potential is a measure of the negative charge or excess of electrons on the surface of particulate matter. The degree of this charge determines the efficacy of the coagulation process. To induce the particles to settle properly, the zeta potential, or zp, should be reduced to close to zero. A zeta potential below -10 indicates very poor settling due to strong repelling forces. A zeta meter is used to measure zp.

6.7.5 Streaming Current Monitors

The stream current in the flow stream of coagulated water is monitored using a stream current detector (STD). This measurement is very similar to the zeta potential discussed above. The optimal SCD reading varies with the pH of the raw water. The advantage of the stream current monitor is that it provides a continuous reading and thus could be used effectively to control the process. If the pH of the water remains constant, it can be used for automatic control.

6.7.6 Particle Counters

Monitoring turbidity at various points in water treatment yields good information about the removal of particulate matter. However, it fails to recognize the size and number of particles removed. This is increasingly important when we want to ensure the removal of crypto and Giardia cysts, two pathogens commonly found in drinking water supplies. Since particle counters can provide size distribution, that is, size and concentration, they can be used for optimizing the coagulation filtration process.

Particle counters emit a beam of laser light, and the reflected or diffused light is sensed by a photodiode. Computer software interprets this information. These instruments are quite expensive and need to be handled by specially trained personnel. However, it has a promising future in water treatment optimization.

Discussion Questions

1. To precipitate turbidity, alum is added that reacts with the alkalinity present in water. Show that during coagulation, alkalinity as $CaCO_3$ consumed is equal to half that of alum dosage.

2. What do you understand by coagulation and flocculation? Explain why they are necessary in potable water treatment.
3. Describe the various type of coagulant and coagulant aids used in the water industry.
4. Discuss the role of coagulant aids and flocculation aids.
5. Compare rapid mixing with slow mixing.
6. Describe the various steps to perform jar testing. What are the disadvantages of applying too high a dosage?
7. Explain the factors that affect flocculation.
8. What process control test can be used to optimize the coagulation/flocculation process?
9. Why is alum the most commonly used coagulant in water treatment?
10. Under what conditions would you need frequent jar testing?
11. Write the chemical reaction between ferric chloride and lime slurry that results in the precipitation of $Fe(OH)_3$.

Practice Problems

1. The raw water of a certain river has an alkalinity of 40 mg/L as $CaCO_3$. Based on jar testing, an optimum alum dosage of 40 mg/L is required to reduce the turbidity to less than 1 NTU. To assure complete precipitation, a minimum of 30 mg/L of residual alkalinity is recommended.
   a. What alkalinity must be added to maintain the desired level of residual alkalinity? (10 mg/L as $CaCO_3$)
   b. Determine the dose of 80% hydrated lime in mg/L that will be needed to add alkalinity? (9.3 mg/L)
   c. What should be the setting on the lime feeder in kg/d while treating a flow of 12 ML/d? (110 kg/d)
2. A water plant is to process water at a rate of 62 ML/d. The dimensions of a flocculation basin are 25 m × 14 m × 4.0 m. It is fitted with four horizontal shafts, each supporting four paddles measuring 14 m × 0.2 m and are centred 1.5 m from the shaft and rotated at 15 rpm. Assume the viscosity of the water is 1.2×10^{-6} m^2/s. Making the appropriate assumptions, calculate the velocity gradient and flocculation time. (10/s, 33 min)
3. In a small water plant, a 15% alum solution is fed at a rate of 200 L/24h when treating a flow of 28 m^3/h. Find the alum dosage applied. (45 mg/L)
4. In a water treatment plant, due to an increase in water demand, the water flow rate is increased to 11 ML/d. To achieve an alum dosage of 30 mg/L, work out the alum pump rate in mL/min of 48% liquid alum with an SG of 1.3. (370 mL/min)
5. A water treatment plant's daily flow rate is 30 ML/d. Select the dimensions of a square rapid mix tank with vertical baffles and flat impeller blades to provide a detention time of 0.50 min. Assume the depth of the mixer is 6.0 m. Determine the power input required to produce a velocity gradient of 900 /s. Assume the viscosity of the water is 1.0×10^{-3} Pa·s. (1.3 m × 1.3 m, 8.2 kW)
6. A flocculation basin that is 20 m long, 14 m wide and 4.5 m deep treats a coagulated water flow of 35 ML/d at a temperature of 15°C (absolute viscosity, $\mu = 1.06 \times 10^{-3}$ Pa·s). The power input to the paddle wheel is 1.2 kW, resulting in a paddle blade-tip velocity of 0.42 and 0.30 m/s for the outer and inner blades, respectively.
   a. Determine the detention time in h. (0.86 h)
   b. Determine the flow through velocity. (6.4 mm/s)
   c. Determine the velocity gradient. (30 /s)

7. A dosage of 60 mg/L of alum is added for the coagulation of river water.
   a. How much natural alkalinity is consumed if 20 mg/L of alkalinity as $CaCO_3$ is added? (10 mg/L)
   b. What is the increase of SO_4 ion in the treated water? (29 mg/L)
   c. What is the reduction in alkalinity as bicarbonate? (12 mg/L)
   d. How is the pH of the water affected? (Drop)
8. A dosage of 30 mg/L of alum and an equivalent dosage of soda ash is added for coagulation of surface water. What changes take place in the ionic character of water? (addition of 14 mg/L of SO_4 ion, 7 mg/L of Na ions and decrease in pH due to CO_2)
9. The coagulation of soft water requires 40 mg/L of alum plus lime to supplement natural alkalinity for proper coagulation and floc formation. If the goal is to react not more than 10 mg/L of natural alkalinity, what dosage of lime as 85% pure CaO is required? (6.6 mg/L)
10. River water is coagulated by dosing ferrous sulphate at a rate of 30 mg/L and an equivalent dosage of lime. The chemical reaction is as follows:

$$2FeSO_4.7H_2O + 2Ca(OH)_2 + 0.5\,O_2 + 2Fe(OH)_3 + 2CaSO_4 + 13H_2O$$

   a. How many kg of hydrated lime are needed per ML of water? (8.0 kg/ML)
   b. How many kg of 88% CaO are needed per ML of water? (6.9 kg/ML)
   c. How many kg of sludge as $Fe(OH)_3$ are needed per ML of water? (12 kg/ML)
11. At a conventional plant, surface water at a rate of 24 ML/d is treated by applying a coagulant dosage of 50 mg/L.
   a. What is the required dosage rate of lime as 85% CaO? (730 kg/d)
   b. How many kg of dry solids as $Fe(OH)_3$ are produced? (790 kg/d)
   c. Assuming wet sludge is 2.0% solids, what is the daily production of inorganic sludge? (40 m^3/d)
12. Determine the settings in percent stroke on a chemical feed pump to apply a dosage of 2.5 mg/L. The water is pumped at a rate of 35 L/s and the strength of the chemical solution fed is 5.0%. The chemical feed pump has a maximum capacity of 300 mL/min at a setting of 100% capacity. (35%)
13. The optimum alum dose as determined by jar testing is 15 mg/L. The label on the liquid alum drum indicates a strength of 45% with a density of 1.3 kg/L.
   a. Determine the setting on the liquid alum feeder in L/24h for treating a flow of 15 ML/d. (390 L/24h))
   b. How long would a container containing 1.5 m^3 of liquid alum last? (3.9 d)

7 Sedimentation

Sedimentation follows coagulation and flocculation to allow the gravity settling of flocs and particulate matter. Settled water has a low turbidity of usually less than 10 NTU, which is removed by filtration. Sedimentation is necessary for the treatment of turbid waters. However, when the turbidity level in the source water is less than 10 NTU, the sedimentation process may be omitted. This process scheme is called direct filtration. The main purpose of sedimentation in the treatment of potable water is to remove solids to minimize the loading on the filters. Experience has shown that, in a well-operated plant, the turbidity of the settled water should be less than 10 NTU.

Pre-sedimentation is practised where the surface water bodies are subjected to heavy loads of solids, such as grit and sand. This usually involves the use of a dam to hold the water to allow sedimentation. A lake is a natural pre-sedimentation tank. In the case of turbid waters, pre-sedimentation is very important since water loaded with grit can damage the mechanical equipment and pipes. Because pre-sedimentation is done before the water enters the plant, it is not normally considered part of the plant process scheme.

7.1 GRAVITY SETTLING

In settling, heavier matter in water is allowed to settle out by gravity under quiescent conditions. For effective settling, water flow velocity needs to be kept low, typically in the range of 0.1–0.5 mm/s. Sedimentation will happen when the settling velocity is more than the surface overflow rate and the sedimentation tank provides sufficient hydraulic detention time. The parameters important for the design and operation of sedimentation tanks include detention time, overflow rate and weir loading.

7.2 SEDIMENTATION BASINS AND TANKS

For efficient operation, sedimentation basins are designed so that water can enter the tank, pass through and leave without creating much turbulence, while preventing any short-circuiting. The operation of a settling tank can be better understood by visualizing it in four distinct zones. In the inlet zone, water entering the tank is distributed across this section by the baffle and slows to a uniform velocity. The settling zone is the main part of the tank, where the water flows slowly and the suspended solids settle. The settled matter or sludge accumulates at the bottom of the tank, referred to as the sludge zone. In the outlet zone, the clarified water collects in suitable channels and leaves the tank.

7.2.1 Rectangular Basins

Rectangular basins are more popular in water treatment plants. The main reasons for their popularity include predictable performance, comparable cost, low maintenance and minimal short-circuiting. They can also cope with sudden increases in loading.

Double deck basins are modified rectangular basins that provide twice the effective sedimentation area of a single basin of equivalent land area. Due to their high operation and maintenance cost, they failed to gain popularity.

DOI: 10.1201/9781003231264-7

7.2.2 Circular and Square

In circular and square basins, the flow radiates from the centre towards the periphery. They are generally more likely to short circuit. The bottom of a circular clarifier is graded towards the centre (typically 1:12) to provide the efficient collection and withdrawal of sludge. Given a circular clarifier with a diameter, D, the volume of water and sludge can be expressed in terms of D. The total volume is the sum of a cylinder of height H equal to the side water depth and a cone made by the sloped part of the clarifier.

$$Capcity,\ V = \frac{\pi D^2 H}{4} + \frac{1}{3} \times \frac{\pi D^2}{4} \times \left(\frac{D}{2 \times 12}\right) = \frac{\pi D^2}{4}\left(H + \frac{D}{72}\right)$$

$$= D^2\left(0.785H + 0.011D\right)$$

7.2.3 Tube Settlers

Conventional sedimentation basins are modified by placing tubes at an upwards angle in the direction of flow through the tank. Water is directed upwards and each tube acts as an individual settler. Tube settlers are basically shallow depth settlers. Together, they provide a high ratio of settling surface area per unit volume of water. High-rate settlers are particularly useful for water treatment applications where the site area is in constrained, packed-type units to increase the capacity of existing units. Since the settling depth is reduced significantly, the time for a particle to settle is also reduced. Due to increased overflow rates, coagulant aids should be used to strengthen the floc.

7.2.4 Solids Contact Units

Solids contact units combine coagulation, flocculation, and sedimentation and sludge removal into a single compartmented tank, as shown in **Figure 7.1**. Such units are common for industrial plants and where softening and turbidity removal are performed simultaneously. The flow is generally in an upward direction through a properly controlled sludge blanket, hence the name solids contact unit or up-flow clarifier.

Raw water and coagulant enter the draft tube just below the mixer. The mixer draws the raw water and the chemical up the draft tube through the rapid mix zone. This phase represents the coagulation

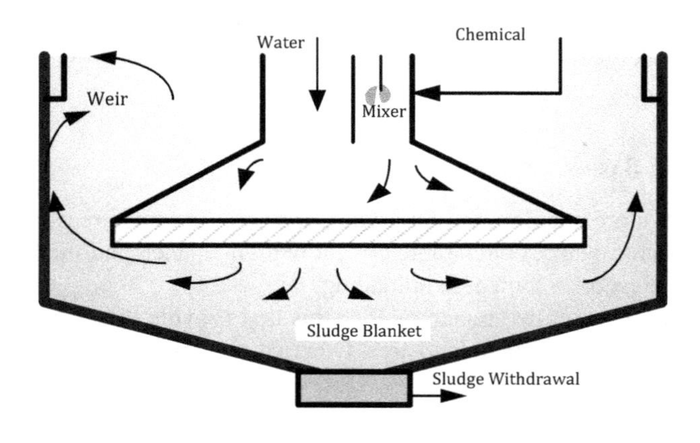

FIGURE 7.1 Solids Contact Unit

process. Water then flows from the top of the draft tube down into a slow mix zone under the skirt, where flocculation takes place. The contact with the sludge solids helps to build better floc. Floc settles to form a layer of sludge at the bottom of the clarifier. Water flows up to the launders from the bottom of the flocculation zone. The upward flow velocity or rise rate decreases in this zone to allow the floc to settle. From the launders, the clarified water flows to the next process, usually filtration.

During the operation of such units, some of the settled sludge is recycled with the incoming raw water, and the remainder is periodically drawn off. There must be a proper balance between the incoming raw water turbidity and the amount of sludge to be drained. The maintenance of the sludge blanket is critical to the operation of the up-flow clarifier. Such devices are very sensitive to changes in influent flow and turbidity.

7.2.5 Pulsator Clarifier

The pulsator clarifier is a vertical-flow type sludge tank. A pulse of upward flow is generated every 30 s to give rapid flow for 5–10 s. This causes the sludge blanket to rise and fall alternately.

7.2.6 Ballasted Flocculation

This is also known as an Actiflo process. Injecting the water with heavy sand enhances the flocculation process. Usually, micro sand with a size range of 60–200 μm is used as ballast. Typically, the sand is applied after coagulation along with a flocculation aid, such as a polymer. Both are added to initiate floc formation for a couple of minutes. In the maturation tank, the polymer allows the floc to gain in size and strength. Typically, hydraulic detention time is about 6 min. Clean sand is separated from the sludge slurry and recycled to the injection tank.

7.3 THEORY OF SEDIMENTATION

Velocity of flow, viscosity of water (ν), size (D) and shape and, more importantly, density or specific gravity (G_S) of the solid influence the settling velocity, v_S, of discrete particles in water. When buoyancy and drag forces counterbalance the gravity forces, the solid particles reach terminal velocity as given by Stokes' law.

$$\text{Stokes, } v_s = \frac{gD^2}{18\nu}\left(G_s - 1\right) \text{ or } v_s = 418D^2\left(G_s - 1\right)\left[\frac{3T + 70}{100}\right]$$

Since viscosity is dependent on temperature, the second equation is a modified form to introduce variation with temperature, T (°C). However, the above equation is valid for particles of size < 0.1 mm and $N_R < 1$, that is, when flow is laminar. If the settling particles are larger than 1.0 mm, settling is turbulent and the settling velocity is determined by Newton's equation.

$$\text{Newton, } v_s = 1.8\sqrt{gD\left(G_s - 1\right)}$$

Example Problem 7.1

Assuming laminar flow conditions, determine the settling velocity of a discrete particle with a diameter of 60 μm. Assume the specific gravity of the solid is 2.65 and the water temperature is 20°C.

Given: T = 20°C

D = 65 μm = 6.0×10^{-5} m = 6.0×10^{-2} mm

ν = 1.01×10^{-6} m²/s

v_s = ?

SOLUTION:

$$v_s = 418D^2\left(G_s - 1\right)\left[\frac{3T + 70}{100}\right]$$

$$= 418 \times \left(6.0 \times 10^{-2}\right)^2 \times \left(2.65 - 1\right)\left[\frac{3 \times 20 + 70}{100}\right]$$

$$= \underline{3.22 = 3.2\ mm/s\ or}$$

$$v_s = \frac{gD^2}{18\vartheta}\left(G_s - 1\right)$$

$$= \frac{1}{18} \times \frac{9.81m}{s^2} \times \left(6.0 \times 10^{-5}\ m\right)^2 \times \frac{s}{1.01 \times 10^{-6}\ m^2} \times 1.65$$

$$= \frac{3.20 \times 10^{-6}\ m}{s} \times \frac{1000\ mm}{m} = 3.20 = \underline{3.2\ mm/s}$$

7.4 DESIGN PARAMETERS

When designing the sedimentation process, engineers use basin guidelines. It is important to understand these guidelines in order to communicate effectively with design engineers and to identify the cause of operational problems.

7.4.1 Detention Time

Hydraulic detention time (HDT) or retention time refers to the time required by a parcel of water to pass through a sedimentation basin at a given flow rate (**Figure 7.2A**). Commonly used values of HDT are two to four hours. The actual flow through time may be less than the calculated value due to short-circuiting.

Actual retention time in a given vessel can be determined by performing a dye test. A dye is placed at the inlet of the basin, and the concentration of the dye, C, is observed at various times after its injection. A general residence time distribution curve is a plot of the concentration of dye in effluent versus time t (**Figure 7.2B**). The mean residence time is the time to the centroid.

$$HDT, t_d = \frac{V}{Q} \qquad t_{50} = \frac{\Sigma(tC)}{\Sigma t}$$

7.4.2 Surface Overflow Rate

Overflow rate (v_0) or surface loading represents the upward velocity with which water is going to rise in a given basin. Particles with settling velocities of less than the overflow rate will not settle out. The overflow rate is determined by dividing the basin flow rate by the basin surface area.

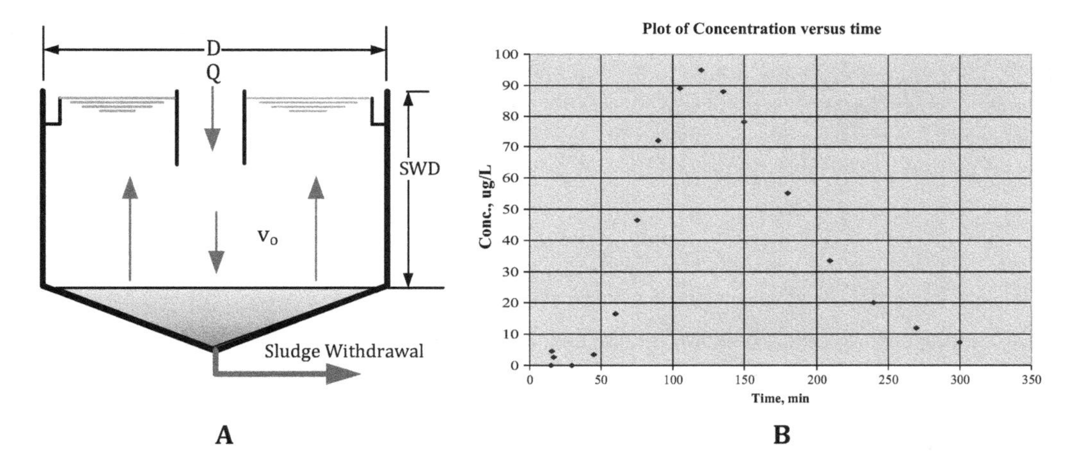

FIGURE 7.2 Detention Time and Residence Time

$$SOR = v_O = \frac{\text{flow rate}}{\text{surface area}} = \frac{Q}{A_S}$$

Typically, the value of SOR ranges from 20 to 33 $m^3/m^2{\cdot}d$. In colder temperatures, settling velocities reduce due to an increase in the density and viscosity of water. Thus, plant capacity is lowered during the winter months. To compensate for the reduced settling rate, weighted agents may be applied, especially in light density floc.

7.4.3 Overflow Rate and Removal Efficiency

Since overflow rate represents the velocity with which water rises, particles with a settling velocity less than equal to the overflow rate will completely settle out in an ideal plug flow system, as in the case of long rectangular tanks. In other words, particles with a settling velocity $\geq v_o$ will be 100% removed. However, particles with lower settling velocities will also settle if introduced at a lesser height. This will only be true in plug flow reactors and not up-flow clarifiers. Following this logic, if particles are uniformly distributed, 50% of particles with a settling velocity of 0.5 v_o should also settle out. Likewise, 25% of particles with a settling velocity of 0.25 v_o will also be removed. Since there is a large gradation of settling particle sizes, this job can be accomplished by performing sieve analysis and a hydrometer test. Based on this analysis, the percentage of particles with $v_s < v_o$, say P_o, is found, and the plot of P versus particle size is plotted to generate a so-called settling velocity curve.

$$PR = \left(100 - P_o\right) + \int_0^{P_o} \left(\frac{v_S}{v_o} \times 100\%\right) dP$$

Since the curve does not follow a specific geometric shape, integration can be performed by selecting strips of area under the curve as follows.

$$PR = \left(100 - P_o\right) + \frac{1}{v_o} \Sigma\, v_s \Delta P$$

7.4.4 Effective Water Depth

Ideally, sedimentation basins should be shallow with a large open area. However, practical considerations, such as depth of sludge blanket, current and wind effects, and desired flow velocity, must be considered in the selection of an appropriate basin depth. The side water depth (SWD), overflow rate (v_0) and hydraulic detention time (HDT) are interrelated parameters.

$$t_d = \frac{side\ water\ depth}{overflow\ rate} = \frac{d_w}{v_O} = \frac{L}{v_H}$$

Since v_0 represents the upward travelling rate, distance divided by velocity gives the time it takes for a given amount of water to rise vertically for discharge through effluent channels.

7.4.5 Mean Flow Velocity

Mean flow velocity is the horizontal flow velocity along the length of the basin. In an ideal basin, it should be uniform. However, due to the non-uniform accumulation of sludge, wind action and density currents, it is not completely uniform throughout the length of the basin. High-flow velocities will tend to scour the settled sludge. It is recommended to limit mean flow velocities to 2.5 mm/s at the design flow rate.

Example Problem 7.2

Size a rectangular sedimentation tank to treat a flow of 4.5 ML/d so as to provide a detention time of 3.0 h without exceeding a flow velocity of 1.5 mm/s.

Given: $Q = 4.5$ ML/d $= 4500\ m^3/d$
$v_H = 1.5$ mm/s
$t_d = 3.0$ h

SOLUTION:

$$Length\ L = t_d \times v_H = 3.0\ h \times \frac{1.5\ mm}{s} \times \frac{3600\ s}{h} \times \frac{m}{1000\ mm}$$

$$= 16.2 = \underline{17\ m\ say}$$

$$Flow\ area,\ A_X = \frac{Q}{v_H} = \frac{4500\ m^3}{d} \times \frac{s}{1.5\ mm} \times \frac{1000\ mm}{m} \times \frac{d}{24\ h} \times \frac{h}{3600\ s}$$

$$= 34.72 = 35\ m^2$$

Assume a working depth of 3.0 m.

$$Width,\ B = \frac{A_X}{H} = \frac{34.72\ m^2}{3.0\ m} = 11.57 = \underline{12\ m\ say}$$

To provide room for sludge accumulation and free board, select the depth of the tank as 3.0 m + 1.0 m + 0.5 m = 4.5 m. Hence, the final design is a rectangular basin of 17 m × 12 m × 4.5 m.

Example Problem 7.3

Size a circular sedimentation tank to treat a flow of 3.0 ML/d so as to provide a detention time of 3.0 h. Assume the bottom of the tank is sloped 1:12 (H:V).

Given: Q = 3.0 ML/d = 3000 m^3/d
t_d = 3.0 h
SWD = 3.0 m (assumed)

SOLUTION:

$$Capacity,\ V = t_d \times Q = 3.0\ h \times \frac{3.0\ ML}{d} \times \frac{d}{24\ h} \times \frac{1000\ m^3}{ML}$$

$$= 375.0 = \underline{375\ m^3}$$

The first estimate of the diameter of the clarifier can be made without considering the conical portion of the clarifier.

$$D = \sqrt{\frac{4V}{\pi H}} = \sqrt{\left(\frac{4}{\pi} \times \frac{375\ m^3}{3.0\ m}\right)} = 12.6\ m$$

The actual required diameter would be less. Assume it to be 12.0 m and check.

$$V = D^2\left(0.785H + 0.011D\right) = \left(12m\right)^2\left(0.785 \times 3.0\ m + 0.011 \times 12\ m\right)$$

$$= 358.12 = 360\ m^3 < 375m^3$$

Hence, the required diameter will be slightly more than 12 m. A diameter of 12.5 m would be able to hold the required volume. Based on this diameter, the height of the bottom cone is:

$$= \frac{D}{2 \times 12} = \frac{12\ m}{24} = 0.50\ m$$

A 12.5 m diameter tank with a total depth of 3.0 m + 0.5 m +0.5 m (free board) = 4.0 m will do the job.

Example Problem 7.4

In a water filtration plant, a rectangular sedimentation tank is 3.0 m deep and 50 m long. What flow velocity would you recommend to effectively remove 30 μm particles at a water temperature of 25°C ($\vartheta = 9.0 \times 10^{-7}$ m^2/s). Assume 2.65 is the specific gravity of the solid particles.

Given: T = 25°C
D = 35 μm = 3.0×10^{-5} m = 3.0×10^{-2} mm
G_s = 2.65
v_H = ?

SOLUTION:

$$v_s = 418D^2\left(G_s - 1\right)\left[\frac{3T+70}{100}\right]$$

$$= 418\times\left(3.0\times10^{-2}\right)^2\times\left(2.65-1\right)\left[\frac{3\times25+70}{100}\right]$$

$$= 0.900 = \underline{0.90\ mm/s}$$

$$v_H = v_S \times \frac{L}{d_w} = \frac{0.90\ mm}{s}\times\frac{50\ m}{3.0\ m} = 15.0 = \underline{15\ mm/s}$$

Hence, to ensure 100% removal of 30 μm particles, the horizontal flow velocity must not exceed 15 mm/s.

7.4.6 Weir Loading Rate

The weir loading rate is the flow rate per unit length of the effluent weir. If the weir loading rate becomes too high, floc will be carried out of the basin. Weir loading is more important in shallow basins than in deeper basins. The maximum weir loading rate is 250 $m^3/m\cdot d$.

7.5 FACTORS AFFECTING OPERATION OF SEDIMENTATION

Some key factors are as follows:

- Large and dense floc will settle faster.
- A lower tank overflow rate is preferred.
- Inlet arrangements should distribute incoming flow over the full cross-section of the tank.
- The outlet design should collect settled water near the top and uniformly across the width of the tank.
- Currents caused by wind action and density differences can cause short-circuiting.
- Particles settle faster in warm water than cold water.
- Usually, two to four hours of residence time is sufficient.
- Short-circuiting results in poor removals.
- Flow rates exceeding design flow can cause solids to carry over.
- The efficiency of the coagulation/flocculation process directly affects the performance of the sedimentation process.
- Sudden changes in raw water quality will affect the sedimentation process.
- In the event of increased dosages of coagulant required for the effective removal of increased suspended solids, the frequency of sludge removal may need to be increased.
- Changes in source water alkalinity and pH caused by storms seriously affect the performance of sedimentation as a result of poor coagulation or flocculation.
- Sudden increases in settled water turbidity will cause the premature clogging of filters and may increase the turbidity of the filter effluent.

7.6 VOLUME OF SLUDGE

Sludge collected from the bottom of a water treatment clarifier is mainly inorganic. As discussed earlier, by knowing the dosage of the coagulant and turbidity removed, the mass of solids removed can be found. By knowing the consistency of wet sludge, the quantity of wet sludge can be estimated. Though inorganic solids are heavier, sludge solids concentration is not more than 2%. For solids concentration up to 10%, it is safe to assume the specific gravity of sludge is 1. The solid concentration of alum sludge ranges from 1 – 3%. When alum is used as the coagulant, the sludge formed is difficult to dewater.

Example Problem 7.5

In a flocculation clarifier, an alum dose of 30 mg/L is used to treat river water at a rate of 4.0 ML/d. Estimate the volume of wet sludge produced as aluminium hydroxide, assuming the solids concentration in the sludge is 0.80% and the specific gravity of dry solids in the sludge, G_s is 1.5. Find the quantity of sludge produced, assuming 15 NTU of turbidity is removed.

Given: Q = 4.0 ML/d
Alum dose = 30 mg/L
Sludge solids, SS = 0.80%
G_s = 1.0
Sludge volume = ?

SOLUTION:

$$\frac{100\%}{G_{sl}} = \frac{SS\%}{G_{ss}} + \frac{w\%}{G_w} \quad or \quad \frac{100\%}{G_{sl}} = \frac{080\%}{1.5} + \frac{99.2\%}{1} \quad or$$

$$G_{sl} = \left(\frac{0.008}{1.5} + .992\right)^{-1} = 1.0026$$

This validates the point earlier that for low solid concentrations, the specific gravity of wet sludge can be safely assumed to be 1.

$$Sludge,\ SS = \frac{0.80\%}{100\%} \times \frac{1000\ kg}{m^3} = 8.0 = 8.0\ kg/m^3$$

$$Mass,\ M_{SS} = Q \times SS = Q \times 0.26\ Alum = \frac{4.0\ ML}{d} \times \frac{0.26 \times 30\ kg}{ML}$$

$$= 31.2 = \underline{31\ kg/d}$$

$$Sludge\ volume = \frac{M_{SS}}{SS_{sl}} = \frac{31.2\ kg}{d} \times \frac{m^3}{8.0\ kg} = 3.90 = \underline{3.9\ m^3/d}$$

In the second case, 15 NTU of turbidity is removed that would increase solids in the sludge.

$$Sludge,\ SS = 0.44 \times Alum\ dose + 0.74 \times Turbidity\ removed\ in\ NTU$$

$$= 0.44 \times (30\ mg)/L + 0.74 \times 15 = 24.3\ mg/L$$

$$M_{SS} = Q \times SS = \frac{4.0\ ML}{d} \times \frac{24.3\ kg}{ML} = 97.2 = \underline{97\ kg / d}$$

$$Sludge\ volume = \frac{M_{SS}}{SS_{sl}} = \frac{97.2\ kg}{d} \times \frac{m^3}{8.0\ kg} = 12.1 = \underline{12\ m^3 / d}$$

That is, roughly three times the wet sludge produced without considering the solids removed due to turbidity.

7.7 SLUDGE DISPOSAL

In many plants, common practice is to discharge the sludge in the sanitary sewer to be treated at the municipal wastewater plant. Recently, water plant residue has been found to have a number of other applications, including as soil conditioner, to name a significant one. Recent applications include cement and brick manufacturing, turf farming, composting with yard waste, road subgrade, forest land application, citrus grove application, landfill and land reclamation.

Discussion Questions

1. Based on the chemical reaction of alum reacting with alkalinity, show that alum floc produced as aluminium hydroxide is roughly about one-fourth of alum dosage.
2. List the conditions suitable for direct filtration.
3. Describe, with a neat sketch, the functional zones of a sedimentation basin.
4. Derive the relationship between detention time and surface overflow rate.
5. Prove that theoretically, surface loading, not depth is a measure of the removal of particles in a sedimentation tank.
6. Explain the working of a circular sedimentation tank with the help of a neat sketch.
7. Define the terms: detention time, flow-through period, overflow rate and weir loading.
8. List the advantages and limitations of solids contact units.
9. Describe the working of a solids contact unit with a neat sketch.
10. Explain how the use of tube settlers can increase plant capacity without expanding its footprint.
11. How does the flow scheme for direct filtration differ from the conventional flow scheme? What limitations of raw water quality are recommended for the adoption of direct filtration?

Practice Problems

1. Calculate the settling velocity of a discrete particle with a diameter of 45 μm, assuming laminar flow conditions. Assume the specific gravity of the solid to be 2.62 and the water temperature to be 20°C ($\nu = 1.0 \times 10^{-6}\ m^2/s$). (1.8 mm/s)
2. A rectangular sedimentation basin that is 16 m long, 10 m wide and 3.2 m deep treats a flow of 4800 m^3/d. Determine the detention time, the overflow rate and the mean flow velocity through the basin. (2.6 h, 30 m^3/m^2.d, 1.7 mm/s)
3. Size a rectangular sedimentation basin to treat a flow of 5.5 ML/d so as to provide a detention time of 3.5 h without exceeding a flow velocity of 1.5 mm/s. Assume a side water depth of 3.3 m. (19 m × 13 × 4.0 m)

4. A sedimentation tank is 3.0 m deep and 50 m long. What flow velocity would you recommend to effectively remove 25 μm particles when the water temperature is 20°C. Assume the specific gravity of the solids particles is 2.65. (9.4 mm/s)
5. In a water filtration plant, water is processed at a daily rate of 5.5 ML/d and turbidity is removed by applying alum at a dosage of 35 mg/L. Estimate the volume of wet sludge produced as aluminium hydroxide, assuming the solids concentration in the sludge consistency is 0.90%. (5.6 m^3/d)
6. In a water plant, coagulated water is fed to two circular settling basins operating in parallel at the rate of 4.0 ML/d. Each basin is 2.5 m deep and has a single peripheral weir attached to the outer wall.
   a. Determine the diameter of each basin based on a surface loading of 18 $m^3/m^2 \cdot d$. (12 m)
   b. What detention time is achieved? (3.3 h)
   c. Calculate the weir loading rate. (53 $m^3/m \cdot d$)
7. Use Stokes' law to calculate the settling velocity of a spherical particle with a diameter of 1.0 mm and a specific gravity of 2.7 in water at a temperature of 25°C. The kinematic viscosity of the water at 25°C is 0.85×10^{-6} m^2/s. (1.1 m/s)
8. What diameter circular clarifier and side water depth is needed to treat a flow of 15 ML/d. The maximum overflow rate must not exceed 16$m^3/m^2 \cdot d$ and a detention time of 4.0 h must be achieved. (35 m, 2.7 m)
9. A spherical particle with a diameter of 0.02 mm and a specific gravity of 2.8 settles in water with a temperature of 20°C. Use Stokes' law equation to calculate the settling velocity. The kinematic viscosity of water at 20°C is 1.0×10^{-6} m^2/s. Convert the settling velocity from metres per second to $m^3/m^2 \cdot d$, which represents the units commonly used for expressing the overflow rate. (3.9×10^{-4} m/s, 34 $m^3/m^2 \cdot d$)
10. Size a circular clarifier for a design flow of 3800 m^3/d.
    a. What diameter tank should be used to allow 100% settling of particles with a settling velocity of 0.24 mm/s? (15 m)
    b. What should be the side water depth of the tank to provide a detention time of 3.0 h? (2.6 m)
    c. What size particles will be 100% removed? Assume viscosity $\nu = 1.0 \times 10^{-6}$ m^2/s, SG = 2.7. (16 μm)
11. Estimate the sludge solids produced when coagulating surface water with a turbidity of 14 NTU and an alum dosage of 45 mg/L. What volume of sludge will there be if the settled waste sludge and filter backwash are concentrated to 1000 mg/L of solids in a clarifier thickener? (30 mg/L, 30 L/m^3)

8 Filtration

The process of filtration removes particulate matter and floc by passing water through a porous bed of sand, coal or other granular material. In conventional filtration, the process scheme consists of coagulation, flocculation, sedimentation and filtration. When raw water turbidity is < 5 NTU, the flocculation step can be eliminated, hence the term direct filtration. When the coagulants are added directly to the filter inlet pipe and flocculation and sedimentation are eliminated, this is called in-line filtration.

8.1 FILTRATION MECHANISMS

Filtration is not just straining but a combination of physical chemical processes and, in some cases, biological processes. Although straining plays a significant role in the overall removal process, it is important to realize that most of the particles removed during filtration are considerably smaller than the pore openings in the filter media. The entire removal process is a complex combination of straining, sedimentation, adsorption and biological action.

Sedimentation occurs when particles settle out on the surface of the media as water flows through the openings or pores of the media. Some media, such as carbon, has absorptive properties. The process of adsorption removes the particles by pulling them onto the surface of the media. As more and more removal takes place, the media is exhausted. Biological action is important when filtration rates are relatively slow, as in the case of slow sand filtration. Biological growth on the media retains impurities from the water, especially organics.

8.2 TYPES OF FILTERS

Earlier filters were slow sand filters consisting of uniform sand with no provision for backwashing. Modern multi-media filters perform at much higher filtration rates and are cleaned by backwashing. Backwashing refers to reversing the flow of water through the bed at a rate higher than the filtration rate.

8.2.1 Slow Sand Filters

The filtration rate, defined as the flow rate per unit surface area, of slow sand filters is extremely low. The removal mechanisms employed in slow sand filters are straining, adsorption and biological action. The main advantage of this method is low maintenance and simplicity of operation. Due to low hydraulic loading, most of the removal takes place in the top portion of the bed. When the bed becomes clogged with turbidity, the top layer is scraped off and replaced with new media. Due to high labour and area requirements, combined with its unsuitability for treating turbid waters (> 10 NTU), slow sand filtration is not very common in larger plants.

8.2.2 Rapid Sand Filters

As the name indicates, in rapid sand filtration, hydraulic loading or filtration rate is 30 to 40 times greater than that of slow sand filtration, as shown in **Table 8.1.** This is achieved by maintaining 2–3 m of head above the media consisting of uniform sand. In addition to straining, removal mechanisms

DOI: 10.1201/9781003231264-8

TABLE 8.1
Comparison of Various Types of Filters

Characteristic	Slow Sand	Rapid Sand	High Rate
Filtration rate, L/s·m²	0.03	1.5	2–5
Media distribution	Unstratified	Fine to coarse	Coarse to fine
Filter run	20–60 d	12–36 h	12–36 h
Loss of head, m initial	0.03	0.3	0.3
Loss of head, m final	1.1	2.5	2.5

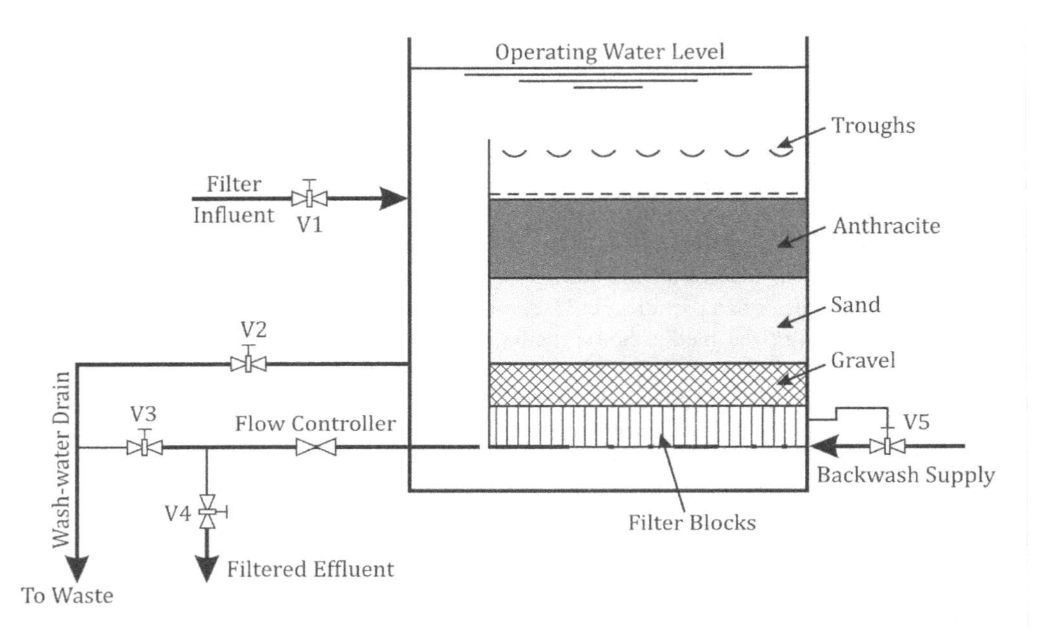

FIGURE 8.1 Gravity Filtration System Components

include sedimentation and adsorption. Because of the rapid sand filter's increased rate of filtration and its ability to be backwashed, chemical coagulation is possible.

8.2.3 High-Rate Filters

In high-rate filters, much higher filtration rates can be afforded. This is achieved by providing multi-media filters with gradation from coarse to fine. Gradation allows in-depth filtration rather than just at the top portion of the filter. In-depth filtration achieves longer filter runs before backwashing is needed. A comparison of various filters is shown in **Table 8.1**.

8.3 COMPONENTS OF A GRAVITY FILTER

The main components of a gravity filtration system, as shown in **Figure 8.1**, include a filter box, filter media, underdrain system, gravel bed, surface wash system, wash-water troughs and control equipment.

8.3.1 Filter Box

A filter box is usually made of concrete and contains all the system's components except for the control equipment. In the majority of the cases, the filter box is a rectangular section. During filtration, water enters above the filter media through an inlet flume. After passing through the granular media and supporting gravel bed, it is discharged through an underdrain pipe to the clear well storage.

8.3.2 Filter Media

The type and depth of filter media greatly affect the removal efficiency of a filter. Filter media can be of single, dual or multi-media type. However, only multi-media filters are able to produce in-depth filtration. After backwashing, the gradation in single-media filters is fine to coarse, thus reducing filter run due to surface plugging. Dual media beds of coarser anthracite overlying the sand filter provide coarser to fine gradation. Mixed media beds can almost be considered the ideal filter since the gradation is coarse to fine, with finer material at the bottom of the media. The various types of filter media are usually classified by the size of grains and specific gravity, as shown in **Table 8.2**.

Effective size is defined as a tenth percentile (D_{10}), which means only 10% of the media grains by mass are smaller than effective size. In a way, effective size represents the finest size in a graded media. Since a larger particle size will produce larger pore openings, coarser media will take more time to reach terminal head loss, but the time taken for turbidity to breakthrough is shortened. The uniformity coefficient (C_u) is defined as the ratio of grain sizes comprising the 60th percentile (D_{60}) and 10th percentile (D_{10}).

$$\text{Uniformity coefficient, } C_U = \frac{D_{60}}{D_{10}} = \frac{D_{60}}{\text{effective size}}$$

A lower uniformity coefficient indicates more uniform materials. Generally, the more uniform the material is, the slower the head loss builds up. In non-uniform (C_u > 2–4) materials, larger pores are filled by smaller particles, thus reducing hydraulic effectiveness.

8.3.3 Underdrain System

The underdrain is the bottom portion of the filter supporting the filter media. Its main function is to evenly distribute the backwash water without disturbing the gravel bed and the media, and to collect the filtered water and feed in of the backwash water. Several different kinds of manufactured filter bottoms are available, including porous plates, filter blocks and nozzles.

Layers of gravel are used in the top portion of an underdrain system. The number and depth of gravel layers, both to prevent the loss of media during filtration and to allow the uniform distribution of backwash water, depend on the size of the opening in the underdrain. Large openings require a

TABLE 8.2
Filter Media Characteristics

Material	Size, mm	Specific Gravity
Conventional sand	0.5–0.6	2.6
Coarse sand	0.7–3.0	2.6
Anthracite coal	1.0–3.0	1.5–1.8
Garnet	0.2–0.4	3.1–4.3
Gravel	1.0–50	2.6

deep gravel layer, hence a deeper filter box, but have reduced head loss. Smaller openings, such as slots in nozzles, require a fine gravel size, hence higher head loss. The gravel bed is typically 200 to 300 mm in thickness with a gradation in the range of 3 mm to 20 mm.

8.3.4 Surface Wash System

The purpose of surface wash equipment is to assist in removing particles trapped in the filter media. The surface washer breaks up the surface of the filter and causes particles to collide and release impurities. Surface wash equipment is positioned directly above the top surface of the filter media.

8.3.5 Wash-Water Troughs

The wash troughs are located above the surface wash equipment. During backwash, the level of water in the filter rises and enters the trough. During filtration, the water trough remains submerged and, as such, does not play a role.

8.3.6 Control Equipment

A flow controller is required to maintain a constant rate of flow. The differential pressure across a venturi or an orifice meter installed in the effluent piping activates the controller. Depending on the impurities caught in the media, head loss across the filter increases during the filter run. When the filter is close to the end of the run, the valve is fully opened since flow is controlled by the head loss across the filter.

8.4 FILTRATION OPERATION

Filtration has three steps: filtering, backwashing, and filtering to waste. A brief discussion follows.

8.4.1 Filtering

During filtration, water passes downward through the filter due to gravity. The maximum operating head is the difference between the water levels above the filter and, in a clear well, is commonly 3–4 m. During filtration, the depth of water above the filter is 0.9–1.2 m. To provide additional head, the underdrain pipe is submerged in the clear well. At the start of the filter run, the media is clean and head loss is very low (about 30 cm), resulting in high operating head conditions. In the absence of a control, the filtration rate will be high. This is the case in declining rate filtration. The head loss at a given point in the filter is the drawdown in water level of the piezometer installed at that point. If at a given point in the filter medium head loss exceeds the operating head, negative pressure will develop.

8.4.2 Constant Rate Control

In constant rate flow control, the flow controller is used to maintain a constant desired filtration rate. A rate of flow controller consists of a valve controlled by a venturi meter. To control the flow at the beginning of the filtration cycle, the flow controller valve is closed, providing additional head loss and therefore maintaining the desired flow rate. As filtration continues, impurities collect in the filter and head loss increases. To maintain the same operating head, the controller valve is gradually opened. When the valve is fully opened, the filtration rate is entirely controlled by the head loss in the filter media. If filtration is continued, the filter rate will drop since the head loss cannot be compensated any further. This is when the filter run is ended (head loss of 2.5 m) and must be

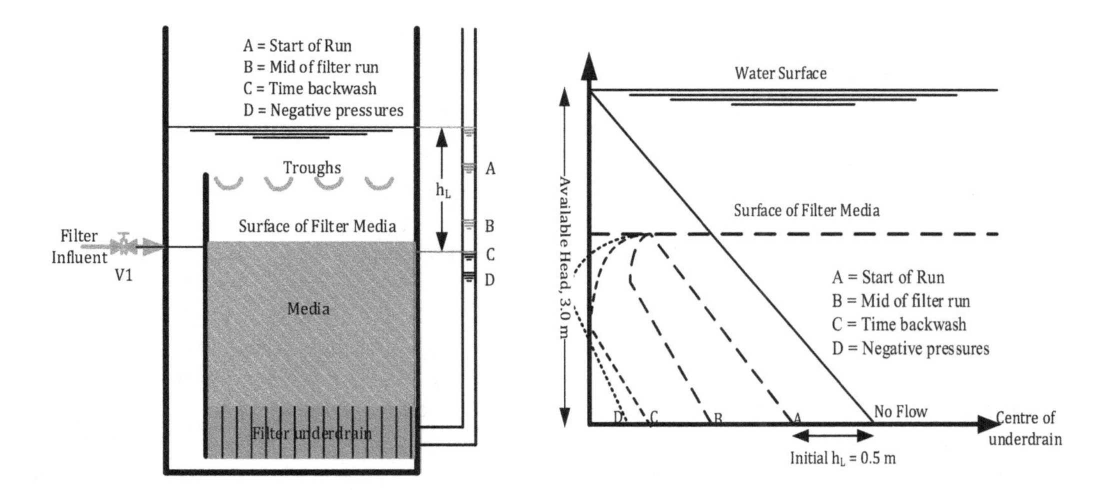

FIGURE 8.2 Head Loss in a Filter

cleaned by backwashing. Referring to **Figure 8.2**, during filtration, only influent and effluent valves are open.

8.4.3 Declining Rate Control

Flow rate is allowed to vary with head loss. Each filter operates at the same water surface level. The filtration rate is highest at the beginning and drops as the filter gets clogged. This system is relatively simple but requires an effluent weir to provide adequate media submergence.

8.4.4 Split Flow Control

As the name indicates, a control weir divides the flow to each filter. With this system, equal flow is automatically distributed to each filter. The filter effluent position is controlled by the water level in the filter. Each filter operates within a narrow water level range.

8.4.5 Backwashing

During filtration, the voids in the filter media become clogged with filtered-out material. The media grains also become coated with the floc and become very sticky. To clean the filter bed, the media grains must be agitated and allowed to rub against each other. Therefore, the backwash rate must be high enough to expand the filter media by about 20%. Surface wash is recommended to provide extra agitation, especially in high-rate filters, due to in-depth filtration. A filter is backwashed when:

1. head loss approaches a set value, usually about 2–2.5 m;
2. floc breakthrough of the filter causes an increase in turbidity; or
3. filter run reaches a set value, usually 36–48 hours, whatever comes earlier.

A decision to backwash the filter should be made based on all these factors. As an example, when the source water is very clean, as in Lake Superior, filter runs based on head loss or turbidity breakthrough can be very long – in the order of four days. However, long filter runs can cause the gradual build-up of organic materials and bacterial populations within the filter. For this reason, filter runs are rarely allowed to exceed four days.

Filter ripening

After the backwash operation, filters need to be rested to allow the media and other particulate matter to settle down. If this filter ripening is not done, filtered water will be of poor quality and turbidity levels may exceed regulatory limits.

8.4.6 Backwash Operation

Referring to **Figure 8.1**, a typical backwash begins by closing the influent (V1) and effluent (V4) valves and opening the drain valve (V2), allowing the water level to drop to a level about 15 cm above the media. The surface washers are turned on and allowed to operate for 1 to 2 minutes. If air agitation is used, compressed air is introduced through the underdrain for scouring. By opening the wash water supply valve (V5), clean water enters the underdrain and passes up through the filter, which is the reverse direction. Before opening valve V5, valve V3 should be opened to allow wash water to go to waste.

During reverse flow, filter media is expanded hydraulically, and dirty wash water is collected in troughs and conveyed for disposal or treatment. The expanded bed is washed for 5 to 15 minutes, depending on how dirty the filter is. The surface washers are usually turned off for about 1 minute before the backwash flow is stopped. To avoid any damage to the underdrain or loss of media, the backwash supply valve should be opened slowly.

8.4.7 Filtering to Waste

At the backwash operation is completed, the first few minutes of the filtered water should be wasted. This is accomplished by opening the filter to the waste valve (V3) when the influent valve (V1) is open. All the other valves remain shut. As the turbidity of the effluent drops to an acceptable level, opening the effluent valve (V4) and closing the filter to the waste valve (V3) starts the next filter run. Common backwashing aids are hydraulic surface agitators, mechanical rakes and compressed air.

8.5 DESIGN AND PERFORMANCE PARAMETERS

8.5.1 Filtration Rate

Filtration rate, v_F, is defined as the flow rate per unit surface area of the filter and is a measure of hydraulic loading rate. The filtration rate basically indicates the velocity of the flow (v_F) at the top of the filter. It can be directly observed by noting the water drop rate after closing off the influent valve. The flow velocity through the filter will be higher as some of the surface area will be occupied by the media grains. Higher filtration rates can be afforded in multi-media filters. Filtration rate is one measure of filter production. Along with filter run time, it provides valuable information for the operation of the filter. Problems can develop when design filtration rates are exceeded. Typical values fall in the range of 2 – 6 L/s.m^2.

Example Problem 8.1

For a slow sand water filtration plant, size six slow sand filters beds to treat a maximum flow of 12 ML/d with a filtration rate not exceeding 4.5 m^3/m^2·d. The filter box is rectangular with a length twice that of its width. Also, assume that one of the six units will be kept as standby.

Given: Q = 12 ML/d
v_F = 5.0 m/d
L = 2B
L, B = ?

SOLUTION:

$$Total\ filter\ surface,\ A_F = \frac{Q}{v_F} = \frac{12\ ML}{d} \times \frac{d}{4.5\ m} \times \frac{1000\ m^3}{ML} = 2666.6\ m^2$$

Since one unit is to be kept as standby, five units should provide the required surface.

$$Width,\ B = \sqrt{\frac{A_F}{2 \times 5}} = \sqrt{\frac{2666.6\ m^2}{10}} = 16.3 = \underline{16.5\ m\ say}$$

Six filter units, each measuring 33.0 m × 16.5 m, will meet the requirements.

8.5.2 Unit Filter Run Volume

Unit filter run volume (UFRV) is a measure of filter performance and is used to compare and evaluate filter runs. To find the UFRV, divide the total volume of water filtered between filter runs by the surface of the filter, A_F. For the majority of filters, UFRV falls in the range of 200–400 m^3/m^2 of the filter area.

$$UFRV = \frac{V_{WF}}{A_F} = \frac{Q \times t_{FR}}{A_F} = v_F \times t_{FR}$$

Example Problem 8.2

A filter with a media surface of 30 m² is operated at a constant flow rate of 90 L/s. The average filter run for this filter is 50 h. Find the filtration velocity and unit filter run volume (UFRV).

Given: A_F = 30 m²
t_F = 50 h
Q = 90 L/s
v_F = ?
UFRV = ?

SOLUTION:

$$v_F = \frac{Q}{A_F} = \frac{90\ L}{s} \times \frac{1}{30\ m^2} = \frac{3.0\ L}{s.m^2} \times \frac{m^3}{1000\ L} \times \frac{3600\ s}{h}$$

$$= 10.8 = \underline{11\ \ m/h}$$

$$UFRV = \frac{V_{WF}}{A_F} = \frac{Q \times t_f}{A_F} = v_F \times t_F = \frac{10.8\ m}{h} \times 50\ h = 540.0 = \underline{540\ m^3/m^2}$$

8.5.3 Backwash Rate

Backwash rates can be calculated in a similar fashion to filtration rates. For a desired backwash rate (v_{BW}), the pumping rate (Q_{BW}) can be calculated as follows:

$$Q_{BW} = v_{BW} \times A_F \quad V_{BW} = Q_{BW} \times t_{BW}$$

The backwash rate represents the rise rate. Backwash rates are 5–10 times that of filtration rates. The volume of water used to backwash is typically in the range of 4–6%.

Example Problem 8.3

For a filter unit ($A_F = 90\ m^2$), determine the backwash-pumping rate in L/s if the desired backwash rate is 15 L/s.m² for a duration of 10 minutes. After backwash, find the drop in water level of the backwash water holding tank with a diameter of 25 m.

Given: A = 90 m²
v = 15 L/s.m²
D = 25 m
t = 10 min

SOLUTION:

$$Pumping\ rate,\ Q_{BW} = v_{BW} \times A_F = \frac{15\ L}{s.m^2} \times 90\ m^2 \times \frac{m^3}{1000\ L}$$

$$= 1.35 = \underline{1.4\ m^3/s}$$

$$Volume\ of\ water,\ V_{BW} = Q_{BW} \times t_{BW} = \frac{1.35\ m^3}{s} \times 10\ min \times \frac{60\ s}{min} = 810.0$$

$$= \frac{810\ m^3}{69500\ m^3} \times 100\% = 1.16 = 1.2\%$$

$$Drop\ in\ level,\ \Delta d = \frac{V}{A} = \frac{810\ m^3}{0.785(25\ m)^2} = 1.65 = \underline{1.7\ m}$$

Example Problem 8.4

Dual media sand filter units have a design filtration rate of 10 m/h. The required water supply is 10 ML/d. Make appropriate assumptions.

Given: Q = 10 ML/d
v_F = 10 m/h
Backwash water = 5.0% (assumed)
Downtime = 0.5 h (assumed)
L = 1.5 × W (assumed)
t_F = 24 h-0.50 = 23.5 h

SOLUTION:

$$A_F = \frac{Q}{v_F} = \frac{10\ ML}{d} \times \frac{1000\ m^3}{ML} \times 1.05 \times \frac{d}{23.5\ h} \times \frac{h}{10\ m} = 44.6 = 45\ m^2$$

Assuming three filter units for flexibility:

$$Width,\ W = \sqrt{\frac{A}{3 \times 1.5}} = \sqrt{\frac{44.6\ m^2}{3 \times 1.5}} = 3.14 = \underline{3.2\ m\ say}$$

Thus three filter units measuring 4.7 m × 3.0 m each will suffice.

8.5.4 Selection of Filter Sand or Other Media

Media commonly available may be too coarse or too fine to meet the design specifications. By washing the fines and screening the course material, you can obtain the required size and uniformity. By performing sieve analysis on the stock material, coarse and fine fractions are based on P_{10} and P_{60} of the stock material.

P_{10} = Percentage of material less than desired effective size (%passing basis)

P_{60} = Percentage of material less than desired $D_{60} = C_u \times D_{10}$

Since the two sizes represent half the desired material, the percentage of useable stock is:

$$P_u = 2\left(P_{60} - P_{10}\right)$$

To meet the specified composition, 10% of the useable percentage can be below effective size. Thus, material that is too fine in the stock can be removed by washing.

$$P_f = P_{10} - 0.10 \times 2\left(P_{60} - P_{10}\right) = 1.2\ P_{10} - 0.20 P_{60}$$

Similarly, the percentage of the material that is too coarse is P_{60} plus 40% of P_u.

$$P_c = P_{60} + 0.40 \times 2\left(P_{60} - P_{10}\right) = 1.8\ P_{60} - 0.80 P_{10}$$

Example Problem 8.5

Sieve analysis was performed on a 2.0 kg sample of stock sand. The results of the analysis are shown in the table below. The desired effective size is 0.40 mm with a uniformity coefficient of 2.0. Determine the particle size of the material to be removed.

SOLUTION:

$$D_{60} = U_c \times D_{10} = 1.5 \times 0.40\ mm = 0.60\ mm\ (\%\text{passing basis})$$

From **Figure 8.3**, P_{10} = 24% and P_{60} = 40%.

TABLE 8.3
Sieve Analysis Data

Opening	Mass Retained			Passing
mm	g	%	Σ	%
2	0	0.00	0.0	100.0
1.5	100	5.00	5.0	95.0
1.2	125	6.25	11.3	88.8
1	150	7.50	18.8	81.3
0.9	210	10.50	29.3	70.8
0.8	115	5.75	35.0	65.0
0.75	100	5.00	40.0	60.0
0.7	170	8.50	48.5	51.5
0.6	235	11.75	60.3	39.8
0.5	168	8.40	68.7	31.4
0.4	159	7.95	76.6	23.4
0.3	165	8.25	84.9	15.2
0.2	95	4.75	89.6	10.4
0.1	120	6.00	95.6	4.4
0.08	51	2.55	98.2	1.8
0.06	33	1.65	99.8	0.2
	4	0.20	100.0	
	2000	100.00		

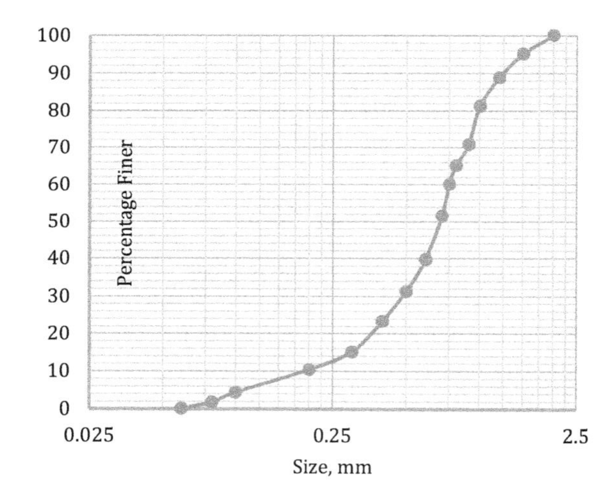

FIGURE 8.3 Distribution Curve (Ex. Prob. 8.5)

$$\textit{Usuable, } P_u = 2(P_{60} - P_{10}) = 2 \times (40\% - 24\%) = 32\%$$

$$\textit{Fine removed, } P_f = 1.2\ P_{10} - 0.20P_{60} = 1.2 \times 24\% - 0.20 \times 40\%$$

$$= 20.8 = \underline{21\%}$$

$$Fine\ size,\ D_f = D_{21} = \underline{0.35\ mm} \text{ as read from the curve}$$

$$Coarse\ removed,\ P_c = 1.8\ P_{60} - 0.80 P_{10} = 1.8 \times 40\% - 0.80 \times 24\%$$

$$= 52.8 = \underline{53\%}$$

$$Corse\ size,\ D_c = D_{53} = \underline{0.75\ mm} \text{ as read from the curve}$$

To prepare the right type of sand, particles of size 0.75 mm < D < 0.35 mm must be removed.

8.6 HYDRAULICS OF GRAVITY FILTERS

As water passes through the media downwards, it encounters resistance, resulting in head loss. As filtration progresses, head losses increase since more and more impurities are caught, and resistance to flow increases, as shown in **Figure 8.2**. The initial head loss as the filter is cleaned primarily depends upon the porosity of the filter media, and to varying degrees on other factors including filtration rate, v_F, the diameter of the grains, D, the depth of the filter media, H, and the drag coefficient, C_D. The Rose equation can be used to estimated head loss through a clean filter or initial head loss.

$$Intial\ head\ loss,\ h_L = \frac{1.067 v_F^2 L}{\varphi g n^4} \Sigma \frac{C_D \times F}{D}$$

Φ = shape factor
1 = rounded particles
F = mass fraction of media particles of diameter, D
L = length of flow path = depth of filter media

$$C_D = 0.4\ for\ N_R > 10^4, \quad C_D = \frac{24}{N_R}\ for\ N_R < 0.5,$$

$$C_D = \frac{24}{N_R} + \frac{3}{\sqrt{N_R}} + 0.34\ for\ 0.5 < N_R < 10^4$$

An initial head loss in excess of 0.6 m indicates that either the filtration rate is too high, the filter media is not backwashed properly or the particle gradation is too fine. The filter box must be at least as deep as the maximum design head loss, after which the filter is backwashed. This value is usually a maximum of 3.0 m.

Example Problem 8.5

After pre-treatment, water is filtered at a rate of 5.0 m/h. The filter media consists of a uniform sand of grain size 0.65 mm, a media depth of 0.70 m, G = 2.65, Φ = 0.85 and porosity n = 40%. Work out the initial head loss through the filter. Assume the temperature of the water is 10°C (kinematic viscosity = $1.51 \times 10^{-6}\ m^2/s$).

Given: $v_F = 5.0$ m/h
$G_s = 2.65$
D = 0.65 mm

v = 1.51 × 10^{-6} m^2/s
n = 40%
T = 10°C
Φ = 0.85
h_L = ?

SOLUTION:

$$N_R = \frac{v\varphi D}{\vartheta} = \frac{5.0\ m}{h} \times \frac{h}{3600s} \times \frac{0.85 \times 0.65\ mm.s}{1.51 \times 10^{-6} m^2} \times \frac{m}{1000\ mm}$$

$$= 0.508 (< 0.5\ Laminar)$$

$$C_D = \frac{24}{N_R} = \frac{24}{0.508} = 47.226 = 47.2$$

$$h_L = \frac{1.067 v_F^2 L}{\varnothing g n^4} \sum \frac{C_D \times F}{D} = \frac{1.067 v_F^2 L}{\varnothing n^4 g} \times \frac{C_D F}{D}$$

$$= 1.067 \times \left(\frac{5.0\ m}{h} \times \frac{h}{3600\ s} \right)^2 \times \frac{0.70\ m.s^2}{0.85 \times 0.40^4 \times 9.81\ m} \times \frac{47.2 \times 1}{0.00065\ m}$$

$$= \underline{0.49\ m}$$

Head loss seems to be okay since the recommended value is 0.60 m.

8.7 OPERATING PROBLEMS

The three major areas in which most filtration problems occur are treatment efficiencies before filtration, the control of the filtration rate and backwashing the filter.

Sudden changes in water quality indicators such as turbidity, pH, alkalinity, threshold odour number, temperature, chlorine residual or colour are an indication of problems in the filtration process or processes preceding filtration. During a normal filtration run, the operator should watch for sudden changes in head loss and turbidity breakthrough.

8.7.1 Filter Breakthrough

Probably the most common filter operation problem is filter breakthrough. Filter breakthrough can be defined as a steady increase in the turbidity of the filtered water. Normally, the turbidity of the filtered water will stay relatively low and constant.

8.7.2 Mud Balls

Mud balls look like small irregularly shaped balls of mud resting near the coal-sand or sand-gravel interfaces. Mud balls are formed due to inadequate backwashing and grow with time if the problem is not corrected. If allowed to remain, mud balls will cause clogged areas in the filter. Generally, proper surface washing will prevent mud ball formation.

8.7.3 Air Binding

Air binding is the result of a negative head condition due to excessive head loss. This is caused by the release of dissolved air in saturated cold water due to a decrease in pressure below atmospheric.

During backwash, there can be violent agitation as the air is released from the filter media. This can damage the filter and result in the loss of filter media.

8.7.4 Media Breakthrough

When media breakthrough occurs, the media starts appearing in the filter effluent or even in the distribution system, and filtration has to be stopped quickly.

8.7.5 Gravel Mounding

When this happens, gravel is blown up and mixes with the filter media. Mounding is often caused by hydraulic surges due to improper backwashing.

8.7.6 Media Boils

Media boils appear during backwash when there is uneven distribution. In filters with nozzle-type underdrains, boils are often the result of nozzle failure.

8.8 OPTIMUM FILTER OPERATION

Getting the best from your filter units requires an understanding of the factors responsible for causing poor efficiency and keeping the system in good working order. Here are some ways to achieve optimum performance.

- Checking the media cleanliness frequently by core sampling or visual inspection.
- Ensuring backwashing is effective so as to operate the filter efficiently and get long filter runs. If the nature of the floc is sticky, it is advisable to have surface wash or air scouring to remove the sticky material from the media grains during backwashing.
- Keeping the filter media clean and maintaining media depth.
- Employing an aggressive backwash with an air-water combination to help wash the media clean with low flow rates and less frequent backwashings.

Maintaining good records of water quality, filter run length, backwash frequency and any changes made helps an appropriate action to be taken in time.

Discussion Questions

1. What is meant by in-depth filtration? How can it be accomplished?
2. Compare slow sand filters with rapid sand filters.
3. With the help of a neat sketch, describe how a dual media filter works.
4. Compare declining rate filtration with constant rate filtration.
5. In a gravity filter, the depth of the water above the filter surface is typically 1.5 m. How can the head loss gauge record a head loss of 2.5 m?
6. Under what conditions can a filter run be terminated?
7. What are the advantages of using a multi-media filter over a single-media sand filter?
8. What is the function of the rate of flow controller in filter operation?
9. Describe the common operating problems encountered in filter operation.
10. What would be the effect of inefficient coagulation/flocculation in filter removal efficiency and operation?
11. Describe the steps for backwashing a gravity filter.
12. Describe the various mechanisms that take place during filtration.

Practice Problems

1. Design five slow sand filter beds to treat a maximum flow of 12 ML/d with a filtration rate not exceeding 6.0 $m^3/m^2 \cdot d$. The filter box is rectangular with a length twice that of its width. Assume that one of the five units will be kept as standby. (32 m × 16m)
2. A filter with a media surface of 11 m × 7.5 m produces a total of 68000 m^3 of filtered water during a 3 days filter run. What is the average filtration rate and unit filter run volume? (11 $m^3/m^2 \cdot h$, 820 m^3/m^2)
3. For a filter unit (A_F = 90 m^2), determine the backwash-pumping rate in L/s if the desired backwash rate is 12 $L/s \cdot m^2$. How deep must the water be in a backwash tank 25 m in diameter if the filter is backwashed for 12 minutes? (1080 L/s, 1.3 m)
4. Design a dual-media sand filter unit for the production of 5.0 ML/d of water supply with all its principal components. The design filtration rate is 10 m/h. Make the appropriate assumptions.
5. After pre-treatment, water is filtered at a rate of 5.0 m/h. The filter media consists of uniform sand of grain size 0.40 mm, a media depth 0.70 m, G = 2.65, Φ = 0.85 and porosity n = 40%. Work out the initial head loss through the filter. Assume a kinematic viscosity of 1.0×10^{-6} m^2/s. (0.86 m)
6. A new water treatment plant is to be considered for treating 80 ML/d.
   a. Estimate the number of filters required if each filter area is not to exceed 60 m^2 and one filter is to be kept as standby. Assume the design filtration rate is 12 m/h. (6 filters)
   b. What are the dimensions of each filter if the length to width ratio is 4:1? (15 m × 3.7 m)
7. After closing the influent valve, it is observed that it took 6 minutes and 30 seconds for the water level to drop by 0.50 m. What is the filtration rate and flow through the filter if the filter measures 8.0 m x 8.0 m? (1.3 mm/s, 7.1 ML/d)
8. A filter measuring 14 m x 7 m produces a total of 72 ML during a 72-hour filter run. What is the average filtration rate in $L/s \cdot m^2$? (2.8 $L/s \cdot m^2$)
9. Calculate the backwash-pumping rate in L/s for a filter measuring 10 m x 10 m if the desired backwash rate is 11 $L/s \cdot m^2$. What volume of backwash water is required if the filter is to be backwashed at this rate for 10 minutes? (1100 L/s, 660 m^3)
10. The average filtration rate during a particular filter run was determined to be 7.8 m/h. If the filter run time was 42.5 hours, calculate the UFRV. (330 m^3/m^2)

9 Disinfection

Disinfection is one of the cleansing processes used to make water hygienic and suitable for consumption. It is the last process before water is pumped into a distribution system. This section focuses on the disease-producing organisms that might be present in water. Disease-producing organisms are called pathogens. These organisms are very small, usually microscopic, ranging from 1–5 microns (μm). Therefore, it is impossible to tell if pathogens are present simply by looking at a water sample.

Pathogenic organisms can be carried and transmitted by water and cause diseases. Dealing with pathogenic organisms is a major concern of water treatment. Disinfection is defined as the selective destruction/inactivation or removal of pathogenic organisms. Sterilization, on the other hand, is the destruction of all living organisms. Sterilization is not necessary for water treatment and is very expensive. Though disinfection is necessary for making water safe, it can also create objectionable tastes and excessive levels of disinfection by-products (DBP). Untreated water from domestic water sources may contain the following organisms:

- Viruses, which could cause diseases such as infectious hepatitis and poliomyelitis.
- Bacteria, which could cause diseases such as cholera, typhoid fever, dysentery and legionnaire's disease.
- Intestinal parasites, such as giant roundworm, which could cause amoebic dysentery or giardiasis.

These three types of organism differ in size and their resistance to disinfection. Of the above-mentioned pathogens, Giardia and crypto are the hardest to destroy.

PRIMARY DISINFECTION

Primary disinfection refers to the removal or inactivation of pathogens by various treatment methods. Thus pre-chlorination, settling and filtration, and post-chlorination all contribute to primary disinfection.

SECONDARY DISINFECTION

Secondary disinfection refers to leaving a disinfectant residual to provide a defence against any possible contamination on its way to the consumer. Thus only disinfectants that create a persistent residual, such as chlorine, chlorine dioxide and chloramines, can be used for secondary disinfection.

- Secondary disinfection is used to protect the water from microbiological re-contamination, reduce bacterial regrowth, control biofilm formation and serve as an indicator of distribution system integrity.
- The loss of disinfectant residual may indicate that system integrity has been compromised.
- UV and ozone do not provide a residual and cannot be used for secondary disinfection.
- Free chlorine residual is more powerful but decays faster compared to combined chlorine residual.

DOI: 10.1201/9781003231264-9

9.1 DISINFECTION METHODS

Pathogens can be removed or inactivated by applying a combination of physical and chemical treatment processes. Disinfection methods are discussed in the following sections.

9.1.1 Removal Processes

Conventional filtration consists of chemical coagulation, rapid mixing, flocculation and sedimentation, followed by rapid-rate sand filtration.

Direct filtration eliminates the sedimentation step from conventional filtration. Direct filtration is suitable for plants with raw water sources of high quality. The turbidity level in raw water is usually less than 10 NTU.

Membrane filtration involves the passing of water through membranes consisting of very fine pores. Depending on their pore size, they are further classified into microfiltration or ultrafiltration membranes. Membrane filtration can remove bacteria, Giardia, and some viruses. They are more suitable for polishing water that has already been treated by other methods. Algae and other solids can quickly clog cartridge filtration. Little information is available concerning the effectiveness of cartridge filters for virus removal. The effectiveness of the removal process is indicated by the turbidity of the filter effluent. For disinfection credits, filter effluent turbidity must be less than 0.5 NTU 95% of the time.

9.1.2 Inactivation Processes

Inactivation involves methods that create a harsh environment for pathogens. Strong oxidizing chemicals are commonly used. These chemicals destroy or impair pathogens by diffusing into the cell wall and impairing or destroying the organism.

9.1.3 UV Light

Ultraviolet light can be used to destroy pathogens, but the process is expensive. However, UV light is gaining popularity since it does not produce any harmful by-products and is more effective in the control of cysts. In the past, this type of disinfection was limited to small or local systems, but nowadays, many municipal plants are being retrofitted with this process. Operators must protect themselves from irradiation. UV light is not very effective against viruses.

9.1.4 Heat

Boiling water for about five minutes will destroy all microorganisms. This method is expensive because of the energy it requires and thus is usually used only in disaster and emergency situations (boil water advisory).

9.1.5 Chemicals

The chemicals that are used to disinfect water have problems associated with them. For example, iodine is expensive and has possible physiological effects. There are difficulties in the handling of bromine. Alkaline chemicals such as sodium hydroxide and lime leave a bitter taste in the water. Excess lime must be removed from water before it is supplied for public use.

9.1.6 Potassium Permanganate

Potassium permanganate is more commonly used for controlling taste and odour. It is also used as a disinfectant, especially for the removal of compounds formed by trihalomethanes (THMs) in raw water, such as humic acid and fulvic acid. Feeding in permanganate as the initial oxidant allows chlorine to be applied later in the treatment process when the precursors have been reduced. When added, permanganate imparts a pink colour to water. It is commonly used as a disinfectant in rural areas.

9.1.7 Ozonation

Ozone is an effective disinfectant and is widely used in some countries. In North America, the process is gaining popularity in conventional water treatment. However, ozonation is too expensive for small systems. The other disadvantage of ozonation is that it fails to provide a measurable residual. Despite its disadvantages, there appears to be renewed interest in ozonation as it does not form THMs, as is the case with chlorination. THMs are considered carcinogenic.

9.1.8 Chlorination

Chlorination offers several advantages for water treatment. It is reliable and relatively low in cost. Also, the slight chlorine residual that stays in the water after purification serves as a tracer that can be used to indicate the presence of the disinfecting agent at any point in the system. Although the primary use of chemical oxidants is for disinfection, these chemicals can serve additional purposes during the disinfection process, such as controlling biological growth in tanks and water mains, controlling tastes and odours, colour removal, and precipitation of iron and manganese.

9.2 CHLORINE COMPOUNDS

Chlorination is by far the most effective way of disinfecting drinking water. Chlorination involves the use of chlorine or chlorine compounds.

1. Chlorine itself can be added to the water to be treated. This method of chlorination is called gas chlorination. Chlorine cylinders containing liquefied gas come in various sizes.
2. Chlorine compound solutions, such as sodium hypochlorite, can be used. This method of chlorination is called hypochlorination. Since only part of the chemical is chlorine, it is relatively safer to use than chlorine gas. This practice is limited to small water systems requiring lesser amounts.
3. Another method of chlorination involves the use of chlorine dioxide. This form of chlorine is becoming more common due to its disinfectant efficiency and lower production of DBPs.

9.2.1 Gas Chlorination

One method of chlorination, especially in larger installations, involves the use of chlorine gas as a disinfectant. This method of chlorination is referred to as gas chlorination. Due to safety considerations and the initial cost of a gas feeder, gas chlorination is not used in small, seasonal drinking water systems.

Chlorine gas is about 2.5 times heavier than air. It has a pungent odour and greenish-yellow colour. The gas is highly toxic. Its odour can be detected at concentrations greater than 0.3 ppm. Chlorine liquid is gas compressed at a high pressure. The liquid is about 99.5% pure chlorine, amber in colour and 50% denser than water. One volume of liquid chlorine yields about 450 volumes of gas.

Chlorine is non-explosive but reacts violently with greases, hydrocarbons, ammonia, and other flammable materials. Chlorine will not burn, but it supports combustion.

9.2.2 Chlorine Safety

Chlorine is very toxic and hazardous. A few breaths of 0.1% chlorine in air can cause death even at low concentrations. It can cause ill health effects. The following table shows the physiological effects of various concentrations of chlorine by volume in air. Handle chlorine carefully and use all the necessary precautions. You need to respect chlorine as it is known as the "green goddess" of water.

- Do not enter a chlorine-containing atmosphere without wearing protective gear.
- Apparatus, cylinder lines and valves should be checked regularly for leaks.
- Since chlorine is 2.5 times heavier than air, store chlorine at the lowest level. For the same reason, do not stoop down if you notice a chlorine smell.

9.3 HYPOCHLORINATION

A common way that chlorine is used for water disinfection in a small water system is called hypochlorination. Because chlorine is dangerous and requires special handling, many smaller water plants use a liquid chlorine compound called hypochlorite instead of chlorine gas. Another advantage of hypochlorination is that it does not need elaborate equipment or a separate room. It is important to note that hypochlorite compounds, which contain an algaecide, should not be used as a disinfectant in potable water systems.

9.3.1 Calcium Hypochlorite

Calcium hypochlorite ($Ca(OCl)_2$) is a dry, white or yellow-white granular material produced from the reaction of lime and chlorine. Calcium hypochlorite is commercially available in granular powder or tablet form. The commercially available compound, high test hypochlorite (HTH), typically contains 65% available chlorine. These products are dissolved in water to form a liquid solution before they are used to disinfect water. While calcium hypochlorite is not as dangerous as chlorine gas, it should be handled according to the recommended procedures.

9.3.2 Sodium Hypochlorite

Household bleach is sodium hypochlorite in a 3–5% concentration. Large systems usually purchase liquid bleach in carboys, drums and railroad tank cars. If needed in very small quantities, one-gallon (4 L) jugs can be purchased. Sodium hypochlorite is alkaline, with a pH ranging from 9 to 11, depending on the strength. Adding chlorine by using this chemical raises pH, unlike gaseous chlorine, which suppresses pH. Due to its poor stability, liquid bleach can lose 2–4% of its available chlorine every month.

9.3.3 Chlorine Dioxide Disinfection

Chlorine dioxide is a greenish-yellow gas at room temperature and is odorous like chlorine.

$$2NaClO_2 + Cl_2 + 2NaCl + 2ClO_2 \uparrow$$

The use of chlorine dioxide as a disinfectant is of interest for a number of reasons:

- Trihalomethanes are not formed when chlorine dioxide is used.
- Chlorine dioxide is about 2–3 times more effective than chlorine in killing bacteria and many times more effective in killing viruses and cryptosporidium. When Giardia or crypto is the problem, treatment by chlorine dioxide followed by chlorine or chloramines is very effective.

- However, because of its higher cost, it is commonly used to treat unappealing taste and odour in water, and to oxidize iron and manganese in difficult-to-treat water.
- The difficulty with chlorine dioxide is that it must be generated on-site by reacting sodium chlorite solution with chlorine solution. Another problem is the handling of sodium chlorite, which is very combustible around organic materials. It is explosive with concentrations exceeding 10% in the air.

9.4 CHEMISTRY OF CHLORINATION

Complex chemical reactions occur when chlorine is added to water, but these reactions are not always obvious. For example, a chlorine taste or odour in water is sometimes the result of too little chlorine rather than too much. When chlorine is added to distilled water it forms hypochlorous acid (HOCl), which, depending on the pH, can ionize to the hypochlorite ion. When pH is below seven, the bulk of the HOCl remains unionized.

$$Gas\ chlorination: Cl_2 + H_2O \rightarrow HCl + HOCl$$

$$HOCl \ \overrightarrow{pH > 8}\ \ H^+ + OCl^-$$

$$Hypochlorination: Ca(OCl)_2 + H_2O \rightarrow Ca^+ + 2\ OCl^- + H_2$$

Chlorine existing in water as hypochlorous acid and hypochlorite ion are two forms of free chlorine residual. Hypochlorous acid as a disinfectant is 100 times more effective than hypochlorite ion. The effectiveness of disinfection increases as pH drops.

9.4.1 Breakpoint Chlorination Curve

When chlorine is added, the amount of free chlorine residual formed is less than the amount of chlorine added since some of the chlorine is used up by reacting with impurities in water. When chlorine is added to natural water containing ammonia and other impurities, the residuals that develop yield a curve similar to that shown in **Figure 9.1**. The reactions that take place can be divided into four groups.

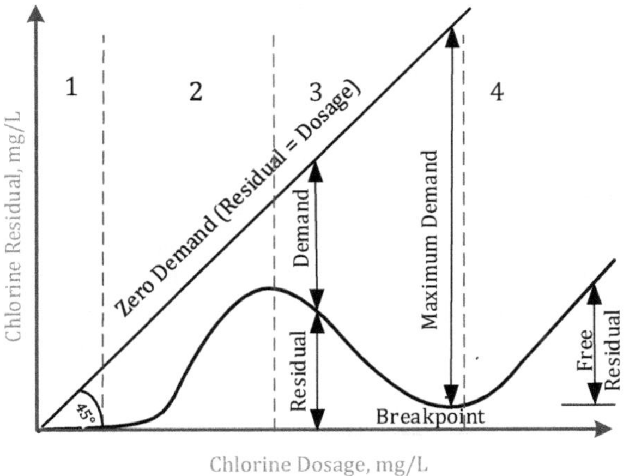

FIGURE 9.1 Breakpoint Chlorination Curve

- **Stage I**: No chlorine residual since all of the chlorine added is used up in reactions with reducing agents such as iron and manganese, nitrites and sulphides. No disinfection happens.
- **Stage II**: As more chlorine is added, it reacts with ammonia and organic water, forming monochloramine and chloroorganic compounds. These compounds have some disinfectant properties and are called combined chlorine residual.

$$NH_3 + HOCl \rightarrow NH_2Cl + H_2O$$

- **Stage III**: Adding more chlorine to water actually decreases the residual. This is because the additional chlorine oxidizes some of the chloroorganic compounds and ammonia.

$$2NH_3 + 3Cl_2 = N_2 + 6HCl$$

The additional chlorine also changes some of the monochloramine to dichloramine and trichloramine.

$$NH_2Cl + HOCl \rightarrow NHCl_2 + H_2O$$

$$NHCl_2 + HOCl \rightarrow NCl_3 + H_2O$$

Chloramine residuals decline as more chlorine is added until they reach a minimum value referred to as the breakpoint. The amount of chlorine added (dosage) minus the chlorine residual is the chlorine demand of water. Chlorine demand is at a maximum at the breakpoint.

- **Stage IV**: Beyond the breakpoint, any chlorine addition will produce free chlorine residual as indicated by the straight line with a one-to-one slope (**Figure 9.1**). Breakpoint chlorine dosage is unique for each sample of water tested. Depending on the number of constituents in water, such as ammonia and other reducing agents, some stages may be absent.

9.4.2 Breakpoint Chlorination

To achieve free chlorine residual, the chlorine dosage must be more than the maximum chlorine demand (breakpoint) by the amount equal to the desired residual. Breakpoint chlorination is a common practice except when THMs are a problem or compounds such as phenol are present in water. Chlorine reacts with phenol to form chlorophenols, which have an intense medicinal odour. In such situations, combining residuals as monochloramine is preferred. Though combined residuals are relatively weaker, they are long-lasting.

Example Problem 9.1

Different chlorine dosages were applied to a water sample, and the chlorine residual was tested after 12 minutes of contact time. For the data shown in the table below, plot the breakpoint chlorination curve and find the breakpoint dosage and residual, maximum chlorine demand and chlorine feeder setting in kg/d for treating a flow of 12 ML/d and for achieving a residual of 0.65 mg/L.

Dosage, mg/L	0.20	0.4	0.60	0.80	1.0	1.2	1.4	1.6	1.8	2.0
Residual, mg/L	0.05	0.15	0.30	0.45	0.30	0.20	0.35	0.62	0.78	1.0

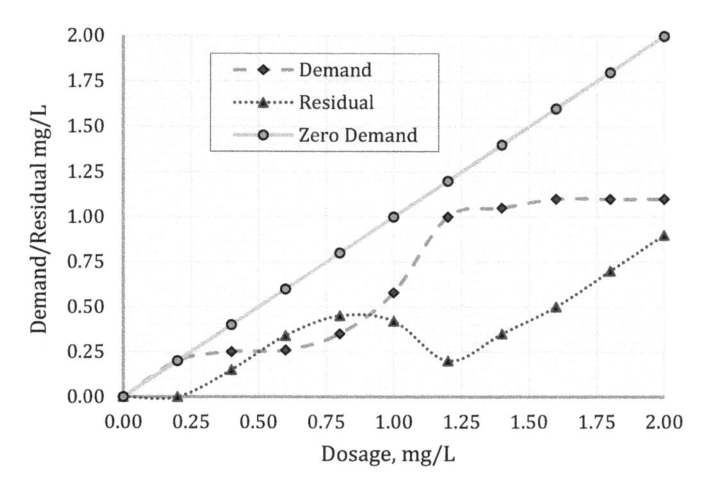

FIGURE 9.2 Breakpoint Chlorination (Ex. Prob. 9.1)

Given: Q = 12 ML/d
Residual = 0.65 mg/L
Dosage, C = ?
Feed Rate = ?

SOLUTION:

The chlorine demand for each dosage is calculated by subtracting chlorine residual from chlorine demand, as shown in the table below:

Dosage, mg/L	0.20	0.40	0.60	0.80	1.00	1.20	1.40	1.60	1.80	2.00
Residual, mg/L	0.05	0.15	0.30	0.45	0.30	0.20	0.35	0.62	0.78	1.00
Demand, mg/L	0.15	0.25	0.30	0.35	0.70	1.00	1.05	0.98	1.02	1.00

From **Figure 9.2**, breakpoint dosage = 1.2 mg/L, max. demand = 1.0 mg/L.

$$Dosage = Demand + Residual = (1.0 + 0.65)mg/L = 1.65\ mg/L$$

$$Feed\ rate = \frac{12\ ML}{d} \times \frac{1.65\ kg}{ML} = 19.8 = \underline{20\ kg/d}$$

Example Problem 9.2

A water pump delivers on average 10 L/s against typical operating heads.

1. If the desired chlorine dosage is 2.0 mg/L, what should the chlorine feeder setting be?
2. If the pump is operated 18 hours per day, how many kg of chlorine will be used in a week?

Given: Q = 10 L/s
C = 2.0 mg/L = 2.0 kg/ML
M = ?

SOLUTION:

$$a)\ Dosage,\ M = Q \times C = \frac{10\ L}{s} \times \frac{2.0\ kg}{ML} \times \frac{1440\ min}{d} \times \frac{60\ s}{min} \times \frac{ML}{10^6 L}$$

$$= 1.72 = 1.7\ kg/d$$

$$b)\ Usage = \frac{1.70\ kg}{24\ h} \times \frac{18\ h}{d} \times \frac{7\ d}{wk} = 9.07 = 9.1\ kg/wk$$

Example Problem 9.3

A chlorinator is set to feed chlorine at a dosage rate of 11 kg/d. This dose results in a chlorine residual of 0.55 mg/L when the average 24-hour flow is 5.0 ML/d. What is the chlorine demand of water? When the feeder setting is increased to 12 kg/d, determine the expected increase in residual.

Given: Dosage rate, M = 11 kg/d
Residual = 0.55 mg/L
Q = 5.0 ML/d
Demand = ?

SOLUTION:

$$Dosage,\ C = \frac{M}{Q} = \frac{11\ kg}{d} \times \frac{d}{5.0\ ML} = 2.20 = 2.2\ kg/ML = 2.2\ mg/L$$

$$Demand = dosage - residual = (2.2 - 0.55)\ mg/L = 1.75 = 1.8\ mg/L$$

$$New\ C = \frac{M}{Q} = \frac{12\ kg}{d} \times \frac{d}{5.0\ ML} = 2.40 = 2.4\ kg/ML = 2.4\ mg/L$$

$$Change\ in\ residual,\ \Delta C = 2.4 - 2.2 = 0.20\ mg/L$$

9.5 CHLORINE PRACTICES

9.5.1 Chloramination

In this method, chlorine residual is combined with natural or added ammonia. This type of residual is less effective but lasts longer. Chloramination is more suitable when the free residual causes taste and odour problems, as with the presence of phenolic compounds. This practice is more commonly used for secondary disinfection.

Chloramination is the adding of ammonia to water to form chloramines. The reason for this practice is the prevention of the formation of THMs and chlorophenols, and for providing a long-lasting residual. The only limitation is that chloramine residual is significantly weaker. The typical dosage of ammonia is one part to three parts of chlorine. Higher dosages of chlorine form dichloramine and will cause taste and odour problems, and may cause the problem of nitrification at high temperatures. Residuals of 2.5 mg/L or greater should be maintained.

9.5.2 Breakpoint Chlorination

As explained earlier, free chlorine residual is produced beyond the breakpoint. Any dosage over and above the breakpoint dosage would result in an equal amount of free chlorine residual. This practice is common, and the dosage depends on the chlorine demand and the level of free chlorine residual to be attained. Another point worth repeating is that free chlorine residual is 25 times more effective compared with combined residual.

9.5.3 Super-Chlorination

Super-chlorination can be thought of as breakpoint chlorination with high levels of free residual. This practice is done in emergencies such as outbreaks of epidemics, breakdowns, water mains repairs or heavily polluted water. Heavy doses of chlorine (10–15 mg/L) are used to effectively destroy resistant organisms and cysts. Excess chlorine is then removed by dechlorination.

9.5.4 Dechlorination

After super-chlorination, water contains high chlorine residuals. These residuals must be brought down to acceptable levels before the water exits the plant. When super-chlorination is used to disinfect tanks and pipes, the water must be dechlorinated before discharge. Reducing compounds, such as sulphur dioxide, sodium thiosulphate and sodium bisulphate, are commonly used to neutralize chlorine. Prolonged storage and adsorption on charcoal or activated carbon is also effective.

9.6 POINTS OF CHLORINATION

Chlorine is used at various stages of a water supply system, starting from raw water pumping to the distribution system. Terms like pre-, post- and re-chlorination are often used to indicate the points of application.

9.6.1 Pre-Chlorination

Pre-chlorination is the application of chlorine to raw water before any other treatment process. It is also called source water chlorination and is mainly done to achieve the following:

- To control biological growth on filters, pipes and basins.
- To promote improved coagulation, iron and manganese removal.
- To prevent the formation of mud balls and slime in filters.
- To reduce taste, odour and colour problems, and minimize the post-chlorination dosage.

In heavily polluted waters, pre-chlorination should be used with caution as chlorine reacts with organic compounds to form THMs as a disinfection by-product (DBP).

THMs are considered carcinogenic and cause taste and odour problems. The formation of THMs can be reduced by eliminating the pre-chlorination process or by moving the injection point to just ahead of the filters. Other methods for reducing THM levels include the use of alternate disinfectants, such as ozone or chlorine dioxide, and the application of activated carbon to adsorb THMs and humic substances.

9.6.2 Post-Chlorination

Post-chlorination is the application of chlorine to treated water to achieve residual for water that exits the plant. This provides protection against any possible contamination in the distribution system or customer plumbing system. Water regulation provides specific guidelines for the type and level of chlorine residual before water leaves the plant.

9.6.3 Re-Chlorination

In large and complex water distribution systems, it may not be possible to maintain the minimum residual of 0.2 mg/L, especially in the remote parts of the system. In such cases, chlorine needs to be applied in stages to boost chlorine residual. This is known as re-chlorination.

9.7 FACTORS AFFECTING CHLORINE DOSAGE

Waters with high turbidity cannot be adequately disinfected. The turbidity particles inhibit disinfection by providing hiding places for bacteria where chlorine cannot reach and by increasing the chlorine demand. The primary factors that determine the disinfectant efficiency of chlorine are as follows.

9.7.1 Chlorine Concentration

The higher the concentration of freely available disinfection chlorine, the more effective and faster is the disinfection.

9.7.2 Contact Time Between the Organism and Chlorine

The longer the contact time, the more effective the disinfection. If the chlorine concentration is decreased, then the contact time, t, must be increased. The disinfecting power, often referred to as the kill, is directly related to two factors: disinfectant residual; and time of contact, known as CT factor.

$$\boxed{Inactivation = C \times T}$$

C = chlorine residual at the end of contact time = strength in mg/L
T = t_{10} = 10th percentile contact time in minutes

Contact time, t_{10}, in the above equation is the length of time during which no more than 10% of the water passes through a given disinfection process. It is calculated using an estimate of baffling in the tank or through a tracer test. Introducing baffling in the clear well increases the length to width ratio. Therefore, a greater contact time for the disinfectant to work in is achieved. In a tracer test, a non-reactive chemical like fluoride or lithium is added to the basin, and the change in concentration of the effluent is measured over time. A curve of C_o/C_t versus time is plotted. The value of time corresponding to a 10% ratio, t_{10}, is used in CT calculations. For a plug flow system, as in the case of a pipe flowing full, contact time can be assumed to be equal to theoretical detention time.

The regulating authority specifies minimum values of CT factor to achieve a given level of disinfection (1–4 Log removals). This is based on the premise that the efficiency of disinfection depends on the strength (concentration) and length of time that the disinfectant remains in contact with the organism. For surface treatment water systems, minimum CT values have to be provided for peak flow conditions to achieve 4-log removals for viruses. In groundwater systems, virus removal is only 2-log since filtration through ground layers significantly contributes to pathogen removal.

Log Removal

Log removal is more commonly used to indicate the inactivation of pathogens. If the concentration is reduced by a factor of 10, that is 90% removal (PR), it amounts to 1-log removal since the log of 10 is 1. In terms of influent (C_i) and effluent (C_e) concentrations, log removal is:

$$\boxed{Log\ removal,\ LR = \log(C_0 / C_e) = -Log(1 - PR / 100\%)}$$

9.7.3 Type of Chlorine Residual

Free chlorine is a much more effective disinfectant than combined chlorine. Combined chlorine residual requires greater concentration acting over a longer period of time to do the same job. When the contact time is short, only a free residual will provide effective disinfection.

9.7.4 Temperature and pH of Water

Usually, the higher the temperature, the more effective the disinfection. The water system operator cannot control the temperature, but they must increase the contact time or dosage at lower temperatures.

Disinfection is generally more effective at a low pH since HOCl is 100 times more effective than the OCl ion. In that sense, a combined residual, such as monochloramine, is least effective. Disinfection is most effective in the pH range of 4 to 7. As an example, the contact time required for 99.9% (3-log) Giardia removal applying 0.6 mg/L of chlorine residual is 100 minutes at a pH of 6 and twice as many minutes at a pH of 8. The pH also affects corrosivity.

9.7.5 Substances in the Water

Turbidity in water can provide shelter for organisms. Therefore, for chlorination to be effective, turbidity must be reduced to the greatest extent possible by the preceding water treatment methods.

Example Problem 9.4

A small community is served by a groundwater supply. Water is pumped into a 1.5 km long water main with a diameter of 400 mm until it reaches the first consumer. The peak hourly pumping rate is 10 m^3/min, and the temperature of the water is 5°C. As the groundwater is under the influence of surface water, regulations demand a 3-log virus inactivation. What free chlorine residual is required at the outlet of the pipeline?

Given: C = ?
L = 1500 m
Q = 10 m^3/ min
D = 400 mm

SOLUTION:

$$Contact\ time,\ t = \frac{A.L}{Q} = 0.785(0.4\ m)^2 \times 1500\ m \times \frac{min}{10\ m^3}$$

$$= 18.84 = 19\ min$$

From the table, the CT factor for a 3-log virus removal at 5°C is 6.0 mg.min/L, hence the minimum free chlorine residual desired to meet the requirement is:

$$Residual,\ C = \frac{CT}{t} = \frac{6.0\ mg.min}{L} \times \frac{1}{18.84\ min} = 0.318 = 0.32\ mg / L$$

Example Problem 9.5

A filtration system is required to provide 0.5-log Giardia and 2-log virus removal by post-chlorination. The t_{10} time for the clear well is evaluated to be 43 minutes and a free residual of 0.45 mg/L is maintained. How much removal is achieved for Giardia?

Given: CT = 19 mg.min/L / 0.5LR for Giardia, 3.0 / 2 LR for virus
C = 0.50 mg/L
t_{10} = 43 min

SOLUTION:

$$For\ Giardia = \frac{CT}{CT\ /\ LR} = \frac{0.45\ mg \times 43\ min}{L} \times \frac{L.0.5LR}{19\ mg.min} = 0.51\ LR$$

$$For\ virus = \frac{CT}{CT\ /\ LR} = \frac{0.45\ mg \times 43\ min}{L} \times \frac{L.LR}{3.0\ mg.min} = 13\ LR$$

9.8 CHLORINATION EQUIPMENT

Because of the toxic nature of chlorine, gas chlorination systems require a different type of chlorine feed system, as well as specialized equipment for health and safety. The main components of a gas chlorination feeding system are as follows.

Weighing Scale

A weighing scale is used to determine the amount of chlorine used or remaining in a container or cylinder. By recording the weights at regular intervals, the chlorine dosage rate can be calculated.

Valves and Piping

Except for when a direct chlorinator is used, an auxiliary valve is connected to the container valve. The auxiliary valve can be used as a shut-off to the supply during emergencies. The valve assembly is connected to the chlorine supply piping or manifold by flexible tubing. The tubing is usually 10 mm copper rated at 3500 kPa.

Chlorinator

A chlorinator can be a simple direct-mount unit or a free-standing cabinet. A variable orifice inserted into the feed line controls the feed rate of chlorine. The reduced pressure downstream of the orifice control allows a uniform gas flow accurately metered by a rotameter or feed rate indicator. This also acts as a safety feature in case a leak develops in the vacuum line.

Injector

An injector (or ejector) is basically a venturi. Water flowing through the venturi creates a vacuum that draws in gas to mix with water to create a strong chlorine solution. This solution can be safely piped to various points in the treatment plant.

Diffuser

The purpose of the diffuser is to disperse the chlorine solution into the main flow of water. To disperse the solution uniformly and quickly, the diffuser pipe is usually perforated.

Chlorine feeders can be operated manually or automatically based on flow, chlorine residual or both.

9.8.1 Feed Control

Manual Control

In this case, the chlorine feed rate is adjusted manually. The chlorine is fed at a constant rate irrespective of flow and the chlorine demand of the water. Adjustments must be made each time the flow rate changes. This type of control is satisfactory in the following situations:

- Where flow and demand are relatively constant.
- Where an operator is available to make manual adjustments.

Automatic Proportional Control

This is recommended when the chlorine demand of the water remains constant. Proportional control adjusts the feed rate to provide a constant pre-established chlorine dosage. This is accomplished by transmitting the flow signal to the feeder, which responds to the transmitted signal by increasing or decreasing the feed rate.

Automatic Residual Control

Automatic residual control provides a set residual rather than dosage, as in the case of proportional control. This is desirable when there are fluctuations in the chlorine demand of the water. A chlorine residual sensor is used in addition to a flow sensor. The feeder receives signals from both the flow meter and the chlorine analyzer.

9.8.2 Hypochlorinators

Hypochlorination is well suited to smaller water supply systems with chlorine uses of less than 1.5 kg/d. Both calcium and sodium hypochlorite systems use hypochlorite solution feeders called hypochlorinators to meter the liquid stock volume into water. Calcium hypochlorite also requires mixing and storage tanks for making up the stock solution from powder.

Solution feeders are more common in small water systems. The hypochlorinators either pump or inject a chlorine solution into the water. When injecting the chlorine solution, a hydraulic device such as a venturi is used to create negative pressure. In this arrangement, the hypochlorite solution is pumped through an injector, which draws in additional water for dilution of the hypochlorite solution. Types of hypochlorinators available include positive displacement feeders, aspirator feeders, suction feeders and tablet hypochlorinators. Positive displacement feeders using diaphragm pumps are most common. Hypochlorination systems consist of a chemical solution tank, a diaphragm-type pump, a pressurized water supply and a mixing tank. The diaphragm pump is normally a positive displacement type and has adjustments for the stroke length of the reciprocating rod and for the speed at which the rod moves.

The positive displacement pump hypochlorinator can be used with any water system. However, it is especially desirable in systems with low, fluctuating water pressure. Most hypochlorinators are controlled electronically from a master control panel. The stopping and starting of the hypochlorinator is synchronized with the pumping unit. A flow switch or other sensing device can be used for this purpose. Having a flow switch avoids operating problems associated with chemicals being fed when the raw water pumps fail to start.

Operation

To achieve desirable results, chlorination equipment must be properly maintained and competently operated. Related activities include mixing solutions, testing chlorine residual, adjusting feed pump rates and calibrating the feed pump.

TABLE 9.1
Troubleshooting Chlorine Equipment

Trouble	Cause and Remedy
Chlorine leak at the cylinder valve packing	Tighten the cylinder valve packing without using excessive force. If this does not eliminate the leak, close the valve and call the chlorine supplier.
Chlorine leak at vent	The pressure relief valve is malfunctioning. The usual cause is dirt on the valve seat. Test as follows: 1. Shut off the water supply to the ejector. 2. Submerge the end of the vent tube in a glass of water. Bubbling indicates a chlorine leak. 3. Before removing the unit from the cylinder, close the cylinder valve, turn the water supply on and operate until the metering ball registers zero flow.
Loss of chlorine feed	1. The nozzle at the ejector is dirty or plugged. Check by removing the chlorine gas line at the ejector and holding a thumb over the fitting. If there is no vacuum or partial vacuum, the ejector nozzle may be plugged. Unscrew the nozzle from the body and clean with a small diameter wire. Take care not to scratch the nozzle edge. 2. There is insufficient water pressure to the ejector. Check by holding a thumb over the ejector fitting to determine whether or not there is a vacuum. Restore water pressure as needed. 3. There is no supply of chlorine. Check the rotameter. 4. The chlorinator filter is plugged. If so, replace with a new one. 5. There is increased backpressure on the injector. The valve is closed or a diffuser is plugged.

Discussion Questions

1. Differentiate between the following:
   a. Disinfection and sterilization.
   b. Primary and secondary disinfection.
   c. Free chlorine residual and combined chlorine residual.
   d. Gas chlorination and hypochlorination.
   e. Pre-chlorination and post-chlorination.
2. Name and explain various types of disinfectants.
3. Explain the breakpoint chlorination curve with a neat sketch.
4. Describe chlorination practices along with application points.
5. What safety measures are taken where chlorination is practised?
6. Briefly discuss the main components of chlorination equipment.
7. Define CT factor and explain how it can be used to determine disinfection efficiency.
8. What is considered the most important water treatment process for preventing the spread of waterborne disease? Is there potential for any harmful effects?
9. What are the characteristics of a good disinfectant?
10. Briefly describe the factors that affect chlorine dosage.
11. What is ozonation? What are its advantages and disadvantages?
12. What are the characteristics of a good disinfectant?
13. Define the meaning of $C \times t$ product. In addition to C and t, what other factors influence the efficiency of disinfection? What kind of pathogens are readily inactivated by free chlorine and which are the most difficult to inactivate?

Practice Problems

1. A chlorine feeder operates on proportional control and is set to provide a chlorine dosage of 2.0 mg/L. At a given hour, the rotameter indicates a reading of 4.0 kg/d. Calculate the hourly flow rate in L/s. (23 L/s)
2. In a well water system, the chlorine feed is controlled manually. What should be the setting on the feeder when the water pumping rate is 55 L/s and the desired chlorine dosage is 2.5 mg/L? (13 kg/24h)
3. A chlorinator is set to feed to filtered water at a dosage rate of 15 kg/d. This dose results in a chlorine residual of 0.65 mg/L when the average 24-hour flow is 270 m^3/h. Determine the chlorine demand of the water. When the chlorine feed rate setting is increased to 18 kg/d, what is the expected increase in free chlorine residual? (1.7 mg/L, 0.46 mg/L)
4. A well water pump delivers on average 15 L/s against typical operating heads.
   a. If the desired chlorine dosage is 2.2 mg/L, what should be the setting on the rotameter for the chlorinator? (2.9 kg/24h)
   b. If the pump is operated 18 hours per day, how many kg of chlorine will be used up over a period of one week? (15 kg/wk.)
   c. How many days will a 45 kg chlorine cylinder last? (21 d)
5. Water pumped from a well is disinfected by a hypochlorinator. A chlorine dosage of 1.3 mg/L is applied to maintain the desired level of chlorine residual. During a one-week period, the flow totalizer indicated 8830 m^3 of water was pumped. A 2.5% sodium hypochlorite solution is stored in a 1 m diameter tank. Determine the expected drop in the level of the hypochlorite tank. (58 cm)
6. How many kg of hypochlorite (65% available chlorine) is required to disinfect 800 m of a 400 mm water main at a chlorine dosage of 100 mg/L? (15 kg)
7. How many litres of chlorine bleach with 12% available chlorine should be used to add to a 2.5 m diameter open well to achieve a dosage of 50 mg/L? Water stands to a height of 3.0 m in the well. (6.1 L)
8. A small community is served by a groundwater supply consisting of a 450 mm diameter main 1400 m in length until it reaches the first consumer. The peak hourly pumping rate is 11 m^3/min and the temperature of the water is 10°C. Regulations state that the groundwater, being under the influence of surface water, demands a 3-log virus inactivation (CT required 6.0 mg.min/L). What free chlorine residual is required at the outlet of the pipeline? (0.20 mg/L)
9. The filtered water at peak hourly flow from a plant with direct filtration has a pH of 7.0 and a temperature of 15°C. The effective detention time in the clear well is 24 minutes, followed by a 1.3 km transmission line at a flow velocity of 1.2 m/s before entering the distribution system. 1-log removal of Giardia is required by chlorination. What chlorine residual is required at the outlet of the clear well and the pipeline? Assume there is no loss of residual in the pipeline. The required value of CT for 1-log Giardia removal at the prevailing conditions is known to be 25 mg.min/L. (0.60 mg/L)
10. A new main is disinfected with water containing 50 mg/L of chlorine by feeding 1.0% chlorine solution to the water entering the pipe.
    a. How many kg of dry hypochlorite powder containing 70% chlorine must be added to 200 L of 1.0% solution? (2.9 kg)
    b. At what rate should a 1.0% solution be applied to water entering the pipe to achieve a dosage of 50 mg/L? (5.0 L/m^3)

11. Chlorine gas, 70% HTH and 12% sodium hypochlorite solution (liquid bleach) are being considered as the primary disinfectants for a water treatment plant with a capacity of 7.5 ML/d. The anticipated chlorine dose is 2.5 mg/L. Calculate the daily cost of each disinfectant to achieve a dosage of 2.5 mg/L where chlorine gas, HTH and liquid bleach cost Rs. 50 /kg, Rs. 150 /kg, and Rs. 95 /kg, respectively? (Rs. 937,4020,14800/d)
12. The results of a chlorine demand test on a raw water sample are as follows:

Dosage	0.20	0.40	0.60	0.80	1.0	1.2	1.4	1.6	1.8	2.0
Residual	0.0	0.15	0.34	0.45	0.42	0.20	0.35	0.50	0.70	0.90

   a. Sketch a breakpoint curve.
   b. What is the chlorine dosage at the breakpoint? (1.2 mg/L)
   c. What is the maximum demand? (1.0 mg/L)
   d. What is the dosage required to achieve a free chlorine residual of 0.50 mg/L? (1.8 mg/L)
13. A chlorine residual of 0.4 mg/L is added to a flow of 4500 m^3/d. What should be the chlorinator setting in kg/d if the chlorine demand is 2.6 mg/L? (14 kg/d)
14. For proper disinfection, a water supply is to be fed at a dosage rate of 2.5 mg/L of chlorine. How much chlorine will be left in a 50 kg chlorine tank after a week if the average daily water flow is 1.5 ML/d? (24 kg)
15. A well water supply is disinfected by feeding a 2.5% sodium hypochlorite solution. Over a period of 12 hours, the total volume of water pumped is 16580 m^3. During the same period, the hypochlorinator level drops by 57 cm in the 0.8 m diameter solution tank. The hypochlorinator is operated continuously over the 12-hour period and a residual of 0.2 mg/L is maintained. Calculate the chlorine demand of the well water and the feed pump rate in mL/min. (0.23 mg/L, 400 mL/min)

10 Water Softening

The hardness of water is due to its calcium and magnesium content. Hardness due to other bivalent and trivalent cations is usually insignificant. Total hardness is the total of calcium and magnesium hardness and is expressed in terms of equivalent calcium carbonate. Water hardness varies considerably in different geographic areas. Since groundwater is in contact with geological formations containing calcium (limestone) and magnesium (dolomite) salts, it is is normally harder than surface water.

Though hardness is not harmful to health, too much hardness in water makes it unsuitable for uses such as heating, cooking, washing, bathing and laundering. Excessively hard water may, for example, deposit scale in water heaters and water piping. Hardness can waste large proportions of the soap used in laundering. As a result of these disadvantages, many municipalities soften their water supplies. **Table 10.1** shows a comparative classification for softness and hardness in water.

Hardness in the range of 80–120 mg/L is quite acceptable. Although relatively soft water is preferable, excessively soft water is corrosive.

10.1 TYPES OF HARDNESS

Hardness can be categorized based on cation and anion content, calcium and magnesium hardness, and carbonate and non-carbonate hardness. Hardness caused by calcium is called calcium hardness regardless of the type of salt associated with it, which may include carbonates, sulphates and chlorides. Likewise, hardness caused by magnesium is called magnesium hardness. Magnesium hardness is relatively hard to treat.

10.1.1 Carbonate Hardness

Carbonate hardness (CH) is primarily caused by bicarbonates and hence is equal to the natural alkalinity of water. When the water is boiled, the bicarbonates break down into carbonates and settle out of the water. Because it can be removed by heating, carbonate hardness is also referred to as temporary hardness.

10.1.2 Non-Carbonate Hardness

Non-carbonate hardness (NCH) is a measure of calcium and magnesium salts other than carbonates and bicarbonates, including sulphates, nitrates and chlorides. Since non-carbonate hardness cannot be removed by prolonged boiling, it is also called permanent hardness. The sum of carbonate hardness and non-carbonate hardness is the total hardness. In most waters, hardness due to other cations, including trivalent, is negligible.

Example Problem 10.1

The chemical analysis of a water sample is shown in **Table 10.2**. Express each concentration as meq/L and determine the various types of hardness.

DOI: 10.1201/9781003231264-10

TABLE 10.1
Classification of Hardness

Classification	mg/L as $CaCO_3$
Extremely soft	0–50
Soft	0–75
Moderately hard	75–100
Hard	100–125
Very hard	125–175
Excessively hard	175–250
Too hard	> 250

TABLE 10.2
Milliequivalent Table (Ex. Prob. 10.1)

Component	mg/L	mg/meq	meq/L	mg/L as $CaCO_3$
CO_2	25	22	1.14	57
Alk	300	61	4.92	246
Ca	135	20	6.75	338
TH	360	50	7.20	360

Given:

Ca = 135 mg/L
TH = 360 mg/L as $CaCO_3$
Alk = 300 mg/L as HCO_3
CO_2 = 25 mg/L

SOLUTION:

$$MgH = TH - CaH = 360 - 338 = 22\ mg/L$$

$$NCH = TH - CH = 360 - 246 = 114 = 110\ mg/L$$

10.2 SOFTENING METHODS

The two common methods of softening water are lime-soda ash and ion exchange. Ion exchange softening is more suitable for treating water high in non-carbonate hardness when the total hardness is below 350 mg/L. It is possible to achieve zero hardness with ion exchange softening. The end products of lime treatment, calcium carbonate and magnesium hydroxide, are soluble in water to some degree. Hence, it is not practical to reduce the hardness of the finished water to less than 25 mg/L by lime-soda ash treatment.

10.2.1 Lime-Soda Ash Softening

In the lime-soda ash process, lime and soda ash are added to the water to form insoluble precipitates that are removed by the sedimentation and filtration processes.

Lime Reactions

Lime added to water increases its pH and reacts with carbonate alkalinity to precipitate as calcium carbonate.

$$Ca(HCO_3)_2 + Ca(OH)_2 \rightarrow 2CaCO_3 \downarrow + 2H_2O$$

$$CO_2 + Ca(OH)_2 \rightarrow CaCO_3 \downarrow + H_2O$$

If sufficient lime is added to reach a pH > 10, magnesium hydroxide is precipitated.

$$Mg(HCO_3)_2 + Ca(OH)_2 \rightarrow CaCO_3 \downarrow + MgCO_3 + 2H_2O$$

$$MgCO_3 + Ca(OH)_2 \rightarrow CaCO_3 \downarrow + Mg(OH)_2 \downarrow$$

Soda Ash Reactions

Soda ash is used to remove non-carbonate hardness.

$$MgSO_4 + Ca(OH)_2 \rightarrow Mg(OH)_2 \downarrow + CaSO_4$$

$$CaSO_4 + Na_2CO_3 \rightarrow CaCO_3 \downarrow + Na_2SO_4$$

In removing magnesium non-carbonate hardness, both lime and soda ash are needed. The lime-soda ash process increases the sodium content of the water, and the pH of the water is high due to residual (insoluble) carbonates and hydroxides.

Dosages

The amounts of chemical required depend on the degree of hardness removal with various types of wastes. According to the chemical reactions shown before, the chemical requirements expressed in equivalents for treating various types of hardness are shown in **Table 10.3**.

Referring to **Table 10.3**, the lime requirement for the removal of carbonate hardness associated with magnesium is twice as much. Non-carbonate hardness associated with magnesium requires one each of lime and soda ash. Based on this, lime and soda ash requirements when expressed in terms of carbonate and non-carbonate hardness are as follows:

$$Lime = CO_2 + Alkalinity + Mg + Excess$$

$$Soda\ ash = NCH = Ca + Mg - Alkalinity$$

TABLE 10.3
Lime and Soda Ash Requirement

Type of Hardness	Component	Lime (eq/eq)	Soda ash (eq/eq)
Carbonate hardness, CH	$Ca(HCO_3)_2$	1	0
	$Mg(HCO_3)_2$	2	0
Non-carbonate hardness, NCH	$CaSO_4$	0	1
	$MgSO_4$	1	1
Carbon dioxide	CO_2	1	0

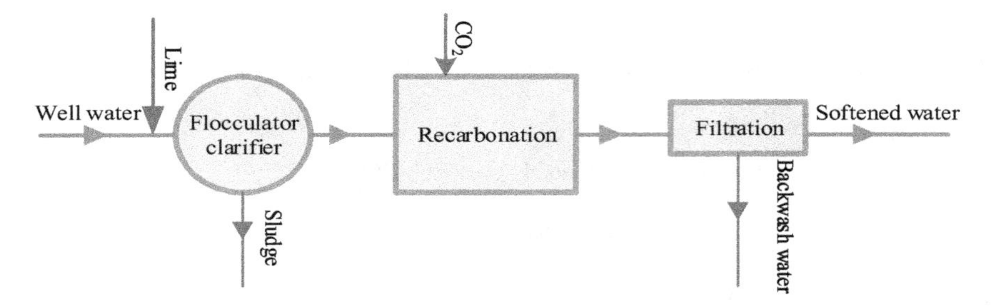

FIGURE 10.1 Single-Stage Lime Softening

It is assumed here that all the alkalinity is due to the presence of bicarbonates, which is mostly true in the case of fresh waters. Lime dosage is commonly expressed in terms of quicklime, CaO. The commercially available quicklime has a purity ranging from 70–96%, with a typical value of 88%. The commercial purity of soda ash is as high as 98–99%.

10.3 TYPES OF LIME-SODA ASH PROCESSES

As indicated by the chemistry of softening, different processes can be used depending on the type and degree of hardness and the quality of the finished water. For example, carbonate hardness or alkalinity can be removed simply by the addition of lime. However, both lime and soda ash are required for the removal of total hardness. The single stage of lime-soda ash process scheme is illustrated in **Figure 10.1**.

Example Problem 10.2

Non-carbonate hardness is to be removed using soda ash. How much 98% pure soda ash is required (kg/d) to remove 40 mg/L of non-carbonate hardness as $CaCO_3$ from a flow of 25 ML/d?

Given: Q = 25 ML/d
NCH = 40 mg/L

SOLUTION:

$$Soda\ ash = NCH = \frac{40\ mg}{L} \times \frac{meq}{50\ mg} \times \frac{53\ mg}{meq} \times \frac{commercial}{0.98\ pure}$$

$$= 43.26 = \underline{43\ mg/L}$$

$$Dosage\ rate,\ M = Q \times C = \frac{25\ ML}{d} \times \frac{43.26\ kg}{ML} = 1081.6 = \underline{1100\ kg/d}$$

Example Problem 10.3

Calculate the lime (88% pure CaO) and soda ash (100% Na_2CO_3) dosage using excess lime treatment at 0.70 meq/L of excess lime for treating hard water described in Example Problem 10.1.

Given: CO_2 = 1.14 meq/L
Alk(CH) = 4.92 meq/L
Mg = 0.45 meq/L
NCH = 2.28 meq/L

SOLUTION:

$$Lime\ dosage = CO_2 + Alk + Mg + Excess = 1.14 + 4.92 + 0.45 + 0.70$$

$$= \frac{7.21\ meq}{L} \times \frac{28\ mg\ CaO}{meq} \times \frac{comm.}{0.88\ pure}$$

$$= 229.4 = \underline{230\ mg / L}$$

$$Soda\ ash\ dosage = NCH = \frac{2.28\ meq}{L} \times \frac{53\ mg}{meq} = 120.8 = \underline{120\ mg / L}$$

10.3.1 Selective Calcium Removal

This process is used when magnesium hardness is less than 40 mg/L as $CaCO_3$. Magnesium hardness exceeding this limit will form scale. The usual process scheme is lime clarification with single-stage recarbonation followed by filtration. Because magnesium is not removed, no excess lime is needed. Soda ash may or may not be required depending on the amount of non-carbonated hardness.

Example Problem 10.4

Water is to be softened by selective calcium hardness removal.

a. Calculate the lime and soda ash requirements.
b. Calculate the hardness and alkalinity in the softened water before and after recarbonation.

Given: Cations: Ca = 4.0 Mg = 1.0 Na = 2 meq/L
Anions: HCO_3 = 3.5 SO_4 = 4.5 CO_2 = 1.0 meq/L

SOLUTION:

Hypothetical combinations.

	Ca	Mg	Na	Σ
HCO_3	2.5	-	-	2.5
SO_4	1.5	1.0	2.0	4.5
Σ	4.0	1.0	2.0	7.0

$$CH = \frac{2.5\ meq}{L} \times \frac{50\ mg}{meq} = 125.0 = 120\ mg / L$$

$$NCH = \frac{(1.5 + 1.0)\ meq}{L} \times \frac{50\ mg}{meq} = 125.0 = \underline{120\ mg / L}$$

$$TH = CH + NCH = 125 + 125 = 250.0 = \underline{250\ mg/L}$$

$$CaH = \frac{(2.5+1.5)meq}{L} \times \frac{50\ mg}{meq} = 200.0 = \underline{200\ mg/L}$$

$$MgH = \frac{1.0\ meq}{L} \times \frac{50\ mg}{meq} = 50.0 = \underline{50\ mg/L}$$

$$Lime = CO_2 + Ca\text{-}CH = 1.0 + 2.5 = \frac{3.5\ meq}{L} \times \frac{28\ mg\ CaO}{meq} \times \frac{comm.}{0.85\ pure}$$

$$= 115.2 = \underline{120\ mg/L}$$

$$Soda\ ash = Ca\text{-}NCH = \frac{1.5\ meq}{L} \times \frac{53\ mg}{meq} = 79.5 = \underline{80\ mg/L}$$

10.3.2 Excess Lime Treatment

To reduce magnesium hardness, excess lime is required to raise the pH above 10.6. The amount of excess lime used is typically 35 mg/L CaO (1.25 meq/L). When this treatment process is used, soda ash is used to remove non-carbonate hardness and recarbonation is used to stabilize the water. A schematic of two-stage excess lime treatment is shown in **Figure 10.2**.

Excess lime treatment can be performed in a single- or double-stage process. In the first stage, carbon dioxide is added to lower the pH to about 10.3 and convert the excess lime into settleable $CaCO_3$ for removal by flocculation and sedimentation. In the second stage, further recarbonation reduces the pH to the range of 8.4 –9.5 to convert the residual carbonate ion to bicarbonate ion to prevent scale formation during filtration and in the distribution of pipelines.

10.3.3 Split Treatment

In split treatment, only a portion of water is treated with excess lime, thus substantially reducing the lime and carbon dioxide requirement (**Figure 10.3**). Excess lime is added to precipitate magnesium

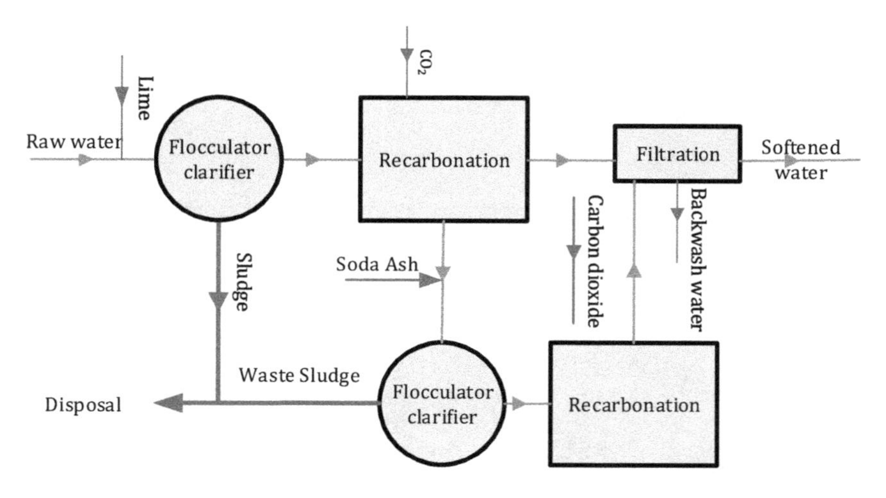

FIGURE 10.2 Excess Lime Softening Treatment

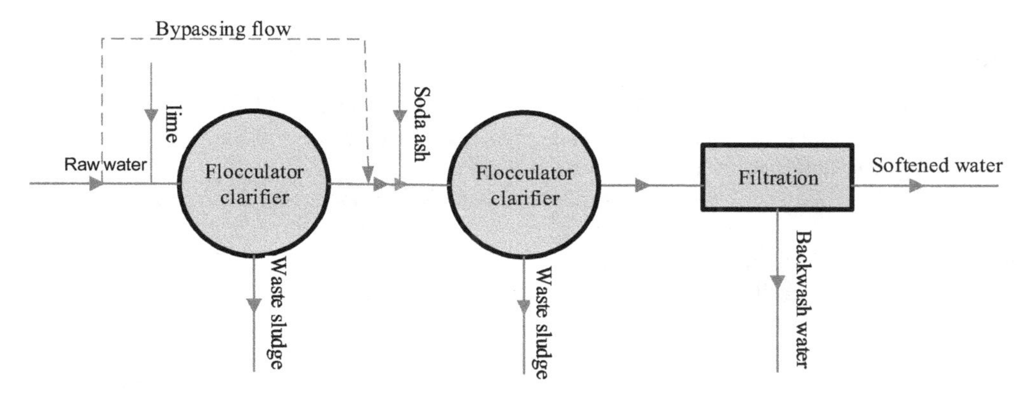

FIGURE 10.3 Split Treatment Softening Plant

in the first stage, which reacts with bypassed water to remove calcium hardness. The quantity of flow split in the first stage depends on the hardness of the water and the desired quality of the finished water. When saving money, the magnesium can be reduced to < 40 mg/L.

Ion-Exchange Softening

Using zeolites, hardness can be removed by replacing the bivalent ions of calcium and magnesium with sodium. Natural zeolites (boiling rock) are minerals that contain a negative charge and have affinity for metal ions. Currently, synthetic zeolites are used in place of natural zeolites. With use, the ion-exchange capacity of the zeolite resin is exhausted. Passing brine through the unit regenerates the resin. The ion-exchange process is especially suitable for small plants and when the majority of the hardness is in non-carbonate form.

10.3.4 Removal Capacity

Removal capacity indicates the total amount of hardness that can be removed by a unit volume of resin. By knowing the removal capacity, the exchange capacity of a softener can be calculated.

Removal capacity can be expressed as g or kg/m^3. Typically, the removal capacity of resins ranges from 15 to 100 kg/m^3.

10.3.5 Water Treatment Capacity

Water treatment capacity is the volume of water that can be softened before the resin must be regenerated. By knowing the exchange capacity and hardness of water, the water treatment capacity can be worked out.

$$Capacity = (Exchange\ capacity) / Hardness$$

Water treatment capacity indicates the volume of water that can be softened with one cycle of operation. By dividing the capacity by the average flow rate, the hours or days of operation can be calculated. Pumping a concentrated brine solution of 5–20% onto exhausted resin regenerates it. The NaCl dosage required for regenerating resin ranges from 80 to 160 kg/m^3 of resin. The feed rate of the solution is typically 40 L/m^2 · min. After regeneration, the medium should be flushed with softened water to remove excess brine.

Discussion Questions

1. What causes hardness in water? Why is it necessary to remove hardness in water?
2. What is water softening? Name and explain the various types of hardness in water.
3. Explain the various water softening methods.
4. Explain lime-soda ash softening methods and their associated chemical reactions.
5. What are the advantages and disadvantages of ion exchange water softening?
6. Under what situations are selective calcium removal recommended? Explain.
7. What is excess lime softening? Why is it necessary to add an excess of lime, and what happens to it in finished water?
8. In what ways is split treatment more economical?
9. In lime-soda ash softening, solids contact type units are commonly used. Explain.
10. If hard water causes problems, why not use soft water?
11. Why is hardness in drinking water supplies not regulated? What is the preferred range of hardness in drinking water supplies?
12. If municipal water is hard, it may be unfit for some industrial uses. Explain. Give three examples of such industries.

Practice Problems

1. A groundwater sample is analyzed and its hypothetical combination table is shown below. Determine TH, CH, NCH, Ca-H and Mg-H. (235, 110, 125, 185, 50 mg/L as $CaCO_3$)

	Ca	Mg	Na	Σ
HCO_3	2.2	-	-	2.2
SO_4	1.5	1.0	2.5	5.0
Σ	3.7	1.0	2.5	7.2

2. The results of the chemical analysis of a water sample are shown below. Make a hypothetical combination table and determine CH, NCH and Mg-H. (246, 104, 37 mg/L as $CaCO_3$)

   Ca = 125 mg/L
   Total hardness, TH = 350 mg/L as $CaCO_3$
   Alk = 300 mg/L as HCO_3
   CO_2 = 25 mg/L

3. The results of the analysis of a water sample are as follows:
   Cations: Ca = 100 mg/L Mg = 41 mg/L Na = 25 mg/L
   Anions: Alk = 180 mg/L SO_4 = 96 mg/L Cl = 136 mg/L (CO_2 = 8.8 mg/L)
   a. Make hypothetical combinations. ($Ca(HCO_3)_2$ = 3.6, $CaCl_2$ = 0.40, $MgCl_2$ = 2.4, $MgSO_4$ = 1.0, $CaSO_4$ = 1.0, NaCl = 1.1 all in meq/L)
   b. Determine TH, CH, NCH, Ca-H and Mg-H.
   (420, 180, 240, 250, 170 mg/L as $CaCO_3$)

4. For the previous problem, determine the required dosages rate.
   a. Calculate the quantity of lime as CaO with 88% purity needed to treat 16 ML/d using excess lime treatment. (4400 kg/d)
   b. Calculate the quantity of 95% soda ash required to remove NCH. (4300 kg/d)

5. A groundwater sample was analyzed and found to have the following constituents:
   Cations: Ca = 70 mg/L Mg = 9.7 mg/L Na = 6.9 mg/L
   Anions: Alk = 115 mg/L SO_4 = 96 mg/L Cl = 11 mg/L (CO_2 = 17 mg/L)
   a. Make hypothetical combinations. ($Ca(HCO_3)_2$ = 2.3 meq/L, $CaSO_4$ = 1.20 meq/L, Mg SO_4 = 0.80 meq/L, NaCl = 0.30 meq/L)
   b. Calculate the quantity of lime as CaO with 85% purity needed to treat 11 ML/d using excess lime treatment. (1860 kg/d)
   c. Calculate the quantity of 98% soda ash required to remove non-carbonate hardness. (1190 kg/d)
6. Calcium hardness is removed by coagulating water with lime (85% CaO) at a dosage of 130 mg/L. What is the concentration of dry solids produced as $CaCO_3$ precipitate? (400 mg/L)
7. A groundwater supply has the following analysis:
   Cations: Ca = 94 mg/L Mg = 24 mg/L Na = 14 mg/L
   Anions: HCO_3 = 317 mg/L SO_4 = 67 mg/L Cl = 24 mg/L
   Calculate the quantities of lime (CaO) and soda ash required for excess lime softening. (240 mg/L, 85 mg/L)
8. Calculate the dry sludge solids ($CaCO_3$ + $Mg(OH)_2$) produced in practice problem 7, above. Assume the practical limits of hardness removal for calcium 30 mg/L and magnesium to be 10 mg/L as $CaCO_3$. How much sludge will be produced for treating every ML of water, assuming the solids concentration of sludge is 9.0%? (630 mg/L, 7.0 m^3/ML)
9. A groundwater supply has the following analysis:
   Cations: Ca = 110 mg/L Mg = 25 mg/L Na = 12 mg/L
   Anions: HCO_3 = 240 mg/L SO_4 = 110 mg/L Cl = 35 mg/L
   a. Make hypothetical combinations to determine TH, CH, NCH, Ca-H and Mg-H. (380, 240, 140, 275, 105 all in mg/L as $CaCO_3$)
   b. Is this water suitable for selective calcium removal? If yes, determine the dosage of lime as CaO. (130 mg/L)
10. A groundwater supply has the following analysis:
    Cations: Ca = 40 mg/L Mg = 14.7 mg/L Na = 13.7 mg/L
    Anions: Alk = 135 mg/L SO_4 = 29 mg/L Cl = 17.8 mg/L (CO_2 = 8.8 mg/L)
    Excess lime and soda ash treatment is used to achieve practical limits of hardness removal in water with the following analysis.
    a. Express concentration in equivalents and make hypothetical combinations to determine the constituents present in this water?
    b. Based on your findings, determine the total hardness, carbonate hardness and non-carbonate hardness. (160 mg/L, 135 g/L, 25 mg/L)
    c. What is the lime (88% CaO) and soda ash (95%) requirement to soften this water? (180 mg/L, 28 mg/L)
11. The analysis of some water (all components measured as meq/L) is as follows:

Ca	Mg	Na	K	HCO_3	SO_4	Cl
3.7	1.0	1.0	0.5	4.0	1.2	1.0

   a) List the hypothetical combinations and find the total hardness. (4.7 meq/L = 235 mg/L)
   b) Calculate the chemical doses of lime (CaO) and soda ash (Na_2CO_3) necessary for excess lime softening. Assume 1.25 meq/L of excess lime. (6.25 meq/L = 175 mg/L as CaO, 0.7 meq/L = 37 mg/L as Na_2CO_3)

12. An ion exchange water softener has a diameter of 3.0 m. It is filled with a resin to a depth of 1.8 m. If the removal capacity of the resin is 21 kg/m^3, what is the total exchange capacity of the softener? (270 kg)
13. The meq/L concentrations of various components in a water sample are reported as follows:

Ca	Mg	Na	K	HCO_3	SO_4	Cl
3.5	0.8	0.6	0.1	4.0	0.6	0.4

    a) Calculate the Ca-H, Mg-H, CH and NCH. (175, 40, 200, 15 mg/L)
    b) Calculate the lime dosage in mg/L of $Ca(OH)_2$ required for selective removal of calcium hardness. (130 mg/L)
    c) Calculate the hardness and alkalinity of the softened water. (70 mg/L, 55 mg/L)
14. $Ca(OH)_2$ (lime slurry) reacts with calcium bicarbonate ($Ca(HCO_3)_2$) in solution to precipitate $CaCO_3$. Calculate the amount of 78% CaO required to react with 185 mg/L of calcium hardness as $CaCO_3$. (133 mg/L)

11 Miscellaneous Water Treatment Methods I

In addition to the water treatment methods already discussed, there are others used for a specific purpose, such as the addition or removal of fluoride and iron, and manganese removal. These are not part of conventional water treatment but are more specific to plants where problems exist.

11.1 FLUORIDATION

The process of fluoridation adds fluoride to water. The fluoridated water reduces tooth decay. It has been documented that fluoridation at a level of 1 mg/L can reduce the incidence of tooth decay among children by 65%. Controlled fluoridation at less than 1.5 mg/L is a safe, effective and economical process. Mottling of the teeth occurs when the fluoride level exceeds 1.5 mg/L. Fluoride occurs naturally in water, and concentrations in excess of 0.1% have been found in water from volcanic regions. Water with fluoride concentrations of 1.4 to 2.4 mg/L should be defluoridated to reduce the concentration to an optimum level of 1 mg/L. The maximum contaminant level (MCL) is 4.0 mg/L.

11.1.1 Fluoride Chemicals

The most commonly used fluoride chemicals in the water industry are sodium fluoride (NaF), sodium silicofluoride (Na_2SiF_6) and hydrofluosilicic acid (H_2SiF_6). These chemicals are refined from minerals found in nature and yield fluoride ions, which dissolve in water. A summary of fluoride chemicals is given in **Table 11.1**.

The fluoride ion content will depend on the chemical formula and commercial purity of the chemical. This is illustrated in the following Example Problem 11.1

Example Problem 11.1

Determine the fluoride content of 30% pure commercial fluosilicic acid, H_2SiF_6.

Given: H = 1 g/mol
Si = 28 g/mol
F = 19 g/mol

SOLUTION:

$$Molecular\ mass = 2 \times 1 + 28 + 6 \times 19 = 144\ g/mol$$

$$Fluoride\ content = \frac{114\ g\ F}{144\ g\ H_2SiF_6} \times \frac{30\%}{100\%} = 0.2375 = 0.24 = \underline{24\%}$$

When selecting a fluoridation chemical, the solubility of the chemical and how well it remains in the solution must be considered. Safety, ease of handling and cost must be given serious

DOI: 10.1201/9781003231264-11

TABLE 11.1
Common Fluoride Compounds

Item	Sodium fluoride	Sodium Silicofluoride	Fluosilicic acid
Chemical formula	NaF	Na_2SiF_6	H_2SiF_6
Commercial form	Powder or crystal	Fine crystal powder	Liquid
Molecular mass, g/mol	42	188	144
Commercial purity, %	90–98	98–99	22–30
Fluoride ion, %	42	61	79
kg / ML for dosing at 1.0 ppm F at indicated purity	2.26 (98%)	1.67 (98.5%)	4.21 (30%)
pH of saturated solution	7.6	3.5	1.2 (1.0%)
Sodium ion contributed at 1.0 ppm F, ppm	1.17	0.40	0.00
Solubility at 25°C, g/L of water	41	7.6	Liquid
Specific gravity	1–1.4	0.9–1.2	1.3

consideration. Hydrofluosilicic acid in liquid form is easy to feed and popular with small and large plants. However, the acid produces toxic fumes that must be vented. Large waterworks use gravimetric dry feeders to apply sodium silicofluoride that is commercially available in various gradations.

11.1.2 Fluoridation Systems

Depending on the natural fluoride level in the source water, there may be four different fluoridation systems:

1. When the natural fluoride level is zero and all the fluoride has to be added to the water supply.
2. When the raw water source has adequate or excessive ions. When water contains excessive fluoride, it must be defluoridated.
3. More commonly, when the natural fluoride level is less than the optimum. Fluoride ions are added to bring the total to the desired level.
4. When there are two sources of water supply with different levels of fluoride. In this instance, the degree of blending must be considered when selecting the fluoride dosage.

Both overdosing and underdosing are undesirable.

Chemical Feeding

The equipment used for fluoridation is similar to that used for feeding other water treatment chemicals. Fluoride ions can be added to water by either solution feeders or dry feeders. Solution feeders are the most economical for small water systems. Whatever the type of feeding system may be, it is important to maintain accurate feeding and prevent overfeeding and siphonage. A given feeding system should provide a means of measuring the fluoride level in the finished water.

It is preferable to add fluoride after filtration to avoid any losses that may occur as a result of reactions with other chemicals, notably alum and lime.

Example Problem 11.2

A liquid feeder applies a 4.0% saturated sodium fluoride solution to treat a flow of 6.0 ML/d with a fluoride dosage of 1.0 mg/L.

a) Express the mass to volume concentration in g/L.
b) What must the feeder pump setting be (L/d) to feed the solution at the desired dosage?

Given:

Parameter	Solution fed	Water treated
C	4.0% NaF	1.0 mg/L
Q	?	6.0 ML/d

SOLUTION:

$$Solution,\ C = \frac{4.0\%\ NaF}{100\%} \times \frac{0.45\ F}{NaF} = 18\ g/L\ F\text{-}ion^{-}$$

$$Dosage\ rate,\ M = Q \times C = \frac{6.0\ ML}{d} \times \frac{1.0\ kg}{ML} = 6.00 = 6.0\ kg/d$$

$$Feed\ pump\ rate,\ Q_1 = \frac{C_2}{C_1} \times Q_2 = \frac{1.0\ kg}{ML} \times \frac{L}{18\ g} \times \frac{1000\ g}{kg} \times \frac{6.0\ ML}{d}$$

$$= 333 = \underline{330\ L/d}$$

Example Problem 11.3

A flow of 8.5 ML/d is to be treated with a 20% solution of hydrofluosilicic acid with a fluoride content of 79%. The relative density (SG) of the acid is known to be 1.2. Calculate the feed pump rate in L/h for applying a 1.5 mg/L dose of fluoride.

Given:

Parameter	Solution fed = 1	Water treated = 2
C	20%, 79% F	1.5 mg/L
SG	1.2	1.0
Q	?	6.0 ML/d

SOLUTION:

$$Solution,\ C = \frac{20\%\ acid}{100\%} \times \frac{0.79\ F}{acid} \times \frac{1.2\ kg}{L} \times \frac{1000\ g}{kg} = 189.6 = 190\ g/L$$

$$Feed\ pump,\ Q_1 = \frac{C_2}{C_1} \times Q_2$$

$$= \frac{1.5\ kg}{ML} \times \frac{L}{189.6\ g} \times \frac{1000\ g}{kg} \times \frac{8.5\ ML}{d} \times \frac{d}{24\ h}$$

$$= 2.80 = \underline{2.8\ L/h}$$

$$Dosage\ rate,\ M = Q \times C = \frac{8.5\ ML}{d} \times \frac{1.5 kg}{ML} = 12.75 = \underline{13\ kg/d}$$

11.2 DEFLUORIDATION

Though an optimum fluoride dose prevents dental caries in children, excessive fluoride in drinking water supplies can cause dental fluorosis (mottling of teeth) and the hardening of arteries and bones in older people. The maximum contaminant level of fluoride is 4.0 mg/L, and water with fluoride concentrations in the range of 1.4–2.4 mg/L should be defluoridated to bring it up to the optimum level. The following are the most commonly used chemical treatments for reducing fluoride in drinking water.

11.2.1 Calcium Phosphate

Bone has a great affinity for fluoride and can be used in the filter media to remove it. The bones are calcinated at high temperatures, followed by mineral treatment. It is then pulverized to a 40–60 mesh and is used in the filter bed. When exhausted, the filter is regenerated with alkali and acid.

11.2.2 Tricalcium Phosphate

Bone charcoal is basically a mixture of tricalcium phosphate and carbon. This material is very successfully used to reduce fluoride content. Calcium triphosphate can also be synthesized by adding phosphoric acid to lime. It has been used in contact filters for the removal of fluoride. Exhausted calcium phosphate is regenerated by adding 1.0% caustic solution followed by dilute hydrochloric acid or carbon dioxide to neutralize the excess of caustic solution.

Fluorex is the trade name for a special mixture of tricalcium phosphate and hydroxyapatite and can be used as filter media. It can be regenerated using a caustic solution followed by a water rinse, and the excess caustic soda can be neutralized with carbon dioxide (carbonic acid).

11.2.3 Ion Exchange

The ion exchange method can be used for the removal of fluoride, for example, by the cation exchange of sulphonated coal type and an amin resin. Alum treated cation exchange resin from avaram bark can be used as an effective material for fluoride removal.

11.2.4 Lime

During the softening process, fluoride can be removed along with magnesium. However, due to residual caustic alkalinity, it must be followed by recarbonation. This process is applicable when treating hard water with fluoride concentrations of less than 4.0 mg/L.

11.2.5 Aluminium Compounds

Alum has a high absorption capacity for fluoride, which can be further enhanced with coagulant aids such as activated silica and clays. Another method utilizing aluminium salts involves the use of contact beds of insoluble materials impregnated with aluminium compounds. Dehydrated aluminium oxide can be used in contact beds.

11.2.6 Activated Carbon

Activated carbon has very high absorptive properties that can be used to remove fluoride. This removal takes place at low pH levels – at pH 8.0 or above, no removal takes place. Exhausted carbon can be regenerated with a weak acid and alkaline solution.

11.3 IRON AND MANGANESE CONTROL

Iron and manganese may be found in water supplies as soluble or insoluble compounds. Both elements are more commonly found in well water supplies than surface water supplies. Groundwater has little or no oxygen; hence, these elements exist in their reduced states as ferrous or manganous. In supply reservoirs, dissolved oxygen is also low, particularly during the winter when reservoir surfaces are covered with ice and snow, and aeration cannot occur. Since iron and manganese are more soluble in the absence of dissolved oxygen (reduced states), high concentrations may sometimes be found in lower portions of reservoirs.

11.3.1 Problems Due to High Iron and Manganese

In their soluble state, these minerals are colourless at the concentrations they are generally found in water supplies. However, when in contact with air during a pumping or aeration process, iron and manganese are oxidized and converted to insoluble forms. In concentrations as low as 0.3 mg/L, these compounds will be noticeable and will stain clothes and fixtures.

Perhaps the most troublesome consequence of iron and manganese in water is that they promote the growth of a group of microorganisms known as iron bacteria. Aesthetic indications of the presence of iron bacteria include the sliming of pipes and fixtures, staining, and taste and odour problems.

When the iron content of water is high, tannic acid in tea or coffee may combine with the iron to darken the beverage like ink. Coffee becomes unpalatable if the iron content exceeds 1.0 mg/L.

Iron and manganese in water can be easily detected by observing the colour of the inside walls of the filters and the filter media. If the raw water is pre-chlorinated, there will be black stains on the walls below the water level and a black coating over the top portion of the sand filter bed. The black colour usually indicates a high level of manganese, while a brownish-black stain develops when water is high in both iron and manganese. As it is, iron and manganese have no adverse effects on health. The generally acceptable limits for iron and manganese are 0.3 mg/L and 0.05 mg/L, respectively. Even if they were available in beneficial amounts, their presence is objectionable.

11.4 CONTROL METHODS

Several methods are available to control iron and manganese in water. Preventive measures are successful when the concentration is low. However, at large concentrations, iron and manganese must be removed.

11.4.1 Phosphate Treatment

Sequestering with phosphate may be a simple solution when the water contains up to 0.3 mg/L of manganese and less than 0.1 mg/L of iron. Chlorine must usually be fed along with the polyphosphate to prevent the growth of iron bacteria. The chlorine dose for phosphate treatment should be sufficient to produce a free-chlorine residual of 0.25 mg/L after 5 minutes of contact time. A minimum of 0.2 mg/L free-chlorine residual should be maintained throughout the system.

11.4.2 Bench-Scale Testing

Any polyphosphate can be used for treatment; however, sodium metaphosphate is effective in lower concentrations. The proper dosage is determined by bench-scale testing in the laboratory. This test is similar to jar testing. In the first step, a series of samples are dosed with chlorine solution to determine the chlorine dose required to produce the desired chlorine residual.

A series of samples are tested with various dosages of polyphosphate and the previously determined chlorine dosage. The chlorine should never be fed ahead of the polyphosphate because the chlorine will oxidize the iron and manganese. Samples are observed daily against a white background, noting the amount of discolouration. The optimum dosage is the lowest dosage that delays noticeable discolouration for a period of four days.

11.4.3 Feed System

In a well water supply system, polyphosphate and chlorine are fed through a polyethylene hose discharging below the suction bowls of the pump. Polyphosphate solutions stronger than 0.5 g/L are very viscous. Stale solutions (> 48 hours) are not good because polyphosphates are hydrolyzed to form orthophosphates, which are less effective.

Polyphosphate treatment to control iron and manganese is usually most effective when the polyphosphate is added before the chlorine. They can also be fed together, but chlorine should never be fed before polyphosphate because chlorine will oxidize iron and manganese.

11.5 REMOVAL METHODS

11.5.1 Oxidation by Aeration

Aerating the water to form insoluble ferric hydroxide can oxidize iron. Aeration followed by sedimentation and filtration, or filtration alone, is used to remove the precipitates. Oxidation by plain aeration is accelerated by an increase in pH. Lime is generally added to raise the pH. If the water contains any organic substances, the oxidation reaction is slower.

$$Fe^{++}\left(soluble\right)+O_2 \rightarrow Fe^{+++}\left(insoluble\right)$$

Plain aeration is not effective in oxidizing manganese. Hence, this method is not suitable for treating water with high manganese concentrations. An advantage of this method is that no chemicals are used except lime or soda ash, which may sometimes be used to raise pH.

11.5.2 Oxidation with Chlorine

Chlorine is a strong oxidant. Maintaining free chlorine residual throughout the treatment process can easily oxidize iron and manganese. The higher the chlorine residual, the faster the oxidation occurs. In some small plants, the water is dosed to maintain a chlorine residual of 5–10 mg/L, filtered, and dechlorinated using sodium bisulphite or other reducing agents.

In smaller plants that remove iron and manganese, a hypochlorite solution may be used instead of chlorine gas. When the water is high in hardness, the calcium carbonate tends to form a coating on the valves in the solution feeder. To avoid this, softened water is used to dilute the commercially available hypochlorite solution.

11.5.3 Oxidation with Permanganate

Much like chlorine, potassium permanganate ($KMnO_4$) is a strong oxidizing agent. For certain water, potassium permanganate oxidation may be advantageous since pH adjustments are not required. The specific dosage required depends on the concentration of metal ions, pH, mixing conditions and other factors. Bench tests should be performed to determine the optimum dosage. A dose that is too small will not oxidize all the manganese in the water, whereas a dose that is too large will leave potassium permanganate residual and may produce a pink colour.

$$Fe(HCO_3)_2 + KMnO_4 \rightarrow Fe(OH)_3 + MnO_2$$

$$Mn(HCO_3)_2 + KMnO_4 \rightarrow MnO_2$$

$$\frac{KMnO_4}{Fe} = \frac{1\ mole \times 158g\ /\ mol}{3\ mole \times 55.6g\ /\ mol} = 0.95$$

Potassium permanganate is a dark purple crystal or powder available commercially at 97.99% purity. Theoretically, the dosages of potassium permanganate required to oxidize each 1 mg/L of iron and manganese, respectively, are 0.95 and 1.92 mg/L. In actual practice, the amount needed is often less than this. When chlorine is used, for each mg/L of iron it takes approximately 0.6 mg/L of $KMnO_4$ and 0.64 mg/L of chlorine. Effective filtration following chemical oxidation is essential. Experience has shown that filtration can effectively remove iron and manganese as long as they are both under 1.0 mg/L.

11.5.4 Ion Exchange with Zeolites

Manganese zeolites, also known as manganese greensand, is a granular material. Greensand is coated with manganese dioxide that removes soluble iron and manganese. The greensand also acts as a filter media. After the zeolite becomes saturated with iron and manganese oxides, it is regenerated using potassium permanganate to remove the insoluble oxides.

In the continuous flow system, potassium permanganate solution is fed into the water. Water containing potassium permanganate passes through a pressure filter that contains a dual-media anthracite and manganese zeolite bed. The upper filter layer removes the metal oxides formed by the chemical reaction with potassium permanganate.

The underlying zeolite layer captures any metal ions not oxidized. Any excess amount of potassium permanganate applied is used to regenerate the greensand. If too much potassium permanganate is applied, the effluent may be coloured pink. A good operating zeolite system can remove 95% of both iron and manganese. However, if the iron content exceeds 20 mg/L, the efficiency of the process drops quickly. A residual of potassium permanganate must be present in the effluent water from the greensand for the zeolite to be effective.

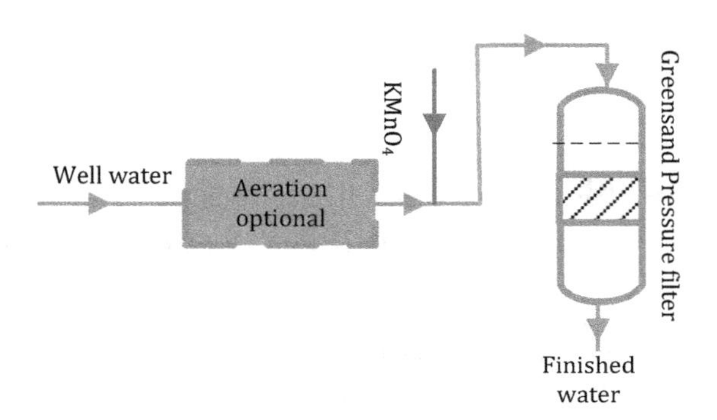

FIGURE 11.1 Process Schematic

Example Problem 11.4

Calculate the theoretical potassium permanganate dosage in mg/L to treat well water with 2.6 mg/L of iron before aeration and 0.3 mg/L after aeration. The manganese concentration is 0.8 mg/L before and after aeration.

Given: Fe = 0.30 mg/L
Mn = 0.80 mg/L
$KMnO_4$ = ?

SOLUTION:

$$Dosage = 0.95Fe + 1.92Mn = 0.95 \times 0.3\ mg/L + 1.92 \times 0.8\ mg/L$$

$$= 1.82 = \underline{1.8\ mg/L}$$

The actual dosage required may be less than 1.8 mg/L. This needs to be checked by bench scale testing.

11.6 ARSENIC REMOVAL

Arsenic is a natural contaminant in some supplies, particularly groundwater, where it is in contact with arsenic-bearing material. It may also be released into aquatic environments by mining and smelting operations. Arsenic is a contaminant of concern near waste remediation sites. The maximum permissible limit in water supplies is 10 μg/L. Water supplies exceeding this limit can use a variety of methods to reduce the level of arsenic present. For most people, the most significant exposure to arsenic is through food. Although found in some surface waters, it is mainly found in groundwater supplies. In soluble form, arsenic primarily occurs as arsenite (AsO_3) and arsenate (AsO_4).

The process of precipitation is the most frequently used method of removing arsenic. Proper coagulants can transform soluble arsenic insoluble, which can then be removed by sedimentation and filtration. However, the inorganic sludge produced would contain arsenic, so the sludge must be treated properly before disposal. Another effective way to reduce arsenic is by adsorption on activated carbon. After adjusting the pH of the water to 5.5, a granular media of activated alumina can provide efficient removal of inorganic arsenic, which is the more predominant form in water. Without pH adjustment, arsenic removal will be relatively low.

Other treatment options for arsenic removal include ion exchange and membrane filtration. Soluble forms of arsenic can be removed by strongly anion exchange resins. Membrane filtration is discussed in the next chapter.

11.7 NITRATE REMOVAL

The maximum limit of nitrates in drinking water is 45 mg/L. Excessive levels of nitrates are damaging to babies. In some groundwaters, nitrates can be excessive and need to be removed. One common method for nitrate removal is anion exchange, much like cation exchange in the case of iron and manganese. In anion exchange, selective nitrate resins are used. These resins are called selective because they are less selective for multivalent anions like sulphates. Sodium chloride or brine is used in the regeneration of both ordinary and selective resins.

$$RCl + NO_3^- \overset{\text{Nitrate removal}}{\Rightarrow} RNO_3 + Cl^-$$

The resins beds are susceptible to plugging caused by hardness, particulates and bacterial growths, resulting in poor removals. To prevent this, water fed to a nitrate exchange unit must be softened to remove excessive harness. If this is not done, backwashing will not be complete since metal oxides and other contaminants have greater resistance to fluidization.

Discussion Questions

1. What is fluoridation and explain when it is beneficial?
2. What are the chemicals used to add fluoride to potable water?
3. What are the problems associated with excess fluoride in drinking water?
4. Define defluoridation. What are the methods of defluoridation used to bring fluoride to a safe level?
5. Briefly explain the various methods used for the removal of excess iron and manganese.
6. What problems are caused by an excess of iron and manganese in potable water?
7. What problems are associated with the growth of iron bacteria?
8. The most common process scheme for the removal of iron and manganese from groundwater is aeration, chemical oxidation and filtration.
   a. What is the purpose of aeration?
   b. Name the oxidation chemicals used.
   c. Why is the majority of iron oxide and manganese dioxide removed in the filter rather than during sedimentation?
9. What is the health concern associated with excessive nitrate in municipal water supplies? What is the common method used for the removal of excessive nitrates?
10. Arsenic in water supplies is not desirable. What are the common methods for its removal in a) larger plants and b) smaller plants?

Practice Problems

1. Hydrofluosilicic acid of 30% purity (SG = 1.3) is fed to fluoridate water. The average daily water flow rate is 12 ML/d and the fluoride dosage is 1.1 mg/L. The acid is pumped from a drum placed on a platform scale. What is the expected daily weight loss of the drum? (43 kg/d)
2. Calculate the kg/d of commercial sodium silicofluoride to be fed if the natural level of fluoride in the water is 0.2 mg/L and the flow to be treated is 56 L/s. The desired level of fluoride is 1.0 mg/L. (6.5 kg/d)
3. A sodium silicofluoride (Na_2SiF_6) solution is prepared by dissolving 5.0 kg of 98% pure commercial salt to make 100 L of solution. Calculate the feed rate of this solution to increase the fluoride content of water by 0.9 mg/L. (30 L/ML)
4. A flow of 24 ML/d is to be treated with a 20% solution of hydrofluosilicic acid with 79% fluoride content and an SG of 1.2. What should the feed rate of the acid be in mL/min to increase the fluoride content of water from 0.2 to 1.0 mg/L? (70 mL/min)
5. A flow of 0.50 ML/d is treated with 1.0 kg/d of commercial salt sodium fluoride (NaF). The commercial purity is 98% and the fluoride content is 45%. Determine the fluoride dosage. (0.88 mg/L)
6. What dosage of potassium permanganate ($KMnO_4$) is needed to treat city well water with 3.5 mg/L iron before aeration and 0.2 mg/L after aeration? The manganese content of 1.2 mg/L remains the same before and after aeration. (2.5 mg/L)

7. What dosage of potassium permanganate is required to oxidize 5.5 mg/L of iron and 1.2 mg/L of manganese? (7.5 mg/L)
8. 0.675 grams of polyphosphate are weighed out. What volume of solution should be made by adding water such that the solution concentration is 0.1% or 1 g/L. How many mL of this solution should be added to a 1.5 L water sample to dose water at the rate of 4.0 mg/L. Calculate how many litres of this solution will be required to dose 1 m^3 of water at a rate of 2.5 mg/L. (675 mL, 6 mL, 2.5 L)
9. Hydrofluosilicic acid is pumped from a shipping drum placed on a platform scale. The recorded weight loss of the drum is 74 kg in processing 15.4 ML of water. What fluoride dosage was applied? (1.5 mg/L)
10. A 5% solution of potassium permanganate is fed into a manganese zeolite pressure filter. The desired dosage is 2.0 mg/L. What should the feed pump rate be in L/h if the well is being pumped at a rate of 15 L/s? (2.2 L/h)
11 Determine the required dosage rate of polyphosphate if 2.6 ML/d of water is treated at a rate of 3.5 mg/L. If the polyphosphate solution strength is 2.0%, what will the daily use of the solution be in L/d? (9.1 kg/d, 460 L/d)

12 Miscellaneous Water Treatment Methods II

A major purpose of water treatment is to produce palatable and aesthetically appealing water. The importance of supplying odour-free water cannot be overemphasized. It should be clear, good-tasting and free of any objectionable odours.

12.1 TASTE AND ODOUR CONTROL

Consumers can judge the quality of water based on what they can see, smell and taste. People would refuse to drink good hygienic water if it is turbid or has an unpleasant taste or odour. On the other hand, they will happily drink toxin-loaded water if it looks clean and is odour-free.

For those who have experience working at a water plant, it is well known that most of the complaints relate to taste and odour problems in water supply. To add to this, any such problems are more likely to become headlines in local newspapers. Although this is ludicrous, a good plant operator would be foolish to ignore it. Any neglect on the part of plant personnel may erode public confidence and force people to seek alternative sources of drinking water, such as bottled water and water from private wells, which may not be safe. In short, a water plant operator must pay serious attention to consumer complaints, especially those related to taste and odour. At no cost should the public lose confidence in the safety and quality of their water.

Adsorption has been used for years to remove organic-causing taste, odour and colour problems. This process has added importance since it removes toxic and carcinogenic organics from drinking water.

12.1.1 Organics in Raw Water

Organic substances such as humic and fulvic acid are produced naturally as a result of decaying vegetation. As a result, surface water usually contains greater concentrations of these substances and has a stronger colour compared with groundwater. In addition to humic substances, a great variety of organics, mainly man-made, are introduced when domestic, agricultural and industrial wastes are discharged into received water bodies.

The presence of organics in water can interfere with other treatment processes such as coagulation and flocculation. During chlorination, THMs are formed when chlorine reacts with organics. As a result, the practice of pre-chlorination is becoming unpopular, and alternate disinfectants, including ozone and chlorine dioxide, are gaining in popularity.

The man-made organics include solvents, hydrocarbons, cleaning compounds, pesticides and herbicides, many of which are toxic to humans. The best solution is the prevention of such substances entering the water systems. Emphasis should be given to watershed management and the implementation of sewer-use control programs.

Algae are the most common cause of taste and odour problems. Their metabolic activities produce odorous compounds. In a eutrophic lake or reservoir, regular copper sulphate applications to the impounded water are effective in controlling algae blooms.

DOI: 10.1201/9781003231264-12

Common sources of taste and odour problems can be summarized as follows:

- Dissolved gases, such as hydrogen sulphide.
- Biological growths, such as algae, protozoa and slimes. Most of the time, the problem arises when an organism dies. This is the reason that water will sometimes develop an odour problem after treatment.
- Product of decaying aquatic vegetation and algae. Urban run-off discharges, especially during spring, are serious problems for communities downstream.
- Industrial waste discharges.
- Problems originating in the distribution system. Half the time, an odour problem will occur after the water leaves the plant. Distributions with dead ends and poor circulation are highly prone to odour problems. Such areas should be identified and the regular flushing of lines should be carried out as a preventive measure.
- Leachate from materials such as plastic, paint and other odour-causing compounds.
- Chemicals used in the processing of water. Although chlorine is used for odour control, it can also contribute to an odour problem if its dosage and residual are not properly controlled. Free chlorine can cause taste and odour problems if the water contains naturally occurring phenolic compounds.

12.1.2 Chemical Dosing

The chemical dose of copper sulphate pentahydrate ($CuSO_4 \cdot 5HO_2$) required partly depends on characteristics including alkalinity, turbidity and temperature. At high levels of alkalinity, citric acid may need to be added to prevent the precipitation of copper. The typical dosage of copper is 2.0 mg/L. Note that for every unit of copper, four units of copper sulphate need to be applied. When treating large reservoirs, sometimes only a depth of perhaps 6 m, or a depth down to the thermocline, may be used in the calculation of volume of water to be treated.

The desired copper sulphate dosage may also be specified as kg/ha. This format is generally used for water with relatively high alkalinity. Under such conditions, algae control is limited to the upper volume of water due to interference of other ions. When using such procedures, the phenomenon of die-off causing odour problems must be considered. In many cases, it may be better to remove the microorganisms by plant processes rather than the chemical dosing of lakes and reservoirs.

12.2 TASTE AND ODOUR REMOVAL

Taste and odour causing organic compounds can be removed by chemical oxidation or by absorption using activated carbon or synthetic resins.

12.2.1 Oxidation

Some taste and odour causing compounds can be removed using chlorine, potassium permanganate to oxidize the compounds by aeration (oxidation and air-stripping), coagulation/sedimentation, or filtration. As discussed earlier, heavy chlorination should be avoided because of THM formation.

12.2.2 Aeration

Aeration brings water in contact with air and allows volatile compounds to escape into the atmosphere. This method of removing volatile compounds is also called degasification. Aeration is also responsible for oxidation. However, oxidation by aeration is more effective for treating inorganic compounds than organic compounds. Therefore, for taste and odour problems, aeration can be considered a physical rather than a chemical process.

Aeration can be achieved by passing diffused air into water or cascading water through the air. A process called air stripping does both. In chemical oxidation, aeration may help prepare the water for treatment, thus reducing chemical dosages required for oxidation.

12.2.3 Chemical Oxidation

Chlorine, potassium permanganate and chlorine dioxide are some of the oxidizing chemicals commonly employed for taste and odour control. Ozone is another strong oxidant. When chlorination is used, dosages are adjusted. Usually, higher dosages than those required for disinfection are needed. It is recommended to add chlorine in the early stages of treatment to get the best results. When super-chlorination is used for odour control, water must be dechlorinated to adjust the residuals. When water contains phenols or similar compounds, chlorine may be the wrong choice for taste and odour control.

Potassium permanganate has been used very successfully for taste and odour control. This is a strong chemical oxidant and produces a purple colour when mixed with water. After oxidation, the colour changes from purple to yellow or brown. Any permanganate not reacted or reduced will result in purple finished water. Permanganate should be added to water as early as possible to maximize contact time. Any excess permanganate can be treated with activated carbon.

Experience at various facilities has shown that permanganate dosage for odour control is in the range of 1–3 mg/L. Since permanganate is an expensive chemical, a cost comparison with alternate methods must be done before making the final choice.

It is important to note that permanganate must never be stored in the vicinity of activated carbon. The two make a combustion mixture. Like most other chemicals, permanganate can be metered using chemical feed pumps, or dry feeders can be used to apply the solid crystals to water. Since permanganate is very corrosive, provisions for dust control must be made.

Chlorine dioxide and ozone are the other two disinfectants that can be used for taste and odour control. Chlorine dioxide is gaining popularity especially where phenols are the major cause of odour problems and there is danger of formation of THMs. Due to the instability of chlorine dioxide, it must be generated on-site and applied immediately. Ozone decreases the formation of THMs and is a strong oxidizing agent to act as a disinfectant as well as a deodorant. However, oxidation is not as successful as adsorption in the removal of taste and odour causing compounds.

Example Problem 12.1

The volume of a reservoir is estimated to be 18 ML. The desired dose of copper is 0.5 mg/L. How many kg of copper sulphate will be needed?

Given: C = 0.50 mg/L as Cu
V = 18 ML

SOLUTION:

$$Molecular\ mass = (63.5 + 32 + 4 \times 16 + 5 \times 18) = 249.5 = \underline{250\ g/mol}$$

$$Fraction\ of\ Cu = \frac{63.5 g/mol}{249.5 g/mol} \times 100\% = 25.4 = \underline{25\%}$$

$$Mass\ applied = 18\ ML \times \frac{0.50\ kg}{ML} \times \frac{100\%\ salt}{25.4\ \%} = 22.5 = \underline{23\ kg}$$

12.2.4 Adsorption

In adsorption, as the name indicates, the organic compounds adhere to the surface of an adsorbing media such as activated carbon. For adsorption to be effective, the adsorbent must provide an extremely large surface for the trapping of the organics. Activated carbon is an excellent adsorbent due to its porous structure and affinity for organics. These pores are created during the manufacturing process by steaming the carbon at high temperatures (800°C). This is known as activation. Each carbon particle consists of many small and large size pores that adsorb the odorous compounds. Once the surface of the pore is covered, the carbon is exhausted and cannot adsorb any more.

The exhausted carbon must then be replaced or reactivated by essentially the same process as activation.

12.2.5 Forms of Activated Carbon

Although activated carbon can be made from a variety of materials such as wood, nutshells, peat and petroleum residue, the activated carbon used in water treatment is usually made from bituminous or lignite coal. There are two common forms of activated carbon, powder and granular. A comparison of the two is shown in **Table 12.1**.

Powder Activated Carbon

Powder Activated Carbon (PAC) is preferred for use in small plants and for controlling periodic taste and odour problems. Powder carbon can be fed dry or as a slurry. Treatment plants that consistently use PAC employ the slurry method.

Dry Method

In the dry method, chemical feeders are specifically designed to handle dry carbon. A hopper with inclined walls sits above the feeder to allow free gravity flow. Feeders should be located in a confined enclosure or area of the plant because of dust problems.

Slurry Method

After its delivery in bulk, the PAC is transferred to the storage tank using an eductor or a carbon slurry pump. At least two tanks are necessary, one for shipment and one for use. All equipment used in PAC application should be lined with corrosion-resistant material.

Slurry is pumped from the storage tank to the day tank, which is equipped with mixers to keep the carbon in suspension. From the day tank, slurry is fed by gravity to the volumetric feeder. The typical concentration of slurry is 100 g/L.

Point of Application

PAC is usually added ahead of coagulation to ensure no residual carbon enters the distribution system. The carbon added as powder ends up as sludge in the sedimentation basin and backwash from filters. It is not practical to reuse the carbon.

TABLE 12.1
Comparing PAC with GAC

Property	PAC	GAC
Particle diameter, mm	< 0.1	1.2–1.6
Specific gravity	0.32–0.72	0.42–0.48
Surface area, m^2/g	500–600	650–1150

Granular Activated Carbon

Granular Activated Carbon (GAC) is typically used when carbon is continuously needed to remove organics. Rather than feeding, as in the case of PAC, water to be treated is passed through a bed of GAC. The bed can be part of the media of the conventional filter or be in separate units called contactors.

Conventional Filter

Partial or complete replacement of the filter media with GAC is economical and suitable for most situations. Granular carbon, being the same density as anthracite, is an equally effective filtering material. The minimum recommended depth of GAC media is 600 mm.

Contactors

Contactors are essentially pressure filters containing deep columns of GAC. Contactors, though expensive, provide longer contact time, flexibility of operation, longer filter runs and longer bed life. Because contactors are used primarily for adsorption and not filtration, they are placed after regular filtration.

12.2.6 Process Control Tests

Process control tests for adsorption range from routine jar tests to more sophisticated water quality analyses. In many situations, total organic carbon removal can be used as a process control parameter. The dosage rate can be determined by a modified jar test.

If PAC is used to remove compounds other than those causing taste and odours, for example, a THM precursor, then a different procedure is needed. In contactors, the influent and effluent samples are analyzed using gas chromatography. The turbidity in the effluent is continuously monitored to determine the passing through of carbon fines and other suspended matter.

Example Problem 12.2

A sample of PAC slurry was collected from a slurry-mixing tank. The one-litre sample was allowed to settle. After six hours of settling, the volume of settled carbon was recorded to be 55 mL. An aliquot of 10 mL of settled carbon was dried and found to weigh 2.2 g. Calculate the concentration of carbon in the slurry and find the dosage of slurry to dose water at 5.0 mg/L.

Given: Settled volume = 55 mL/L
Settled carbon V = 10 mL
m = 2.2 g

SOLUTION:

The bulk density of powdered carbon is relatively small since a large portion of the volume is occupied by air. The particle density of the powder carbon is typical, about 500 g/L.

$$Slurry = \frac{55\ mL\ sludge}{L\ slurry} \times \frac{2.2\ g\ solids}{10\ mL\ sludge} = 12.1 = \underline{12\ g/L}$$

$$Slurry\ dosage,\ \frac{V_1}{V_2} = \frac{C_2}{C_1} = \frac{5.0\ g}{m^3} \times \frac{L}{12.1\ g} \times \frac{1000\ m^3}{ML} = 413 = \underline{410\ L/ML}$$

12.3 MEMBRANE FILTRATION

Membrane filtration has become more common recently. This is because firstly, the technology has improved, and secondly, because of its smaller footprint, the capacity of a plant can be increased when space is limited for expansion.

Membrane filtration is a process in which hydrostatic pressure is applied to force the water through semipermeable membranes to filter out contaminants. To avoid the premature plugging and fouling of membranes, feed water must be treated adequately by conventional methods.

12.3.1 Microfiltration and Ultrafiltration

The terms micro and ultra are used to indicate the size of the particles that can be removed. Microfiltration (MF) membranes are generally considered to have an average size of 0.1 μm and are capable of filtering out suspended solids and some bacteria. Ultrafiltration (UF) membranes have pores of smaller size, typically 0.01 μm, and are capable of removing colloidal particles, mostly bacteria, viruses and large molecules. Membranes in these cases are usually made of plastic hollow fibre material. Several such tubes are wrapped together in a fibreglass tube, several of which are assembled together in the form of modules. UF and MF membrane filters are operated at a feed pressure of 700 kPa or less. Membrane systems are normally completely automated to control feed rate and frequency of backwashing.

12.3.2 Nanofiltration and Reverse Osmosis

Nanofiltration (NF) and reverse osmosis (RO) systems are capable of removing small molecules and dissolved solids like those causing colour and hardness. Pore sizes in NF membranes are on average 1 nm, and the RO process uses even smaller pore sizes than this. Semipermeable membranes are used in these processes. Because of smaller pore sizes, operating pressures for reverse osmosis are in the range of 1–5 MPa, depending on the recovery of product water.

RO membranes are more susceptible to plugging, scaling and fouling by dissolved compounds, chemical precipitates and bacterial growths.

12.4 DESALINATION

Water with a high level of dissolved minerals or salts is unfit for domestic, industrial or agricultural uses. Whereas the salt concentration of brackish water is typically 1.0 g/L or 0.1%, sea water has 3–4 times that concentration. To make these waters fit for various uses, salt needs to be removed by a process called desalination. Reducing the salt content to less than 500 mg/L or less makes it potable.

12.4.1 Membrane Technology

Reverse osmosis and electrodialysis are two membrane treatment processes used for desalination, although electrodialysis is limited to treating brackish water. In electrodialysis, a voltage, rather than pressure, is applied across the salty water, causing ions to migrate towards the electrode of opposite charge. Plastic membranes that are selectively permeable to either cations or anions are used to separate fresh water from salty water.

In a reverse osmosis process, a semipermeable membrane separates salty water of two different concentrations. The normal osmotic flow of water through a semipermeable membrane is for the low concentration water to pass through the membrane to dilute the saline water of a higher concentration. The pressure created due to differences in concentrations is called osmotic pressure. Reverse osmosis is the forced passage of water through a semipermeable membrane against the natural osmotic pressure, hence the name. The osmotic pressure of sea water is about 1 MPa. However, brackish water, having a lower concentration of salts, has a significantly lower osmotic pressure. A basic reverse osmosis system consists of pre-treatment units, high-pressure pumps, post-treatment tanks and appurtenances for cleaning and flushing, and a disposal system for the reject brine.

12.4.2 Distillation of Sea water

Multistage flash distillation is the most common process used for desalination. In distillation, fresh water is separated from sea water by heating, evaporation and condensation, much like nature separates fresh water from the ocean to fall as rain. Water starts boiling at a lower temperature if the air pressure is lowered. This fact is used in flash distillation to save energy. This process is carried out in a series of closed vessels, set progressively at lower pressures. When preheated water enters a vessel that is at low pressure, some of the water rapidly boils into vapour. The water vapour is condensed into fresh water in heat-exchange tubes. The remaining salt water flows to the next stage set at an even lower pressure. Some facilities can have as many as 40 stages.

12.5 WATER STABILIZATION

One of the several objectives of water treatment is the production of stabilized water that is neither corrosive nor scale forming. Unstable water can cause problems related to health, aesthetics and economics. Some of the problems due to unstable water are briefly described in the following sections.

Colour

Iron corrosion stains bathroom vanities and white laundry with a familiar rust colour. In homes, copper corrosion results in blue-green stains that show up in sinks or baths or the hair of blond individuals.

Taste and Odour

Corrosion can cause a metallic taste and sometimes a musty odour. The corroded iron acts as a food for iron bacteria, resulting in a serious taste and odour problem.

Economic Costs

Unstabilized water can increase the operating cost of a water distribution system. Build-up of corrosion products (tuberculation) or uncontrolled scale deposits can seriously reduce pipeline capacity and increase resistance to flow. This could lead to the premature replacement of mains water. Scaling can also cause the operation of hot water heaters and boilers to become more expensive due to reduced volume and heating capacity.

Health Problems

Corrosive water can cause the leaching of toxic metals such as lead and copper into a water supply. The iron deposits or tubercles provide shelter for microorganisms. Pressure or velocity changes in the main can dislodge these deposits, thus releasing the organisms into the water.

12.5.1 Classifying Water Stability

The stability of treated water depends on the chemical composition of the water. Basically, water can be classified into the following three categories: scaling (supersaturated), neutral (stabilized) and corrosive (aggressive).

Scaling typically indicates hard water that is supersaturated with calcium carbonate. Other scale-forming compounds include magnesium carbonate, calcium sulphate and magnesium chloride. Scaling tendencies are easily noticed in hot water heaters.

Neutral water is in equilibrium or stabilized. It is neither scale-forming nor corrosive. However, any change in the chemical composition or temperature can make it aggressive or scale-forming. By adjusting the water to slightly scaling, a protective eggshell crust forms on the interior pipe surface.

Corrosive water tends to dissolve pipe material. Deposits of tuberculation in a cast iron system are typical by-products.

It is important to remember that the same water can be scaling at one point of use and aggressive at another point due to changes in physical or chemical characteristics. Mildly corrosive water can become scaling after passing through the hot water heater. In other cases, the water may become corrosive in the home. This can result from high flow velocity or the use of a home water softener or a reverse osmosis unit.

12.5.2 Chemistry of Corrosion

Localized corrosion occurs due to galvanic corrosion. It is called cell corrosion because the corrosion generates an electric current that flows through the metal being corroded. Due to minor impurities and variations in metal, one spot on the pipe starts acting like an anode in relation to another spot that acts like a cathode. At the anode, atoms of iron break away and go into the solution. As each atom breaks away, it ionizes by losing two electrons, which travel to the cathode. Chemical reactions can occur within the water balance, with electrical and chemical reactions at the anode and the cathode.

The formation of $Fe(OH)_2$ leaves an excess of H^+ near the anode, and the formation of H_2 leaves an excess of OH^- near the cathode. This imbalance in the H^+ and OH^- is the reason for the localized corrosion that causes pitting and tuberculation.

$$Anode: Fe \rightarrow Fe^{++} + 2\,e$$

$$Cathode: 2\,H + 2\,e \rightarrow H_2 \uparrow$$

Fe^{++} is further oxidized to Fe^{+++} in the presence of oxygen to form $Fe(OH)_3$ or rust.

$$2\,Fe^{++} + 5\,H_2\,O + 0.5\,O_2 \rightarrow 2\,Fe\,(OH)_3 \downarrow + 4\,H^+$$

The rust precipitates, forming deposits called tubercles. As the reaction indicates, the rate of corrosion is controlled by the concentration of dissolved oxygen. Scale-forming water can protect the pipe from corrosion by depositing a thin scale that bars the flow of the current. However, uncontrolled scale deposits can significantly reduce the flow-carrying capacity due to a reduction in the effective flow area and an increase in flow resistance.

12.5.3 Factors Affecting Corrosion

Corrosion is the result of a combination of physical, chemical and biological factors.

Physical Factors

Physical factors include flow and temperature. Low-flow, dead-end areas typically have higher rates of corrosion. This is why tuberculation problems tend to be worse in dead-end areas. On the other hand, erosion may cause corrosion in areas where water moves at extremely high velocities and at elbows.

Corrosion tends to increase with higher temperatures. This is why higher corrosion rates can result, especially in surface water supplies.

Chemical Factors

- Chemical factors include alkalinity, hardness, conductivity, dissolved oxygen, and the presence of sulphates or chlorides.
- Low alkalinity limits the formation of a calcium carbonate coating and provides little resistance to pH change.
- Low hardness offers little protection in the form of calcium carbonate.
- Conductivity means the ability to transmit electricity. Some types of water are more conductive (transmit electricity more easily) than others. Because corrosion is an electrochemical process, the more conductive the water, the greater the potential for corrosion.
- Dissolved oxygen at levels greater than 3 to 5 mg/L can encourage corrosion.
- High levels of sulphates and chlorides can increase the rate of corrosion. Chlorides are a common problem where there is salt water intrusion. Chlorides form soluble corrosion by-products that can cause pitting problems.

Biological Factors

Iron bacteria and sulphate-reducing bacteria can speed both corrosion and the formation of corrosion by-products. These problems are usually noticed because of slime growths in the system or a musty or stale taste.

Microbiologically influenced corrosion commonly refers to corrosion from iron-related bacteria and sulphate-reducing bacteria. Iron-related bacteria use iron as an energy source; sulphate-reducing bacteria use sulphate.

12.5.4 Stability Index

The corrosive or scaling tendency of water can be determined with a Baylis curve or by calculating the Langelier Saturation Index. The Baylis curve (**Figure 12.1**) shows the relationship between pH values and alkalinity or the solubility of $CaCO_3$. Note how the pH and the alkalinity can be adjusted to make the water non-aggressive. Low alkalinity combined with low pH makes water very corrosive.

Langelier Index

A corrosivity index is a way of determining the tendency of water to form scale or to corrode. The Langelier Saturation Index is the most common and is applicable in the pH range of 6.5 to 9.5. The Langelier Index (LI) is defined as follows:

$$\boxed{Langlier\ index,\ LI = pH - pH_S,\ pH_S = A + B - \log\left(Ca \times Alk\right)}$$

Constant A depends on temperature and constant B depends on total dissolved solids. The values of constants A and B can be read from **Table 12.2** and **Table 12.3**.

A positive (> 1) LI indicates that the water is supersaturated with $CaCO_3$ and will tend to form scale. If the actual pH is less than the pH_S (LI < 1), the water is corrosive. If the pH and pH_S are equal (LI = 0), the water is stable. Alk is the alkalinity and Ca is the calcium hardness, both expressed in mg/L as $CaCO_3$. This calculation is accurate enough up to a pH_S value of 9.3.

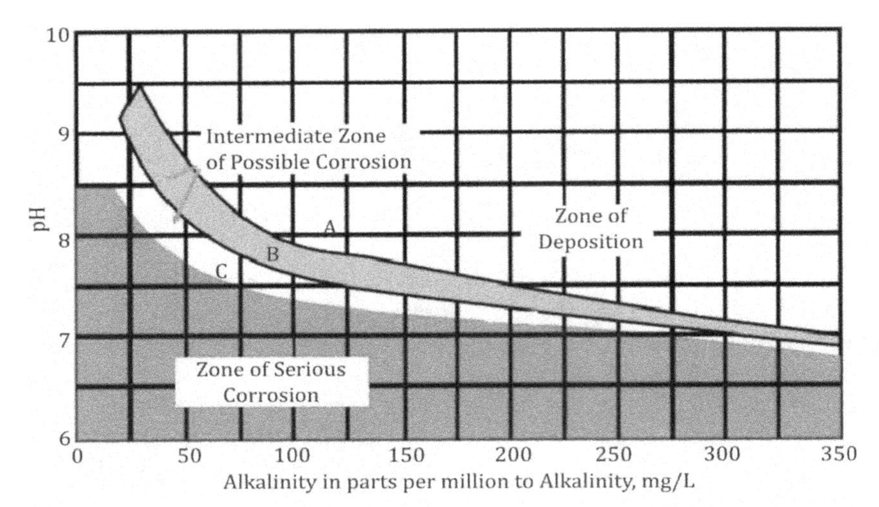

Note :
A = Curve of values necessary to produce a coating of calcium carbonate.
B = Curve of calcium carbonate equilibrium.
C = Curve of values of necessary to prevent iron stains.

FIGURE 12.1 Baylis Curve

TABLE 12.2
Values of Constant A

T°C	0	5	10	15	20	25	30
A	2.34	2.27	2.20	2.12	2.05	1.98	1.91

TABLE 12.3
Values of Constant B

TDS, mg/L	0	50	100	200	400	800	1000
B	9.63	9.72	9.75	9.80	9.86	9.94	10.04

Example Problem 12.3

Calculate the Langelier Index of water at 10°C having a total dissolved solids concentration of 210 mg/L, and an alkalinity and calcium content of 110 and 40 mg/L as $CaCO_3$, respectively.

Given: Ca = 40. mg/L
Alk = 110 mg/L
TDS = 210 mg/L
pH = 8.4 at 10°C

SOLUTION:

Constant A, from **Table 12.2** T = 10°C, A = 2.20
Constant B, from **Table 12.3** TDS = 210 mg/L, B = 9.80

$$pH_S = A + B - \log(Ca \times Alk) = 2.2 + 9.8 - \log(40 \times 110) = 8.35 = 8.4$$

$$LI = pH - pH_S = 8.4 - 8.35 = 0.05 = \underline{0.1}$$

Example Problem 12.4

Calculate the Aggressive Index of water at 10°C having a total dissolved solids (TDS) of 100 mg/L, alkalinity of 80 mg/L and Ca hardness of 40 mg/L as $CaCO_3$. The actual pH at the prevailing temperature is 8.4.

Given: TDS = 100 mg/L
T = 10°C
Alk = 80 mg/L
pH = 8.4
Ca = 40 mg/L as $CaCO_3$

SOLUTION:

$$AI = pH + \log(Ca \times Alk) = 8.4 + \log(40 \times 80) = 11.9 = \underline{12}$$

12.5.5 Measuring Corrosion

Corrosion rates can be measured using corrosion coupons or by testing the water.

Coupons

A coupon is a piece of metal, the same type of which is used in the water system. The coupons are set up in test racks, and the racks are inserted into the pipes or tanks at various points in the distribution system. After a certain period of time, they are removed and tested or examined for types and amounts of corrosion.

While coupons provide a useful way to judge and compare rates of corrosion, they do not measure what is actually happening inside a 45-year-old main that doesn't at all resemble the coupon.

Testing

Analyzing water samples from the distribution system provides an accurate and practical way to monitor actual corrosion rates. By sampling and analyzing the water regularly, its metal content can be monitored. The metal content indicates the rate of corrosion.

When lead leaching is the concern, obtaining first-draw, customer-service and distribution samples is the best way to monitor.

Marble Test

This test is used to determine the degree to which a given water is saturated with calcium carbonate ($CaCO_3$). The test is based on the principle that water in contact with powdered calcite ($CaCO_3$) will approach saturation. During the test, the temperature of the water should not be allowed to rise, and the water being tested should not be exposed to atmospheric carbon dioxide. After saturating the sample with $CaCO_3$, the change in pH is recorded. This change in pH closely represents the Langelier Index (LI).

If this value is less than 0.2, the water is very near saturation and, hence, relatively stable. After recording the final pH, the sample is filtered, and hardness and alkalinity tests are performed on the filtrate (water that passed through the filter). The $CaCO_3$ precipitation potential can be calculated using the following equation:

TABLE 12.4
Precipitation Potential

LI	CaCO3 Precipitation Potential	
> 0 Positive	Supersaturated	Scale forming
< 0 Negative	Unsaturated	Corrosive
= 0 Zero	Equilibrium	Stable

If the change in temperature is more than one degree, the test becomes invalid as the solubility of $CaCO_3$ is very strongly affected by temperature. Both the LI and precipitation potential will have the same sign in a given case.

Example Problem 12.5

Results from a Marble Test performed on a water sample are listed in the following table. Calculate calcium carbonate precipitation potential and LI.

Given:

	Temp. °C	pH	Hardness, mg/L
Initial	11.0	8.50	55
Final	11.4	8.95	58

SOLUTION:

$$LI = pH - pH_S = 8.50 - 8.95 = -0.45 = -0.45$$

$$Precip.\ potential = intial - final = (55 - 58) mg/L$$

$$= -3.0 = \underline{-3\ mg/L}$$

A negative sign indicates that the water is unsaturated with $CaCO_3$.

12.5.6 Corrosion Control

There are two ways to inhibit, or slow down, corrosion: adjusting the pH or adding a corrosion inhibitor. Sometimes, both are used.

Adjusting pH

If the water is corrosive and lead leaching is a problem, consider raising the pH to make the water more likely to form scale. **Table 12.5** lists several chemicals that can be used to raise pH. Using this chart in combination with the Baylis curve or the Langelier Saturation Index, an operator can see how to change the water quality.

While each chemical listed raises the pH, each provides different sources of hydroxide (OH) or carbonate (CO_3) alkalinity. For example, only lime provides calcium, and it is the least expensive chemical on a per kg basis. However, it is also the most troublesome to feed. Liquid caustic soda can be fed easily and safely. However, long-term shortages of caustic soda will make it expensive. Before adjusting the pH, other factors may need to be considered, such as trihalomethanes (THM)

TABLE 12.5
Proper Use of pH Adjustment Chemicals

Chemical	Dosage mg/L	Alkalinity Increase	Equipment
Caustic (NaOH)- 50% solution	1–29	1.25	Chemical feed pump
Soda ash (Na_2CO_3)	1–40	0.94	Feed pump, solution tank
Lime ($Ca(OH)_2$)	1–20	1.35	Lime slaker
Sodium bicarbonate ($NaHCO_3$)	5–30	0.59	Feed pump, solution tank

formation at higher pH levels, and the fact that disinfection by chlorination is most efficient between pH 7.2 and pH 7.5.

Sodium phosphates have been used for film formation, sequestration and dispersion for more than 100 years. They can also be used to stabilize and buffer pH-adjusted water. These phosphates generally work over a wide range of pH and under a variety of conditions.

Products such as sodium hexametaphosphate (SHMP) and sodium tripolyphosphate are two traditional mainstays. Other powders and powdered blends are also available. Recently, many liquid-blended phosphates have shown the best performance and ease of application. For water below a pH of 8.0, zinc phosphates can be added for film formation. Zinc phosphates can be made of zinc orthophosphate or zinc polyphosphate and come in liquid and powder forms. The maximum contaminant levels for heavy metals, such as for zinc in treatment sludge, have limited the use of zinc phosphates in some applications. Silicates, more commonly known as water glass, are known for the glass-like coatings they form in distribution systems. They can be blended with phosphates for corrosion reduction.

Corrosion Inhibitors

Chemical inhibitors may be used alone or as part of a pH adjustment program. Besides forming a film on the inside of distribution piping, some inhibitors also sequester minerals and buffer the water to help stabilize pH changes.

Discussion Questions

1. How does cathodic protection prevent the internal corrosion of a steel water tank?
2. Explain how unprotected iron pipes are corroded. What is the most common method to protect the pitting of ductile iron pipes?
3. Compare tuberculation with scaling.
4. What is water stabilization?
5. Chlorine can be used for taste and odour control. However, if the water contains phenols, chlorination may worsen things. Explain.
6. Briefly describe the common methods used for taste and odour control.
7. Briefly describe the various types of index used to measure the stability of water.
8. Why is the passage of water through a semipermeable membrane for the removal of dissolved salts called reverse osmosis? What other process is used for desalination and how does it work?
9. How does the use of acid prevent scale formation on reverse osmosis membranes?
10. How does cathodic protection prevent the internal corrosion of pipes and tanks?
11. What are corrosion inhibitors? List the chemicals commonly used as corrosion inhibitors.

Practice Problems

1. A one-litre sample of PAC slurry was collected from a slurry-mixing tank and allowed to settle. After six hours of settling, the volume of settled carbon was recorded to be 44.8 mL. An aliquot of 10 mL of settled carbon was dried and found to weigh 2.52 g. Calculate the concentration of carbon in the slurry and find the dosage of slurry to dose water at 4.0 mg/L. (11g/L, 360 L/ML)
2. The volume of a reservoir is estimated to be 28 ML to prevent algae growth. The desired dose of copper is 0.65 mg/L. How many kg of copper sulphate are needed to treat reservoir water? (72 kg)
3. Calculate the Langelier Index of water at 10°C with a TDS of 100 mg/L, an alkalinity of 80 mg/L and a calcium hardness of 40 mg/L as $CaCO_3$ and pH of 8.40. Determine the corresponding value of the Aggressive Index. (-0.04, 12)
4. A water sample has the following characteristics: Alkalinity = 80 mg/L as $CaCO_3$, Ca = 72 mg/L, TDS = 400 mg/L, T = 12°C, pH = 7.80. Calculate the Langelier Saturation Index. (-0.14)
5. Some finished water has the following characteristics: TDS = 300 mg/L, T = 10°C, pH = 7.5 Alk = 60 mg/L, Ca = 30 mg/L as $CaCO_3$.
   a) Based on LI, determine whether the water is corrosive or not? (Yes)
   b) Express the water stability as Aggressive Index. (10.8)

13 Water Distribution

Water distribution is the second most important step in potable water supply systems. In many cases, a poorly operated and maintained water distribution is responsible for degradation in water quality. It may be bacterial contamination due to lack of chlorine residual, or taste and odour problems caused by poor circulation of water in the distribution system. Under severe water demands, water pressures can fall below atmospheric, causing backflow that can lead to contamination or collapsing of the water mains.

The operating personnel of water distribution systems must ensure that their clients have access to clean, safe water while maintaining sufficient pressure and volume to meet the users' demands. It should be capable of meeting water requirements during emergencies, including fighting fires. The design and operation of the system must provide safeguards against backflow contamination.

As a rule of thumb, water pressure in a distribution system should not drop below 250–300 kPa in order to provide adequate pressure. However, in developing countries, water pressures are usually less than 200 kPa. Maximum pressures are usually kept below 550 kPa and are not allowed to exceed 700 kPa, to reduce the chances of leaks and water breaks.

A well-designed system must put in place devices that protect the system against backflow. All distribution operating personnel should always be aware of the consequences of cross-connections between a water distribution system and an unapproved source of water. Cross-connections have led to serious public health problems in many cities.

In summary, a properly operated and maintained water distribution system must be able to: meet normal water demands and demands during emergencies (e.g. firefighting); provide positive pressure of more than 250 kPa at all times except during emergencies and firefighting (minimum pressure of 150 kPa); maintain water quality; ensure infrequent interruptions; and minimize the costs of maintenance and operation.

13.1 SYSTEM COMPONENTS

Water distribution systems consist of water mains, pumping stations, storage reservoirs, valves, hydrants, meters and other ancillary devices. People responsible for the operation have to be aware of the strengths and weaknesses of their system so as to always be on guard to maintain the safety of the water supply. The operation of a distribution system includes: the operation of pump stations; the location and repair of water main breaks; the flushing and cleaning of distribution lines; the chlorination of water distribution lines, reservoirs or storage containers; and the thawing of distribution lines during winter.

As discussed above, distribution systems consist of piping, pump stations, and storage facilities such as reservoirs, elevated tanks and standpipes. In addition, there are numerous appurtenances, such as valves, fire hydrants, water meters and service connections. It is essential that distribution systems are operated to minimize supply interruptions, protect the water from contamination and maintain a positive pressure in the system at all times. A lack of positive pressure can cause contaminated groundwater to enter pipes through cracks, and negative pressure may draw contaminated water from an unapproved source of water. This phenomenon is known as backflow due to back siphoning.

Water distribution systems must also be prepared to operate under unusual conditions. These can include excessive demands caused by firefighting or broken or frozen water mains. Pipe sizes are based on peak hourly demand or fire flow demands, whichever is largest. In addition, pump

DOI: 10.1201/9781003231264-13

failures can create serious problems. Plans need to be prepared and taught to all staff before an emergency occurs.

13.2 EQUALIZING DEMAND

Water distribution systems are designed to meet peak flows and firefighting needs while maintaining a uniform pumping and treatment rate equal to the maximum daily demand. This is accomplished by providing equalizing or balancing storage to meet peak hourly demand. Generally, the volume of water needed to equalize the peak demand is about 20% of the average daily demand. The location of the reservoir is selected based on the following considerations:

- Where the areas of high demand are most likely to occur (load centre).
- The comparative cost of pumping and pipelines.
- The equalizing pressure in the water system, which is determined by the topographic features of the area served.
- The hydraulic grade line developed by the reservoir and associated pumping facilities.
- If the topography permits, storage is located at a central location and at high elevation.

13.2.1 EQUALIZING STORAGE CAPACITY

An extremely important element in a water distribution system is water storage. System storage has a far-reaching effect on a water system's ability to provide adequate and reliable water supplies for domestic and commercial use, and especially firefighting. The main function of storage is to equalize demand and storage during emergencies and for fighting fires. However, its primary function is to balance distribution storage to meet fluctuating demand with a constant rate of water supply from the treatment plant.

Balancing storage can be worked out by knowing the maximum cumulative deficit (supply minus demand) either by using the mass curve or an analytical method. The latter is gaining more popularity as it is easier to set up a table using a spreadsheet computer program like Microsoft Excel. Using a tabular method, hourly rate supply and demand are entered, and cumulative values are calculated. A column indicating the difference in the cumulative supply and cumulative demand is then created. A negative value would indicate excess demand and a positive value excess supply. The total equalizing storage capacity is then worked out by adding the maximum excess demand and maximum excess supply. This procedure is further illustrated by the example problems shown below.

Example Problem 13.1

A medium-size city has variable demand, as shown in **Table 13.1**. Determine the storage capacity required to equalize the demand against a constant rate of pumping at 24 ML/d and an intermittent rate of pumping from 5 am to 11 am, and from 2 pm to 8 pm at 48 ML/d.

SOLUTION:

As seen in **Table 13.1**, the rate of demand in column 3 is converted to volume in column 4, and the cumulative value is shown in column 5. The same is done for the supply, and the difference in cumulative supply and cumulative demand is shown in column 9. Storage capacity is calculated by adding the maximum excess supply and maximum excess demand.

For a constant supply, the computations are shown in **Table 13.1**

TABLE 13.1
Excel Worksheet (Constant Supply)

1	3	4	5	6	7	8	9	10	11
hour	Deman, Q_o			Supply, Q_i			Q_i–Q_o		
	Rate ML/d	Volume m^3	Σ m^3	Rate ML/d	Volume m^3	Σ m^3	Volume m^3	Excess m^3	
1	12.0	499	120	26	1067	1000	1067	1067	Supply
2	11.4	476	596	26	1067	2067	1471	1471	Supply
3	9.8	408	1004	26	1067	3133	2129	2129	Supply
4	7.6	317	1321	26	1067	4200	2879	2879	Supply
5	7.1	295	1616	26	1067	5267	3651	3651	Supply
6	6.5	272	1888	26	1067	6333	4445	4445	Supply
7	10.9	453	2341	26	1067	7400	5059	5059	Supply
8	19.0	793	3135	26	1067	8467	5332	5332	Supply
9	27.2	1133	4268	26	1067	9533	5265	5265	Supply
10	32.6	1360	5628	26	1067	10600	4972	4972	Supply
11	34.8	1451	7079	26	1067	116670	4588	4588	Supply
12	38.1	1587	8665	26	1067	12733	4068	4068	Supply
13	35.9	1496	10161	26	1067	13800	3639	3639	Supply
14	34.8	1451	11612	26	1067	14867	3255	3255	Supply
15	34.3	1428	13040	26	1067	15933	2893	2893	Supply
16	34.8	1451	14491	26	1067	17000	2509	2509	Supply
17	34.8	1451	15941	26	1067	18067	2125	2125	Supply
18	36.4	1519	17460	26	1067	19133	1673	1673	Supply
19	40.3	1677	19137	26	1067	20200	1063	1063	Supply
20	50.0	2085	21223	26	1067	21267	44	44	Supply
21	45.7	1904	23127	26	1067	22333	–793	793	Demand
22	27.2	1133	24260	26	1067	23400	–860	860	Demand
23	17.4	725	24985	26	1067	24467	–519	519	Demand
24	15.2	635	25620	26	1067	25533	–87	87	Demand
Σ	624	25999		614					

Max excess supply = 5332 m^3
Max excess demand = 860 m^3
Reqd. storage capacity = 6192 m^3

As can be read from **Table 13.1**, the maximum excess supply (8th hour) is 5332 m^3 and the maximum excess demand is 860 m^3. The required storage capacity is the sum of these two.

$$\textit{Storage capacity} = 5332 + 860 = 6192 = 6200\ m^3$$

For the intermittent supply, the computations are shown in **Table 13.2**.

It is worth noting that the storage capacity required in the second case is less since the supply rate is twice as much and most of the demand is met by the supply. Hence, the required balancing storage is relatively smaller. However, the supply rate, though needed for only half the duration, is double. Thus, unit devices and pipe sizes need to be larger to produce water at twice the rate. The total storage capacity must include the volume of water required for firefighting.

The plot of demand and supply for both cases is shown in **Figure 13.1**.

TABLE 13.2
Excel Worksheet (Intermittent Supply)

1	3	4	5	6	7	8	9	10	11
hour	Deman, Q_o			Supply, Q_i			Q_i–Q_o		
	Rate ML/d	Volume m^3	Σ m^3	Rate ML/d	Volume m^3	Σ m^3	Volume m^3	Excess m^3	
1	12.0	499	120		0	0	−120	120	Demand
2	11.4	476	596		0	0	−596	596	Demand
3	9.8	408	1004		0	0	−1004	1004	Demand
4	7.6	317	1321		0	0	−1321	1321	Demand
5	7.1	295	1616		0	0	−1616	1616	Demand
6	6.5	272	1888	52	2167	2167	279	279	Supply
7	10.9	453	2341	52	2167	4333	1992	1992	Supply
8	19.0	793	3135	52	2167	6500	3365	3365	Supply
9	27.2	1133	4268	52	2167	8667	4399	4399	Supply
10	32.6	1360	5628	52	2167	10833	5205	5205	Supply
11	34.8	1451	7079	52	2167	13000	5921	5921	Supply
12	38.1	1587	8665		0	13000	4335	4335	Supply
13	35.9	1496	10161		0	13000	2839	2839	Supply
14	34.8	1451	11612		0	13000	1388	1388	Supply
15	34.3	1428	13040	52	2167	15167	2127	2127	Supply
16	34.8	1451	14491	52	2167	17333	2843	2843	Supply
17	34.8	1451	15941	52	2167	19500	3559	3559	Supply
18	36.4	1519	17460	52	2167	21667	4207	4208	Supply
19	40.3	1677	19137	52	2167	23833	4696	4696	Supply
20	50.0	2085	21223	52	2167	26000	4777	4777	Supply
21	45.7	1904	23127		0	26000	2873	2873	Supply
22	27.2	1133	24260		0	26000	1740	1740	Supply
23	17.4	725	24985		0	26000	1015	1015	Supply
24	15.2	635	25620		0	26000	380	380	Supply
Σ									

Max supply = 5921 m^3
Max demand = 1616 m^3
Storage capacity = 7537 m^3

13.2.2 Other Purposes of Storage

Distribution storage tanks and reservoirs are used to provide the community with emergency storage capabilities. They also allow the treatment facility to produce excess water during slow periods for use during periods of high demand. In this way, the water treatment facility can be designed for the average demand rather than the peak demand. Storage facilities improve system flows and pressures.

Another advantage of storage is that it provides a location for blending water from various sources and for extended chlorination. Pumps operate at lower and uniform pumping rates, resulting in cost savings.

13.2.3 Types of Storage

Common types of storage facilities include ground storage, underground reservoirs, elevated tanks and standpipes. An underground storage reservoir built as part of a water treatment plant is called a

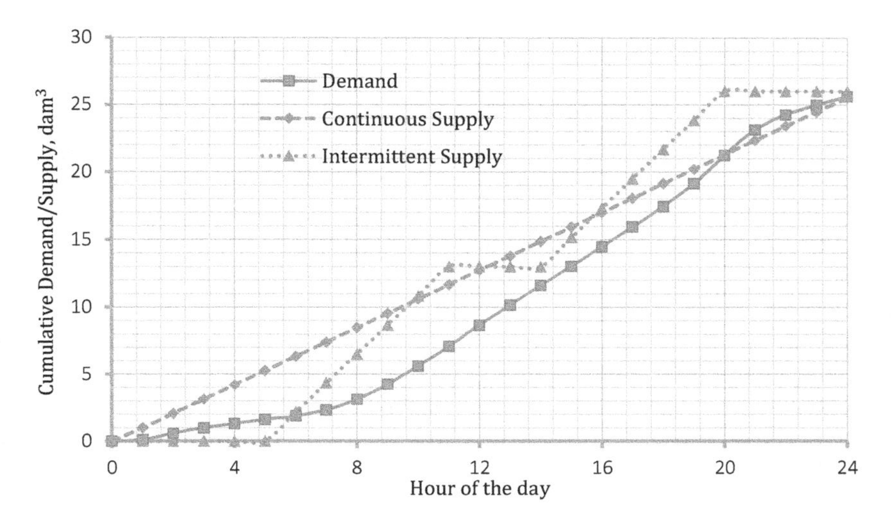

FIGURE 13.1 Mass Curve (Ex. Prob. 13.1)

clear well. An elevated storage tank's main purpose is to maintain system pressure. Level switches control flow in and out of the tank. Hydropneumatic tanks are used for small-scale and individual household water systems. Storage reservoirs need to be protected against corrosion, contamination and vandalism. To maintain the quality of water, about two-thirds of the storage's volume should be renewed every day. Procedures for operating reservoirs depend on the type of facility and the magnitudes of system demands. However, replenishing the reservoir after the peak demand period and maintaining system pressure should be the main consideration in any operation scheme.

Ground-Level Storage

Since water is stored at ground level, it must be pumped to the point of use. This limits water system effectiveness for fire protection in three ways:

1. There must be excess pumping capacity to deliver the peak demand for normal use as well as any firefighting demands.
2. Standby power sources and pumping units must be maintained at all times.
3. Water mains in the system must be oversized to handle peak demand plus fire flow, no matter where a fire occurs.

Elevated Storage

Properly sized elevated water tanks provide dedicated fire storage and are used to maintain constant system pressure. Where elevated tanks are used, ground storage tanks may still be required depending on the type of system design. However, the size of this storage would be much less. On a daily basis, water from the top 3–5 m is used for supply and the remaining 70–75% is held as reserve. When domestic consumption is minimum during the day, water in the elevated tank is recycled with fresh water to eliminate the ageing of water in the holding tank.

Standpipe

Standpipes are normally constructed using steel or concrete. They are cylindrical in shape with heights exceeding their diameters. Standpipes provide more storage than elevated tanks. But this is only useful for equalizing purposes if the water surface is above the elevation required for maintain minimum pressure. The water stored below the minimum elevation can be used for firefighting with pumper trucks or during other emergencies.

Location of Storage Tanks

Using several small storages tanks or water wells near the major centres of water withdrawal is preferred to using one large tank near the pumping station. It is best to locate tanks on the opposite side of the demand centre from the pumping station. This allows for more uniform pressures throughout the system and smaller diameter mains and pumps.

Pumping Stations

In most distribution systems, the topography of the land requires the use of pumping stations to lift the water to a higher elevation and to maintain a positive pressure within all sections of the system. Several types of pumping arrangements are common. Stations may elect to use a pump large enough to handle the maximum flow rate, although this tends to be inefficient during low demand periods. Stations may also use a variable speed motor to decrease the pump's capacity during low demand periods. Most systems have an elevated storage tank or reservoir on high ground floating on the system. For larger municipalities, the pumping station generally consists of two or more pumps of different sizes to match the required rate of flow. Backup pumps are usually available in case of failure.

13.3 PIPELINE LAYOUT

The main distribution pipes within a system are known as the trunk mains. Smaller diameter pipes branch off and distribute water to individual streets. Water mains are generally not less than 150 mm in diameter. There are, in general, four different types of networks commonly used, either singly or in combination. A brief description is given in the following paragraphs.

13.3.1 Dead-End Systems

Also called a tree system, water from the source goes to the main, which is further divided into many submains. Each submain then divides into several branch pipes called laterals. Service connections are provided from laterals to serve users. This type of system may be adopted for older towns and cities where development took place haphazardly. The main advantages are:

- Modelling of the system is easy.
- A low number of shut-off valves are required.
- Shorter pipe lengths are needed.
- It is easy to expand or extend the system.

One of the serious drawbacks of this system is the presence of dead ends that prevent recirculation, thus making water stale and vulnerable to contamination. The other disadvantages of the system include:

- Since there is only one path, larger parts of the system need to be shut down during repairs, thus causing inconvenience to consumers and risking failing to attend to any emergency during this period.
- To maintain the quality of water, dead-end hydrants need to be flushed frequently, resulting in wastage of treated water.
- Since there is only one route for water to reach a given point, it is not possible to increase the water supply to meet higher demand during emergencies.

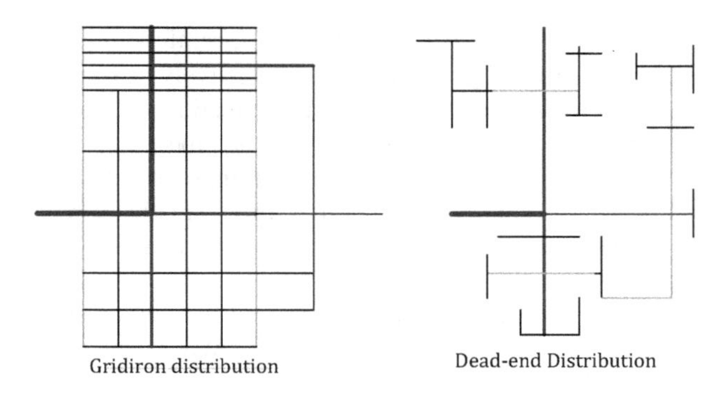

FIGURE 13.2 Pipe Networks

13.3.2 Grid-Iron System

A grid-iron arrangement of pipes is preferred to a layout that has many dead-end branches. Dead ends are extremely susceptible to taste and odour problems and require frequent flushing. In a grid-iron pattern, valves are used to isolate the broken section of the pipes, and water can still reach consumers from the other side of the loop. The advantages of a grid-iron system are as follows:

- Since there is more than one route for water, the flow carried by each pipe is relatively small, thus the size and friction losses are small too.
- In case of repairs, it is easy to shut down or isolate part of an area without affecting the supply to other users.
- Dead ends are eliminated and water remains in circulation.
- During emergencies such as fires, more water can be diverted to the demand area.

However, such a system requires more hardware and is thus expensive to construct. The design of such a system is complex, and the use of computer models is almost a necessity.

13.3.3 Ring System

This is also called a circular system. In a ring system, a ring of pipeline runs around the served area. The distribution area is divided into rectangular or circular blocks. This system is well suited to well-planned and laid out towns and cities.

13.3.4 Radial System

This system is recommended for towns with radial roads emerging from different centres. Distribution reservoirs are placed at such centres, and water is then supplied through radially laid distribution pipes. This method ensures high pressures and different water distribution. The design of such a system is relatively simple.

13.4 PIPE MATERIAL

Water pipes in the distribution system flow under high pressure. Pressure pipes are available in pressure rating from 100 psi (700 kPa) to 250 psi (1750 kPa). Class 150 (pressure rating) pipe is the most widely used pressure pipe. AWWA C900 is the most commonly used type of pipe, which adapts easily to existing piping systems. For larger diameter pipes, reinforced concrete pipes are usually used. Ductile iron is extremely strong, lightly flexible and relatively heavy, and needs corrosion protection. Steel pipes are generally used for high-pressure applications. However, corrosion protection must be provided for the interior as well as the exterior of the pipe.

Pipes used in water distribution systems must have the strength to withstand internal and external pressures and abnormal pressures due to water hammer. Pipes should be resistant to corrosion and must not be detrimental to the potability of water. Of lesser consideration is the ease of handling and jointing during installation.

The pipe materials most commonly used in water supply and distribution include: polyvinyl chloride (PVC), high-density polyethylene (HDPE), asbestos cement (AC), ductile iron (DIP) and reinforced concrete pressure pipe (RCP). Plastic pipes are inexpensive, flexible, lightweight, corrosion-free, easily installed and low in friction. The smoothness of a pipe, as indicated by the friction factor, increases the flow-carrying capacity or yields less pressure loss while carrying the same flow.

13.4.1 Plastic

This is currently the most common type of pipe used in distribution systems. It is available in lengths of 3 m, 6 m and 12 m. One main advantage is that it is light, allowing for easy installation. One disadvantage is its ability to withstand shock loads. The National Sanitation Foundation (NSF) in the US currently lists most brands of PVC pipe as being acceptable for potable use. Plastic pipe has several advantages over metallic pipes, including being resistant to rupture from freezing and corrosion-proof, and it can be installed above or below ground.

Plastic pipes are made of different strengths. Sometimes, polyvinyl chloride is further chlorinated to obtain greater strength, a higher level of impact resistance and a greater resistance to extreme temperature. Chlorinated blend pipes are designated as CPVC. High-density polyethylene (HDPE) pipes are flexible and suitable for household fittings. They are used for minor domestic works. Polythene and PVC pipes can withstand pressures of up to 1000 kPa.

13.4.2 Cast Iron

Cast iron (CI) pipes have been in use for a long time, although they are less common these days, especially in developed countries. However, because of their long life, CI pipes are found in most cities. CI pipes are sufficiently resistant to corrosion, durable, reasonably cheap and relatively easy to join. They are available in nominal pipe sizes (NPS) of 80 mm to 1200 mm and can withstand pressures of up to 2400 kPa. CI pipes lose their flow capacity over time due to tuberculation and other deposits. They are heavy and uneconomical in larger sections. In addition, they are brittle and likely to break during transportation or handling. Although they are not currently the material of choice, there is still a lot of it in the ground.

13.4.3 Ductile iron

Ductile iron (DI) pipes were developed to overcome the breakage problems in cast iron pipes. The main advantage of DI pipes is that they can withstand high pressure both internally and externally. They can be protected from highly corrosive soils by wrapping them in plastic sheeting.

13.4.4 Steel

Steel pipe is used mostly for transmission mains where internal pressures are usually low. They have a smooth interior and can withstand pressures of up to 1700 kPa. Steel pipes need to be protected from internal and external corrosion. To protect the pipe, internal coal tar epoxy linings and cement linings are used; for outside protection, polyethylene linings are used.

Steel pipes are either welded or rivetted steel sections. Welded pipes are smoother and stronger than rivetted ones and hence generally preferred. As steel is strong in tension, even pipes with diameters of up to 6 m can be fabricated out of thin steel shells and they can resist very high internal pressures.

Galvanized iron steel pipes with corrugations on their circumferences are stronger than plain steel pipes. They are manufactured in diameters ranging from 200 mm to 2000 mm. They are comparatively light and can be transported to large distances.

If they are to be placed above the ground, expansion joints are necessary to counteract temperature stress. Steel pipes are liable to quick rusting. They last for around 40 years. Steel pipes cannot withstand high negative pressures or vacuums created inside them, or the combined effect of an internal vacuum and external stresses due to backfill and traffic loads above. Hence, plain steel pipes are not used much in India.

13.4.5 Cement Concrete and RCC

Plain cement concrete pipes are manufactured up to a maximum diameter of 600 mm. For greater diameters, reinforced cement concrete (RCC) is used. The diameters of RCC pipes may vary from 1.5 m to 4.5 m. They may be precast or cast in situ. Plain concrete pipes are used for pressures of up to 150 kPa, whereas RCC pipes are used for pressures of up to 750 kPa. For greater pressures, prestressed concrete pipes are used.

As per IS 458-1971, there are three categories of concrete pipes: P1, P2 and P3. P1 is available in diameters from 80 mm to 1200 mm with a test pressure of 200 kPa; P2 in diameters from 80 mm to 400 mm with a test pressure of 400 kPa; and P3 in diameters from 80 mm to 400 mm with a test pressure of 600 kPa.

Concrete pipes resist corrosion inside and outside and can resist external compressive loads. They do not fail under traffic loads and nominal vacuum and are durable, with a lifespan of about 75 years. Having a low coefficient of expansion, when placed over ground, no expansion joints are required. If groundwater contains acids, alkalis or sulphur compounds, RCC pipes may get affected by corrosion. Repairs are difficult and connections are not easy. Shrinkage cracks and porous texture may result in leakage.

13.4.6 Asbestos Cement

Asbestos cement (AC) is a homogenous material prepared out of three ingredients: asbestos, silica and cement. It has high strength and is quite dense and impervious. Asbestos cement pipes are available in diameters typically ranging from 100 mm to 900 mm and length 4 m. Four grades of asbestos cement pipes are manufactured to suit pressures varying from 350 kPa to 1400 kPa. To join these pipes, a special coupling called a simplex joint is used.

AC pipes are light, and so transportation and handling are easy. They are very resistant to corrosion and hydraulically efficient due to their smooth surface. They are suitable for use as small distribution pipes.

AC pipes are not very strong and so are liable to be damaged during transportation and handling, and from overlying loads or lateral earth pressures. They are quite costly. Asbestos fibre is dangerous to health, therefore safety precautions must be taken when handling and joining these pipes.

13.5 PIPE JOINTS

Pipes are manufactured in standard sizes and various types of joints are used to connect them. The most common joints in use are: flanged, socket and spigot, flexible, mechanical, expansion and simplex.

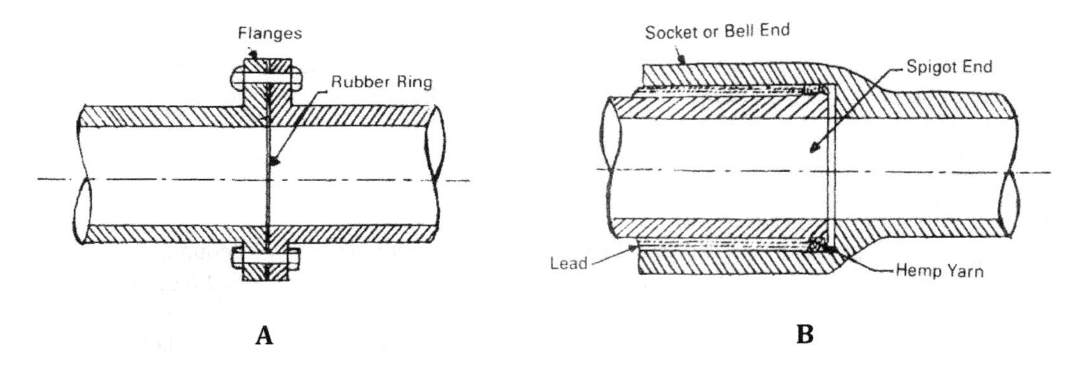

FIGURE 13.3 Flanged Joint and Socket Joint

13.5.1 Flanged Joint

These are adopted in pumping stations, filter plants, etc., where occasional disjointing is necessary. Two flanges are brought together with a rubber gasket between them to create water tightness. They are fixed together with nuts and bolts (**Figure 13.3A**). These joints are rigid, strong and expensive. They cannot be used where there are vibrations and deflections. They are quite costly.

13.5.2 Socket and Spigot Joint

These are also called bell and spigot joints and are usually used for cast iron pipes. The joining ends of the pipes are made into the shapes of wider bells or sockets in such a way that the socket and spigot fit properly (**Figure 13.4B**). The gap between the socket and spigot is filled with jute and molten lead to make it watertight. This joint is flexible and can be used for pipes laid on flat curves.

Bell and spigot joint systems are more common in plastic, ductile iron and reinforced concrete pipes. The pipes should be installed so that the bell is in the direction of the pipe laying. Steel pipes are joined using mechanical joints or dresser coupling. Bedding is an important consideration as plastic pipes have low resistance to crushing. After laying the pipe, it should be properly backfilled, tested, flushed and disinfected before being put in operation.

13.5.3 Flexible Joint

Flexible joints are used where pipes are to be laid in rivers with uneven beds and large-scale settlements are likely to occur, or pipes are to be laid on curves. Under these circumstances, the joints are likely to break if there is not sufficient flexibility.

The socket is made spherical. The spigot is provided with a bead at the end. The special rubber gasket is held in position by a retainer ring. On that is placed a split cast iron "gland ring". They are tightened by bolts and nuts. The spigot end can be moved to tackle the desired deflection and the nuts are tightened over the gland ring.

13.5.4 Mechanical Joint

This is also called a dresser coupling. This joint is adopted when it is necessary to join the plain ends of cast iron pipes. A metallic collar is fitted and tightened over the abutting ends of the pipes, forming a mechanical joint. The usual type adopted is called a dresser coupling.

Another type of mechanical joint is the Victaulic coupling. In this joint, an iron ring and a gasket are slipped over the abutting ends of the pipes. An iron sleeve is introduced between the gaskets.

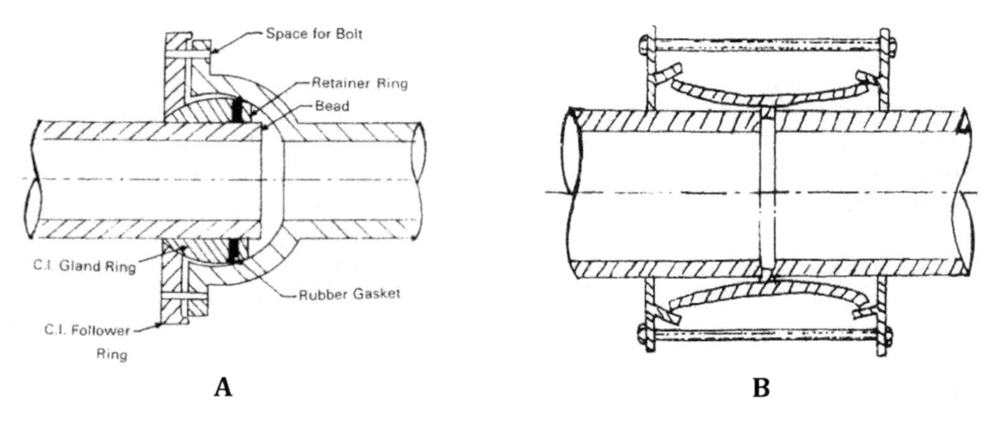

FIGURE 13.4 Flexible Joint and Mechanical Joint

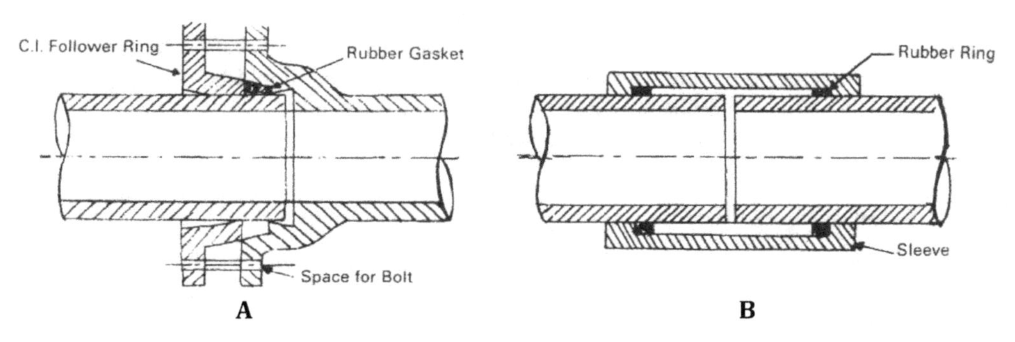

FIGURE 13.5 Expansion Joint and Simplex Joint

These iron rings are tightened with bolts. Being rigid and strong, these joints can withstand heavy vibrations. They are used for pipes carried over or below bridges.

13.5.5 Expansion Joint

Expansion joints are necessary to neutralize the effects of temperature stresses. They are provided at suitable intervals in pipelines. The socket end is cast to have a flange and the spigot end remains plain. The socket end is fixed rigidly to an annular ring that can slide freely over the spigot end. A small gap is allowed between the face of the spigot and the inner face of the socket. A rubber gasket fills up the annular space between the socket and the spigot. The flanges are then tightened with bolts and nuts. When the pipes expand, the socket moves forward and the gap gets closed. Similarly, when the pipes contract, the socket contracts, leaving a gap. Meanwhile, the annular ring follows the movements of the socket and keeps the gasket in position, thus keeping the joint watertight.

13.5.6 Simplex Joint

This joint is used to connect two asbestos cement pipes. A sleeve is provided over the abutting ends that it fits over them both. Two rubber rings are compressed between the sleeve and the pipe ends.

13.6 PIPE-LAYING AND TESTING

The various operations to be done for the laying of water mains are discussed briefly.

1. After a detailed and accurate survey, road maps must be prepared. These must indicate the positions of curbs, sewers, water pipes, electrical and telephone conduits, etc.
2. To mark the trench to be excavated, a central line is marked by driving spikes at 30 m intervals on straight portions and 7.5 m to 15 m intervals on curves.
3. The depth and width of the excavated trenches should be adequate to permit the laying and joining of pipes. The width must be about 40 cm more than the outer diameter of the pipe, and there should be a clear ground cover of about 90 cm above the top of the pipe barrel.
4. The bottom of the trench should be carefully prepared so that the barrel can be bedded true to the line and gradient over its entire length on a firm surface. If the excavation is in rock, a 15 cm bed is provided on the bottom of the trench so that adequate protection is given against any possible settlement. For jointing the pipes, joint holes are provided. Pipes must be introduced into the trench carefully so that they do not undergo any damage.
5. Pipes should be laid in an "uphill" direction so that joint making is easy. They should not be laid with a flat slope parallel to the hydraulic gradient. They should be placed either with a continuous rise to high points or a continuous fall to low points.
6. For jointing pipes, suitable joints are adopted depending upon the type of pipe.

13.6.1 Anchoring of Pipes

Thrust refers to the force exerted due to changes in flow direction as dictated by the law of mechanics. Thus, at all positions where there are bends, toes, valves, branch connections, etc., there is hydraulic thrust. At such positions, to transmit this thrust over a wider area of soil, "thrust blocks" of concrete are provided. In the case of pipes laid on sloping grounds, there will be upward hydraulic pressures. In such cases, "anchor blocks" of concrete are provided at regular intervals, pipes being rigidly secured to them with steel straps.

Thrust blocks must be placed on the opposite side of the thrust exerted and must bear against undisturbed soil, since disturbed soil is subject to compression upon loading. The size of the thrust block depends on the water pressure in the pipe, the type of fitting and the bearing capacity of the soil.

13.6.2 Backfilling with Earth

This is the operation of filling up the trench with excavated stuff. The material used around the pipe should be soft and should not contain any rock pieces, lumps, stone, etc. The filling may be done in layers, each 15 cm to 30 cm thick and well consolidated so that no pipe movement is possible inside the trench.

The remaining upper portion of the trench may be filled with excavated debris and the top of the trench brought to a level so that after some time, due to consolidation by traffic moving above, the level of the top of the trench will become flush with the surrounding road.

13.6.3 Testing of Pipes

After the pipes are laid and joined, before backfilling, the pipes must be tested under pressure. The pipe is filled with water so that all the air inside is drawn out. It is kept in this condition for some time and a test pressure of about 500 kPa or 50% of the maximum working pressure, whichever is greater, is applied. The pressure may be applied by a manually operated test pump or a power-driven test pump in the case of large pipes or longer lengths of mains.

After the test pump registers the test pressure it is stopped, and any fall in pressure is recorded. The pipe is considered to be okay if the pipe maintains the test pressure without any loss for at least 30 minutes. This test is carried out in sections as the laying of pipe proceeds. In other cases, the test pressure is much greater, typically 1000 kPa, and the volume of water needed to be pumped into the water main to maintain pressure for one hour is recorded. The volume of water used per joint is worked out and compared with the allowable water loss.

13.6.4 Flow Velocity

Most pipe sizes are sized to carry both fire flows and maximum day demands. Flow velocity usually ranges from 0.5 to 1.5 m/s. The principles of hydraulics dictate that head losses increase in much greater proportions compared to the increase in flow velocity. A very large flow will cause excessive head losses and hence a significant pressure drop. For this reason, if there is a break in the main or too many hydrants are open, pressure at critical points may fall below atmospheric. This can result in the contamination of water due to back siphoning. Very low velocities are also not desirable since there is greater risk of contamination and biological growth due to stagnation.

13.7 VALVES

Valves are used within a distribution system to isolate portions of the system for repair, cleaning, maintenance or adding additional lines. Valves may also be used to regulate the flow or pressure of water within a pipe.

Different types of valves may be used for different applications. Most valves are of the gate type, which allows the full passage of water without restricting flow. The main purpose of gate valves is isolation and should never be used for throttling. Gate valves are either completely open or closed and should never be operated in a partially open position. Distribution valves generally suffer from lack of use rather than wear. It is recommended practice to exercise the valves on a scheduled basis. Based on the movement of the closure element, valves can be categorized as linear or rotary.

13.7.1 Linear Valves

The closure element in this kind of valve has linear movement when closed or opened. Each turn of the valve handle will move the valve in its seat up or down in pitch. Thus, the stem of the valve might be turned several times to a fully open or closed position. A motorized valve operator is typically installed in large valves. Valves falling into this category include globe, gate and pressure and flow control valves.

Unlike gate valves, globe valves allow for throttling and pressure regulation. Globe valves are used exclusively on pipes with diameters of 100 mm or smaller. In a diaphragm valve, a flexible piece inside the valve's body can be moved up or down to adjust the opening size. The diaphragm is made of flexible material such as rubber or leather.

Air and vacuum relief valves consist of a float-operated valve that allows the opening of an orifice to break the vacuum or allow air to escape to the atmosphere when draining or filling a water main. These valves serve three functions: they allow air to escape during filling; they allow air to enter the pipeline when draining; and they allow entrained air to escape when a line is operating under pressure. Unprotected vacuum conditions can result in the pipe collapsing. High points in water systems are more susceptible to air entrapment. Air bubbles in flow streams can cause significant resistance to flow. Reduced buoyancy due to entrapped air causes the float to open smaller orifices for bubbles to escape.

13.7.2 Rotary Valves

As the name indicates, a rotary valve in a pipe rotates. Usually, this kind of valve can be opened and closed by a one-quarter turn. Thus, they are also known as quarter-turn valves. The most common valves in this category are butterfly, plug and ball.

Butterfly valves have a movable disk large enough to fit in the inside of the pipe completely. The disk rotates on a spindle or shaft in only one direction to either close or open position. Butterfly valves can be used for throttling as well. However, as the body of the valve is in the pipe, the valve needs to be removed when swabbing the pipe.

Plug valves may have a tapered or cylindrical plug with an opening to the side that can be turned to open, restrict or close the flow. In smaller sizes, these valves are commonly used as corporation stops on service lines.

Ball valves are similar in design to the plug valve except the plug is ball-shaped with a cylindrical hole bored through it. When the valve is in a fully open position, the borehole is in parallel to the direction of flow. When rotated through 90 degrees from the fully open position, the opening moves opposite to the direction of flow, thus completely stopping the flow. Such valves are not suitable for throttling.

13.7.3 Special Function Valves

Some valves are designed to perform a special function. The valve used to allow flow in the flow direction only is called a check valve. There are many designs of check valves, including swing check, slanting disk, double-door and foot valve.

A foot valve is put on the end of the suction pipe of a pump to maintain priming when the pump stops. The discharge side of valves is equipped with check valves to prevent backflow when the pump is turned off.

Altitude valves are used in reservoirs to prevent overflow. As the water reaches a set altitude or level in the reservoir, the pressure generated activates the valve to close. This valve is also used in tanks where the full system pressure might cause them to overflow. Altitude valves help maintain a constant water level in a tank or reservoir as long as the pressure in the distribution system is adequate.

Pressure sustaining valves (PSV) are usually globe valves that are spring-loaded to adjust the opening to maintain a set pressure on the downstream side regardless of the fluctuating demand downstream. As the pressure downstream of the valve increases, it compresses the spring and makes the element move to close the opening further and decrease the flow and pressure. When the downstream pressure decreases, the spring will open and allow more flow.

Pressure regulating valves (PRV) are usually globe valves that control pressure and operate by restricting flow. The pressure downstream of the valve regulates the flow. They are used to deliver water from a high-pressure zone to a low-pressure zone and are spring-loaded to adjust the opening to maintain a set pressure on the downstream side regardless of the fluctuating demand downstream. As the pressure downstream of the valve increases, it compresses the spring and makes the element move to close the opening further and decrease the flow and pressure. When the downstream pressure decreases, the spring will open and allow more flow.

13.7.4 Exercising of Valves

Most of the problems with valves are due to their non-use over a long period. A valve stuck due to corrosion, rust or freezing becomes a bigger problem when it needs to work in an emergency. Hence, it is very important to exercise valves.

Valve exercising should be done once a year, especially in main lines as part of a preventative maintenance program. A valve inspection should include drawing valve location maps to show distances (ties) to the valves from specific reference points.

13.8 CROSS-CONTAMINATION

Cross-connection means any actual or potential connection or structural arrangement between a drinking water system and any non-potable water source or system through which it is possible to introduce into the distribution system contaminated water, industrial fluid, gas or substances other than the intended drinking water with which the system is supplied. Cross-connections constitute a serious public health risk. There are numerous well-documented cases of cross-connections that contaminated drinking water and resulted in serious illness.

Cross-contamination in water distribution can occur due to the backflow of non-potable water into the water main line. Backflow occurs in two ways: back siphoning and back pressure.

13.8.1 Back Pressure

Back pressure happens when contaminants enter the drinking-water system when the pressure of the contaminant source exceeds the pressure of the water system. The lower the system pressure or greater the instances of leakage in the piping network, the higher the probability of contaminant ingress.

13.8.2 Back Siphoning

Back siphoning occurs when vacuum conditions occur in the water pipeline and allow water from a contaminated source or system to enter the drinking water supply. Reduction in potable water supply pressure or partial vacuum conditions occurs in situations such as water line flushing, firefighting, or breaks in water mains. In addition to physical faults in the distribution system, the backflow of contaminants can come from connections to non-potable systems, tanks, receptors, equipment, or plumbing fixtures where inadequate cross-connection controls, including backflow prevention devices, have been installed or where maintenance has been inadequate.

13.8.3 Backflow Prevention

Although backflows through cross-connections have caused a broad and varied range of outbreaks of illnesses associated with drinking water, surveys of water utilities have found that many do not have inspection programmes or have programmes that are insufficient to provide protection against cross-connections.

There are many examples of cross-connection that speak to the considerable need to protect against cross-contamination. In June 1983, yellow gushy stuff poured from faucets in the town of Woodsboro, Maryland, USA. This contamination occurred due to the backflow of a powerful agricultural herbicide. In 2007–2008, approximately 5,000 cases of gastrointestinal illness attributed to Campylobacter and Salmonella were reported in a population of about 30,000 in Nokia, Finland. The outbreak was caused by the cross-connection of the sewage system and the drinking water system.

Cross-connections must be either physically disconnected or have an approved backflow prevention device installed to protect the public water system. Approved devices include: air gap (method); atmospheric vacuum breaker; pressure vacuum breaker; double-check valve; and reduced pressure backflow preventer.

13.9 HYDRANTS

Fire hydrants are used to fight fires and to flush lines and may be used to obtain water for construction purposes. In addition, they provide locations for adding chlorine to disinfect pipes, inserting pigs and swabs for pipe cleaning, and expelling air from the system. Hydraulic performance (friction factor) testing, or fire flow testing, can be facilitated at hydrant locations. It is important for operators of hydrants to be properly trained. Improper use may result in damage, which could jeopardize the community's firefighting abilities.

There are two basic types of hydrants: dry barrel and wet barrel. Most hydrants are of the dry barrel compression type. The operating valve on the dry barrel type is located at the bottom and is provided with a drain hole. On closing the valve, water is drained. Thus, the barrel remains dry when not in use. This prevents water from freezing during winter. On wet barrel hydrants, the operating valve is at the outlet, and for this reason, these hydrants are limited to climates where freezing is less likely.

Fire hydrants should be inspected and maintained at least twice a year, normally in the spring and fall. Most hydrants have two nozzles and a pumper connection. The typical size of the nozzle is 65 mm.

13.9.1 Hydrant Testing

During hydrant testing, a reduction in pressure at the residual hydrant can be observed by opening the hydrant in the vicinity of the hydrant being studied. This test allows for the estimation of the flow capacity of a given hydrant during firefighting or other similar emergencies. Hydrant discharge from a flowing hydrant is found by observing the trajectory of the water jet or by reading the velocity pressure of the water jet, as explained earlier.

$$\boxed{Q = 0.785D^2\sqrt{2g \times \Delta h} = D^2\sqrt{\Delta p}}$$

$$p = kPa,\ D = m,\ Q = m^3/s$$

Example Problem 13.2

During a fire flow test, the pitot gauge reads a velocity pressure of 55 kPa. Applying the principle of hydraulics, estimate the discharge rate from a 60 mm hydrant nozzle. The coefficient of discharge can be assumed to be 85%.

Given: D = 60 mm = 0.060 m
Q = ?
C = 0.85
Δp = 55 kPa

SOLUTION:

$$Q_F = D^2\sqrt{\Delta p} = (0.060)^2\sqrt{55} = 2.66\times10^{-2}\,m^3/s = 26.6 = \underline{27\ L/s}$$

13.9.2 Hydrant Flow Capacity

A flow test is carried out to estimate the flow capacity of a given hydrant. Pressure readings at the hydrant are read during static (no flow) and flow conditions. The discharge rate during flow conditions is estimated as discussed earlier. The hydrant flow capacity at the residual pressure, which is usually 150 kPa, is calculated as follows:

$$\boxed{Q_R = Q_F \times \left(\frac{\Delta p_R}{\Delta p_F}\right)^{0.54} = Q_F \times \left(\frac{p_S - p_R}{p_S - p_F}\right)^{0.54}}$$

Q_F = total flow from all the flowing hydrants during testing
Δp_F = drop in pressure at the residual hydrant = $p_S - p_F$
Δp_R = static pressure minus the residual pressure = $p_S - p_R$
p_R = residual pressure = 140–150 kPa

Hydrants are colour-coded to indicate capacity. Hydrants with maximum flows > 100 L/s are painted blue; 30–65 L/s range are orange; 65–100 L/s range are red; and < 30 L/s are yellow.

Example Problem 13.3

In a fire flow test, four hydrants were opened to produce a total flow of 140 L/s. During the test, the pressure at the residual hydrant dropped from 480 to 310 kPa. What hydrant flow can be expected at a residual pressure of 150 kPa?

Given: $\Delta p_F = 480 - 310 = 170$ kPa
$Q_F = 140$ L/s
$\Delta p_R = 480 - 150 = 330$ kPa

SOLUTION:

$$Q_R = Q_F\left(\frac{\Delta p_R}{\Delta p_F}\right)^{0.54} = 140\ L/s \times \left(\frac{330\ kPa}{170\ kPa}\right)^{0.54} = 200.3 = \underline{200\ L/s}$$

13.10 SERVICE CONNECTIONS

Water from a distribution main reaches the property line of individual consumers through a service pipe, usually made of copper or plastic, with a minimum diameter of 20 mm. Water connections are made initially when the main is installed (dry tap) or later when the main is in service (wet tap). A special fitting called a corporation stop is employed to make the service connection. At the property line there is an underground shut-off valve known as a curb stop. This valve can be operated from the ground surface.

13.11 WATER METERS

Water meters are used to measure flows. Most meters read cumulative flow volume. Differences in readings over a period of time gives the amount of water used. In developed countries, service connections are all metered, preventing the wastage of water. In countries like India, municipal connections are not metered. Water meters should be accurate, rugged, durable and prevent backflow. Water meters used in water distribution systems are usually velocity meters or displacement type meters.

Velocity or inferential type water meters measure the flow velocity across a cross-section and are suitable for measuring high flows. Rotary and turbine meters fall into this category. Displacement meters are more suitable for low flow applications, as in the case of residential services. In this type of meter, the quantity of water passing through it is measured by filling and emptying a chamber of known capacity. Types of displacement meters in use include reciprocating, oscillating and mutating disc.

13.12 DUAL WATER SYSTEMS

A dual water distribution system is one that provides two independent pipeline networks within the same municipal service area: one system carries potable water and a second system carries non-potable water. Non-potable water is usually recycled wastewater that can be used for the irrigation of lawns and gardens, firefighting and street cleaning.

The potable network of a dual water system requires smaller water mains and pumps as the water requirement is reduced. This also means less water needs to be treated at the water treatment plant, and it may be able to meet the need for future demand without expanding the facility. Dual water systems make sense where water shortage is a recurring problem. Although dual water systems are uncommon at present, they are a viable option for the future. With advancements in water treatment technology such as nanofiltration, it is now possible to reclaim wastewater to drinking water standards.

Discussion Questions and Problems

1. Discuss the advantages and disadvantages of the various types of pipe materials used in water supply systems?
2. Define and explain the following terms:
   a. Tuberculation.
   b. Cross-connection.
   c. Exercising valves.
   d. Thrust.
   e. Equalizing demand.
3. Describe the various types of joints used in joining water pipes with the help of sketches.
4. Compare the following:
   a. Intermittent supply and continuous supply.
   b. Elevated storage and ground storage.
   c. Inferential meters and displacement meters.
   d. Dry barrel hydrants and wet barrel hydrants.
   e. Back pressure and back siphoning.
   f. Linear valves and rotary valves.
   g. Dry tap and wet tap.
   h. Grid-iron and dead-end systems
   i. RCC pipes and prestressed pipes.
   j. Curb stop and corporation stop.
5. Describe the various methods of preventing cross-contamination.
6. Briefly describe the steps involved in laying down water pipes.
7. Why is it important to install water meters in service connections? Describe the desired features of water meters.
8. What is the most common type of valve used in a water distribution system? Describe the common types of valves used in a water system.

TABLE 13.3
Storage Capacity Problem

Hour	1	2	3	4	5	6	7	8	9	10	11	12
Demand, ML/d	2.9	2.9	3.6	4.8	6.0	8.4	18	29	46	52	36	24
Supply, ML/d						48	48	48	48	48	48	48
Hour	13	14	15	16	17	18	19	20	21	22	23	24
Demand, ML/d	1.0	0.8	0.6	1.1	1.5	1.8	1.8	1.6	1.4	0.91	0.35	0.25
Supply, ML/d			48	48	48	48	48	48				

9. Though the grid-iron layout of water pipes is most desirable, it is not used everywhere. Explain why.
10. A medium-size city has variable demand, as shown in **Table 13.3**. Determine the storage capacity required to equalize the demand against:
    a. a constant rate of pumping at 24 ML/d. (9.0 ML)
    b. an intermittent rate of pumping (**Table 13.3**) (6.0 ML)
11. Briefly discuss variations in water demand over time. Sketch a graph that illustrates hourly variations.
12. What is a dual water system? What are its advantages and disadvantages?
13. What is meant by the term "equalizing storage"? What benefits does equalizing storage provides in water distribution systems?

14 Pipeline Systems

To discuss the design of pipeline systems, it is important to understand the application of flow equations. Details of flow equations can be found in any standard text on fluid mechanics. A brief discussion of flow equations follows.

14.1 FLOW EQUATIONS

A flow equation relates pipe characteristics including diameter, roughness and length to the flow rate and head loss due to friction. The three equations commonly in use are Darcy–Weisbach, Hazen–Williams and Manning's equations.

14.1.1 Darcy–Weisbach Flow Equation

The Darcy–Weisbach flow equation is theoretical and can be used for any consistent units. Friction factor depends on the Reynolds number and relative roughness; thus, it yields more precise results and is applicable to all kinds of fluids and flow conditions.

$$\text{Frictional loss, } h_f = \frac{f}{1.23g} \times L \times \frac{Q^2}{D^5} \left(\text{consistent units}\right)$$

$$= \frac{f}{12} \times L \times \frac{Q^2}{D^5} \quad (SI)$$

$$\text{Flow capacity, } Q = \sqrt{\frac{1.23\,g}{f} \times \frac{h_f}{L} \times D^5} = \sqrt{\frac{1.23\,g}{f} \times S_f \times D^5}$$

$$= \sqrt{\frac{12}{f} \times S_f \times D^5} \quad (SI)$$

In turbulent flow conditions, the friction factor, f, is dependent on the Reynolds number and the relative roughness of the pipe, RR. Absolute roughness of a pipe is measured in terms of the average thickness of irregularities on the surface of the pipe and is denoted by the symbol ε (epsilon). Typical values of absolute roughness for commonly used pipe materials are shown in **Table 14.1**. Relative roughness is the ratio of the diameter of the pipe to the absolute roughness.

$$N_R = \frac{\text{velocity} \times \text{diameter}}{\text{kinematic viscosity}} = \frac{vD}{\vartheta}$$

$$RR = \frac{\text{diamater}}{\text{asolute roughness}} = \frac{D}{\varepsilon}$$

DOI: 10.1201/9781003231264-14

TABLE 14.1
Absolute Roughness

Pipe Material	Roughness ε μm
Glass, plastic	Smooth
Copper, brass	1.5
Commercial steel	46
Cast iron, 1 uncoated	240
2 Coated	120
Concrete	1200
Riveted steel	1800

14.1.2 Moody Diagram

Most of the usable data available for evaluating friction factor has been derived from experimental data. This data presented in graphical form is called a Moody diagram and is widely used. The diagram shows the friction factor f plotted against N_R with a series of parametric curves, each representing a specific value of D/ε. Some authors use the inverse relationship, that of ε/D, for representing relative roughness. Of course, in a modern world of calculators and computers, not many of us are used to reading graphs as it is more cumbersome and not suited to computer use. An explicit relationship like the one shown below is gaining popularity.

$$f = 0.0055 + 0.0055 \times \sqrt[3]{\frac{20000}{D/\epsilon} + \frac{1000000}{N_R}}$$

14.1.3 Hazen–Williams Flow Equation

The Darcy–Weisbach equation is cumbersome for solving problems where there are more than two unknowns. In such cases, empirical flow equations, which provide direct solutions, are more commonly used. The most common pipe flow formula used in the design and evaluation of water supply systems is the Hazen–Williams equation. This equation relates the flow-carrying capacity (Q) with the size (D) of pipe, slope of the hydraulic gradient (S_f) and a coefficient of friction C, which depends on the roughness of the pipe. The value of C is chosen based on the judgement of the designer. Results based on this equation would be less accurate compared with the Darcy–Weisbach equation.

$$Q(\text{m}^3/\text{s}) = 0.278CD^{2.63} \times S_f^{0.54}$$

$$h_f = 10.7L \times (Q/C)^{1.85} \times D^{-4.87}$$

C = friction coefficient
D = diameter of pipe, m
S_f = friction slope = $h_f/L = \Delta h/L$

The Hazen–Williams equation is valid only when the fluid is water, the flow velocity is less than 3 m/s and the pipe diameter is larger than 2 cm. Users of this equation must be aware that the value of the frictional coefficient varies over a broad range for a given surface and the effect of N_R is not

TABLE 14.2
Hazen–Williams Coefficient, C

Pipe Material	C
Asbestos cement	140
Cast iron	
- Cement, lined	130–150
- New, unlined	130
- 5-year old, unlined	120
- 20-year old, unlined	100
Concrete	130
Copper	130–140
Plastic	140–150
Commercial steel	120
New riveted steel	110

considered. However, if the value of C is chosen judiciously, this equation provides good results. The values of the roughness coefficient C for selected pipe materials are given in **Table 14.2.**

14.1.4 Manning's Flow Equation

The Manning flow equation is also empirical in nature. It can be applied both to closed-pipe and open-channel flow systems, though it is more commonly used for open-channel flow conditions.

14.2 SERIES AND PARALLEL

If a pipeline system is arranged so that the fluid flows in a continuous line without branching, it is referred to as a series system. Conversely, if the system causes the flow to branch out into more than one path, it is called a parallel system.

Pipelines connected in series will have the same flow rate to maintain the continuity of flow. However, the head loss in each length of pipeline will be different depending on its diameter, length and roughness. The total head loss in the system will be the sum of the individual head loss components.

For parallel pipeline systems, head loss in each pipeline is the same; however, flow-carrying capacity varies. Hence, the total flow carried is the sum of all the flows carried by individual pipelines. For analysis purposes, different branch lines can be replaced by a single pipe of uniform diameter and of a length that will pass the discharge, Q, with the head loss, h_f. This is called an equivalent pipe.

14.3 EQUIVALENT PIPE

In engineering analysis, quite often, if a series of pipes are replaced by a single diameter pipe with the same head loss and discharge rate, the pipe is called an equivalent pipe and the diameter is called the equivalent diameter. Since head loss in the equivalent pipe is the same as that of all the pipes in series, equating the two, we can express equivalent diameter in terms of individual diameters as follows:

$$\frac{L_e}{D_e^5} = \frac{L_1}{D_1^5} + \frac{L_2}{D_2^5} + \ldots (series) \quad \sqrt{\frac{D_e^5}{f}} = \sqrt{\frac{D_1^5}{f_1}} + \sqrt{\frac{D_2^5}{f_2}} + \ldots (parallel)$$

$$\sqrt{D_e^5} = \sqrt{D_1^5} + \sqrt{D_2^5} + ..(\ same\ f)$$

Example Problem 14.1

Two water reservoirs are connected by a pipeline system consisting of three pipes connected in series. The three pipes are 300 m of 30 cm diameter, 150 m of 20 cm diameter, and 250 m of 25 cm diameter, respectively, of new cast iron. If the elevation difference between the water levels in the two reservoirs is 10 m, find the rate of flow.

Given: $L_1 = 300m$
$L_2 = 150m$
$L_3 = 250m$
$D_1 = 30\ mm$
$D_2 = 200\ mm$
$D_3 = 250\ mm$
$\Delta Z = 10\ m$
$C = 120$
$Q = ?$

SOLUTION:

Writing the energy equation between water levels in the two reservoirs and noting that points 1 and 2 are in open reservoirs and minor losses are negligible, the energy equation reduces to:

$$h_f = Z_1 - Z_2 = 10\ m = 10.7\left(\frac{Q}{C}\right)^{1.85}\left(\frac{L_1}{D_1^{4.87}} + \frac{L_2}{D_2^{4.87}} + \frac{L_3}{D_3^{4.87}}\right)$$

$$Q^{-1.85} = \frac{10.7}{10\times120^{1.85}}\left(\frac{L_1}{D_1^{4.87}} + \frac{L_2}{D_2^{4.87}} + \frac{L_3}{D_3^{4.87}}\right)$$

$$Q^{-1.85} = \frac{10.7}{10\times120^{1.85}}\left(\frac{300}{0.30^{4.87}} + \frac{150}{0.20^{4.87}} + \frac{250}{0.25^{4.87}}\right) = 106.5$$

$$Q = 106.5^{-0.54} = \frac{0.0803m^3}{s}\times\frac{1000\ L}{m^3} = 80.3 = \underline{80\ L/s}$$

14.4 SYSTEM CLASSIFICATION

When analyzing a fluid flow system, the primary parameters involved are energy additions (h_a) or subtractions (h_r or h_L), fluid flow rate (Q) or velocity, diameter (D) and length (L) of the pipe, wall roughness and fluid properties including density (γ) and viscosity (ν).

As shown earlier, when applying the energy equation, only one unknown can be determined at a time. Basic algebraic principles suggest that we need an equation (relationship) for each unknown. Usually, one of the first three parameters, Q, h_f and D, is the true unknown, while the remaining items are known or specified by the analyst.

14.4.1 Class I Systems

By knowing the pipeline characteristics (L, D, ε), head losses can be determined for a given flow, Q. Knowing h_l, the head added by the pump, h_a, can be determined by applying the energy equation. If the flow is all due to gravity ($h_a = 0$), h_l can be found by applying the energy equation.

Example Problem 14.2

A hydraulic gradient test is performed in the field by opening the end hydrant. In the direction of flow, pressure readings at hydrant 1 and 2, respectively, were observed to be 413 kPa and 395 kPa. From the map of the area, the elevations of hydrants 1 and 2 are 112.4 m and 111.8 m, respectively. Given that the two hydrants are connected by 610 m of 300 mm diameter line, what is the hydraulic gradient for the test conditions?

Given:

Variable	Hydrant 1	Hydrant 2
Pressure, kPa	413	395
Elevation, m	112.4	111.8
Head loss	?	
Length, m	610	

SOLUTION:

$$h_f = (Z_1 - Z_2) + \frac{(p_1 - p_2)}{\gamma} = (112.4 - 111.8)m + (413 - 395)kPa \times \frac{m}{9.81\ kPa}$$

$$= 2.43 = \underline{2.4\ m}$$

$$S_f = \frac{h_f}{L} = \frac{2.43\ m}{610\ m} \times 100\% = 0.398 = \underline{0.40\%}$$

Example Problem 14.3

A hydraulic gradient test is performed in the field. Two hydrants 750 m apart were chosen to observe the hydraulic head. For the maximum flow conditions, the pressure readings at hydrants 1 and 2 were observed to be 477 kPa and 495 kPa, respectively. From the map of the area, the elevations of hydrants 1 and 2 are 112.4 m and 111.9 m, respectively. Find the hydraulic head at each of the hydrants and the head loss in the pipeline connecting the two hydrants.

GIVEN:

Variable	Hydrant 1	Hydrant 2
Pressure, kPa	477	495
Elevation, m	112.4	111.9
Head loss, h_L	?	
Length, m	750	

SOLUTION:

$$h_l = (Z_1 - Z_2) + \frac{(p_1 - p_2)}{\gamma} = (112.4 - 111.9)m + (477 - 495)kPa \times \frac{m}{9.81\ kPa} = -1.33 = \underline{-1.3\ m}$$

$$S_f = \frac{h_f}{L} = \frac{1.33\ m}{750\ m} \times 100\% = 0.177 = \underline{0.18\%}$$

14.4.2 Class II Systems

There is added difficulty here due to more than one unknown in solving Class II-type problems. If the unknown is Q, N_R cannot be determined. Hence, h_f cannot be estimated. To reduce the number of unknowns to one, the other unknown quantities need to be guessed. Based on the assumed value (educated guess), the flow rate is determined. This is called trial or iteration number one. Note, our answer for Q is based on an assumed value of f, so it needs to be checked. Based on the computed value of Q, N_R is calculated to find f. The new value of friction factor f is compared with the assumed value. If the two values are close, the guess was right; otherwise, computations must be started over with a newly computed value of f and another trial run. Each trial will get closer to the correct value. The number of trials will depend on how much precision is required in the analysis.

Example Problem 14.4

Calculate the water flow rate in a 1 m diameter commercial steel pipe for an allowable head loss of 4 m per km length of pipeline.

Given: D = 1.0 m
S_f = 4m/km = 4.0×10^{-3} m/m

SOLUTION:

This is essentially a Class II-type problem. To solve this by applying the Darcy–Weisbach flow equation, one has to resort to the trial-and-error technique. The Hazen–Williams equation provides a direct solution. For new pipes, $C = 120$ (old pipes get rougher due to corrosion and incrustation)

$$Q = 0.278\ C\ D^{2.63} S_f^{0.54} = 0.278 \times 120 \times (1.0)^{2.63} \times (4.0 \times 10^{-3})^{0.54} = 1.69 = \underline{1.7\ m^3/s}$$

Example Problem 14.5

At night, water is pumped from a water treatment plant reservoir to elevated storage through a 1.5 km long, 250 mm diameter pipeline with absolute roughness of 1.0 mm. Calculate the discharge pressure required to supply 60 L/s of water to the tank. Assume there is no withdrawal from the system. The elevations are: water surface in the supply reservoir = 3.0 m; pump = 6.0 m; and water level in the elevated tank = 42 m. Assume kinematic viscosity is 1.1×10^{-6} m²/s.

Given: L = 1.5 km = 1500 m
D = 250 mm = 0.25 m
ε = 1.0 mm
Q = 60 L/s = 0.06 m³/s

$Z_1 = 3\ .0$ m (supply)
$Z_2 = 6.0$ m (pump)
$Z_3 = 42$ m (storage)
$\nu = 1.1 \times 10^{-6}$ m²/s

SOLUTION:

$$v_2 = \frac{Q}{A_2} = \frac{4Q}{\pi D^2} = \frac{4}{\pi} \times \frac{0.06\ m^3}{s} \times \frac{1}{(0.25m)^2} = 1.22\ m/s$$

$$N_R = \frac{vD}{\vartheta} = \frac{1.22\ m}{s} \times 0.25\ m \times \frac{s}{1.1 \times 10^{-6} m^2} = 2.77 \times 10^{-5}$$

$$f = 0.0055 + 0.0055 \times \sqrt[3]{(20000 \times \frac{1.0\ mm}{250\ mm} + \frac{10^6}{2.77 \times 10^5})} = 0.0295$$

$$h_f = f \times \frac{L}{D} \times \frac{v^2}{2g} = 0.0295 \times \frac{1500\ m}{0.25\ m} \times \left(\frac{1.22m}{s}\right)^2 \times \frac{s^2}{2 \times 9.81m}$$

$$= 13.45 = 13.5\ m$$

$$p_2 = \gamma\left(\Delta Z + h_f - h_{v2}\right)$$

$$= \left((42\ m - 6.0\ m) + 13.45\ m - 0.076\ m\right) \times \frac{9.81\ kPa}{m} = \underline{480\ kPa}$$

Example Problem 14.6

Water at the filtration plant is pumped into a 350 mm diameter water main (C = 110) at a discharge pressure of 350 kPa. The 1.5 km long water main connects to the load centre where the residual pressure during the peak demand period is known to be 170 kPa. The elevations of the pump and the load centre, respectively, are 102.0 m and 105.0 m.

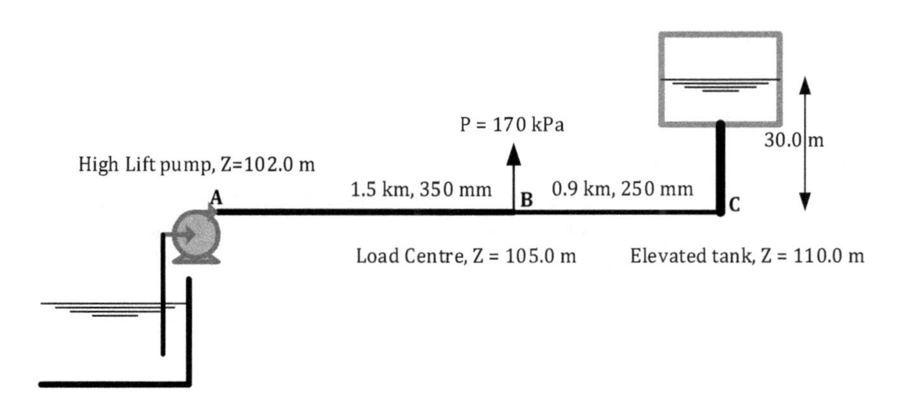

FIGURE 14.1 Simplified Water Distribution System

i. Calculate the percent hydraulic gradient in the water main A-B.
ii. Applying the Hazen–Williams equation ($Q = 0.278C \times D^{2.63} \times S_f^{0.54}$), find how much demand in L/s is served by the clear well at the water plant.
iii. Calculate the flow contribution from the elevated tank.

Given:

Parameter	A	B	C
Elevation, Z, m	102.0	105.0	110.0
Pressure, kPa	350	170	294
Head, m	35.7	17.3	30.0
Hyd. head m	137.7	122.3	140.0
Parameter	**AB**	**CB**	
Length, km	1.5	0.90	
Diameter, mm	350	250	
Coefficient, C	110	110	
Head loss, m	15.37	17.7	
Friction slope, S_f, %	1.022	1.96	

SOLUTION:

$$S_{AB} = \frac{h_f}{L} = \left(\frac{(350-170)\,kPa.m}{9.81\,kPa} - 3.0\,m\right) \times \frac{1}{1500\,m} = 0.0102 = \underline{1.0\%}$$

$$Q_{AB} = 278C \times D^{2.63} \times S_f^{0.54} = 278 \times 110 \times 0.35^{2.63} \times 0.0102^{0.54}$$

$$= 162.8 = \underline{160\ L/s}$$

$$S_{CB} = \frac{\Delta h}{L} = \left(140\,m - \frac{170\,kPa.m}{9.81\,kPa} + 105\,m\right) \times \frac{1}{900\,m} = 0.0196 = \underline{2.0\%}$$

$$Q_{CB} = 278C \times D^{2.63} \times S_f^{0.54} = 278 \times 110 \times 0.25^{2.63} \times 0.0196^{0.54}$$

$$= 95.46 = 95\ L/s$$

$$Q_B = Q_{AB} + Q_{CB} = 163 + 95 = 258 = \underline{260\ \ L/s}$$

14.4.3 Class III Systems

This category represents a true design problem because we are required to determine the size of the pipe needed to carry a given flow, Q, without exceeding the allowable friction slope (head loss per unit length). Without knowing the friction slope D, v, N_R and f cannot be found. This leaves us with more than one unknown, so a direct solution is not possible. Much like Class II systems, the iteration technique is used to solve problems of this type.

14.5 COMPLEX PIPE NETWORKS

Modern pipe networks consist of pipe loops, nodes and junctions. The advantage of this type of system is that the flow direction in a given pipe of a loop can change depending on the demand at a given node. As compared to a single pipeline, part of the system can be separated without affecting the supply of water in other sections. The residual pressure at a given node will be determined by the demand and the pipeline characteristics, that is, length, diameter and roughness. Based on the principles of hydraulics, the following two conditions must be satisfied.

Continuity of Flow

Based on the principle of continuity of flow, at a given junction, the flow entering must equal to the flow exiting. That is to say, there must not be any storage at a junction. In mathematical terms, the algebraic sum of all the flows at every junction must be zero.

Continuity of Pressure

This principle applies to a pipe loop in a given network. According to this principle, the algebraic sum of all the head losses in a loop must be zero. In other words, the pressure at a given junction must be the same, whether the calculation is done clockwise or anticlockwise. By applying these two principles, pipe networks can be analyzed for pressures and flows.

14.5.1 Hardy Cross Method

This method is based on successive iterations such that the correction to flow is almost negligible. Here the correction is based on the value added or subtracted from the assumed flow for carrying out the next iteration.

$$\Delta Q = Q - Q_{actual} \text{ and } h_L = KQ^n$$

The exponent n is 2 for a Darcy–Weisbach equation and 1.85 for a Hazen–Williams equation, and constant K is based on the other terms in the flow formula, such as length, friction factor and pipe diameter. Expanding the terms, the flow correction at the end of iteration can be expressed as follows:

$$\boxed{\Delta Q = -\frac{\Sigma h_L}{n\Sigma\left(\frac{h_L}{Q}\right)}}$$

- The flow in at least one pipe in a loop will have a negative sign.
- Head loss, h_L, and flow, Q, in a pipe have the same algebraic sign.
- If $\sum h_L = 0$, then the correction is zero – that is to say, the assumed flows are the actual flows.
- If $\sum h_L > 0$, correction $\Delta Q < 0$ and vice versa.
- When there are two or more loops, there is always a common pipe. For the common pipe, two corrections need to be applied.
- Since the advent of computers, analyzing a pipe network has been made very easy.

The entire procedure is further illustrated by a couple of example problems. The first example problem is a very simple loop; hence, only one iteration needs to be carried to find the correct flows in each segment. In the second example, there are many nodes, though only one circuit. Further, all the calculations are done using a spreadsheet program such as Microsoft Excel.

Example Problem 14.7

Solve the single loop of a pipe network given below using the Hardy Cross method, balancing the head loss.

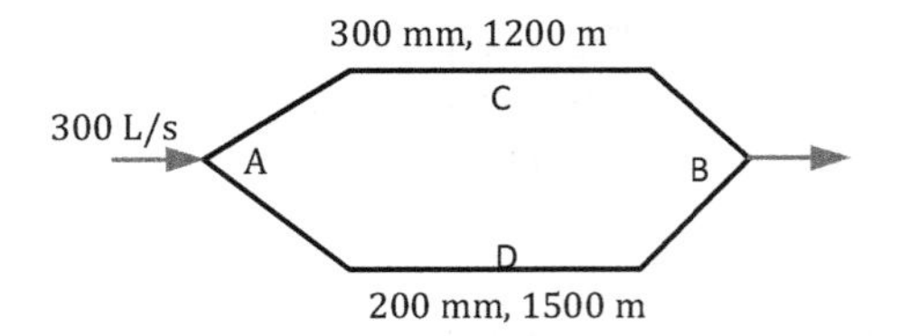

FIGURE 14.2 Pipe Network (Ex. Prob. 14.7)

TABLE 14.3
Excel Worksheet (Ex. Prob. 14.7)

Pipe	Length L, m	Coefficient C	Diameter D, mm	Flow Q, L/s	Loss h_L m	Ratio hL/Q, m·s/L
Trial 1						
ACB	1200	100	300	240	64.32	0.268
ADB	1500	100	200	–60	–44.57	0.743
				Σ =	19.75	1.011
$\Delta Q = -1.85\Sigma h_L/\Sigma(h_L/Q) = -10.56 L/s$						
Trial II						
ACB	1200	100	300	229.44	59.18	0.258
ADB	1500	100	200	–70.56	–60.16	0.853
				Σ =	–0.98	1.111
$\Delta Q = -1.85\Sigma h_L/\Sigma(h_L/Q) = 0.48$ L/s						

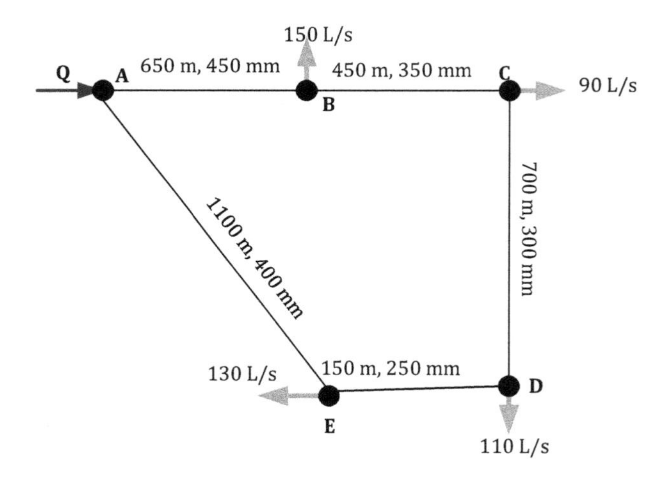

FIGURE 14.3 Pipe Network (Ex. Prob. 14.8)

Example Problem 14.8

An elevated tank at location A maintains a pressure head of 42 m and the minimum desired pressure at remote location D is 210 kPa. Work out the flows in all the pipes based on the demands shown in **Figure 14.3**. Check the available pressure at remote point D and other junctions.

TABLE 14.4
Excel Worksheet (Ex. Prob. 14.8)

Pipe Designation	Length L, m	Coefficient C	Diameter D, mm	Flow Q, L/s	Loss h_L, m	Ratio h_L/Q, m.s/L
Trial 1						
AB	650	110	450	300	6.1	2.0E-02
BC	450	110	350	150	4.0	2.7E-02
CD	700	110	300	60	2.4	4.0E-02
DE	150	110	250	–50	–0.9	1.8E-02
EA	1100	110	400	–180	–7.2	4.0E-02
				Σ =	4.5	1.5E-01
	$\Delta Q = -\Sigma h_L/(1.85\Sigma(h_L/Q)) = -16.75$ L/s					
Trial II						
AB	650	110	450	283	5.5	1.9E-02
BC	450	110	350	133	3.2	2.4E-02
CD	700	110	300	43	1.3	3.1E-02
DE	150	110	250	–67	–1.5	2.3E-02
EA	1100	110	400	–197	–8.4	4.3E-02
				Σ =	0.1	1.4E-01
	$\Delta Q = -\Sigma h_L/(1.85\Sigma(h_L/Q)) = 0.3$ L/s					
Trial III						
AB	650	110	450	283	5.5	1.9E-02
BC	450	110	350	133	3.2	2.4E-02
CD	700	110	300	43	1.3	3.0E-02
DE	150	110	250	–67	–1.5	2.3E-02
EA	1100	110	400	–197	–8.5	4.3E-02
				Σ =	0.0	1.4E-01
	$\Delta Q = -\Sigma h_L/(1.85\Sigma(h_L/Q)) = 0.0$ L/s					

SOLUTION:

Flows are assumed such that there is no net flow at a junction. Applying the Hardy Cross method, flows and corresponding head losses are computed as shown in the Excel worksheet (**Table 14.4**). Flows are balanced after three trials.

14.6 COMPUTER APPLICATIONS

Municipal water distribution systems are, of course, more complex than one or two loops, as illustrated in the previous two example problems. Network analysis is made using a spreadsheet computer program. Previously, the same problems were solved by hand. One of the advantages of using a computer is that it can do in seconds repetitive calculations that once took hours by hand.

Mathematical models of distribution networks are now readily created and analyzed using computers. Modern network software combines hydraulic calculations with other software, like CAD and GIS programs. A designer can simulate various scenarios to choose an alternative that best serves a purpose. Furthermore, new models can be created to simulate the quality of water in various parts of a water distribution system. This allows operating personnel to make necessary changes so that there is no threat to water quality under various demands. Engineers can spend more time in planning and emergency management rather than just number crunching.

Discussion Questions

1. Modify the Darcy–Weisbach equation to find flow capacity Q for SI units.
2. Comparing Darcy–Weisbach equation with Manning's flow equation, find the relationship between n and f.
3. Describe the three types of flow equations and their suitability in the hydraulic analysis of pipeline systems.
4. Discuss how the principles of continuity of flow and continuity of pressure are used in pipe network analysis.
5. Describe the steps involved in the application of the Hardy Cross method.
6. When sizing water mains using a Darcy–Weisbach flow equation, multiple iterations may be required. However, using the Hazen–Williams equation, a direct answer can be achieved. Why not then always use the Hazen–Williams equation?
7. Search the internet to find commonly used network software and comment.
8. What is an equivalent pipe? What is the purpose of determining equivalent pipes in water distribution systems?
9. Discuss pipes connected in parallel and in series. Using the Hazen–William flow equation, derive the expressions for equivalent pipes in both cases.
10. Professional bodies recommend the use of the Darcy–Weisbach equation over the Hazen–Williams flow equation. Comment.

Practice Problems

1. Calculate the water carrying capacity of a 150 mm diameter pipe (C = 100) for an allowable friction loss of 5.0 m per km length of pipe. (11 L/s)
2. What size pipe should be used to supply 100 L/s so that the head losses do not exceed 2.5 m per km length of pipe, assuming C = 100? (400 mm)
3. A pump installed at an elevation of 170 m delivers 220 L/s of water through a horizontal pipeline to a pressurized tank (**Figure 14.4**). The water in the tank is at a height of 1.73 m. The suction side of the pump reads 70 kPa vacuum, and at the 30 cm diameter discharge side the pressure is 260 kPa. The cast iron discharge pipe is 73 m long. Determine the air pressure in the tank assuming the coefficient of friction, C, is 120? (220 kPa).
4. Determine the rate of flow of water that a 100 mm cast iron pipe (C = 110) can carry for an allowable head loss of 10.5 m/km. (6.1 L/s)
5. Estimate the flow capacity of a 300 mm diameter main for an allowable friction slope of 0.1%. Assume the Hazen–Williams coefficient of friction is 120, a relatively smooth pipe. (34 L/s)

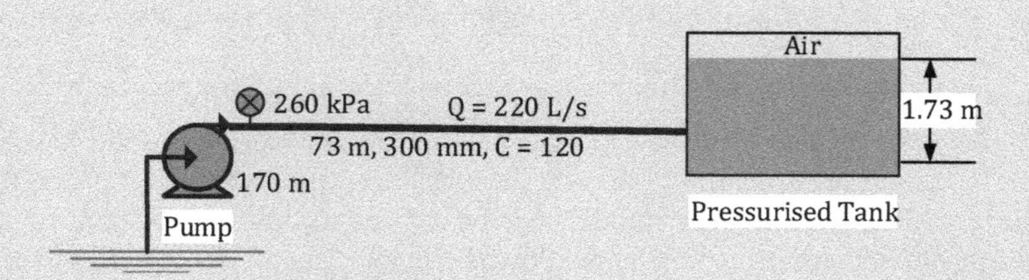

FIGURE 14.4 Water Supply to a Pressurized Tank

6. The water supply system of a town has started experiencing a serious problem of tuberculation. It has been estimated that the effective pipe diameter of a 200 mm water main has been reduced by 3.0 mm and the coefficient of friction, C, has dropped from 110 to 90. What is the percent increase in head loss? (56%)
7. What diameter cast iron pipe ($\varepsilon = 0.36$ mm) is required to carry water at a rate of 250 L/s for an allowable head loss of 2.0%? Assume the kinematic viscosity of water at the prevailing temperature of 10°C is 1.3×10^{-6} m²/s. (350 mm)
8. What diameter cast iron pipe (C = 110) is required to carry water at a rate of 250 L/s for an allowable head loss of 2.0%? (400 mm NPS)
9. What frictional slope can be expected in a 300 mm diameter water main with a roughness coefficient of 120 when carrying a flow of 80 L/s? (5.0 m/km)
10. A 600 mm diameter and 12 km long water main carries water from the pumping station to the load centre. Calculate the hydraulic gradient/frictional slope and hence the head loss in the water main with a coefficient C of 110 when carrying a flow of 200 L/s? (13 m)
11. Water at the filtration plant is pumped into a 350 mm diameter water main (C = 120) at a discharge pressure of 350 kPa (**Figure 14.5**). The 1.5 km long water main connects to the load centre where the residual pressure during the peak demand period is known to be 170 kPa. A 600 m long and 200 mm in diameter pipeline connects the water tower to the load centre. The elevations of the pump, the load centre and the water tower are 102.0 m, 105.0 m and 106.5 m, respectively. What is the peak demand when the water level in the tower is 30 m above ground? (280 L/s)
12. Two pipelines are connected in parallel from junction A to junction B. The short line is 500 m long with a diameter of 300 mm, and the long branch line is 1500 long with a diameter of 200 mm. Determine the equivalent diameter of a single 500 m long pipeline that could replace the two. (320 mm)
13. Solve the one loop of pipe network given below using the Hardy Cross method, balancing the head loss.
14. An elevated tank at location A maintains a pressure head of 42 m and the minimum desired pressure at remote location D is 210 kPa. Design the pipe network based on per capita demand of 200 L/c.d and assume the peak hour demand is 2.6 times the average demand. The population served at various nodes is shown in **Figure 14.6**. (290 kPa)

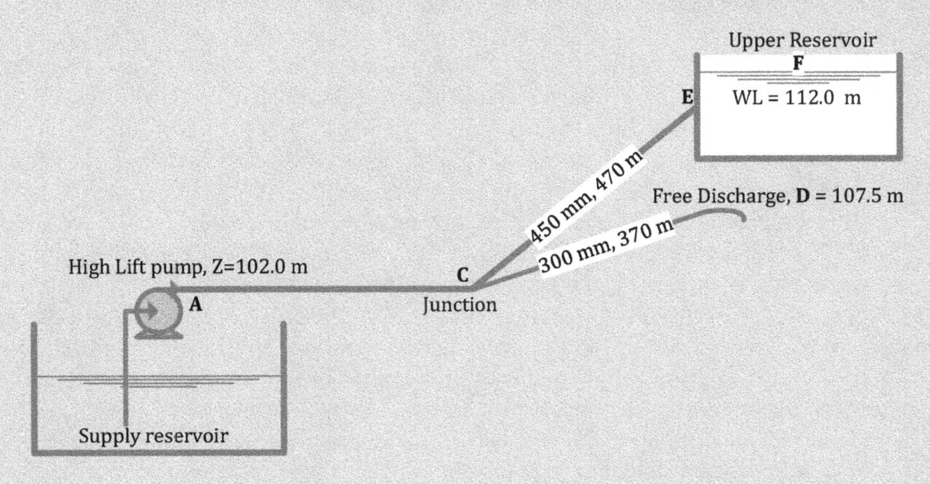

FIGURE 14.5 Simple Water Distribution System

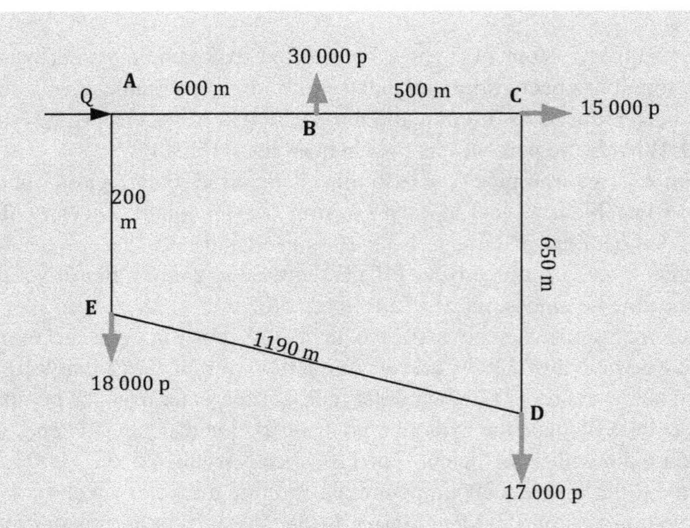

FIGURE 14.6 Pipe Network Problem

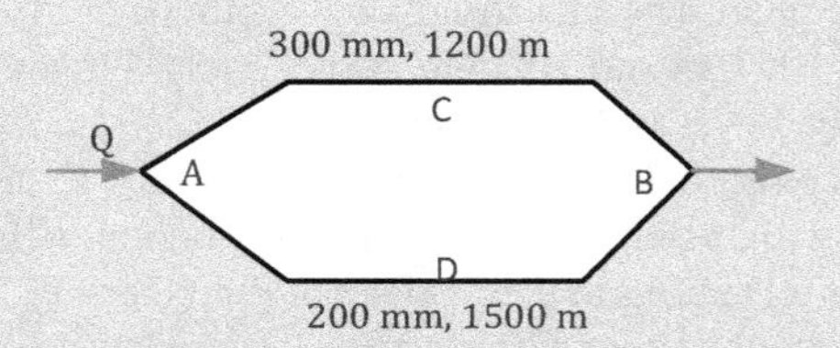

FIGURE 14.7 Single Loop Problem

15. Solve the one loop pipe network shown in **Figure 14.7** using the Hardy Cross method. The flow entering at A is 250 L/s. Determine the flow in each pipe. (192 L/s, 52 L/s)
16. Repeat Example Problem 14.8, assuming the head loss coefficient for all pipes in the network is 100. (290 kPa)
17. A 50 mm diameter and 8.0 m long pipeline is connected to a reservoir with a square entry. At the end it is connected to a 100 mm diameter and 45 m long pipe. What must the level of water in the reservoir be if it is to achieve an exit velocity of 1.5 m/s? Assume friction factor $f = 0.026$. (11 m)
18. Two water reservoirs are connected by a pipeline system consisting of three pipes connected in series. The size of the three cast iron pipes ($C = 110$), respectively, are: 300 m long and 30 cm in diameter; 150 m long and 20 cm in diameter; and 250 m long and 25 cm in diameter. If the elevation difference between the water levels in the two reservoirs is 10 m, what is the rate of flow? (450 L/s)
19. Water is pumped from a lower reservoir to an upper reservoir with the water level F at an elevation of 112 m. At a point along the main line, pipeline CD of 300 mm in diameter and 370 m in length branches of at junction C. Pipeline CD discharges water at 190 L/s freely into the atmosphere at point D with an elevation of 107 m. From junction C, main pipeline CE connects to the upper reservoir at point E. This line is 450 mm in diameter and 470 m in length. For both pipelines, assume a friction factor of 0.029.

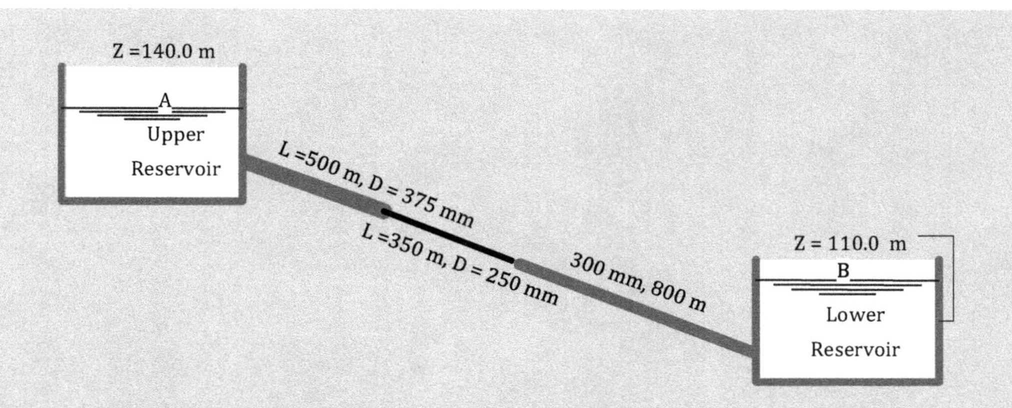

FIGURE 14.8 Two Connected Reservoirs

a. Calculate the head loss due to friction in pipeline CD and pressure/head at junction C. (17 m)
b. Calculate the friction slope in line CE and hence flow carried to the upper reservoir. (4.2%, 570 L/s)
c. Calculate the rate of pumping from the lower reservoir. (760 L/s)

20. Two water reservoirs are connected by three pipelines, as shown in **Figure 14.8**. The lengths and diameters of the pipelines are indicated in the figure. Assuming friction factor f = 0.02 for all three pipes, determine the flow rate at the elevation difference water levels shown. (150 L/s)

15 Pumps and Pumping

In water distribution, a wide variety of pumps are used to transport water. Though centrifugal pumps are common, other types of pumps are also used in special applications. The two broad categories of pumps are positive displacement pumps and kinetic or velocity pumps.

15.1 POSITIVE DISPLACEMENT PUMPS

These pumps are designed to deliver a fixed quantity of fluid for each revolution of the pump rotor. Therefore, except for minor slippage, the delivery of the pump is unaffected by changes in the delivery pressure. In general, these pumps will pump against high pressure, but their capacity is low. These pumps are well-suited for pumping high viscosity liquids and can be used for metering since output is directly proportional to rotational speed. Reciprocation pumps, piston pumps and rotatory pumps are all examples of positive displacement pumps.

15.2 VELOCITY PUMPS

As the name indicates, this category of pump adds to the kinetic energy of the fluid, which is later converted to pressure energy. A centrifugal pump is the most commonly used velocity pump. Non-positive displacement pumps are generally used to transfer large volumes of liquids at relatively low pressures. However, if pressure is increased, pumping rate drops. This is in strong contrast with positive displacement pumps.

15.2.1 Types of Centrifugal Pumps

A centrifugal pump has a radial flow or mixed type impeller. The casing of a centrifugal pump may be either volute type or turbine type, also called diffuser casing. In a volute pump, flow velocity is reduced as the water enters the outlet, thus increasing the pressure. In a turbine pump, velocity is reduced by stationary guide vanes before the water enters the casing, thus giving a better transfer of flow energy or efficiency.

Turbine Pumps

Turbine pumps are usually multistage and used for pumping from deep wells. In multistage pumps, each stage adds the same head. The prime mover is kept at ground surface level and impellers are attached to the bottom of a vertical shaft suspended in the borehole.

Submersible Pumps

As the name indicates, both motor and pump are placed underwater, so no vertical shaft is needed. Water rises through a riser pipe to which the whole assembly is attached. This kind of pump can be used for both domestic and municipal water supplies.

15.3 PUMPING HEAD

Pumps are used to add energy to the water in the form of pressure or lift. The head added or total dynamic head (TDH) can be calculated by writing an energy equation for two points such that the pump falls between the two points.

DOI: 10.1201/9781003231264-15

$$\text{Head added, } h_a = \frac{v_2^2}{2g} - \frac{v_1^2}{2g} + h_l + Z_2 - Z_1 + \frac{p_2}{\gamma} - \frac{p_1}{\gamma}$$

If points 1 and 2 refer to the inlet and outlet of the pump, then $(Z_1 - Z_2) = 0$, $h_l = 0$ and the term $(v_2^2 - v_1^2)/2g$ is negligibly small. By knowing the pressure readings before and after the pump, the head added is simply the difference in pressure head readings.

The head added by pump h_a is usually called the pumping head or total dynamic head (TDH). By knowing the water pumping rate and efficiency from the head capacity curve, the power input to the pump can be figured out. Conversely, if the power input to the pump is being monitored, the actual efficiency of the pump can be calculated and compared to the one read from the performance curves. If a noticeable difference is found, the pump needs to be repaired or replaced. Knowing the pumping head makes it possible to ascertain if the pump is operating in the vicinity of the best efficiency point (BEP). By knowing the pumping rate and head, the waterpower can be worked out. If the power being delivered to the pump is known, the efficiency of the pump can be calculated.

$$\text{Power added, } P_a = W \times h_a = Q \times \gamma \times h_a = Q \times p_a$$

15.3.1 Positive Displacement Pump Characteristics

The capacity of positive displacement pumps is independent of delivery pressure. At high pressures, a small decrease in capacity may occur due to internal leakage. The power required to drive the pump varies linearly with pressure. Therefore, it becomes necessary to protect the positive displacement pumps with relief valves to prevent damage by over-pressurization.

15.3.2 Performance Curves of Centrifugal Pumps

The pumping capacity of centrifugal pumps is strongly affected by the operating head. There is an inverse relationship between capacity, Q, and total dynamic head, H (h_a). As the head increases, the capacity decreases. Performance curves or characteristic curves show the relationships of head, efficiency and power versus capacity at a given speed of the pump. A typical head capacity curve (H versus Q) is shown in **Figure 15.1**. The pump will never operate at a point that does not lie on this line.

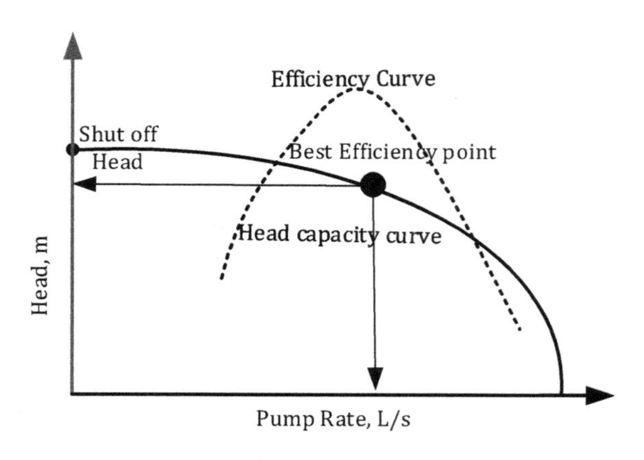

FIGURE 15.1 Pump Performance Curves

In the case of centrifugal pumps, if the operating head changes, the capacity will change, and so efficiency and power will change. Keep in mind, though, for small drops in pumping head, flow capacity may increase in greater proportions. In fact, a pump will fail to pump any water if it equals or exceeds the shut-off head. The operating point corresponding to peak efficiency is known as the best efficiency point (BEP). To get more out of a pump, it is a good practice to operate a rate of discharge within a range of 60–120% of BEP.

Example Problem 15.1

Sault City's water treatment plant gets its water from Lake Superior. The intake is below the water surface at an elevation of 230.4 m. The lake water is pumped to the plant influent at an elevation of 242.4 m. The total head losses are estimated to be 5.5 m when water is drawn at a rate of 340 L/s. Calculate the pumping head, waterpower and pump power, assuming the pump is 72% efficient.

GIVEN:

Variable	Suction = 1	Discharge = 2
Pressure, p	$p_1 = 0$	$p_2 = 0$
Elevation, Z	$Z_1 = 230.4$ m	$Z_2 = 242.4$ m
Head loss, h_L	5.5 m	

SOLUTION:

$$h_a = \frac{p_2}{\gamma} - \frac{p_1}{\gamma} + Z_2 - Z_1 + h_l = 0 + (242.4 - 230.4)m + 5.5\ m$$

$$= 17.5\ m = 18\ m$$

$$P_a = Q \times \gamma \times h_a = \frac{0.340\ m^3}{s} \times \frac{9.81\ kN}{m^3} \times 17.5\ m \times \frac{kW.s}{kN.m}$$

$$= 58.3 = 58\ kW$$

$$P_P = \frac{P_a}{E_P} = \frac{58.3\ kW}{72\%} \times 100\% = 80.9 = \underline{81\ kW}$$

Example Problem 15.2

A city is served partly by a deep well. This well is pumped at a constant rate of 35 L/s to an overhead storage tank through a 200 mm diameter and 460 m long rising main. The difference in water levels, including drawdown, is 32 m. Assume the coefficient of friction f = 0.02. What power is required by an electric motor assuming the overall efficiency of the pumping unit is 70%?

Given: D = 200 mm
L = 460 m

f = 0.02
Q = 35 L/s
ΔZ = 32 m
E_o = 70%
P_I = ?

SOLUTION:

$$v_2 = \frac{Q}{A_2} = \frac{4Q}{\pi D^2} = \frac{4}{\pi} \times \frac{0.035\ m^3}{s} \times \frac{1}{(0.20\ m)^2} = 1.114 = 1.1\ m/s$$

$$h_f = f \times \frac{L}{D} \times \frac{v_2^2}{2g} = 0.02 \times \frac{460\ m}{0.20\ m} \times \left(\frac{1.11 m}{s}\right)^2 \times \frac{s^2}{2 \times 9.81 m}$$

$$= 2.86 = 2.9\ m$$

$$h_a = \frac{p_2}{\gamma} - \frac{p_1}{\gamma} + Z_2 - Z_1 + h_l = 0 + 32\ m + 2.87\ m = 34.87\ m = 34.9\ m$$

$$P_a = Q \times \gamma \times h_a = \frac{0.035 m^3}{s} \times \frac{9.81\ kN}{m^3} \times 34.87\ m \times \frac{kW.s}{kN.m}$$

$$= 11.97 = 12\ kW$$

$$P_M == \frac{P_a}{E_o} = \frac{11.97\ kW}{0.70} = 17.1 = 17\ kW = 23\ hp$$

Example Problem 15.3

Water is pumped at a rate of 180 L/s from a lower reservoir to an upper reservoir. The elevations of the water surfaces in the reservoirs are 110.0 m and 120.0 m, respectively. Two reservoirs are connected via pipeline BC of 450 mm diameter and length of 1.2 km. At junction C it splits into two pipelines, CD and CE. Pipeline CD is 300 mm and 830 m long, while pipeline CE is 250 mm and 710 m long. For all three pipelines, assume a friction factor of 0.025.

i. Calculate the head loss due to friction in pipeline BC carrying a flow of 180 L/s.
ii. Calculate the flows carried by line CD and CE.
iii. Calculate the pumping head and pump power, assuming a pump efficiency of 80%.

GIVEN:

Parameter	A	B	C	F
Elevation, Z, m	110.0	?	Same as B	120.0
Head, m	0	?	?	0
Hyd. head, m	110.0	?	?	120.0
Parameter	**BC**	**CD**	**CE**	
Length, m	1200	830	710	

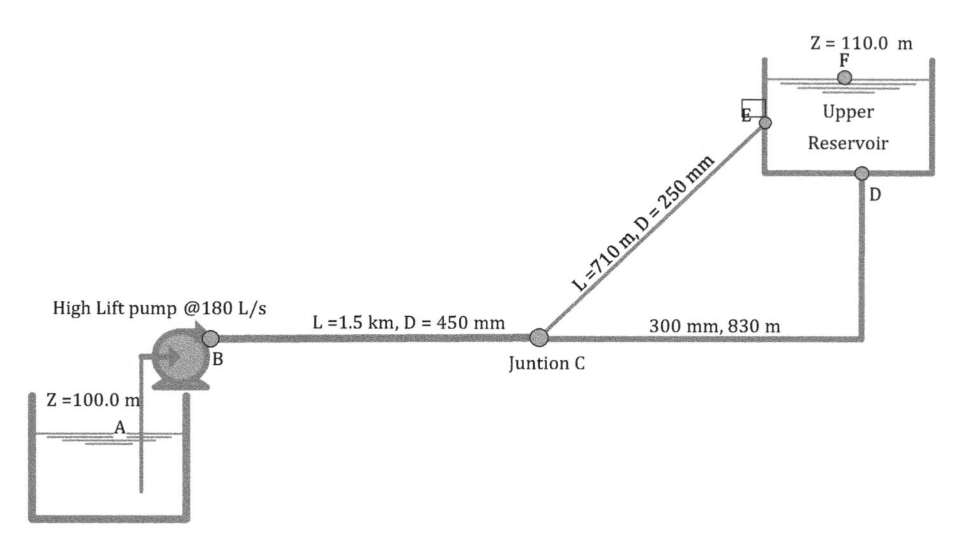

FIGURE 15.2 Simple Distribution Pumping System

Parameter	**BC**	**CD**	**CE**
Diameter, mm	450	300	250
Friction factor, f	0.025	0.025	0.025
Head loss, m	?	?	Same as CD
Friction slope, S_f, %	?	?	Same as CD

SOLUTION:

$$h_f(BC) = \frac{fL}{1.23g} \times \frac{Q^2}{D^5} = \frac{0.025\ s^2}{12.1\ m} \times \frac{1200\ m}{(0.45\ m)^5} \times \left(\frac{0.18\ m^3}{s}\right)^2$$

$$= 4.35 = 4.4\ m$$

Since head loss in CD = head loss in CE.

$$\frac{Q_{CD}}{Q_{CE}} = \sqrt{\frac{L_{CE}}{L_{CD}} \times \left(\frac{D_{CD}}{D_{CE}}\right)^5} = \sqrt{\frac{710}{830} \times \left(\frac{300}{250}\right)^5} = 1.458\ \ or\ Q_{CD} = 1.46 Q_{CE}$$

$$Q_{CD} + Q_{CE} = 180\ L/s\ \ 1.458 Q_{CE} + Q_{CE} = 180\ L/s$$

$$Q_{CE} = \frac{180\ L/s}{2.458} = 73.2 = 73\ L/s \qquad Q_{CD} = 180 - 73.2 = 106.8 = \underline{107\ L/s}$$

$$h_f(CD) = \frac{fL}{1.23g} \times \frac{Q^2}{D^5} = \frac{0.0250\ s^2}{12.1\ m} \times \frac{830\ m}{(0.300\ m)^5} \times \left(\frac{0.107\ m^3}{s}\right)^2$$

$$= 8.07 = 8.1\ m$$

$$h_f(Check) = \frac{fL}{1.23g}(Check) = \frac{fL}{1.23g} \times \frac{Q^2}{D^5} = \frac{0.0250\ s^2}{12.1\ m} \times \frac{710\ m}{(0.250\ m)^5} \times \left(\frac{0.073\ m^3}{s}\right)^2$$

$$= 8.005 = \underline{8.0\ m}$$

$$h_a = \frac{p_2}{\gamma} - \frac{p_1}{\gamma} + Z_2 - Z_1 + h_l = 0 + 10\ m + 0 + 4.35\ m + 8.07\ m$$

$$= 22.42 = 22.4\ m$$

$$P_a = Q \times \gamma \times h_a = \frac{0.180 m^3}{s} \times \frac{9.81\ kN}{m^3} \times 22.42\ m \times \frac{kW.s}{kN.m}$$

$$= 39.58 = 40\ kW$$

$$P_P = \frac{P_a}{E_P} = \frac{39.58\ kW}{80\%} \times 100\% = 49.48 = \underline{49\ kW}$$

15.4 SYSTEM HEAD

The operating point of the pump is determined by the system it is serving. For a given water system, an increase in discharge will cause an increase in head loss since the head loss in a straight pipe is proportional to the flow rate squared. Hence, as shown in **Figure 15.3**, the shape of the system head curve is concave upwards. System head consists of two components: fixed or static head, and a variable component due to head losses.

$$System\ head,\ h_{sys} = Fixed\ head,\ h_0 + Variable\ head,\ h_l$$

$$h_{sys} = h_1 - h_2 + h_L = \frac{p_2 - p_1}{\gamma} + z_2 - z_1 + h_L$$

Points 1 and 2 refer to open reservoirs, $p_1 = p_2 = 0$, and the fixed head equals the total lift. The second term in the system head is the head loss, which can be described using a flow equation such as the Darcy–Weisbach equation. The system head is equal to the static head plus the head losses. Major head losses are calculated using the flow formula, and minor losses are usually neglected or considered by applying the equivalent length technique. The relationship between system head and flow rate is called a system head curve. By superimposing the system curve on the pump curve, the point of intersection is the operating point of the pump serving the system (**Figure 15.3**).

By varying the head (opening or closing valves), the pump operating point will shift along the head capacity curve. For a given system, the pump that operates in the vicinity of BEP should be selected. If the system head or demand varies over a wider range, a variable speed pump or a combination (parallel and series) of pumps may be the answer.

Example Problem 15.4

Pump and system characteristics are given in **Table 15.1**. Plot the pump curve and system head curve and read the operating point.

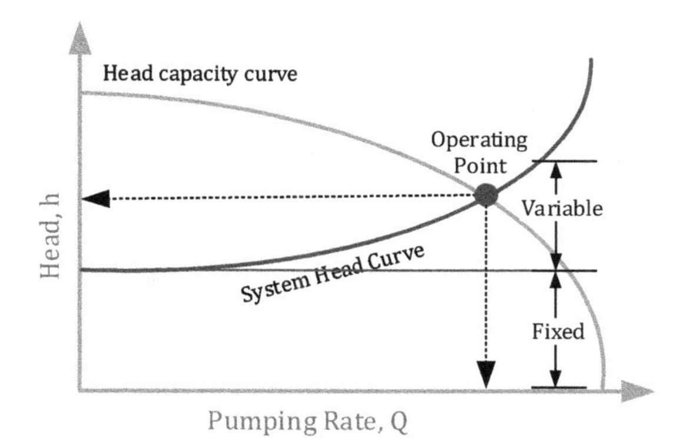

FIGURE 15.3 Operating Point of a Pump

TABLE 15.1
Pump System Head (Ex. Prob. 15.4)

Pump Characteristics			System Head	
Flow, L/s	**Head, m**	**Efficiency, %**	**Flow, L/s**	**Head, m**
15	29	22	13	17
27	28	37	33	18
40	27	49	47	20
53	27	57	57	21
67	26	63	65	23
80	24	64	72	24
93	22	62	80	26
107	20	59	87	27
120	15	54		

SOLUTION:

The curves are shown in **Figure 15.4**. The operating point is where both curves meet (74 L/s, 24.5 m, 65%).

15.4.1 System Head Equation

Both fixed head and variable head can be expressed in terms of flow rate. The system head equation can be expressed in terms of flow rate as follows:

$$H = h_{sys} = Z_2 - Z_1 + \frac{fLQ^2}{1.23gD^5} = \Delta Z + KQ^2;\ K = \frac{fL}{1.23gD^5}$$

15.4.2 Operating Point

The pumping rate for the operating point can be found by equating the system head to the pumping head described earlier.

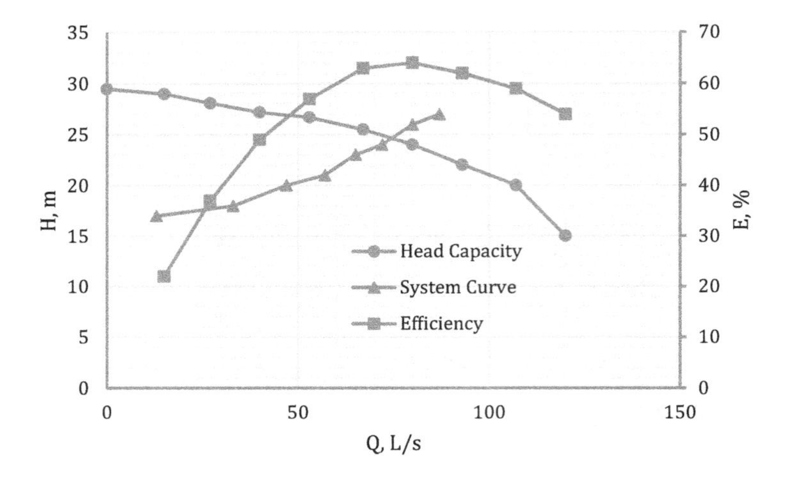

FIGURE 15.4 Pump Operating Point (Ex. Prob. 15.4)

$$H = h_a\left(pump\right) = AQ^2 + BQ + C \quad \text{Comparing the two}$$

$$aQ^2 + bQ + c = 0$$

Where: a = A-K, b = B, c = C - ΔZ; solving the quadratic equation. Q can be found.

$$Q = \frac{-B - \sqrt{\left(B^2 - 4ac\right)}}{2a} = \frac{-B - \sqrt{\left(B^2 - 4\left(A - K\right)\left(C - \Delta Z\right)\right)}}{2\left(A - K\right)}$$

Example Problem 15.5

A pump delivers water from a lower reservoir to an upper reservoir via a 400 mm diameter and 1500 m long pipeline. Assuming a friction factor f = 0,02, and knowing that the difference in elevation of the water levels of the two reservoirs is 39 m, find the system head equation. Knowing the pump head equation is $H = 52 - 230Q^2 + 11Q$, find the operating point of the pump.

Given: Q = ?
f = 0.02
L = 1500 m
D = 400 mm
ΔZ = 39 m

SOLUTION:

$$K = \frac{fL}{1.23gD^5} = \frac{0.02}{12} \times \frac{1500}{0.4^5} = 244.1 = 244 \qquad H = h_{sys} = 39 + 244Q^2$$

$$Q = \frac{-B - \sqrt{\left(B^2 - 4\left(A - K\right)\left(C - \Delta Z\right)\right)}}{2\left(A - K\right)}$$

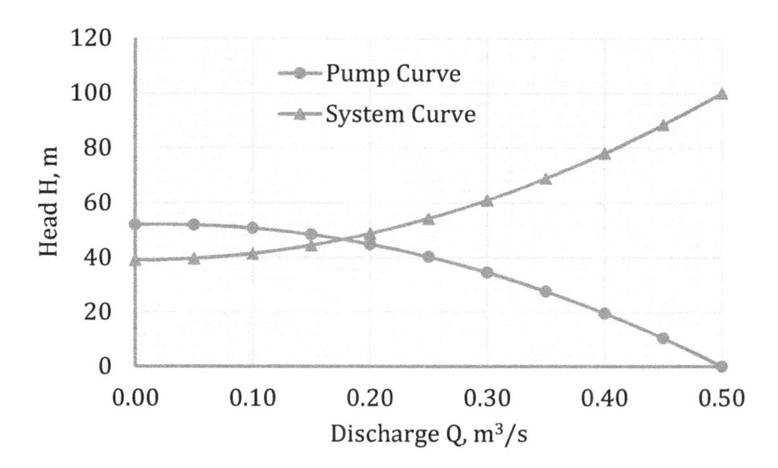

FIGURE 15.5 Operating Point (Ex. Prob. 15.5)

$$= \frac{-11 - \sqrt{\left(11^2 - 4(-230-244)(52-39)\right)}}{2(-230-244)} = 0.1776 = 0.18\ m^3/s$$

This value is the same as the value read from the intersection point of the pump and system head curves shown in **Figure 15.5**.

15.5 AFFINITY LAWS

The performance of centrifugal pumps can be adjusted by varying the rotative speed, N. To some degree, this can also be accomplished by trimming down the pump impeller. Affinity laws determine how pump characteristics change in either speed or diameter. For a given pump, the impeller diameter is fixed or D = constant. When the speed is changed from N_1 to N_2 and ratio N_2/N_1 = constant, C, the pump characteristics change according to the following relationships.

$$\frac{Q_2}{Q_1} = \frac{N_2}{N_1} \qquad \frac{H_2}{H_1} = \left(\frac{N_2}{N_1}\right)^2 \qquad \frac{P_2}{P_1} = \left(\frac{N_2}{N_1}\right)^3$$

When the diameter, D, of the impeller is reduced (trimmed) from D_1 to D_2 such that ratio D_2/D_1 = *constant, C* and speed remain the same, and similar relationships apply except you use a new constant value. At various pump speeds, the factor H/Q^2 remains constant, thus the pump speed can be adjusted to produce desired characteristics.

15.6 SPECIFIC SPEED

In order to compare the performance of different pumps, the term specific speed is used, which is common to all centrifugal pumps. Also called type number, specific speed is another performance factor widely used for the preliminary design and selection of pumps.

$$N_s(English) = \frac{N(rpm)\sqrt{Q(gpm)}}{\left(H(ft)\right)^{3/4}} \qquad N_s(Metric) = \frac{N(rpm)\sqrt{Q(L/s)}}{\left(H(m)\right)^{3/4}}$$

The values of Q and H correspond to the best efficiency point (maximum efficiency) for the shaft speed used. The value in metric units is 61% of the value in English units. Unfortunately, the units shown above do not use a dimensionless form for N_s. In order to write this equation in consistent units and make it a dimensionless number, the denominator has to contain the value of acceleration due to gravity, as shown below:

$$N_s = \frac{N(rps)\sqrt{Q(m^3/s)}}{\left(gH(m)\right)^{3/4}}$$

The values of Q and H correspond to the BEP (maximum efficiency) for the shaft speed used. Most of the published data obtained are based on the first form of the equation. Multiplying the dimensionless number by a factor of 17,200 gives the specific speed in the old fashion. For a multistage pump, the value of H should be the same as that for a one-stage pump. A given specific speed refers to a certain combination of head, speed and capacity, which is typical of a given type of pump. Specific speed can tell us the combination of factors that are both possible and desirable. Low specific speeds refer to centrifugal pumps, whereas axial flow (propeller type) pumps have high specific speeds. Best efficiencies are obtained when specific speeds are in the range 1500–3000.

15.7 HOMOLOGOUS PUMPS

Dynamically similar pumps are called homologous pumps. In addition to geometrical similarity, homologous pumps have the same specific speed, the same operating efficiency and a similar flow pattern. When two pumps are dynamically similar, the capacity is proportional to diameter cubed and head is proportional to diameter squared.

Multiple Pumps

When the desired capacity is beyond the range of a single suitable pump, two pumps operating in parallel may be used. Similarly, for high head requirements, two pumps connected in series will be more efficient. When two or more pumps are used in series or in parallel, a combined pump curve should be developed to determine the operating point. When multiple pumps are running in parallel, the head remains the same and discharge rates are additive. On the other hand, when pumps are in series, heads are additives and the discharge rate remain the same.

In medium to large communities, pumps are usually a combination of small and large pumps connected in parallel to meet the varying demand. Some pumps run all the time, while other pumps come on when demand increases.

In smaller communities, sometimes a variable speed pump, or variable frequency drive (VFD), may suffice. As discussed earlier, pump characteristics change when there is a change in pump speed in accordance with affinity laws. In a VFD pump, an adjustable speed motor is used. Using a VFD motor, pump impeller speed is reduced from its designed operating point, effectively lowering the position of the discharge capacity curve and allowing the pump to operate at a low discharge rate without wasting electrical energy. Electronic sensors installed in the system at locations remote from the pump can control VFD pumps so that the pump speed matches the system requirements.

Newer models of VFD pumps are equipped with microprocessors that adjust both speed and pressure to meet system requirements without placing remote sensors.

Example Problem 15.7

The head capacity data of a centrifugal pump is shown in **Table 15.2**. The static head component of system head is 20 m and the variable head is 7620 Q^2, where Q is in m^3/s. Find the operating head if two pumps are connected in a) series and b) parallel.

TABLE 15.2
Pumps in Series and Parallel

Pump Head, m (Series)		Flow, L/s (Parallel)		System Head
H_1	H_{1+1}	Q_1	Q_{1+1}	H_{sys}
26	52	0	0	20
23	46	10	20	23
21	42	15	30	27
16	32	25	50	39
9	18	35	70	57
5	10	40	80	69

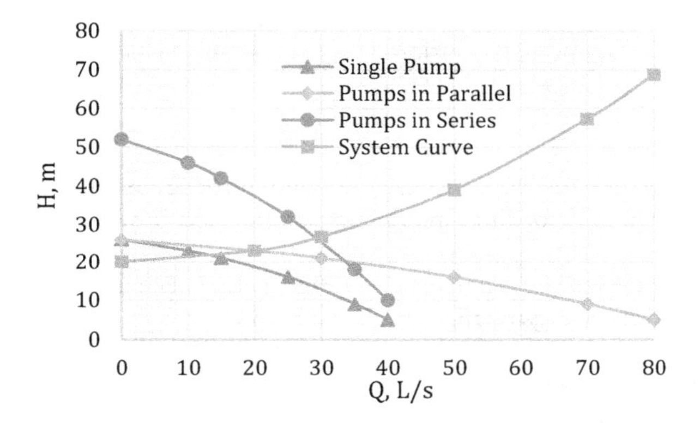

FIGURE 15.6 Two Identical Pumps in Series and Parallel

SOLUTION:

Pump and system data are shown in **Table 15.2**. When the pumps are connected in series, the heads are cumulative. Since the pumps are identical, for each discharge value the head is doubled, as shown in **Table 15.2**, column 2.

When the pumps are connected in parallel, the flows are cumulative. Since the pumps are identical, for each head value, flow is doubled, as shown in column 4 of **Table 15.2**. The system head is the sum of the fixed head (static) and variable head. For example, when Q = 20 L/s, the system head is:

$$H_{sys} = 20 + 7620Q^2 = 20 + 7620\left(\frac{20\ L}{s} \times \frac{m^3}{1000\ L}\right)^2 = 23.04 = 23\ m$$

System head values are shown in the last column of the table. A plot of pump curves and system curve is shown in **Figure 15.6**. The operating point is where the system head curve crosses the pump curves. The operating point for the single pump is 22 L/s, 15 m.

a) In series operation, the pump curve moves up since the heads are cumulative, and thus the operating point moves. In this case, the operating point as read from **Figure 15.6** is 29 L/s, 26 m.
b) For parallel operation, the pump curve moves towards the right since the flows are cumulative.
c) From **Figure 15.6**, the operating point for pumps running in parallel is 20 L/s, 22 m.

Example Problem 15.8

In the following table, the head capacity data of a centrifugal pump is as follows:

Q_1, L/s	0	60	75	85	100	110	120
H_1, m	75	60	55	52	46	40	30

The static head component of the system head is 30 m, 35 m and 40 m for low, half-full and full-tank levels, respectively. The total system head for the three tank levels and flow rates ranging from 140 L/s to 350 L/s are shown in **Table 15.3**.

a) Plot the head capacity curves and system head curves for all the combinations.
b) Read the operating point for a single-pump, tank-low scenario.
c) Read the operating point for a double-pump, tank-half-full scenario.
d) Read the operating point for a triple-pump, tank-full scenario.

SOLUTION:

For pumps operating in parallel, the flows are cumulative. Thus, for two identical pumps, the flows are doubled, and for three pumps in parallel, the flows are tripled, as shown in **Table 15.4**.

Plotted curves for all scenarios are shown in **Figure 15.7**. The operating points for various combinations as read from **Figure 15.7** are shown below:

Single pump pumping into a low tank: Q = 115 L/s, H = 36 m
Double pump pumping into a half-full tank: Q = 195 L/s, H = 46 m
Triple pump pumping into a full tank: Q = 225 L/s, H = 54 m

TABLE 15.3
System Head (Ex. Prob. 15.8)

Q, L/s		**0**	**140**	**180**	**225**	**270**	**350**
Tank Level	**Static, m**	**System Head, m**					
Low	30	30	37	40	45	50	60
Half	35	35	42	45	50	55	65
Full	40	40	47	50	55	60	70

TABLE 15.4
Identical Pumps in Parallel (Ex. Prob. 15.8)

H_1, m	75	60	55	52	46	40	30
Q_{1+1}, L/s	0	60	75	85	100	110	120
Q_{1+1}, L/s	0	120	150	170	200	220	240
Q_{1+1+1}, L/s	0	180	225	255	300	330	360

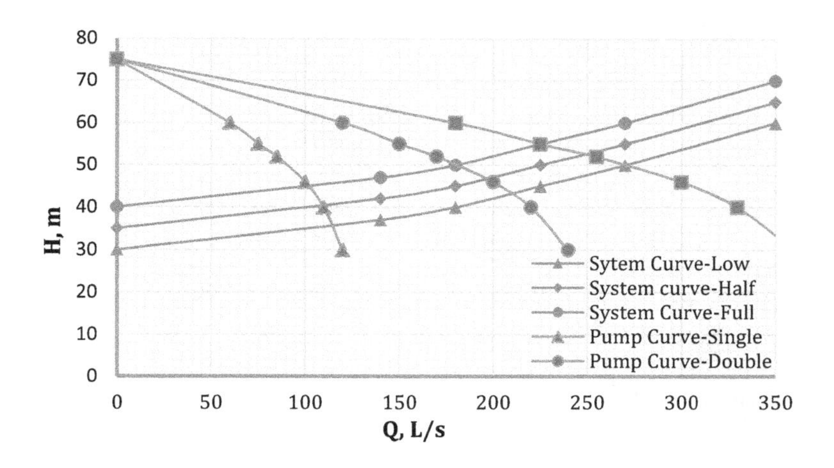

FIGURE 15.7 Pumps Operating in Parallel (Ex. Prob.15.8)

Example Problem 15.9

A pump is to supply water against a static head of 35 m. The suction pipe is 300 mm in diameter, with a friction factor f = 0.020 and 48 m in length, including equivalent length for minor losses. The discharge pipe is 250 mm in diameter, f =0.025 and is 1.3 km long. Add 75 m of length to account for minor losses in the discharge pipe. The pump head capacity relationship is described by the following equation: $h_p = 95 - 5600\ Q^2$, in SI units. Find the operating point of the pump.

GIVEN:

Pipe	D. mm	L, m	f	$h_f = \frac{f}{12.1} \times \frac{L}{D^5} \times Q^2$
Suction	300	48	0.020	$\frac{0.020}{12.1} \times \frac{48}{0.30^5} \times Q^2 = 32.6Q^2$
Delivery	250	1300+75=1375	0.025	$\frac{0.025}{12.1} \times \frac{1375}{0.250^5} \times Q^2 = 2909Q^2$
			Σ	$2942Q^2$

SOLUTION:

$$System\ h_{sys} = h_L + z_2 - z_1 = 2942Q^2 + 35$$

$$2942Q^2 + 35 = 95 - 5600Q^2$$

$$8542Q^2 = 60\ \ or\ Q = \sqrt{60/8542} = 0.0838\ m^3/s$$

$$h_p = 95 - 5600\ Q^2 = 95 - 5600 \times 0.0838^2 = 55.66 = 56\ m$$

$$Check\ h_P = 2942Q^2 + 35 = 2942 \times 0.0838^2 = 55.66 = 56\ m$$

15.8 CAVITATION

The vapour pressure of the liquid refers to the pressure at which liquid transforms into vapour. For any liquid, there is a definite relationship between the vapour pressure and the temperature of the fluid. Vapour pressure increases with an increase in temperature. At normal atmospheric pressure, water starts to boil at 100°C. In other words, at 100°C, the vapour pressure of water is one atmosphere, or about 100 kPa (abs). When operating a pump, the pressure at the pump inlet is lower than the intake due to losses. If the inlet pressure is allowed to drop close to the vapour pressure, bubbles of vapour will start forming. These bubbles suddenly collapse as they enter the high-pressure region of the pump body. This will create sudden noise and pitting of the metal surface due to the explosion of bubbles. This phenomenon is called cavitation.

15.8.1 NET POSITIVE SUCTION HEAD

The basic measure for protecting a pump against cavitation is to avoid pressures that are similar to the vapour pressure of the liquid. Pumps are operated under lift conditions or positive head conditions, and are more susceptible to cavitation when operated under lift conditions. The head available at the pump inlet, called the net positive suction head (NPSH), is less than at the intake due to losses and lift. To prevent cavitation, pumping systems are designed and operated such that NPSH remains above the vapour pressure head. Using the pump centreline as a reference, the net positive suction head available is given by:

$$\boxed{NPSH_A = h_{atm} - h_{vap} - (Z_2 - Z_1) - h_l}$$

h_{atm} = atmospheric pressure as head based on altitude
h_{vap} = vapour pressure head
h_l = head loss due to friction
Z_2 = pump elevation
Z_1 = pumping surface elevation
$(Z_2$-$Z_{1)}$ = suction lift

Note: In suction head conditions, $Z_1 > Z_2$ and the term $(Z_2 - Z_1)$ is negative. In suction lift conditions, $Z_1 < Z_2$ and the term $(Z_2 - Z_1)$ is positive.

Example Problem 15.10

Determine the available NPSH for a pumping system pumping water at 20°C from a well when the pumping water level is 2.5 m below the pump. The atmospheric pressure is at 101 kPa. Total head losses in the suction line are estimated to be 0.45 m.

Given: p_{atm} = 101 kPa (abs)
Z_2-Z_1 = Suction lift = 2.5 m
h_l = 0.45 m
T = 20°C
p_{vap} = 2.34 kPa (abs)
γ = 9.79 kPa/m

SOLUTION:

$$NPSH_A = h_{atm} - h_{vap} - (Z_2 - Z_1) - h_l$$

$$= (101 - 2.34)\,kPa \times \frac{m}{9.79\ kPa} - 2.5\ m - 0.45\ m = 7.05 = \underline{7.0\ m}$$

15.8.2 Permissible Suction Lift

Manufacturers specify the NPSH required (NPSHR) for the efficient operation of their pumps. As long as the machine is operated above this value, the operation will be satisfactory. When the pumping level is open to the atmosphere and there is a positive suction lift (negative head), it is important to check that NPSHA > NPSHR. The maximum lift allowed for a given pump can be found by knowing the NPSHR and the vapour pressure of the liquid being pumped. For a given suction lift, SL (the height of the pump above the pumping level), the net positive suction head available is given by the following expression.

$$NPSH_A = h_{atm} - h_{vap} - SL - h_l$$

The maximum permissible suction lift (MPSL) that does not cause cavitation corresponds to the minimum NPSHA = NPSHR. By making these substitutions, the equation for calculating the maximum permissible suction lift is as follows:

$$\boxed{MPSL = h_{atm} - h_l - NPSH_R - h_{vap}}$$

Example Problem 15.11

Find the permissible suction lift for a pump that requires 3.5 m of NPSH. The total head losses in the suction line are estimated to be 0.7 m. The temperature of the water is 15°C and the atmospheric pressure head is 10.2 m.

Given: $h_l = 0.7$ m
$NPSH_R = 3.5$ m
$T = 15°C$
$h_{vap} = 0.21$ m
$h_{ams} = 10.2$ m

SOLUTION:

$$MPSL = h_{atm} - h_l - NPSH_R - h_{vap} = 10.2\ m - 0.21m - 3.5\ m - 0.70\ m = 5.78 = \underline{5.8\ m}$$

If the pump is placed more than 5.8 m above the pumping water level, it will cause cavitation.

15.9 OPERATION AND MAINTENANCE

Proper operation and maintenance (O&M) procedures must be followed to obtain satisfactory service from centrifugal pumps. Pumps must be properly lubricated using the recommended lubricant (both under- and over-lubrication can be damaging). The rotating part is the only moving part in the casing. A packing gland or seal is used where the pump shaft protrudes to stop air from leaking in or water leaking out. Never overtighten the packing. To ensure a seal is working properly, tighten it to allow some leakage of about 15–20 drops per minute. In the case of mechanical seals, no leakage is allowed. Never overtighten the seal if too much water starts coming out.

A centrifugal pump must be primed or filled with water when it is started. Many pumping units have self-priming units attached to them. The foot valve is designed to prevent the suction line from emptying and to keep the pump primed. If the pump is placed below the pumping water level or suction head conditions, the pump is always primed. The pump should be started with the suction valve open and the discharge valve closed. As the motor picks up speed, the discharge valve is opened slowly. In many cases, this is an automatic operation. When shutting off the pump, the discharge valve must be closed slowly to avoid water hammer. Water hammer refers to tremendous transient pressures that can damage pipes or pumps.

Routine inspection of pumps is important to check for noise, vibrations, alignment, excessive heat, and leakage from gland packing. Vibrations are an indication that the pumping unit is out of alignment and, if not corrected, can lead to premature failures.

Discussion Questions

1. Explain why the efficiency of diffuser pumps is better than volute pumps.
2. Briefly describe the principle on which centrifugal pumps work.
3. Explain why centrifugal type pumps are more common in water systems.
4. Compare reciprocating pumps with centrifugal pumps.
5. How would pump characteristics change if two identical pumps were coupled in series or in parallel?
6. How would you go about choosing the right pump for a given water pumping system?
7. In suction lift conditions, it is recommended to use minimal fittings and preferably a larger suction pipe. Discuss.
8. Describe the relationship between a specific speed and type of velocity pump.
9. How does the operation of a pumping system affect its operating efficiency?
10. What steps can be taken to prevent pump cavitation?
11. Cavitation problems are common when pumping relatively warm liquids and at high altitudes. Comment.
12. What types of pumps are commonly used to withdraw water from water wells?
13. What are the two components of system head? Discuss the shape of a system head curve and the factors affecting it.
14. Sketch a typical head capacity curve and an efficiency curve for a centrifugal pump, and use these curves to explain shut-off head and best efficiency point.
15. Compare two identical pumps connected in series and in parallel. In water distribution systems, what arrangement is common to meet varying demand?
16. Discuss how a change in impeller speed of a centrifugal pump affects head, capacity and power requirements.
17. What is VFD? What is the principle on which it works, and where is it commonly used?
18. Make a sketch showing the difference between suction lift and suction head for a centrifugal pump. Wherever practical, suction head conditions are preferred. Explain why.

Practice Problems

1. A prototype pump has an impeller diameter exactly three times that of a model test pump. The model pump delivers 0.1 L/s when operating against a head of 7.5 m at the best efficiency point. Predict the head and capacity of the prototype operating at the same rpm as the test pump. (2.7 L/s, 68 m)
2. Find the NPSHA when pumping water at 80°C from a well. The pumping water level is 2.5 m below the pump, and assume the atmospheric pressure is 101 kPa. Head losses in the suction line are estimated to be 0.45 m. (2.7 m)

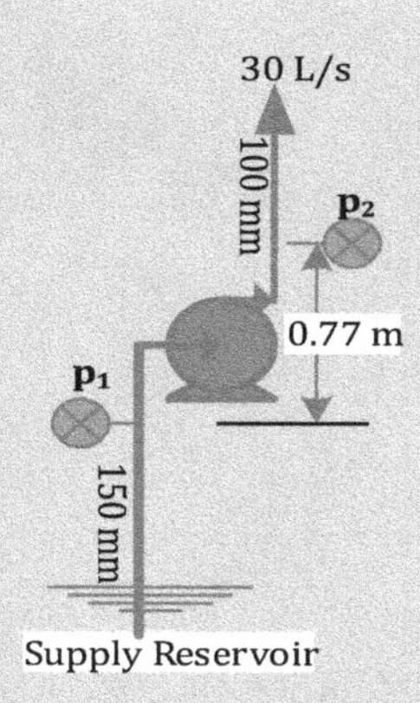

FIGURE 15.8 Pumping System

3. Find the maximum permissible suction lift (MPSL) for a pump that requires 3.5 m of NPSH for the following conditions. Assume total head losses in the suction line are 0.7 m.
   a. The temperature of the water is 60°C and atmospheric pressure is normal. (4.1 m)
   b. The atmospheric pressure is 80 kPa and water temperature is 20°C. (3.7 m)
4. In a pumping system, the suction gauge reads 42 cm of mercury column (vacuum) and the discharge gauge reads 310 kPa. Assuming there is no elevation difference between the two gauges, how much head is added to the pump? (37 m)
5. As shown in **Figure 15.8**, the gauges attached to the suction side (150 mm) and discharge side (100 mm) of a pump read a vacuum of 250 mm of mercury and 140 kPa, respectively. The gauge on the suction side is placed 0.60 m below the pump and the discharge pressure gauge 0.17 m above the pump. Calculate the power required to drive the pump when pumping water at 30 L/s, assuming the pump is 60% efficient. (9.0 kW)
6. A water treatment plant gets its water from a lake. The intake is below the water surface at an elevation of 230.4 m. The lake water is pumped to the plant influent at an elevation of 242.4 m. The total head losses are estimated to be 5.5 m when water is drawn at a rate of 340 L/s. Calculate the pumping head, waterpower and pump power, assuming the pump is 72% efficient. (18 m, 58 kW, 81kW)
7. A centrifugal pump running at 1400 rpm has the following characteristics:

Q, L/s	13	19	25	31	38	44	50
H, m	28	28	26	25	23	21	18
E, %	65	70	73	74	72	69	63

   Plot the performance curves and determine the best efficiency point. At the maximum efficiency, find the power required to drive the pump. (31 L/s, 25 m, 74%, 10 kW)
8. How many stages of a multistage pump are required to pump water at 70 L/s against a total head of 185 m? The speed of the pump is 750 rpm and the specific speed of the pump is not to exceed 700. (10)
9. A centrifugal pump with an efficiency of 65% discharges 100 L/s into a system that includes 900 m of 250 mm diameter pipe with a C of 100. If the total static head is 28 m, compute the required pump power. (77 kW)
10. A centrifugal pump is required to discharge water at 55 L/s against a total static head of 32 m through a discharge pipeline 150 mm in diameter and 100 m long. Assume a

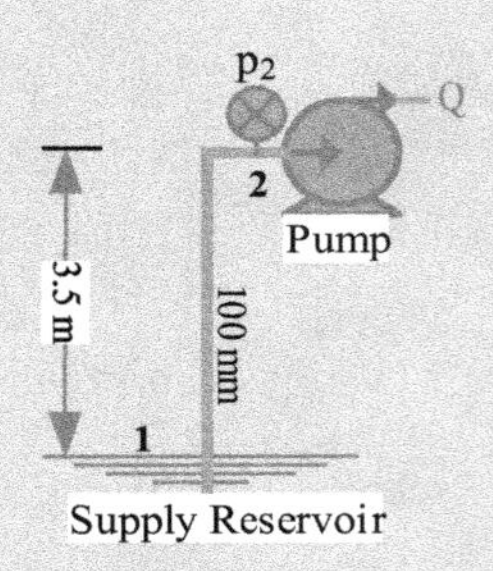

FIGURE 15.9 Pumping with Lift

friction factor f = 0.02 and minor losses equivalent to 10 m of length. Determine the pump power assuming the pump is 65% efficient. (33 kW)

11. As shown in **Figure 15.9**, a centrifugal pump delivers water at a rate of 15 L/s when static suction lift is 3.5 m. The pressure gauge on the suction side just before the water enters the pump indicates a reading of 61 kPa absolute. Find the head losses on the suction side of the pump. (0.60 m)
12. A single-stage centrifugal pump is discharging 7.5 L/s producing a head of 15 m. A tachometer registers the speed of the pump at 1250 rpm and the pump power required to run the pump is 6.0 kW. If the rotational speed of the pump is increased to 1450 rpm, find the new discharge, head and power? (8.7 L/s, 20 m, 9.4 kW)
13. At a certain location, atmospheric pressure and vapour pressure are 95 kPa and 3.0 kPa, respectively, expressed as absolute pressure. If the NPSH required to run the pump is 3.0 m, what maximum suction lift can be afforded, assuming head losses on the suction side to be 0.40 m. (6.0 m)
14. An 8.5 km long, 600 mm diameter water main carries water from the pumping station to a reservoir with a water level of 35 m above the supply reservoir. Assuming f = 0.022, find the system head equation. ($H = 35 + 200Q^2$)
15. A pump with a head capacity equation $H = 52 - 230Q^2 + 11Q$ (SI) serves a system with a system head equation $H = 38 + 180Q^2$. What is the pumping rate? (200 L/s)
16. In a water pumping system, the discharge pressure gauge is 0.50 m above the suction gauge. During pumping, the suction gauge reads 36 cm of mercury column (vacuum) and the discharge gauge reads 370 kPa. What is the pumping head? (43 m)
17. A pumping unit draws 22 kW of power when pumping at 2.7 m^3/min against a pressure of 310 kPa. What is the overall efficiency of the pumping unit? (63%)
18. A pump delivers 30 L/s at 1000 rpm against a total head of 15 m. Determine its performance at 1100 rpm. (33 L/s, 18 m)
19. A water treatment plant gets its water from a reservoir. Water is pumped to the plant influent at an elevation of 24. 2 m above the water level in the reservoir. The total head losses are estimated to be 6.5 m when water is drawn at a rate of 360 L/s. Calculate the pumping head, waterpower and pump power, assuming the pump is 75% efficient. (31 m, 110 kW, 140 kW)
20. A centrifugal pump is pumping into a 4500 m long water pipe that has an inside diameter of 305 mm and a friction coefficient, C, of 100. The system has a total static head of 50 m. The pump characteristics are shown in the table below. Plot the pump head capacity curve and system head curve to determine the operating point. (90 L/s, 83 m)

Q, L/s	0	25	50	75	100
H, m	145	140	126	104	65

21. A pump lifts water from a lower reservoir to an upper reservoir with a static lift of 11 m. The pipeline is 250 mm in diameter and 1500 m in length. The pump characteristics are given in the following table.

Q, L/s	0	8	16	24	32	40	48
H, m	20	19	17	14	11	7.4	3.6
E, %	0	48	69	75	70	60	47

Assuming the friction factor f = 0.04 and minor losses are five times the velocity head, find the system head equation and operating point of the pump. ($H_{sys} = 11 + 5230Q^2$) (24 L/s, 14 m, 75%)

16 Wastewater Collection System

The sewage collection system is made up of several different components, all working together to carry waste to the wastewater treatment plant. The main components of the sewage collection system are building services (building sewers), sewer mains, lift stations, force mains and a wastewater treatment plant. Sewage moves from the point of origin to the wastewater treatment plant in the following steps shown in **Figure 16.1**.

The sewage drains from the building sewer to the sewer main. This is a pipe at least 8 in (200 mm) in diameter. It is graded and slopes to a low point in the town where the sewage is collected in a lift station. Depending on the size of the city, sewer lines may consist of laterals, submains, mains, trunk sewers and intercepting sewers. The lateral is the first line to no other joins except the building sewer. Submains can receive flow from more than one lateral or connect directly to the building sewer. Trunk sewers receive wastewater from a large area.

The lateral, being the smallest in size, is subjected to greater variations of flow, and as we go up in the hierarchy pipe, size increases and flows get smoothed out. For this reason, laterals are designed to carry peak flow flowing half full, and mains and trunk sewers are designed to flow more than half full.

Manholes for servicing are located along the sewer mains at distances of 100–150 m. At low points in the collection system, a lift station is installed. All nearby sewers drain into these low areas. Pumps located in the station then move the collected sewage via a force main to the wastewater treatment plant or main sewer. A force main is an isolated line linking the lift station and the disposal area. It is called a force main because it flows under pressure, while other sewer mains are designed to flow partially full under gravity or open-channel flow conditions.

A wastewater treatment plant is either a lagoon system or a mechanical treatment plant. Lagoons consist of holding ponds where the sewage is treated by natural aerobic and anaerobic biodegradation processes brought by bacteria, sunlight and wind. Mechanical treatment plants treat the wastewater through a combination of settling, aeration and chemical treatment. To produce effluents of acceptable quality, biological treatment (secondary treatment) is required as a minimum level of treatment.

COMBINED SEWERS

Most communities have two sewer systems, one for carrying sewage and one for carrying storm water. However, some older communities have only one sewer system that collects both wastewater and storm water together. This is called a combined sewer system. Combined sewers save costs on sewer installation; however, they need to be much larger in diameter to accommodate storm flow, and the wastewater treatment plant is also larger than necessary to treat just sewage. Most communities are switching to separate sewer systems due to the fact that during wet weather, combined discharges exceed the capacity of the plant, thus flows are bypassed, resulting in pollution of the receiving water body.

STORM SEWERS

The purpose of storm sewers is to carry drainage water or storm run-off. Since combined sewer systems lead to pollution and contamination, separate sewer systems are the norm in all developed

DOI: 10.1201/9781003231264-16

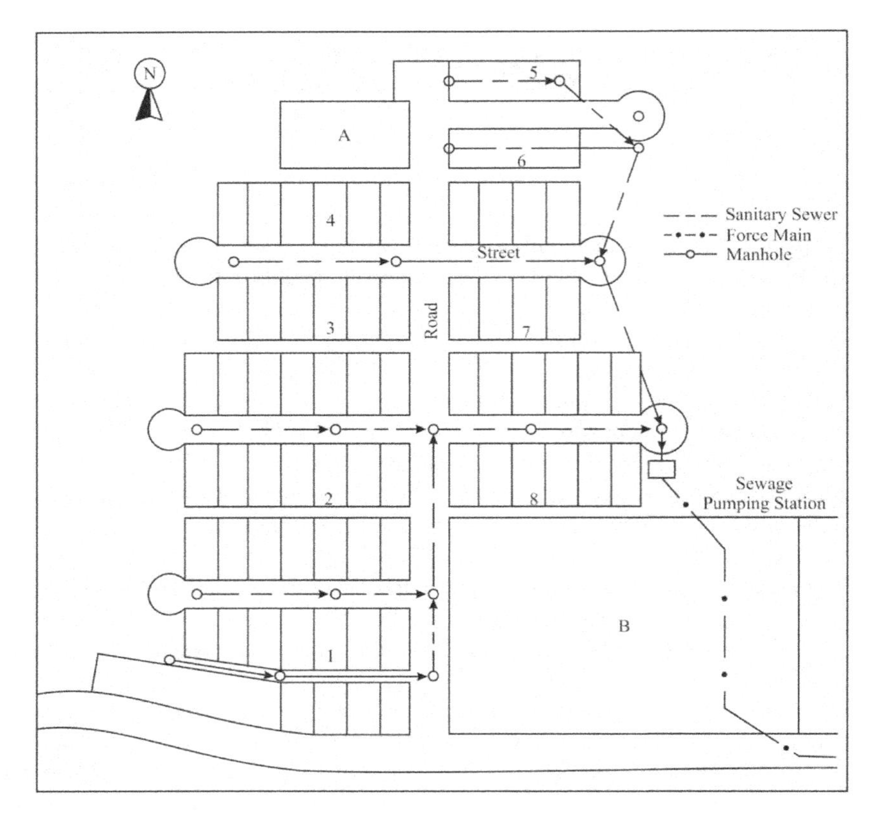

FIGURE 16.1 Sewage Collection System

TABLE 16.1
Storm Sewers versus Sanitary Sewers

Parameter	Sanitary Sewers	Storm Sewers
Size	Relatively smaller	Much larger
Depth	Much deeper	Shallow
Flow and season	Flow year around	During run-off season
Design	Based on 3–4 × avg. daily	Peak run-off rate
Flow depth	Partial	Full
Flow velocity	> 0.5 m/s	> 1.0 m/s
Surcharging	Not permitted	In extreme events

areas. Storm sewers are different in many aspects compared to sanitary sewers. Since storm sewers are supposed to carry peak storm flows, they are larger and not as deep as sanitary sewers. Storm sewers are designed to flow full and are allowed to surcharge during extreme events. Some of the striking differences between the two types of sewer systems are listed in **Table 16.1.**

16.1 INFILTRATION AND INFLOW

Infiltration and inflow (I&I) are used to describe the flow of non-wastewater sources into the wastewater collection system. Infiltration refers to groundwater that enters the collection system through building connections, defective pipes, pipe joints and seals, manhole walls and other parts of the collection system that are not well sealed.

Inflow refers to water that enters the system from illegal connections such as sump pumps, roof, eavestrough drains, catch basin connections and storm water run-off over improperly sealed manhole covers or other access points. Exfiltration is the term used to indicate the wastewater leaking out into the ground during dry weather when the groundwater level is below the sewer pipe. If uncontrolled, exfiltration can lead to contamination.

I&I can cause significant problems to the wastewater collection system if it is not sized to handle the extra flow. Under normal conditions, not more than 10% of the daily flow should come from I&I. In a newer system, this should be less than 10%. It may also cause a problem at the wastewater treatment plant if too much water is coming through the plant to be treated efficiently. I&I may also dilute the wastewater, i.e. lower the strength of the sewage, if its source is relatively clean. Part of the wastewater collection system personnel's job will be to monitor the wastewater collection system at different times of year for I&I and mitigate it where possible.

16.2 WASTEWATER FLOWS

The rated capacity of a wastewater treatment plant is usually based on the average annual daily flow rate. However, from a practical point of view, it makes sense to understand how flow hydrographs change over the course of a day, week or even year to maximize treatment efficiency. Wastewater flow rates will vary depending on the use, the source of the wastewater and the time of day. A properly designed and operated collection system must be able to handle all these variables. If a wastewater collection system cannot handle all the volume generated, it may overflow to the surface or back up the building sewers into basements or floor drains.

Typical residential flow patterns can be described as bimodal, which means the pattern has two main peaks over the course of a day (**Figure 16.2**). The first peak usually occurs in the morning, as people get up and get ready to go to work. The second peak usually occurs in the evening, between 7 and 9 pm, when people return to their homes. Since a wastewater collection system not only collects and transports residential wastewater but industrial, commercial and institutional wastewater as well, it is important to recognize that different design factors will have to be considered for different parts of the collection system.

A similar flow pattern is experienced at the plant. However, the peaks are a bit smaller compared to the collection system, and the timings are lagged depending on the extent of the collection network. In a typical dry weather flow hydrograph, as shown in **Figure 16.2**, minimum flows at the plant are experienced during the early morning hours and peak flow just during afternoon hours. Flow peaks are smaller due to the levelling out effect as wastewater flows through various sections of the collection system. **Figure 16.2** shows a pollutograph of biochemical oxygen demand (BOD). It follows a similar pattern as wastewater flows, however it is slightly lagged. It also shows that BOD loading on the plant is minimal during early morning hours and high in the afternoon hours.

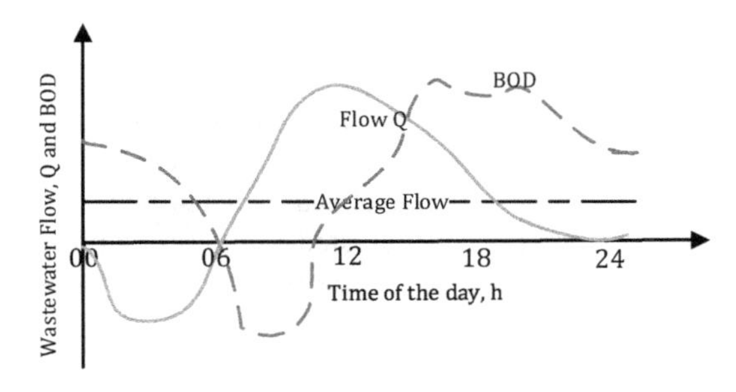

FIGURE 16.2 Dry Weather Wastewater Flow Hydrograph

Excessive I&I, as experienced in poorly maintained and older systems, can run havoc both on the collection system and plant operations. The difference in volume, as indicated by the differences in wet weather and dry weather flow hydrographs, is indicative of the degree of I&I. When describing the variable flow conditions in a collection system, several terms are used.

Base flow is the average of the daily flows sustained during dry weather periods with limited infiltration. It is the minimum flow reaching the plant.

Average wet weather flow is the average of the daily flows sustained during wet weather periods when infiltration is significant.

Average annual daily flow is the average flow rate over a 24-hour period based on annual flow data.

Instantaneous flow represents the highest recorded flow rate for any period. It should be noted that the recorded peak might, in some cases, be considerably lower than the actual peak due to recording equipment limitations.

Sustained flow is the flow rate value that is maintained for any given period.

Wastewater collection systems are usually designed according to peak flow rates. Peak flow rates can be described in terms of instantaneous peaks, or they can be based on an hourly, daily or weekly value. Peak hourly flow rates are typically used to size most components of a wastewater collection system, including lift station components.

16.3 SEWER MAINS

Sewer pipes, also called sewer mains, are installed as either gravity mains or force mains. In gravity mains, the sewage moves through the pipe by gravity alone. When gravity sewers are installed, they must be graded or sloped so that the sewage can be moved by gravity alone, with no other force or pressure required. In force mains, the sewage is pushed, or forced, through the sewer by a pump.

16.3.1 Pipe Size

Minimum pipe sizes (diameters) have been established for different parts of the collection system. Sewer mains pipes must be a minimum of 200 mm (8 in), house connection or building service pipes are a minimum of 100 mm (4 in) and commercial building pipes (such as in office buildings or apartments) are 150 mm (6 in) or larger.

Circular sewer pipe is manufactured with an inside diameter ranging from 100 mm (4 in) to as much as 3.5 m. In the range from 100 mm (4 in) to 300 mm (12 in), available pipe sizes are at an interval of 50 mm in diameter. In the middle range, 300 mm to 900 mm at a 75 mm interval is used. In larger pipe sizes with a diameter range of 900 mm to 3500 mm, pipe sizes are available at 150 mm intervals. Maximum pipe sizes vary with the material of the pipe. For example, the maximum pipe size of a concrete pipe is 1050 mm in diameter, whereas reinforced cement concrete (RCC) pipe is available in large sizes too.

16.3.2 Pipe Slope

Sewer lines are usually graded so that sewage will flow at a rate of at least 0.6 m/s when the sewer pipe is half full or more. This is called the minimum scour velocity for gravity sewer pipes. To achieve the minimum scour velocity, the pipe must be laid on a minimum slope or grade. If the sewer or building service line has less slope than is indicated in the sewer pipe grading (**Table 16.2**), the solids will sink to the bottom of the line and build up there. This process will eventually block the sewer line.

TABLE 16.2
Gravity Sewer Pipe Grading and Size

Pipe Size mm (in)	mm/ 30 m	%
100 (4")	312.5	1.0
150 (6")	180	0.6
200 (8")	120	0.4
250 (10")	85	0.3
300 (12")	65	0.2

Too much slope will cause the solids and the liquids in the sewage to separate, resulting in loss of scrubbing action in the pipes. Paper and other debris will be left behind in the pipe, causing eventual blockage of the sewer line. To prevent erosion of pipe surface, flow velocities of more than 2.5 m/s should be avoided.

16.3.3 Pipe Flow Velocity and Capacity

The sewer pipe is graded to provide a scouring velocity of 0.6 m/s when flowing full. The previous table defines the minimum slopes needed to achieve this velocity. Flow velocity and rate can be found by applying Manning's equation.

$$v = \frac{1}{n} \times R_h^{2/3} \times \sqrt{S} \quad - \quad SI$$

In this equation, R_h is the hydraulic radius. In circular pipes flowing full, the hydraulic radius is one-fourth the diameter of the pipe. Thus, Manning's equation for full flow becomes:

$$v_F = \frac{0.4}{n} \times D^{2/3} \times \sqrt{S} - SI$$

$$Q_F = \frac{0.312}{n} \times D^{8/3} \times \sqrt{S} - (SI)$$

This formula yields the flow velocity for full-flow conditions. A pipe flowing half full would carry exactly half of this full-flow capacity. The factor n is the roughness factor and is typically 0.013. Manning's equation is discussed in more detail in the chapter on the design of sewers. Some simple problems further illustrate the use of this equation.

Example Problem 16.1

Work out the full-flow velocity and flow capacity of a 300 mm sewer pipe laid on a slope of 0.20%.

Given: v = ?
D = 300 mm
S = 0.20%

SOLUTION:

$$v_F = \frac{0.4}{n} \times D^{2/3} \times \sqrt{S} = \frac{0.4}{0.013} \times (0.30)^{2/3} \times \sqrt{\frac{0.20\%}{100\%}} = 0.643 = \underline{0.66\, m/s}$$

$$Q_F = A_F \times v_F = \frac{\pi(0.30\ m)^2}{4} \times \frac{0.643\ m}{s} \times \frac{1000\ L}{m^3} = 43.5 = \underline{44\ L/s}$$

Example Problem 16.2

What is the maximum population that can be served by a 200 mm sanitary sewer laid on a grade to provide a full-flow velocity of 0.6 m/s? Assume the design flow to be 1600 L/c.d.

Given: v = 0.60 m/s
D = 200 mm
PE = ?

SOLUTION:

$$Q_F = A_F \times v_F = \frac{\pi(0.60\ m)^2}{4} \times \frac{0.60\ m}{s} \times \frac{1000\ L}{m^3} = 18.8 = 19\ L/s$$

$$p = \frac{18.8L}{s} \times \frac{60\ s}{min} \times \frac{1440\ min}{d} \times \frac{p.d}{1600L} = 1017 = \underline{1000\ people}$$

Example Problem 16.3

Determine the flow-carrying capacity of a 200 mm sewer pipe (n = 0.012) laid on a 0.33% slope.

Given: D = 0.20 m
n = 0.012
S = 0.33%

SOLUTION:

$$Q_F = \frac{0.312}{n} \times D^{8/3} \times \sqrt{S} = \frac{0.312}{0.012} \times 0.20^{8/3} \times \sqrt{\frac{0.33\%}{100\%}}$$

$$= 0.0204\ m^3/s = \underline{20.L/s}$$

Example Problem 16.4

The desired full-flow capacity of a sewer main 30 L/s and available grade is 0.10%. Select the size of the sewer main.

Givn: D = ?
n = 0.013
Q = 30 L/s = 0.03 m^3/s
S = 0.10% = 0.001

SOLUTION:

$$D_{min} = \left[\frac{Q_F \times n}{0.312\sqrt{S}}\right]^{3/8} = \left[\frac{0.03 \times 0.013}{0.312 \times \sqrt{0.001}}\right]^{3/8} = 0.297\ m$$

$$= 297\ mm = \underline{300\ mm(12\ in\ NPS)}$$

16.3.4 Gravity Sewer Mains

Gravity sewer mains are large diameter pipes that carry sewage from the building service to a lift station. These are called gravity sewers since the flow in them is due to gravity. Gravity sewer mains have a minimum slope and diameter that must be maintained along the length of the pipe to keep the sewage flowing by gravity. There are no pumps in these sections to move sewage along.

A building sewer is a sewer that connects the house or business to the main sewer in the street in front of the building. A building service sewer line is usually a gravity line and has a minimum diameter of 100 mm for residential applications and 150 mm for commercial applications.

16.3.5 Force Mains

A force main is any part of a sewage collection system that is under positive pressure. Pumps are employed to create pressure in the pipe. A force main is always downstream of a lift station and sometimes connects the lift station and the wastewater treatment plant. The hydraulic grade line (HGL) in a force main is higher than the pipe. However, the pressure is much lower compared to water distribution systems. All sections of a collection system that are under pressure will be indicated on the as-built drawings as force mains.

Force mains have additional features that gravity mains do not have, such as vents and shut-off valves. Since a force main is under pressure, it always flows full, and the pipe material is usually ductile iron or steel. Force mains are sized such that, during low flow pumping, a minimum flow velocity of 0.60 m/s is achieved. This flow velocity will allow the flow stream to keep solids in suspension and prevent the deposition of solids in the force main.

The shut-off valves will be located at each end of the force main – at the lift station and just before the discharge end of the force main – and are normally open. These valves can be used to

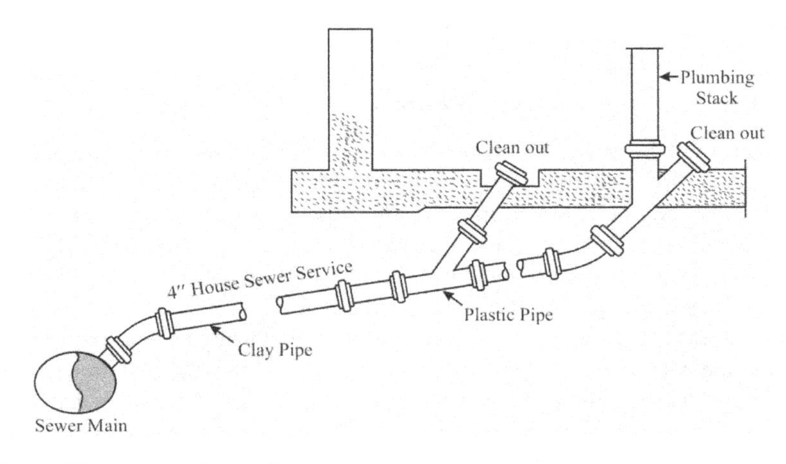

FIGURE 16.3 Building Sewer Connection

shut off flow through the force main when maintenance is needed. The valves should be opened and closed (exercised) twice per year to ensure that they will operate when needed. The force main route should be inspected periodically after digging or construction activity, which might endanger the line.

16.4 OPERATION AND MAINTENANCE

The main activities that form part of wastewater collection system maintenance include: detecting obstructions in a sewer main by the inspection of the manholes in the sewage collection system; removing obstructions in a sewer main; undertaking repairs of collapsed or separated sewer mains; and repairing frozen or blocked building service sewer lines.

16.4.1 Crown Corrosion

If anaerobic conditions exist in sewer pipes, sulphates in water will be reduced by bacteria to produce hydrogen sulphide gas (H_2S). As gas rises to the top, it mixes with moisture or condensation to produce sulphuric acid. This acid corrodes the crown section of the pipe and results in weakening of the pipe. To control crown corrosion, the pipe should be ventilated and septic conditions avoided.

$$H_2S + 2O_2 \xrightarrow{Bacteria} H_2SO_4$$

The accumulation of hydrogen sulphide is known to cause at least three detrimental effects. Firstly, it is odorous and hazardous to people who work in the vicinity and can be fatal at high concentrations. Secondly, it is flammable and explosive. Finally, it is acidic and can cause corrosion problems in sewers and sewage treatment works. Some of the methods that can be used to reduce crown corrosion are listed below.

1. Lining pipes with corrosion-resistant material.
2. Ventilating sewers.
3. Making the sewer run full.
4. Prohibiting the entry of wastes containing sulphides.
5. Aerating and chlorinating septic sewage.
6. Neutralizing sulphides by the addition of chemicals.

16.4.2 Repairing Collapsed/Broken Sections

Sewer main collapses or breaks can be detected with diligent manhole inspections. When a sewer main has collapsed or separated, the operator will see stones, sand and gravel in the manholes downstream of the problem area. Collapsed mains can be caused by freezing, which shatters the pipe, or by poor bedding or rock under the pipe. Separated mains are usually caused by water washing the support bedding away from under the pipe, allowing several sections of the main to drop. This separates the pipe sections by pulling them apart at the joints.

To repair the main, an operator must locate the problem spot using a sewer tape or rod to measure the distance from the manhole to the problem area. A backhoe may then be used to excavate down to the broken or collapsed section of the pipe, and new pipe may be laid as required. All safety precautions, including appropriate signage and barricading on the street, and shoring of the excavation, must be followed when conducting a pipe repair.

16.4.3 Force Main Maintenance

Materials such as sand or grit will build up in the forced main over time. Forced mains should be flushed out in the spring and fall of each year to clean out this accumulation. The force main may be flushed out as follows:

1. Locate a 2 in nipple and valve fitted on the discharge header in the lift station.
2. Run a fire hose from a fire truck or hydrant to the nipple and attach the hose securely.
3. Shut off the lift station pumps.
4. Start water flow through the hose to flush the force main. Flushing must continue for at least one hour to be of any value. CAUTION: When using a fire truck, be sure not to exceed the pressure rating of the pipe in the force main. Piping in some force mains will not withstand pressures of more than 400–450 kPa (60 to 70 psi). A fire truck can quickly and easily develop a pressure of 3500 kPa (500 psi).
5. Check the air vents (if any). Some systems have air vents or air relief valves installed in manholes located at the highest point between the lift station and the disposal area. Air relief valves seldom require maintenance but should be checked occasionally for flooding (especially during spring run-off), vandalism or accidental damage.

16.5 INSPECTION

As part of the inspection of sewer systems, manholes and illegal connections are checked frequently. Two tests are common: smoke test and dye test.

16.5.1 Smoke Test

Smoke testing is done to determine the following:

- Illegal connections.
- The source of surface flow.
- To confirm a building is connected to the sewer system.
- To locate broken manholes and lost sewers.

A smoke test is usually performed when the groundwater table is low. This allows the smoke to rise through openings and soil. Smoke is introduced in the blocked section of the sewer to be tested. The two end manholes of the test section are plugged and a smoke bomb is fired in the middle. An important consideration is to establish airflow before the smoke bomb is fired. To achieve this, a blower is run for 10–15 minutes before firing. Residents in the area must be notified to avoid panic. Smoke emanating from the ground surface is indicative of infiltration or exfiltration, and confirms the connections, legal or illegal, discharging into the sewer system.

16.5.2 Dye Test

A dye test is carried out by introducing a dye at the point of discharge. This can test if a said point is connected to the sewer collection system. It can help detect if non-sanitary water is being discharged into the collection system. If wastewater is overflowing or discharging into a nearby water body, a dye test can be used to locate the location of exfiltration.

If the point of entry is flowing, the dye is introduced directly; if it is running dry, it is first diluted and then poured. Confirmation is made if the dye appears in the downstream sewer pipe. When performing multiple dye tests in each area of a system, start at the downstream end and progress upstream. If the test starts at the upstream end, dye from the first test will interfere with the following test.

The time it takes for the dye to travel from the application point to the observation point must be estimated by assuming the appropriate flow velocity and distance along the sewer pipe. This calculation helps to figure out the approximate time the dye appearance can be expected at the observation point.

16.5.3 Closed-Circuit Television

Closed-circuit television is the best method for inspecting sewers, especially small diameter pipes. This method has become more common lately since the prices of equipment have come down. After a sewer is cleaned, a skid-mounted video camera is pulled through the pipe and a continuous picture is transmitted to the receiver in the service truck. This technique allows for visual inspection of the location of structural defects, infiltration inflow, illegal connections, root growth and grease build-up. In addition, video can be replayed for further examination and record-keeping.

16.6 INVERTED SIPHON

An inverted siphon (**Figure 16.4**) is a depressed sewer that drops below the hydraulic grade line. Pressure conditions in these sewer lines are of positive pressure since atmospheric conditions exist at both the inlet and the outlet, and the water level at the outlet is below the inlet water level. It is called inverted since the pipe is below the inlet and the outlet. Flow in the siphon occurs due to the difference in water levels at the inlet and the outlet. In other words, flow velocity in the siphon is such that head loss is equal to the difference in water levels at the two ends.

Inverted siphons are needed to cross obstructions such as a stream, railway line or depressed highway. In inverted siphons, flow velocity in the pipes must be high to prevent the solids from depositing. Usually, inverted siphons are designed to provide flow velocities of 1.0 m/s or more. This is accomplished by constructing an inlet splitter box that directs the flow to two or more siphon pipes placed in parallel. Depending on the flow, one or more pipes are in operation to maintain the minimum flow velocity. In addition to the control of flow, the inlet and outlet structures provide access for cleaning sewer lines. The difference in water level between the inlet and outlet is the operating head and equals the head loss.

16.6.1 Design of Inverted Siphon

One of the key requirements in the design of an inverted siphon is achieving a self-cleansing velocity of 1.0 m/s even during minimum flow conditions. If this is not achieved, because the pipe is

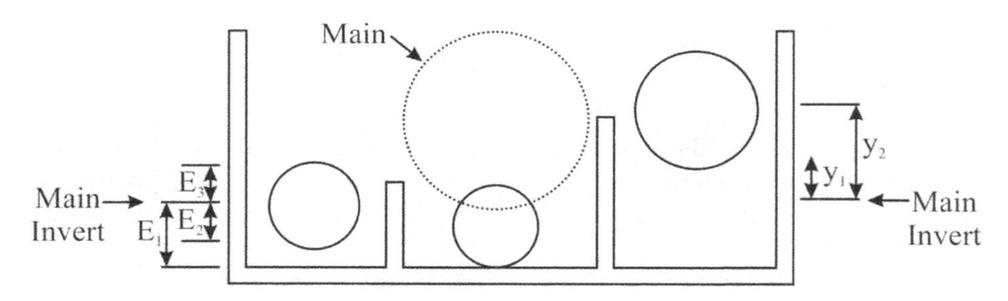

FIGURE 16.4 Inverted Siphon

depressed, it will clog and will not function. Moreover, it is very difficult to clean or unclog siphon pipes. For this reason, a siphon usually consists of three pipes laid side by side: one for carrying minimum flow, one for carrying maximum flow and one for carrying combined flow. If the siphon is to carry only sanitary flow, two pipes may suffice. In addition, the following points should be kept in mind when designing an inverted siphon.

1. If the length of the pipe is extra long, hatch boxes should be provided every 100 m to facilitate rodding. Hatch boxes should be provided with vent pipes to prevent the formation of airlocks.
2. Any changes in the direction of a sewer pipe should be gradual to minimize head losses and clogging.
3. Since the barrels of inverted siphons flow under pressure, head losses due to bends and other structures must be considered when selecting pipe size.
4. The minimum pipe size is taken to be 150 –200 mm in diameter.
5. Manholes or access points should be provided at each end of the siphon.
6. A diversion for the sewage flow should be provided in case the siphon gets fully clogged.
7. The inlet chamber should be provided with screens to prevent the entry of any debris.

Example Problem 16.6

Design an inverted siphon to cross a stream. The total length of the siphon including the slopes is 75 m. For minor head loss, assume an equivalent length of 12 m. The available head is 0.50 m and the average flow is 0.35 m³/s. Assume maximum and minimum flow to be 250% and 50% of average flow, respectively.

Given: $h_f = 0.50$ m
$n = 0.013$
$L = 75 \text{ m} + 12 \text{ m} = 87 \text{ m}$
$Q_{avg} = 0.35 \text{ m}^3/\text{s}$
$Q_{min} = 0.5 \times 0.35 = 0.175 \text{ m}^3/\text{s}$
$Q_{max} = 2.5 \times 0.35 = 0.875 \text{ m}^3/\text{s}$
$v = 1.0 \text{ m/s}$

SOLUTION:

$$D(min.\ Q) = \sqrt{\frac{1.27Q_{min}}{v}} = \sqrt{1.27 \times \frac{0.175\ m^3}{s} \times \frac{s}{1.0\ m}} = 0.471\ m = 450\ mm\ select$$

$$h_f = \left[\frac{Qn}{0.312 \times D^{8/3}}\right]^2 \times L = \left[\frac{0.175 \times 0.013}{0.312} \times \frac{1}{0.45^{8/3}}\right]^2 \times 87\ m = 0.33\ m < 0.50\ m\ okay$$

$$Q_{excess} = Q_{max} - Q_{min} = 2 \times Q = 2 \times 0.35\ m^3/s = 0.70\ m^3/s$$

$$D = \sqrt{\frac{1.27Q}{v}} = \sqrt{1.27 \times \frac{0.70\ m^3}{s} \times \frac{s}{1.0\ m}} = 0.942\ m = 900\ mm\ select$$

$$h_f = \left[\frac{Qn}{0.312 \times D^{8/3}}\right]^2 \times L = \left[\frac{0.7 \times 0.013}{0.312} \times \frac{1}{0.90^{8/3}}\right]^2 \times 87\ m = 0.129 = 0.13\ m < 0.50\ m$$

Example Problem 16.7

Design a sag pipe system using the following given conditions:

Minimum flow velocity in sag pipe = 0.9 m/s.
Length of sag pipe = 100 m.
Minimum flow = 80 L/s, average flow = 300 L/s, maximum flow = 650 L/s.

Design three sag pipes from the inlet chamber, (1) to carry minimum flow; (2) to carry flows from minimum to average (maximum dry weather flow); and (3) to carry all flows above the average flow.

Available hydraulic grade line = 1.0%.
Roughness coefficient n = 0.015 (ductile-iron pipe).

Given: $S_f = 1.0\%$
$n = 0.015$
$L = 100$ m
$Q_{avg} = 0.30\ m^3/s$
$Q_{min} = 0.08\ m^3/s$
$Q_{max} = 0.65\ m^3/s$
$v_{min} = 0.90$ m/s

SOLUTION:

$$D(min.\ Q) = \left[\frac{Qn}{0.312 \times \sqrt{S_f}}\right]^{0.375} = \left[\frac{0.08 \times 0.015}{0.312} \times \frac{1}{\sqrt{0.01}}\right]^{0.375} = 0.294\ m = 300\ mm$$

$$Check,\ v_F = \frac{0.4}{n} \times D^{2/3} \times \sqrt{S} = \frac{0.4}{0.015} \times (0.30)^{2/3} \times \sqrt{\frac{1.0\%}{100\%}} = 1.19\underline{m/s > 0.9\ m/s}$$

$$Q_{excess} = Q_{avg} - Q_{min} = 0.30 - 0.08 = 0.22\ m^3/s$$

$$D(2nd) = \left[\frac{Qn}{0.312 \times \sqrt{S_f}}\right]^{0.375} = \left[\frac{0.22 \times 0.015}{0.312} \times \frac{1}{\sqrt{0.01}}\right]^{0.375} = 0.430\ m = 450\ mm$$

$$v_F = \frac{0.4}{n} \times D^{2/3} \times \sqrt{S} = \frac{0.4}{0.015} \times (0.450)^{2/3} \times \sqrt{\frac{1.0\%}{100\%}} = 1.56\ \underline{m/s > 0.9\ m/s}$$

$$Q_{excess} = Q_{max} - Q_{avg} - Q_{min} = 0.65 - 0.30 - 0.08 = 0.27\ m^3/s$$

$$D(3rd) = \left[\frac{Qn}{0.312 \times \sqrt{S_f}}\right]^{0.375} = \left[\frac{0.27 \times 0.015}{0.312} \times \frac{1}{\sqrt{0.01}}\right]^{0.375} = 0.465\ m = 500\ mm$$

$$v_F = \frac{0.4}{n} \times D^{2/3} \times \sqrt{S} = \frac{0.4}{0.015} \times (0.50)^{2/3} \times \sqrt{\frac{1.0\%}{100\%}} = 1.67\ m/s \underline{> 0.9\ m/s\ okay}$$

16.7 MANHOLES

Sewer manholes are one of the most important appurtenances of sewer systems. They consist of an opening constructed on the alignment of a sewer and facilitate a person to gain access to the sewer for inspection, testing, cleaning and removal of obstructions from the sewer line. Manholes, more appropriately called maintenance access points (MAP), are built into a sewage collection system at every change of alignment, gradient or diameter, at the head of all sewers and branches, and at every junction of two or more sewers. Here is a further list of reasons for which manholes are provided:

- To provide access to the sewage mains for maintenance and inspection.
- To allow a change of pipe size in the sewer main.
- To allow for a change in the direction of the sewer main.
- To provide a crossing point or collection area for pipes leading from two or more directions.
- To allow for a change in the grade or slope of the sewer main.

Figure 16.5 shows some of these different configurations.

16.7.2 Ordinary Manhole

Manholes are generally constructed of concrete rings placed one on top of the other. The rings are usually wide enough in diameter to allow an operator to comfortably descend into the manhole. They are equipped with steel ladder rungs on one side for accessibility. Manholes are spaced along sewer lines at distances of not more than 150 m. This spacing allows the operator to reach obstructions in the sewer main with the tools used to clear them.

16.7.3 Spacing

As mentioned above, sewers are typically spaced about 100–150 m on straight sections of sewer pipes. The spacing of manholes on large diameter sewers is dictated by the method and tools used to clean the sewers. For straight runs of large sewers exceeding 1.5 m, spacing > 150 m may be allowed.

16.7.4 Constructional Details

Manholes are generally constructed directly over the centreline of the sewer. They are usually circular but, in some cases, a rectangular shape may be used. Circular sewers are stronger, so they are preferred. Manhole size should be such that it allows for the necessary cleaning and inspection of sewers. As shown in **Figure 16.6**, manholes are straight down in the lower portion and narrowed at the top cover, with an opening equal to the internal diameter of the manhole cover.

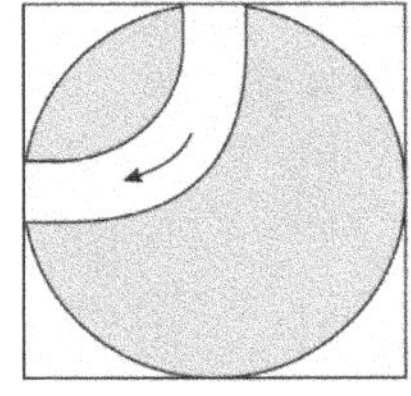

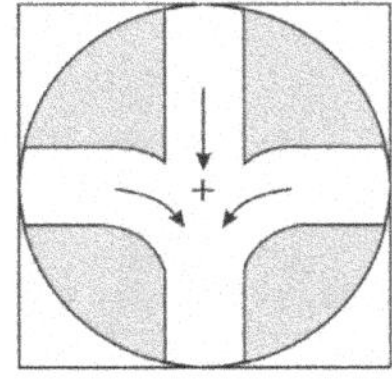

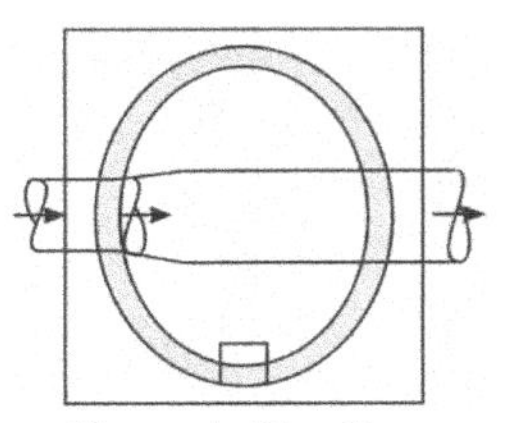

FIGURE 16.5 Manhole Configurations

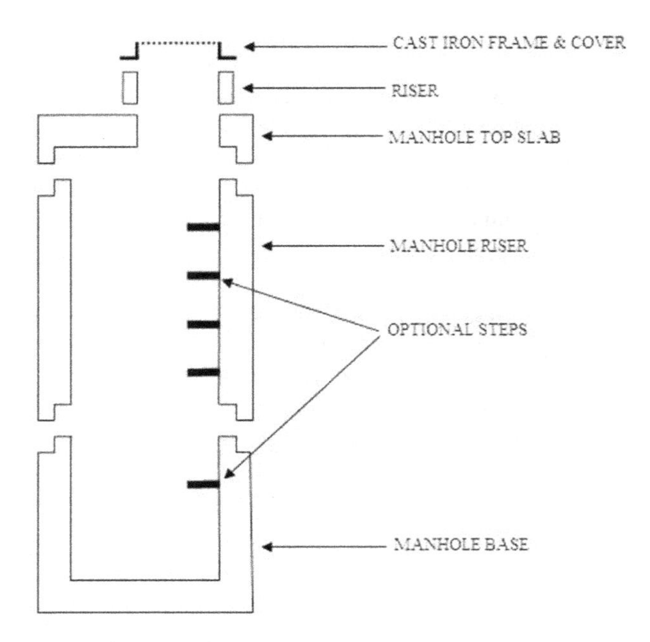

FIGURE 16.6 Components of a Manhole

The diameter of a manhole changes depending on the depth. The opening to the manhole should be of such minimum size as to allow a workman to gain access without difficulty. A minimum opening of 600 mm is usually provided. Manhole covers and frames are usually cast iron with a minimum clear opening of 540 mm (21 in). Solid covers are used on sanitary sewers and open covers are used on storm sewers. Steps or ladder rings are placed for access. The material of the rings needs to be corrosion proof. Walls may be constructed of precast rings, concrete block, brick or poured concrete.

Wastewater is conveyed through the manhole in a smooth U-shaped channel formed in the concrete base. In junction manholes, where more than one sewer enters the manhole, the flowing through-channels should be curved to merge the flow streams. If the sewer changes direction without a change in size, a drop of 50–75 mm is provided to account for head loss. When a smaller sewer joins a larger sewer, the bottom of the larger sewer should be lowered sufficiently to maintain uniform flow transition, and so as to not cause any backup in the smaller sewer. To achieve this, either 80% of the depth of both the sewers should be at the same elevation, or the crowns of both sewers should be matched in elevation.

16.7.5 Lamp Hole

In sewer systems of the 19th century, one can still sometimes find lamp holes fixed to the crown of a section of a sewer between two manholes. They were used as a replacement for a manhole and allowed for the lowering of a lamp when the condition of the sewer was to be inspected from the neighbouring shaft by means of mirroring. If the reflected light emitted from one lamp is received at the other end's mirror, it would indicate the line is clear of any obstructions.

16.7.6 Drop Manholes and Dead-End Manholes

Drop manholes are used where laterals join a deep main line or where the slope of the ground makes the sewer grade too steep. Drop manholes may be used to bring a sewer main down a hill in steps to avoid sewer grades, which are too steep for proper sewage flow. **Figure 16.7** illustrates this concept.

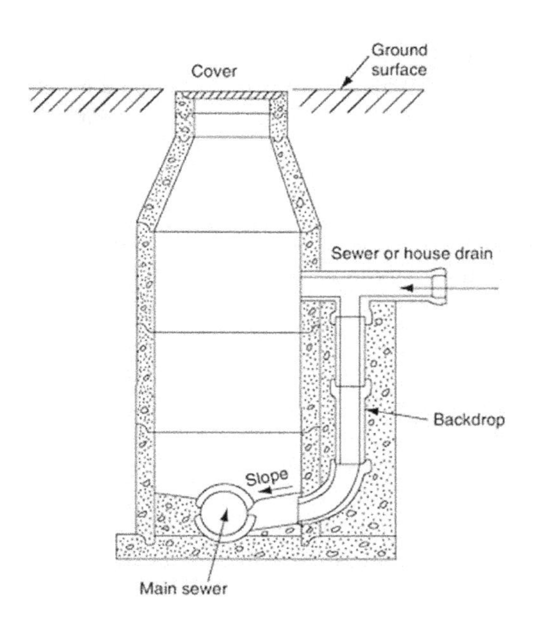

FIGURE 16.7 A Drop Manhole

Dead-end manholes are located at the end of a sewage collection line. Although these manholes may be clean and dry in appearance, they can be deadly to a careless operator. Poisonous or explosive gases can build up in these manholes. Because of its location, the dead-end manhole is seldom inspected. The cover may have several inches of dirt sealing the vent holes, thus creating the ideal gas trap.

16.7.7 Manhole Safety

A manhole is a hazardous place to work in so every care must be taken to prevent injury and loss of life. Hazards include oxygen deficiency, presence of toxic gases likes hydrogen sulphide, carbon monoxide and combustibility. Before entering a manhole, one must check for the presence of obnoxious gases and the deficiency of oxygen with a gas meter. If needed, allow for natural ventilation or use forced air to help make the environment in the manhole safer to work. Since it is a case of confined safe entry, all regulations and approved procedures must be followed. The operator is responsible for his or her safety. If working conditions are unsafe, workers have the right to refuse to work. Here are some key points one should keep in mind while working in manholes during the inspection and maintenance of sewers.

- Before starting any work, place proper traffic signs.
- Before entering, lower a gas probe and check for oxygen deficiency and the presence of toxic gases.
- Lift the cover of the manhole with care as it is heavy and can cause back injury if not done properly.
- Work in a team of two people so that one person can wait on the ground to pull the harness in case it becomes dangerous to work in the manhole.
- The person entering the manhole should have a self-contained breathing apparatus (SCBA) if there is danger of oxygen deficiency.
- Working tools must be lowered in a bucket.
- Keep all materials away from the opening to prevent them from falling into the hole.

- Check the condition of the rungs of the ladder.
- Only non-sparking tools and explosion-free lighting should be used.
- There must be no smoking or open flames near the work area.

16.7.8 Sewer Ventilation

As discussed above, sewer lines contain acidic vapours, combustibles, noxious gases and strong odours. The stack connected to the building sewer usually provides sewer ventilation. Where there is a need, venting shafts connected to the sewer manhole may be provided. In some cases, forced ventilation may also be done.

Example Problem 16.8

You need to enter a 2.5 diameter manhole. Before entering, you need to make sure that, for safety reasons, you ventilate the manhole using an air blower such that all the air is replaced by fresh air. If the air blower capacity is 2.0 m^3/min and the depth to water in the manhole is 3.0 m, calculate the minimum time for which the blower should be run.

$$t = \frac{V}{Q} = \frac{\pi(2.5\ m)^2}{4} \times 3.0\ m \times \frac{min}{2.0\ m^3} = 7.35 = \underline{7.4\ min\left(say\, 8\ min\right)}$$

16.7.9 Manhole Inspection and Maintenance

The operators of wastewater collection systems should establish a daily maintenance inspection schedule for the manholes in a system. To establish a maintenance schedule, the as-constructed drawings for the sewage collection system will need to be consulted. These drawings show the location and number of each manhole in the collection network. The total number of manholes should be divided into groups for daily inspections (a different group will be inspected each day). If properly scheduled, most of the manholes in a system can be inspected in a week.

Carefully written records should be kept of all manhole inspections, including the date and time of inspection, the number of the manhole inspected, the condition of the manhole, including the flow in and out of the manhole, and the name of the inspector. Any build-up of solids in the manhole must be cleaned out so that sewage flow through the manhole is not slowed down. With experience and by keeping good records, it will soon be possible to identify which manholes need regular attention and which are trouble-free. The daily inspection schedule can then be changed to give more attention to manholes that need frequent checking and less to the more trouble-free manholes.

In severe winter conditions, inspection notes should include any build-up of ice on the walls of the manholes in the sewage collection system. Ice build-up in some manholes can completely fill the upper part of the structure. When thawing begins in spring, the ice will melt away from the walls of the manhole, and large chunks may fall to the bottom of the manhole and cut off flow, causing a backup of sewage. The ice should be removed in these manholes before the thaw by steaming or using an ice chisel to avoid this situation.

Manholes located in low areas of the collection system or in areas where storm run-off is likely to flow along the street and into it may present a problem in rainy season. When run-off flows down the street and into a manhole, it carries sand and gravel with it, which may freeze to the sewer mains and to the bottom of the manhole.

16.8 SAMPLING AND FLOW MEASUREMENT

Monitoring sewer use requires composite sampling and flow measurements at selected points in the sewage collection system. Flow measurement and the sampling of wastewater are important to monitor industrial discharges and inflow and infiltration studies. For medium-to-large industries, sampling stations are built to accommodate automatic sampler and flow measuring devices.

16.8.1 Flow Measurement in Sewers

The first estimates of flow in a sewer can be achieved by observing the depth of flow. A flow calculation can be made by applying Manning's formula. Alternatively, flow velocity can be observed with a float, current meter, or dye float and then multiplied by the wetted area.

The installation of a flume is necessary for accurate flow measurement and automatic composite sampling. The Palmer-Bowlus flume, with a trapezoidal section, is most common. By observing the head upstream of the flume, flow is read from the rating equation or table for the flume provided by the manufacturers. Downstream conditions should be free of discharge. The rating equation for a Parshall flume changes with the size of the flume and is accurate over a certain range of flow. As an example, the following equation is for a 4 in Palmer-Bowlus flume in imperial units.

$$Q = 1.73(H + 0.00588)^{1.9573} \quad Q = cfs \quad H = ft$$

Palmer-Bowlus flumes are constructed for temporary installation in the half section of the sewer in a manhole. These flumes are prefabricated with materials such as fibreglass or plastic and are available in various sizes to match the diameter of the sewer pipe. For more permanent installation, the flume is built and embedded in poured concrete. Weirs can also be used but are not preferred because of silting problems.

Various types of depth sensors, including ultrasonic, submerged probe and bubbler, are common. An ultrasonic sensor measures based on the delay in ultrasonic pulse as it is reflected back from the water surface. It is installed above the water surface and is not affected by grease or other floating stuff, so flow is smooth.

A submerged probe works on the pressure-differential principle. Floating debris, grease or foaming does not affect the accuracy of the meter.

FIGURE 16.8 Small Palmer-Bowlus Flume in a Manhole

A bubbler gauge measures depth by sensing the pressure required to force air to bubble through the water column. Many flow meters are programmable and can record the depth of the water and display flow rate and totalized flow. Using a keypad, conversion formulae can be entered.

16.8.2 Sample Collection

Samples can be collected manually or using automatic samplers. Irrespective of the technique used, the sample must be representative. A sampling location should be selected where the flow stream is smooth and well mixed. A sample scoped from the surface or the bottom would be a biased sample. Automatic sampling is becoming very common. Manual sampling is usually done for spot-checking and similar uses.

Though composite samples are more common, to inspect and audit an industrial user, a grab sample is sometimes collected. However, analysis of a grab sample represents the point value with respect to time and place. A composite sample test result would give an average value over the sampling period. In some cases, like pH and dissolved oxygen (DO) measurements, only grab samples are not composited, since this would change the accuracy of the results due to chemical changes.

Discussion Questions

1. Modify Manning's equation to find the flow capacity of a circular sewer of diameter D in SI units.
2. Prove that for a circular sewer flowing full or half full, the hydraulic radius is 1/4th of the diameter of the sewer pipe.
3. Define crown corrosion and describe the measures that can be taken to prevent it.
4. What are the functions of a drop manhole and a lamp hole?
5. Differentiate between sewage and sewerage.
6. State the merits and demerits of (i) a separate system of sewerage and (ii) a combined system of sewerage.
7. Discuss diurnal flow variation in sewage flow. How is this variation considered in the design of sewers?
8. What are the various types of storm water regulators used in a sewerage system?
9. Sketch and explain the components of an ordinary manhole.
10. What are drop manholes, and where are they used in a sewerage collection system?
11. Why is ventilation important in sewers? How is it done?
12. Explain the construction details of a manhole.
13. Describe the different methods for the sampling of wastewater.
14. Name common flow measurement devices used in a sewer collection system.
15. Compare combined sewers with storm sewers.

Practice Problems

1. Design an inverted siphon to cross a river. The total length of the siphon including the slopes is 85 m. Assume 0.30 m minor head losses and 0.012 for roughness coefficient. The available head is 0.75 m and the average flow is 450 L/s. Assume maximum and minimum flows to be 250% and 50% of average flow, respectively. (500 mm, 900 mm, head loss = 0.55 m and 0.50 < 0.75 m)
2. You need to enter a 3.0 diameter manhole. Before entering, you need to make sure that, for safety reasons, you ventilate the manhole using an air blower such that all the air is

replaced by fresh air. If the air blower capacity is 2.0 m^3/min and the depth to water in the manhole is 3.0 m, calculate the minimum time for which the blower should be run. (11 minutes)

3. Work out the full-flow velocity and flow capacity of a 400 mm sewer pipe laid on a slope of 0.20%. (0.75 m/s, 93 L/s)
4. Design an inverted siphon to cross a canal. The length of the sewer pipe including slopes is 60 m. Consider an equivalent length of 9.0 m to account for minor losses. The available head is 0.90 m, and the average flow is 350 L/s. The maximum and minimum rates of flow may be taken as 300% and 50% of average flow, respectively. (450 mm, 0.26 m, 750 mm, 0.27 m)
5. Work out the full-flow velocity and flow capacity of a 500 mm sewer pipe laid on a slope of 0.30%. (1.1 m/s, 210 L/s)
6. A 300 mm stoneware sewer is laid on a slope of 1.0%. Assuming n = 0.013, compute the full-flow velocity and discharge. (1.4 m/s, 97 L/s)
7. What diameter sewer pipe (n = 0.012) is required to carry wastewater with a flow velocity of 0.5 m/s flowing full. The sewer pipe is laid on a gradient of 1 in 500. (200 mm NPS)
8. To carry solids, a flow velocity of 1.4 m/s is required in a 300 mm concrete sewer pipe flowing full. Assuming n = 0.013, what is the minimum grade on which the pipe should be laid? (1 in 100)

17 Design of Sewers

Sewer pipes, also called sewer mains, are installed as gravity mains or force mains. In a gravity main, the sewage moves through the pipe by gravity alone; no other pressure or force is required. In a force main, the sewage is pushed, or forced, through the sewer by a pump. In comparison to closed-conduit flow, open-channel flow has one surface that is free and open to the atmosphere. In closed conduits (pipe flow), the flow is primarily due to pressure, whereas in open conduits it is all due to gravity; therefore, open-channel flow is also called gravity flow. In partially filled conduits, as in sewer mains, the flow is due to gravity and open-channel flow conditions prevail except in a force main, which flows full.

17.1 FLOW CLASSIFICATION

Flow may be classified as being steady (independent of time) or unsteady (time-dependent) and based on the depth of the flow. Therefore, flow velocity is uniform and non-uniform.

Steady uniform flow occurs when the flow rate (discharge) remains constant with time (steady) and the depth of the liquid does not vary along the length of the channel. Steady uniform flow occurs in channels that are long and straight and whose depth and slope are constant (prismatic) along the length of the channel. In this type of flow, the slope of the free water surface (S_w) is parallel to the slope of the bed of the channel (S_o), as shown in **Figure 17.1**.

Steady non-uniform (varied) flow occurs where the depth of flow varies along the length of the channel. This will occur if the channel is non-prismatic. Varied flow can be further classified as rapidly varied flow (RVF) or gradually varied flow (GVF) depending on the rate of change in depth along the channel. Unsteady non-uniform flow occurs when the depth of the fluid varies both with time and along the channel bed. It is not practical to achieve unsteady uniform flow.

17.2 HYDRAULIC SLOPE

Channel bed slope, S_o, refers to the change in elevation of the channel bed per unit horizontal length of the channel. This can be expressed as a fraction, a percent, or the angle the channel bed makes with the horizontal plane. For example, a bed slope of 0.1% can also be reported as follows:

$$0.1\% = 0.001 = \frac{0.1m}{100m} = \frac{10cm}{100m} = \frac{1m}{1km}$$

$$Sin\ \theta = 0.001\ or\ \ \theta = \sin^{-1}(0.001) = 0.057°$$

Water (liquid) surface, S_w is the slope of the water surface. In the case of open-channel flow, flow is all due to gravity, therefore the hydraulic head at a given section is equal to the elevation of the water surface. It can also be said that the water surface represents the hydraulic grade line (HGL) in open-channel flow conditions. Therefore, the hydraulic slope, S_w is a fall in water surface elevation per unit horizontal length of the channel.

DOI: 10.1201/9781003231264-17

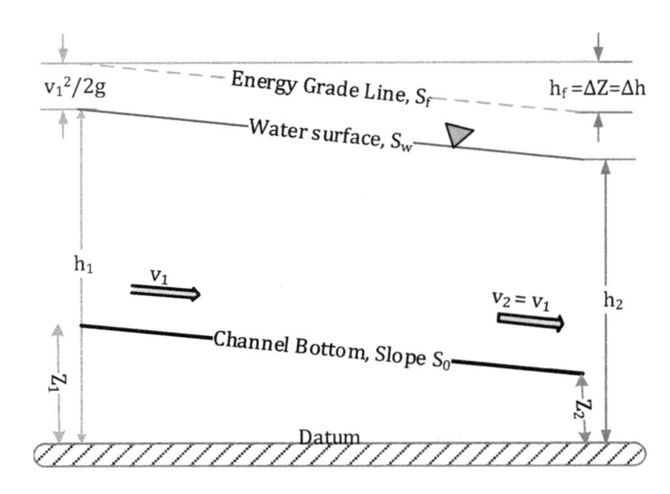

FIGURE 17.1 Open-Channel Flow Energy

- Energy slope, S_f is the slope of the energy grade line. In open-channel flow conditions, the energy grade line is vertically above the liquid surface by an amount equal to velocity head.
- In most channels, the bed slope is small ($\theta < 5°$), therefore $\sin \theta = \tan \theta$, or the length of the channel is approximately equal to the horizontal distance.
- For uniform flow conditions, depth y remains constant; therefore, the velocity of flow is the same at every section.
- The energy grade line, the water surface and the channel bed are all parallel for uniform flow conditions.

Therefore frictional force, which depends on the roughness of the channel, balances the driving force provided by gravity. Equating gravity forces to frictional forces, an expression for the average velocity of uniform flow can be derived. However, the most widely used formula for uniform flow conditions is empirical in nature and is popularly known as Manning's equation.

17.3 MANNING'S EQUATION

Manning's flow equation is briefly discussed in the previous chapter. Because Manning's equation is empirical in nature, its units must be consistent. In SI units, Manning's flow equation is:

$$v = \frac{1}{n} \times R_h^{2/3} \times \sqrt{S} \quad or \quad Q = A \times v = \frac{A}{n} \times R_h^{2/3} \times \sqrt{S}$$

The average velocity of flow v will be in m/s when the hydraulic radius, R_h is in m. Hydraulic radius is defined as the ratio of the liquid section area to the wetted perimeter, P_w, and wetted perimeter is the length of the wetted surface of the channel. It is important to note that the wetted perimeter is equal to the total perimeter of the liquid section minus the length of the free liquid surface. Some examples are shown in **Figure 17.2**.

The term n is a roughness factor, commonly called Manning's n. The friction slope, S, is dimensionless and equal to the slope of the channel bed for uniform flow conditions. The flow rate (discharge) for which uniform flow will occur is referred to as normal discharge. The corresponding depth of flow is called normal depth.

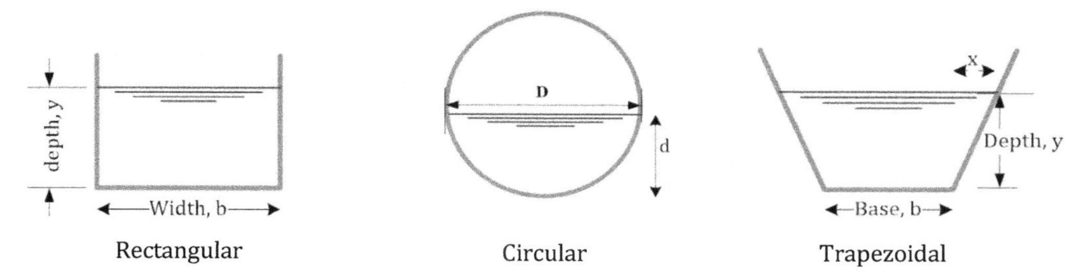

FIGURE 17.2 Common Flow Sections

17.3.1 Section Factor versus Conveyance Factor

The section factor of a channel section is solely dependent on the geometry of the channel section. The conveyance factor is the section factor divided by the roughness coefficient.

$$\textit{Section factor},\ A \times R_h^{2/3} \quad \textit{Conveyence factor} = \frac{A}{n} \times R_h^{2/3}$$

In a circular section flowing half full, both the flow area and the wetted perimeter are one-half of the full section. Hence, the hydraulic radius remains one-fourth that of diameter.

17.3.2 Laminar and Turbulent Flow

For circular pipes flowing full, $R_h = D/4$. Based on this, the expression for the Reynolds number in open-channel flow becomes:

$$N_R = \frac{vD}{\nu} = \frac{v \times 4R_h}{\nu}$$

For $N_R < 2000$, laminar flow occurs, and turbulent flow occurs when $N_R > 4000$.

17.3.3 Circular Pipes Flowing Full

For circular pipes flowing full, $R_h = 0.25D$ and $A = 0.785D^2$. Making these substitutions, Manning's flow equation becomes:

$$v_F = \frac{0.4}{n} \times D^{2/3} \times \sqrt{S} \quad \textit{and} \quad Q_F = \frac{0.312}{n} \times D^{8/3} \times \sqrt{S}$$

Example Problem 17.1

Determine the flow-carrying capacity of a 300 mm sewer pipe (n = 0.013) running half full and laid on a slope that drops 1 m over a run of 1 km.

Given: D = 0.3 m
n = 0.013
S = 1m/km = 0.001

SOLUTION:

$$Q_F = \frac{0.312}{n} \times D^{8/3} \times \sqrt{S} = \frac{0.312}{0.013} \times 0.3^{8/3} \times \sqrt{0.001} = 0.0306\ m^3/s$$

$$Q_{0.5F} = 0.5 \times \frac{0.0306\ m^3}{s} \times \frac{1000\ L}{m^3} = 15.2 = 15\ L/s$$

Example Problem 17.2

Calculate the minimum slope on which a rectangular channel 1.2 m in width should be laid to maintain a velocity of 1.0 m/s while flowing at a depth of 0.60 m. The channel is made of unfinished concrete (n = 0.017).

Given: B = 1.2 m
d = 0.60 m
v = 1.0 m/s
n = 0.017
S = ?

SOLUTION:

$$R_h = \frac{A}{P_w} = \frac{1.2m \times 0.60m}{1.2m + 2 \times 0.6m} = 0.30m \quad \mathrm{R_h} = \frac{\mathrm{A}}{\mathrm{P_w}} = \frac{1.2\mathrm{m} \times 0.60\mathrm{m}}{1.2\mathrm{m} + 2 \times 0.6\mathrm{m}} = 0.30\mathrm{m}$$

$$S = \frac{n^2 v^2}{R_h^{\ 4/3}} = 0.017^2 \times \frac{1.0^2}{0.30^{4/3}} = 1.44 \times 10^{-3} = \underline{0.14\%}$$

Example Problem 17.3

A finished concrete rectangular trench is to be built to carry a discharge of 9.0 m³/s. The available slope is 1.0%. Determine the size of the trench if the depth is assumed to be half the width.

Given: d = b/2
S = 1.0% = 0.01
Q = 9.0 m³/s
n = 0.013
b = ?

SOLUTION:

$$R_h = \frac{bd}{b+2d} = \frac{b \times 0.5b}{b+b} = 0.25b\mathrm{R_h} = \frac{\mathrm{bd}}{\mathrm{b+2d}} = \frac{\mathrm{b \times 0.5b}}{\mathrm{b+b}} = 0.25\mathrm{b}$$

$$A \times R_h^{2/3} = 0.5b^2 \times (0.25b)^{2/3} = 0.198b^{8/3} = \frac{Qn}{\sqrt{S}} = \frac{9.0 \times 0.013}{\sqrt{0.01}} = 1.17$$

$$0.198b^{8/3} = 1.17,\ b = (65/11)^{3/8} = 1.946 = 1.95m,$$
$$\mathrm{d = 0.5b = 0.5 \times 1.95 = 0.98 = \underline{1.0m}}$$

The width of the channel should be 2.0 m. The design depth of the channel will be more than 1.0 m to provide a freeboard.

For rectangular and trapezoidal sections, the hydraulic radius of the most efficient section is one for which R = y/2. This corresponds to a rectangle whose depth is one half the width, or b = 2y.

17.4 MINIMUM FLOW VELOCITY

A velocity that does not permit solids to settle down and even scours deposited particles of a given size is called a self-cleansing velocity. This minimum velocity should develop at least once a day to prevent any deposition in the sewers. If deposition does take place, it will obstruct free flow, causing further deposition and eventually leading to the sewer being completely blocked. Shields suggested the following expression for self-cleansing velocity based on particle size, specific gravity, grade, roughness coefficient and hydraulic radius of the sewer pipe.

$$\text{Self cleansing } v_s = \frac{R_h^{1/6}}{n} \times \sqrt{kD(G_s - 1)}$$

Where k depends on particle properties. The typical values of k for clean inorganic and organic matter present in sewage are 0.04 and 0.60, respectively. From the above equation, the self-cleansing velocity for a sand particle of diameter 1.0 mm, specific gravity 2.65 and sewer pipe of size 200 mm is shown below.

$$v_s = \frac{R_h^{1/6}}{n} \times \sqrt{kD(G_s - 1)} v_s = \frac{R_h^{1/6}}{n} \times \sqrt{kD(G_s - 1)}$$

$$= \left(\frac{0.2}{4}\right)^{1/6} \times \frac{1}{0.013} \times \sqrt{0.04 \times 0.001 \times 1.65} = 0.379 = 0.4 \text{ m/s}$$

Based on this, to achieve self-cleaning, flow velocity during minimum flow conditions is kept at around 0.45 m/s or a flowing full velocity of about 1.0 m/s. Hence, when finalizing the sizes and gradients of sewers, they must be checked for the minimum velocity that would be generated at minimum discharge, i.e. about one-third of the average discharge. When designing sewers, the flow velocity at full depth is generally kept at about 0.8–1.0 m/s. Since sewers are generally designed to be one-half to three-quarters full, the velocity at the designed discharge will be even more than the full depth velocity.

17.5 MAXIMUM VELOCITY OR NON-SCOURING VELOCITY

The interior surface of the sewer pipe gets scored due to continuous abrasion caused by suspended solids present in sewage. The scoring is more pronounced at a higher velocity than can be tolerated by pipe materials. This limiting or non-scouring velocity mainly depends upon the material of the sewer. Thus, to prevent abrasion and scouring of the pipe surface, maximum flow velocities should not exceed 4.0 m/s for vitrified clay, PVC and cast iron, 3.0 m/s for cement concrete pipes, and 1.5–2.5 m/s for brick-lined sewers. The limiting velocity for different sewer materials is provided in **Table 17.1**.

The problem of maximum or non-scouring velocity is severe in hilly areas where the ground slope is very steep and is overcome by constructing drop manholes at suitable places along the sewer line.

TABLE 17.1
Limiting Velocity for Sewer Pipes

Sewer Material	Non-Scouring Velocity, m/s
Vitrified tiles	4.5–5.5
Cast iron sewer	3.5–4.5
Cement concrete	2.5–3.0
Stoneware sewer	3.0–4.5
Brick lined sewer	1.5–2.5

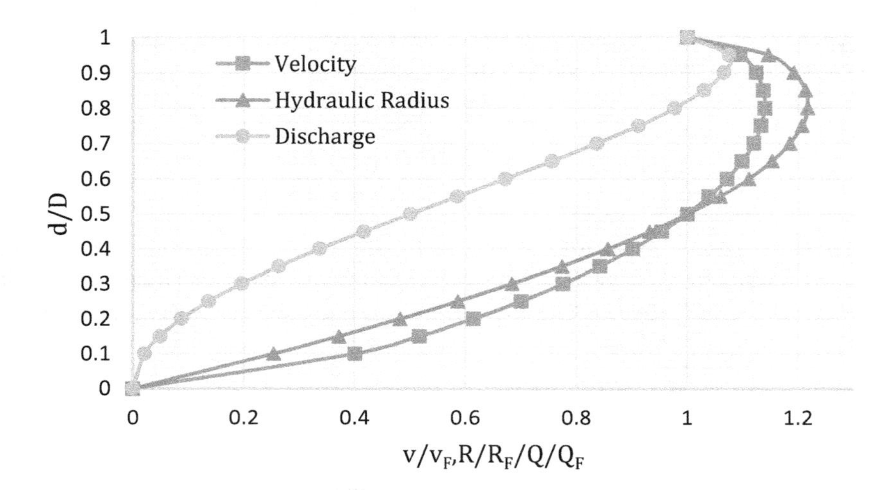

FIGURE 17.3 Standard Chart for Proportionate Elements

17.6 PARTIALLY FULL PIPES

In circular pipes, flow frequently occurs at partial depth. A good example of this is storm and sanitary sewer systems. As noted earlier, the hydraulic radius for pipes flowing full or half full is equal to one-fourth the diameter. For other depths of flow, this is a complex relationship. The simplest way to handle partly full pipes is to compute the velocity or flow rate for the pipe's full condition and adjust to partly full conditions by using a chart as given in **Table 17.2**, or a graphical relationship as shown in **Figure 17.3**.

As an example, when the depth of flow is 30% of the maximum depth (diameter), the corresponding values of flow, area and velocity, as read from **Table 17.2**, are:

$$\frac{d}{D} = 0.30, \quad \frac{Q}{Q_F} = 0.20, \quad \frac{A}{A_F} = 0.25, \quad \frac{v}{v_F} = 0.78$$

It is important to note that in partially filled pipes, maximum velocity is generated when the pipe is flowing at d/D = 0.81 not full. Similarly, a sewer pipe can carry maximum discharge when flowing at a depth equal to 95% of full depth or diameter.

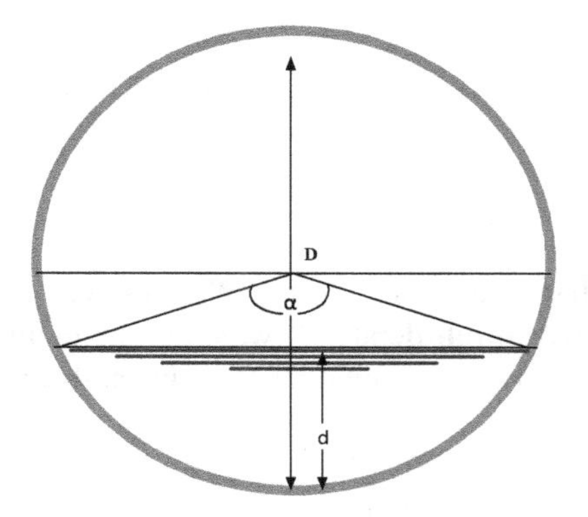

FIGURE 17.4 Partially Full Pipes

TABLE 17.2
Partial Flow and Equivalent Hydraulic Elements

d/D	rad	Deg	P/P_F	R/R_F	A/A_F	v/v_F	Q/Q_F	S_S/S_F	v_S/v_F	Q_S/Q_F
0.10	1.287	73.7	0.205	0.254	0.052	0.401	0.021	3.936	0.796	0.041
0.15	1.591	91.1	0.253	0.372	0.094	0.517	0.049	2.692	0.848	0.080
0.20	1.855	106.3	0.005	0.482	0.142	0.615	0.088	2.073	0.886	0.126
0.25	2.094	120.0	0.006	0.587	0.196	0.701	0.137	1.705	0.915	0.179
0.30	2.319	132.8	0.006	0.684	0.252	0.776	0.196	1.462	0.939	0.237
0.35	2.532	145.1	0.007	0.774	0.312	0.843	0.263	1.292	0.958	0.299
0.40	2.739	156.9	0.008	0.857	0.374	0.902	0.337	1.167	0.975	0.364
0.45	2.941	168.5	0.008	0.932	0.436	0.954	0.417	1.073	0.988	0.431
0.50	3.142	180.0	0.009	1.000	0.500	1.000	0.500	1.000	1.000	0.500
0.55	3.342	191.5	0.009	1.060	0.564	1.039	0.586	0.944	1.010	0.569
0.60	3.544	203.1	0.010	1.111	0.626	1.072	0.672	0.900	1.018	0.638
0.65	3.751	214.9	0.010	1.153	0.688	1.099	0.756	0.868	1.024	0.705
0.70	3.965	227.2	0.011	1.185	0.748	1.120	0.837	0.844	1.029	0.769
0.75	4.189	240.0	0.012	1.207	0.804	1.133	0.912	0.829	1.032	0.830
0.80	4.429	253.7	0.012	1.217	0.858	1.140	0.977	0.822	1.033	0.886
0.85	4.692	268.9	0.013	1.213	0.906	1.137	1.030	0.824	1.033	0.936
0.90	4.996	286.3	0.014	1.192	0.948	1.124	1.066	0.839	1.030	0.976
0.95	5.381	308.3	0.015	1.146	0.981	1.095	1.075	0.873	1.023	1.004
1.00	6.283	360.0	0.017	1.000	1.000	1.000	1.000	1.000	1.000	1.000

Example Problem 17.4

A 450 mm sewer pipe (n = 0.013) is laid on a slope of 0.25%. At what flow depth is the flow velocity 0.6 m/s?

Given: D = 450 mm = 0.450m
n = 0.013
S = 0.25% = 0.0025
v = 0.6m/s

SOLUTION:

$$v_F = \frac{0.40}{n} \times D^{2/3} \times \sqrt{S} = \frac{0.40}{0.013} \times 0.45^{2/3} \times \sqrt{0.0025} = 0.90\text{m/s}$$

$$\frac{v}{v_F} = \frac{0.60}{0.90} = 0.67 \text{ From Table 17.2, } \frac{d}{D} = 0.23, d = 0.23 \times 450\,mm = 103 = \underline{100\ mm}$$

17.6.1 Mathematical Relationships

In this era of computers and fast calculators, the use of tables and charts is becoming obsolete. For any flow depth, d (as compared to full depth, D, which is the diameter of the pipe), partial-flow relationships can be expressed in terms of angle α subtended at the centre by the wetted arc. This is shown in **Figure 17.4**.

In terms of angle α in degrees, proportional depth, area and hydraulic radius can be determined using the following relationships. Similarly, by knowing the proportionate depth, the angle can be found.

$$\frac{d}{D} = 0.5\left(1 - \cos\left(\frac{\alpha}{2}\right)\right) \text{ and } \alpha = 2\cos^{-1}\left(1 - \frac{2d}{D}\right)$$

For example, when the proportional depth is 30% of the full depth, D:

$$\alpha = 2\cos^{-1}\left(1 - \frac{2d}{D}\right) = 2 \times \cos^{-1}(1 - 2 \times 0.30) = 132.84 = 133°$$

Angle α < 180° indicates the pipe is less than half full. On the same note, α > 180° means the pipe is more than half full. Once the angle is known, other proportional variables can be worked out using the following relationships.

$$\frac{A}{A_F} = \left[\frac{\alpha}{360} - \frac{\sin\alpha}{2\pi}\right], \quad \frac{R}{R_F} = \left[1 - \frac{360}{\alpha} \times \frac{\sin\alpha}{2\pi}\right], \quad \frac{v}{v_F} = \left(\frac{R}{R_F}\right)^{\frac{2}{3}}$$

$$= \left[1 - \frac{360}{\alpha} \times \frac{\sin\alpha}{2\pi}\right]^{2/3}$$

$$\frac{Q}{Q_F} = \frac{\alpha}{360}\left[1 - \frac{360}{\alpha} \times \frac{\sin\alpha}{2\pi}\right]^{5/3} \text{ or } \frac{Q}{Q_F} = \frac{n}{n_F} \times \frac{A}{A_F} \times \left[\frac{R}{R_F}\right]^{2/3}$$

Example Problem 17.5

A 600 mm diameter sanitary sewer (n = 0.020) is laid on a grade of 1.5 m per km. When the minimum flow is 40 L/s, what is the unused capacity of the pipe as a percentage of full flow?

Given: n = 0.02

D = 600 mm = 0.60 m

S = 1.5 m/km

Q = 40 L /s

SOLUTION:

$$Q_F = \frac{0.312}{n} \times D^{\frac{8}{3}} \times \sqrt{S} = \frac{0.312}{0.02} \times 0.60^{\frac{8}{3}} \times \sqrt{\frac{1.5\,m}{km} \times \frac{km}{1000\,m}}$$

$$= 0.1547\ m^3/s = 150\ L/s$$

$$\frac{Q}{Q_F} = \frac{40}{154.7} = 0.2585 = 26\%\ Unsed = 100\% - 26\% = 74\%$$

Example Problem 17.6

Select a sewer main for carrying a flow of 280 L/s when running half full. The allowable slope is 0.1% and n = 0.013.

Given: n = 0.013
D = ?
S = 0.1% = 0.01
Q = 280 L /s = 0.30 m^3/s

SOLUTION:

$$Q_F = 2 \times Q = 2 \times 0.28 = 0.560\,m^3/s = 5.6 \times 10^{-1}\,m^3/s$$

$$D = \left(\frac{Qn}{0.312\sqrt{S}}\right)^{0.375} = \left(\frac{0.56 \times 0.013}{0.312 \times \sqrt{0.001}}\right)^{0.375}$$

$$= 0.892\ m = 900\ mm(36\,in\,NPS))$$

$$v_F = \frac{0.40}{n} \times D^{2/3} \times \sqrt{S} = \frac{0.40}{0.013} \times 0.90^{2/3} \times \sqrt{0.001} = 0.906 = 0.91\,m/s$$

Example Problem 17.7

Design a sanitary sewer flowing 70% full to carry wastewater from a community with a population of 85 000. Assume daily per capita wastewater production is 150 L/c.d. The sewer main is lined, n = 0.012, and the available slope is 0.16%. Check for self-cleansing velocity at peak flow conditions.

Given: d/D = 0.70 n = 0.012
D = ?
S = 0.16%
Q = 85 000 people at 150 L/c.d

SOLUTION:

$$Q_{max} = 3 \times Q_{avg} = 3 \times 85000\,p \times \frac{150\,L}{p.d} \times \frac{m^3}{1000\,L} \times \frac{d}{1440\,min} \times \frac{min}{60\,s} = 0.443\,m^3/s$$

$$\alpha = 2\cos^{-1}\left(1-\frac{2d}{D}\right) = 2\cos^{-1}(1-2\times 0.7) = 227.1° = 227°$$

$$\frac{Q}{Q_F} = \frac{\alpha}{360}\left[1-\frac{360}{\alpha}\times\frac{\sin\alpha}{2\pi}\right]^{5/3} = \frac{227.1}{360}\left[1-\frac{360}{227.1}\times\frac{\sin(227.1)}{2\pi}\right]^{5/3} = 0.8368$$

$$\frac{v}{v_F} = \left(\frac{R}{R_F}\right)^{\frac{2}{3}} = \left[1-\frac{360}{\alpha}\times\frac{\sin\alpha}{2\pi}\right]^{\frac{2}{3}} = \left(1-\frac{360}{227.1}\div\frac{\sin 227.1}{2\pi}\right)^{\frac{2}{3}}$$

$$= 1.119 = 1.12$$

$$Q_F = \frac{0.443\,m^3}{s}\times\frac{1}{0.8368} = 0.5289 = 0.53\,m^3/s$$

$$D = \left(\frac{Qn}{0.312\sqrt{S}}\right)^{0.375} = \left(\frac{0.5289\times 0.012}{0.312\times\sqrt{0.0016}}\right)^{0.375}$$

$$= 0.776\,m = 776\,mm\,(36\,in\,NPS)$$

$$v_F = \frac{0.40}{n}\times D^{2/3}\times\sqrt{S} = \frac{0.40}{0.012}\times 0.90^{2/3}\times\sqrt{0.0016} = 1.24\,m/s$$

$$v = v_F\times 1.12 = 1.24\times 1.12 = 1.4\,m/s \gg 0.60\,m/s$$

17.7 EQUIVALENT SELF-CLEANING VELOCITY

As seen in **Figure 17.5**, the velocities produced in circular pipes are equal to or more than flowing full velocity when d/D is > 80%. Nevertheless, sewers flowing with depths of 50% to 80% of diameter do not need to be laid at steeper grades to be as self-cleansing as flowing full. This is because velocity and discharge are functions of tractive force intensity, which further depends upon the friction and flow velocity. The required ratio of various hydraulic elements to have an equivalent self-cleansing velocity as full flow is given by the following relationship:

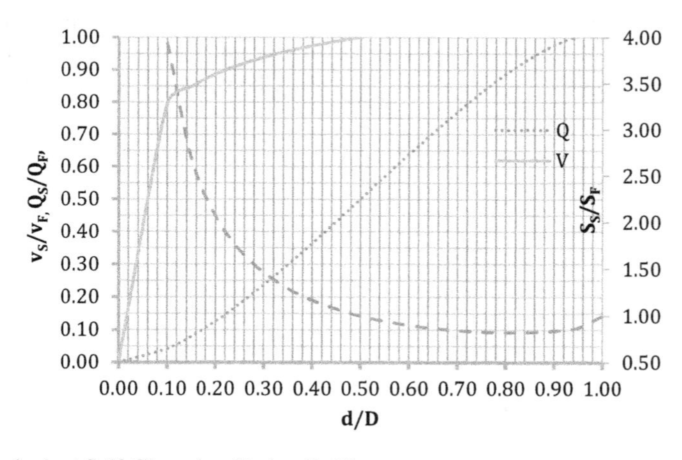

FIGURE 17.5 Equivalent Self-Cleansing Hydraulic Elements

$$\frac{v_s}{v_F} = \frac{n}{n_F} \times \left(\frac{R}{R_F}\right)^{1/6} \quad and \quad \frac{Q_s}{Q_F} = \frac{n}{n_F} \times \frac{A}{A_F} \times \left(\frac{R}{R_F}\right)^{1/6}$$

Example Problem 17.8

A 300 mm sewer pipe is to flow at 30% of full depth on a slope ensuring a self-cleaning equivalent to full-flow velocity of 1.0 m/s. What grade is required if variation of n with depth is ignored and assumed to be 0.012?

Given: d/D = 0.30
n = 0.012
D = 300 mm
S = ?

SOLUTION:

$$\alpha = 2\cos^{-1}\left(1 - \frac{2d}{D}\right) = 2\cos^{-1}(1 - 2\times 0.3) = 132.84° = 133°$$

$$\frac{R}{R_F} = \left[1 - \frac{360}{\alpha} \times \frac{\sin\alpha}{2\pi}\right] = \left[1 - \frac{360}{133} \times \frac{\sin 133}{2\pi}\right] = 0.684$$

$$S_F = \left[\frac{n}{0.4} \times \frac{v_F}{D^{2/3}}\right]^2 = \left[\frac{0.012}{0.4} \times \frac{1.0}{0.30^{2/3}}\right]^2 = 4.48\times 10^{-3} = 0.45\%$$

$$S_s = S_F \times \frac{R_F}{R} = 4.48\times 10^{-3} \times \frac{1}{0.684} = 6.549\times 10^{-3} = 0.65\%$$

Example Problem 17.9

A 450 mm diameter sewer is to carry a flow of 60 L/s at a flow velocity that ensures self-cleansing equivalent to that at a full depth velocity of 0.90 m/s. Find the depth and velocity of flow. Assume n = 0.015 for all depths.

Given: v_F = 0.90 m/s
n = 0.015
D = 450 mm
Q = 60 L/s
d/D, vs., S_s = ?

SOLUTION:

$$S_F = \left[\frac{n}{0.4} \times \frac{v_F}{D^{2/3}}\right]^2 = \left[\frac{0.012}{0.4} \times \frac{0.90}{0.450^{2/3}}\right]^2 = 3.30\times 10^{-3} = 0.33\%$$

$$Q_F = \frac{0.312}{n} \times D^{\frac{8}{3}} \times \sqrt{S} D^{8/3} \times \sqrt{S} = \frac{0.312}{0.015} \times 0.450^{\frac{8}{3}} \times \sqrt{3.3 \times 10^{-3}}$$

$$= 0.450^{8/3} \times \sqrt{3.3 \times 10^{-3}} = 0.142 m^3 / s = \underline{140 L / s}$$

$$\frac{Q}{Q_F} = \frac{60}{142} = 0.422 = 0.42 \, From \, table \frac{Q}{Q_F} = 0.42, \frac{v}{v_F} = 0.5, \frac{d}{D} = 0.45$$

$$d = D \times 0.45 = 450 \, mm \times 0.45 = 202.5 = \underline{200 \, mm}$$

$$\alpha = 2\cos^{-1}\left(1 - \frac{2d}{D}\right) = 2\cos^{-1}(1 - 2 \times 0.45) = 168.5^\circ = 169^\circ$$

$$\frac{R}{R_F} = \left[1 - \frac{360}{\alpha} \times \frac{\sin\alpha}{2\pi}\right] = \left[1 - \frac{360}{169} \times \frac{\sin 169}{2\pi}\right] = 0.935$$

$$S_s = S_F \times \frac{R_F}{R} = 3.3 \times 10^{-3} \times \frac{1}{0.935} = 3.52 \times 10^{-3} = 0.35\%$$

$$v_s = v_F \times \left(\frac{R}{R_F}\right)^{1/6} = \frac{0.90 \, m}{s} \times (0.935)^{1/6} = 0.889 = 0.89 \, m / s$$

$$v = v_F \times 0.95 = \frac{0.90 \, m}{s} \times 0.95 = 0.855 = 0.86 \, m / s$$

Actual flow velocity is very close to equivalent self-cleansing flow velocity.

17.8 STORM DRAINAGE

Providing adequate drainage for storm water is one of the many jobs of an environmental engineer. This job has gained new dimensions since the drainage works changed from primitive ditches to complex networks of curbs, gutters and underground conduits. The simple rules of thumb and crude empirical formulas are generally inadequate. In addition, demands by society for better environmental control require that water quality considerations be superimposed on estimates of quantity so that management of the total water resource can be affected. To design a system with sufficient capacity to yield the adequate drainage of water, for most of the time, a complete understanding of the rainfall–run-off relationship for a given watershed system is very important.

Several design methods for estimating peak discharges are available and vary from a rule of thumb approach, macroscopic approach, microscopic approach and continuous simulation. Each method should be applied only after its strengths and limitations are understood. The most popular methods for estimating peak flows are the Rational Method and the US Natural Resources Conservation Service (NRCS) method. One of the most common methods used is the Rational Method.

17.8.1 Rational Method

The most commonly used peak discharge design method is the Rational Method (Emil Kuichling, 1889). A peak flow is associated with certain watershed parameters, such as area, topography, soil texture, vegetation and surface storage, and with the storm characteristics of duration, intensity and frequency. The Rational Method for computing peak flow is formulated as:

$$Q_p = C \times I \times A$$

C = run-off coefficient, the proportion of the total rainfall which runs off
I = the maximum rainfall intensity for a duration equal to the time of concentration, tc and for a storm of design return period
A = the basin area contributing to run-off

Run-Off Coefficient

The run-off coefficient, C, is estimated based on knowledge of the watershed characteristics. Run-off coefficients can be ascertained from **Table 17.3** and **Table 17.4**. When not sure, it is recommended to select the mid-range.

Time of Concentration

Time of concentration (t_c) is the time it takes for water to travel from hydraulically remote catchment areas to the outlet or drain inlet. The estimation of time of concentration can be calculated by dividing the distance of travel by the average flow velocity as found by Manning's flow equation. For rural catchments, the time of concentration can be estimated by the empirical formula. One such formula for bare earth is:

$$t_c = \frac{0.02L^{0.77}}{S^{0.385}} = \frac{0.02L^{1.2}}{H^{0.385}} = t_i + t_f = \frac{0.02L^{0.77}}{S^{0.385}} + \frac{L}{v}$$

t = time in min
L = length in m
H = drop in m
S = slope as fraction
t_i = inlet time
t_f = flow time

TABLE 17.3
Run-Off Coefficients for Rural Areas

Topography and Vegetation	Soil Texture		
	Sandy Loam	**Clay and Silt Loam**	**Tight Clay**
Woodland			
Flat (0–5% slope)	0.10	0.30	0.40
Rolling (5–10% slope)	0.25	0.35	0.50
Hilly (10–30% slope)	0.30	0.50	0.60
Pasture			
Flat	0.10	0.30	0.40
Rolling	0.16	0.36	0.55
Hilly	0.22	0.42	0.60
Cultivated			
Flat	0.30	0.50	0.60
Rolling	0.40	0.60	0.70
Hilly	0.52	0.72	0.82
Developed areas	30% impervious	50% impervious	70% impervious
Flat	0.40	0.55	0.65
Rolling	0.50	0.65	0.80

TABLE 17.4
Typical C Values (Urban Areas)

Description of Area	Run-Off Coefficients
Business	
Downtown areas	0.70–0.95
Neighbourhood areas	0.50–0.70
Residential	
Single-family areas	0.30–0.50
Multi units, detached	0.40–0.60
Multi units, attached	0.60–0.75
Residential (suburban)	0.25–0.40
Apartment dwelling areas	0.50–0.70
Industrial	
Light areas	0.50–0.80
Heavy areas	0.60–0.90
Parks, cemeteries	0.10–0.25
Playgrounds	0.20–0.35
Railroad yard areas	0.20–0.40
Unimproved areas	0.10–0.30
Streets	
Asphaltic	0.70–0.95
Concrete	0.80–0.95
Brick	0.70–0.85
Drives and walks	0.75–0.85
Roofs	0.70–0.95
Lawns; Sandy Soil:	
Flat, 2%	0.05–0.10
Average, 2–7%	0.10–0.15
Steep, 7%	0.15–0.20
Lawns; Heavy Soil:	
Flat, 2%	0.13–0.17
Average, 2–7%	0.18–0.22
Steep, 7%	0.25–0.35

For covered surfaces, this value can be modified. For example, for an area fully covered with short grass, the time of concentration will be twice as much. In urban catchments, the time of concentration would consist of two components: inlet time, or overland flow time, which is the time to reach the inlet; and flow time, which is the time to travel through the drain or gutter. Flow time, t_f, is found by estimating the flow velocity from Manning's equation, as discussed earlier. Inlet time is found by using empirical formula, as discussed earlier.

Rainfall Intensity

For the time of concentration estimated and a selected rainfall frequency, the design rainfall intensity, I, is obtained by plotting rainfall intensity–duration–frequency (IDF) curves or equations. Standard drainage systems and small spillways are normally designed for rainfall events of 5-year to 10-year frequency.

$$I = \frac{850T^{0.2}}{(D+15)^{0.75}} \quad \textit{and weighted} \quad \bar{C} = \frac{\Sigma C_i A_i}{\Sigma A_i} = \frac{\Sigma C_i A_i}{A}$$

T = return period
D = duration in min
I = rainfall intensity, mm/h,

17.8.2 Limitations of the Rational Method

The following limitations should be recognized when applying the Rational Method.

1. The run-off coefficient *C* is difficult to establish consistently and objectively, particularly for agricultural watersheds where infiltration is highly variable with time and space.
2. The time of concentration t_c is not constant as the formula suggests and may vary with the storm characteristics. It is an extremely difficult parameter to assess meaningfully but plays an important role in estimating the peak discharge.
3. For rural watersheds, the run-off frequency is not equal to the rainfall frequency. In other words, a 10-year storm does not produce a 10-year flood. Therefore, the Rational Method does not yield run-off estimates of selected frequencies but run-off amounts, which might occur because of a rainfall event of a given return period.
4. The rainfall intensity–duration–frequency information is for point rainfall. The probability that such rates will occur uniformly over an area decreases as the area of interest increases.
5. The method is not recommended for agricultural drainage basins larger than 500 ha or 5.0 km^2.
6. The value of run-off coefficient C is assumed to be constant during the progress of an individual storm and from storm to storm. The estimation of the coefficient is based on the degree of imperviousness and the infiltration capacity of the drainage surface. But as infiltration capacity varies with time, depending on initial moisture conditions, C must vary. It is for this reason that run-off frequency is not always equal to rainfall frequency.
7. This method yields a point value Q_p and says nothing about the nature of the rest of the hydrograph.

In urban watersheds where most of the area is impervious, infiltration does not play a major role in the production of run-off. This is the main reason the Rational Method has been used more successfully in its application to small urban watersheds.

Example Problem 17.10

Calculate the peak run-off rate from a 3.5 ha area parking lot. The time of concentration and run-off coefficient are 20 minutes and 0.75, respectively. The rainfall intensity for a 20-minute duration is 75 mm/h.

Given: D = 20 min
I = 75 mm/h
A = 3.5 ha

SOLUTION:

$$Q_p = CIA = 0.75 \times \frac{75mm}{h} \times 3.5ha \times \frac{10\ m^3}{ha.mm} \times \frac{h}{3600\,s}$$

$$= 0.546 = \underline{55\ m^3 / s}$$

Example Problem 17.11

A roof drain has a maximum flow-carrying capacity of 1.25 L/s. Determine how much maximum roof area it can drain without overflowing. Assume the rainfall intensity for a 10-year storm to be 120 mm/h and C = 0.85.

Given: Q = 1.25 L/s
I = 120 mm/h
A = ?

SOLUTION:

$$A = \frac{Q_p}{CI} = \frac{1.25L}{s} \times \frac{1}{0.85} \times \frac{h}{120mm} \times \frac{3600s}{h} \times \frac{m^3}{1000L} \times \frac{1000mm}{m} = 44.12 = \underline{44\ m^2}$$

Example Problem 17.12

A 1.5 km² drainage area is being proposed for development. Pre-development and post-development conditions suggest the following hydrologic parameters.

Development	Surface	C	D = t_c, h
Pre-development	Natural	0.3	3.0
Post development	Partially paved	0.6	2.0

Calculate the peak run-off for pre-development conditions using a 5-year storm.

GIVEN:

IDF function: I (mm/h) = 850 T $^{0.2}$/ (D + 15)$^{0.75}$
T = return period in y
D = rainfall duration in minutes.

SOLUTION:

$$Pre\,development,\ I = \frac{850T^{0.2}}{(D+15)^{0.75}} = \frac{850 \times (5.0)^{0.2}}{(180+15)^{0.75}}$$

$$= 22.7 = 22.5\,mm\,/\,h$$

$$Q_p = CIA = 0.3 \times \frac{22.47\,mm}{h} \times 1.5\,km^2 \times \frac{10^3\,m^3}{km^2.mm} \times \frac{h}{3600\,s}$$

$$= 2.81 = \underline{2.8\ m^3\,/\,s}$$

$$Post\,development,\ I = \frac{850T^{0.2}}{(D+15)^{0.75}} = \frac{850 \times (5.0)^{0.2}}{(120+15)^{0.75}}$$

$$= 29.61 = 29.6\ mm\,/\,h$$

$$Q_p = CIA = 0.6 \times \frac{29.61\,mm}{h} \times 1.5\,km^2 \times \frac{10^3\,m^3}{km^2.mm} \times \frac{h}{3600\,s}$$

$$= 7.40 = \underline{7.4\ m^3 / s}$$

Example Problem 17.13

Compute the peak flow rate from a 5-ha catchment with light vegetation and fine-textured soils. The IDF function for the area is $I = \frac{760T^{0.2}}{(D+15)^{0.75}}$

Given: A = 5 ha
T = 10 y
L = 205 m
S = 0.9% = 0.009

SOLUTION:

$$t_c\,(bare\,earth) = \frac{0.02L^{0.77}}{S^{0.385}} = \frac{0.02 \times 205^{0.77}}{0.009^{0.385}} = 7.39\,min$$

The roughness of a surface covered with short grass is relatively high compared to bare earth. The time of concentration is multiplied by a factor of two: 7.39 × 2 = 14.8 = 15 min, T = 10 years, and D = 15 min I C = 0.36 for clay and silt loam (**Table 17.3**).

$$I = \frac{760T^{0.2}}{(D+15)^{0.75}} = \frac{760 \times (10)^{0.2}}{(15+15)^{0.75}} = 93.9 = 94\,mm / h$$

$$Q_p = CIA = 0.36 \times \frac{93.9\,mm}{h} \times 5.0\,ha \times \frac{10\,m^3}{ha.mm} \times \frac{h}{3600\,s} \times \frac{1000L}{m^3} = 469.8 = \underline{470\,L / s}$$

Example Problem 17.14

Two adjacent fields contribute run-off to a collector whose capacity is to be determined. It is estimated that in 25-minute areas from both the fields start contributing. The rainfall intensity in mm/h for the 15-year storm is given, $i = \frac{1500}{(D+15)^{0.8}}$ where D is the duration of the storm in minutes.

Given: $C_1 = 0.35$
$C_2 = 0.65$
$A_1 = 1$ ha
$A_2 = 2$ ha
$D = t_c = 25$ min

SOLUTION:

$$I = \frac{1500}{(D+15)^{0.8}} = \frac{1500}{(25+15)^{0.8}} = 78.4\,mm\,/\,h$$

$$C = \frac{\Sigma C_i A_i}{\Sigma A_i} = \frac{0.35\times1+0.65\times2}{1+2} = 0.55$$

$$Q_p = CIA = 0.55\times\frac{78.4\,mm}{h}\times3.0\,ha\times\frac{10\,m^3}{ha.mm}\times\frac{h}{3600\,s}\times\frac{1000L}{m^3}$$

$$= 35.6 = \underline{36\ L\,/\,s}$$

17.8.3 Urban Catchments

Application of the Rational Method to urban catchments ($1 < A < 2.5$ km^2) requires special techniques. Flow varies widely along the length of the main channel (sewer) and is generally small at the upstream and larger at the downstream. It may be difficult to determine average t_c. An alternative is to apply the Rational Method incrementally. This method requires the sub-division of the catchment into several sub-areas, as shown in **Figure 17.6**. Water moves through a watershed as sheet flow, shallow concentrated flow, open-channel flow or some combination of these before it exits the sewer line. This is called inlet time.

Referring to **Figure 17.6**, the time of concentration for inlet 1 is inlet time for run-off from catchment area A to reach the inlet. This is the time taken by the overland flow to reach point 1. Based on flow in reach AB of the sewer line, travel time from A to B, t_{AB} is calculated using Manning's equation. The time of concentration for inlet 2 is equal to the inlet time for area A plus travel time from 1 to 2 or inlet time for catchment area B, whichever is greater. This process is continued until the peak flow at the outlet is worked out.

Except for the first inlet, more than one area contributes to run-off. Thus, it becomes a case of composite catchment. Various flow paths are tried and the one producing maximum flow is selected, as shown in the following example.

Example Problem 17.15

A storm drainage system comprises the four areas shown in **Figure 17.6**. Determine the 5-year design flow for each section of the sewer line.

GIVEN:

Catchment	A	B	C	D
Area A, ha	5	5	12	6
Coefficient, C	0.8	0.8	0.6	0.9
Inlet time, t_i, min	10	11	20	8

Sample of Calculations (inlet 2)

$$\Sigma C_i A_i = 0.8\times5+0.8\times5 = 8.0\,ha$$

$$t_{A-2} = 10\,min + 100\,m\times\frac{s}{1.5\,m}\times\frac{min}{60\,s} = 11.1\,min$$

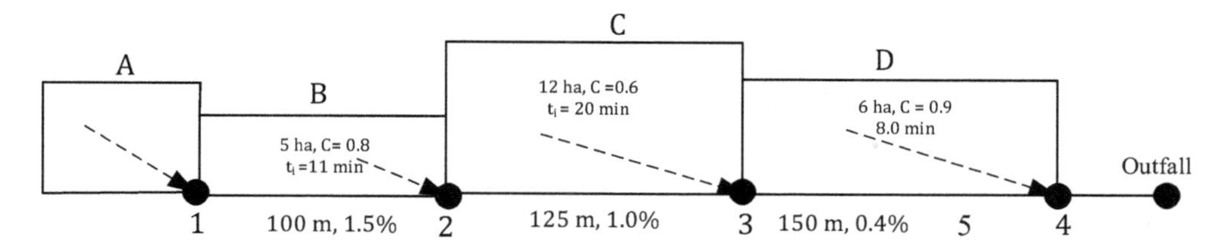

FIGURE 17.6 Urban Draining System (Ex. Prob. 17.15)

TABLE 17.5
Table of Computations

inlet	C	A ha	CA ha	ΣCA ha	Time of concentration, min: Path	Inlet, t_i	L, m	v, m/s	t_f	Tot	I mm/h	Q_p m^3/s
1	0.8	5	4	4	A–1	10				10.0	100.0	1.1
2	0.8	5	4	8	A–2	10	100	1.5	1.1	11.1	95.7	2.1
3	0.6	12	7.2	15.2	C–3	20			0.0	20.0	71.4	3.0
4	0.9	6	5.4	20.6	C–4	20	150	1.5	1.7	21.7	68.2	3.9

Note: For a given outlet, the longest path is used to calculate the time of concentration, which is used to calculate rainfall intensity.

$$I = \frac{2500}{D+15} = \frac{2500}{11.1+15} = 95.7\,mm/h$$

$$Q_p = I \times \Sigma C_i A_i = \frac{95.7\,mm}{h} \times 8.0\,ha \times \frac{10\,m^3}{ha.mm} \times \frac{h}{3600\,s}$$

$$= 2.10 = \underline{2.1 m^3/s}$$

Example Problem 17.16

Find the size of a combined sewer pipe to carry combined flow from an area of 110 ha with a population density of 250 p/ha without exceeding a flow velocity of 3.0 m/s. Assume a water supply rate of 220 L/p·d of which 80% becomes sewage water. The weighted run-off coefficient is 0.45, and inlet time and flow time are 13 minutes and 22 minutes, respectively. The rainfall intensity for the design storm is given by the following IDF function: $I = \frac{1500}{(D+15)^{0.8}}$

Given: C = 0.45
A = 110 ha
Q = 220 L/p·d
Population = 250 p/ha

SOLUTION:

$$Q_{ww} = 0.80 \times \frac{220\,L}{p.d} \times 110\,ha \times \frac{250\,p}{ha} \times \frac{m^3}{1000\,L}$$

$$= \frac{4840\,m^3}{d} = \frac{4840\,m^3}{d} \times \frac{d}{24h} \times \frac{h}{4600\,s} = 0.056\,m^3 / s$$

$$t_c = t_i + t_f = 13\,min + 22\,min = 35\,\text{min},$$

$$I = \frac{1500}{(D+15)^{0.8}} = \frac{1500}{(35+15)^{0.8}} = 65.6\ mm / h$$

$$Q_p = CIA = 0.45 \times \frac{65.6\,mm}{h} \times 110\,ha \times \frac{10\,m^3}{ha.mm} \times \frac{h}{3600\,s}$$

$$= 9.02 = 9.0\ m^3 / s$$

$$Q_c = Q_{ww} + Q_{sw} = 0.056 + 9.02 = 9.076 = 9.1\,m^3 / s$$

As the data shows, wastewater flow is relatively negligible compared to storm flow. Thus, during dry weather, the sewer may not be able to achieve self-cleansing velocities.

$$D_{min} = \sqrt{\frac{4}{\pi} \times \frac{Q_p}{v_{max}}} = \sqrt{\frac{4}{\pi} \times \frac{9.076}{s} \times \frac{s}{3.0\,m}} = 1.96\,m = \underline{2.0\,m}$$

Discussion Questions

1. Comparing the Darcy–Weisbach flow equation with Manning's flow equation, develop the relationship between f and n.
2. Show that for circular pipes flowing full or half full, the hydraulic radius is one-fourth of the diameter of the pipe.
3. Chézy's flow equation is $v = C\sqrt{R_h S}$. Find the relationship between C and n.
4. Explain why the Rational Method is more successful for small urban catchments.
5. What are the limitations of the Rational Method?
6. Define time of concentration and the factors affecting it.
7. Describe the methodology for sizing storm sewers in a given catchment area.
8. Compare section factor with conveyance factor.
9. Modify Manning's equation (SI) for circular sewer pipes flowing full.
10. What is normal depth? Compare gradually varied flow with rapidly varied flow using examples.
11. As it is, Manning's equation is applicable for uniform flow conditions. When applying this equation to GVF, what modifications need to be made?
12. A section with a minimum wetted perimeter is hydraulically most efficient to carry flow. Based on this, prove that a rectangular section is most efficient when the depth of flow is half of the width.
13. Define self-cleansing velocity and non-scouring velocity and the accepted range of values for this in sewer pipes.

14. Explain why it is necessary to achieve this flow velocity at least for some time during a 24-hour period.
15. Describe the relationship between equivalent self-cleansing slope and velocity at partial depth.
16. A rectangular sewer with a width twice its depth is equivalent to a circular one. Derive the relationship between diameter and depth in the rectangular section.
17. What are the different hydraulic elements, and the relationships between them, that govern the discharge in sewer pipes?
18. What considerations should be made when designing sanitary sewer pipes?
19. What are the factors that affect the hydraulics of sewer lines?
20. Wastewater flows have diurnal variation. How does this affect the flow velocity in circular sewer pipes?

Practice Problems

1. Determine the size of a sanitary sewer to serve a population of 50 000 when flowing at a depth of 70% of the diameter. The peak hourly flow is 2.5 times that of the average flow and the available grade is 1/1200. Assume a roughness coefficient n of 0.013. (450 mm)
2. Determine the size of a combined sewer pipe to carry combined flow from an area of 90 ha with a population density of 250 p/ha without exceeding a flow velocity of 3.0 m/s. Assume a water supply rate of 250 L/p·d, of which 80% becomes sewage water. The weighted run-off coefficient is 0.45, and inlet time and flow time are 11 minutes and 18 minutes, respectively. The rainfall intensity for the design storm is given by the following IDF function: $I = \frac{1500}{(D+12)^{0.8}}; I = \frac{mm}{h}, D = min.$ (2.0 m)
3. A town with a population of 40 000 is spread over a drainage area of 75 ha with a run-off coefficient of 0.70 and a time of concentration of 40 minutes. The average daily water consumption is 120 L/d, 70% of which reaches the sewer. Calculate the peak discharge rate. The rainfall intensity function is as follows: $I(mm/h) = \left(\frac{80}{1+t_c(h)}\right)$ ($7.0\ m^3/s + 0.04\ m^3/s \times 3 = 7.12\ m^3/s$)
4. What is the required flow capacity of a drainage sewer pipe that drains two adjacent fields? It is estimated that in 28-minute areas from both the fields start contributing. The rainfall intensity in mm/h for the 15-year storm is given by the IDF function as in Example Problem 17.12. (280 L/s)

Area, ha	Coefficient, C
1.5	0.45
0.8	0.60

5. A 400 mm sewer pipe is to flow at 30% of full depth on a slope ensuring a self-cleaning equivalent to full-flow velocity of 0.80 m/s. What grade is required if variation of n with depth is ignored and assumed to be constant 0.013? (0.34%)
6. A storm drainage system is shown in **Figure 17.7**. For each inlet, catchment area, run-off coefficient and inlet time are shown. For each reach, length and slope are also

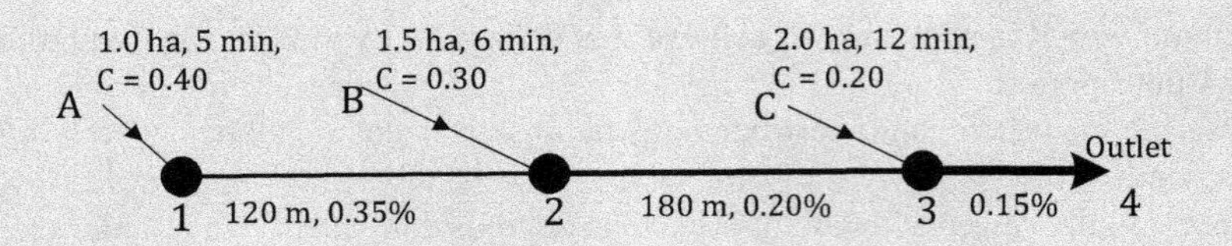

FIGURE 17.7 Storm Drainage System

indicated. For a 10-year return period, the maximum rainfall intensity for various durations are shown below:

Duration, min	5	10	15	20
Intensity, mm/h	150	138	125	110

Using the Rational Method, compute the design flow and required pipe size for each reach assuming maximum velocity of 3.0 m/s. (300 mm, 450 mm, 400 mm NPS).

7. Design a sanitary sewer flowing 70% full to carry wastewater from a community with a population of 55 000. Assume daily per capita wastewater production is 150 L/c.d. The sewer main is lined, n = 0.012, and the available gradient is 0.20%. Check for self-cleansing velocity at peak flow conditions. (750 mm NPS, 1.2 m/s okay)
8. Design a sewer for a maximum discharge of 450 L/s running half full. Consider Manning's roughness coefficient n = 0.012 and the gradient of sewer S = 0.0001. (1.6 m)
9. Work out the capacity of a 1250 mm diameter sewer laid in a slope of 1:400 when flowing half full. Assume Manning's n of 0.011. (1300 L/s)
10. A 450 mm sewer pipe of n = 0.013 is laid on a slope of 0.0025. At what depth of flow is the flow velocity 0.60 m/s? (110 mm)
11. A 300 mm diameter sewer is to flow at a depth of 30% of the diameter on a grade ensuring a degree of self-cleansing equivalent to that obtained at full depth at a velocity of 0.90 m/s. Find the required grade and associated velocity. Assume Manning's n = 0.013. The variation of n with depth may be neglected. (0.63%, 0.85 m/s)
12. A 350 mm diameter sewer (n = 0.014) is to flow at a depth of 35 mm on a grade such that it generates self-cleansing equivalent to a full-depth velocity of 0.80 m/s. Find the required grade and associated velocity at partial depth. The variation of n with depth may be neglected. (1.3%, 0.64 m/s)
13. A 600 mm diameter sewer is to carry 70 L/s at a self-cleansing velocity equivalent to a sewer flowing full at 0.85 m/s. Assuming a uniform value of n = 0.015, find the depth, velocity and required grade. (190 mm, 0.81 m/s, 0.27%)
14. A 1.5 ha residential area is comprised of 20% roof area (C = 0.90), 25% paved area (C = 0.85), 50% open ground and lawn area (C = 0.10) and 5% wooded are (C = 0.05). Assuming a maximum rainfall intensity of 65 mm/h, what is the peak run-off rate for this area? (610 L/s)
15. Design a plastic sewer pipe (n = 0.012) to carry a flow of 65 L/s flowing full. The available slope is 0.13%. (375 mm, 15-in NPS)

16. Find the full-flow capacity of the sewer pipe mentioned in Problem 15. Also find the discharge and flow velocity when flowing partially full at 60% of the full depth. (69 L/s, 46 L/s, 0.67 m/s)
17. A 900 mm diameter sewer carries a certain flow when flowing full. What will the discharge rate and flow velocity be when the depth of flow is 30% of full depth? (133°, 18%, 78%)
18. A sewer main has a diameter of 300 mm and n = 0.013. Applying Shields equation, determine the minimum cleaning velocity required to transport:
    a. a 0.75 mm sand particle (SG = 2.65); (0.35 m/s)
    b. a 4.0 mm organic particle (SG = 1.1); (0.77 m/s)
19. Design an outfall sanitary sewer to carry the sewage flow from a population of 95 000. Assume that per capita wastewater production is 160 L/c.d. The maximum flow can be assumed to be three times the daily average flow. The sewer pipe is lined (n = 0.012) and laid on a grade of 1 in 1000. A self-cleansing velocity of 0.75 m/s needs to be developed. Dry weather flow can be assumed to be one-third of the maximum. (900 mm NPS, 0.98 m/s, 0.78 m/s)
20. A sewer main is to be designed to carry peak flow while flowing 75% full at a velocity of 1.0 m/s. If the ratio of the maximum/average flow is 3.0 and average/minimum flow is 2.5, determine the proportionate depth of flow and flow velocity at:
    a. average flow; and (0.38, 0.78 m/s)
    b. minimum flow. (0.24, 0.57 m/s)
21. A town has a population of 82 000 with a water supply demand of 250 L/c.d. Design a sewer main running at a partial depth of 75% of the diameter. The available slope is 0.18%. Assume a peaking factor of 2.5 and n = 0.013. (825 mm NPS)

18 Construction of Sewers

Sewer pipes can be as small as building sewers and as big as intercepting and trunk sewers. No one material suits all kinds of sewer pipes. The important factors to be considered in the selection of materials for sewer construction include available sizes, lengths and fittings, life expectancy, resistance to scour and chemicals, type of joint, strength, ease of handling and cost.

18.1 MATERIALS FOR SEWERS

The most common materials used to convey sewage include vitrified clay pipe (VCP), asbestos-cement pipe (ACP), plastic pipe and concrete pipe. There are two main types of plastic pipes available: rigid plastic pipe, which can be either polyvinyl chloride (PVC) or acrylonitrile butadiene styrene (ABS); and soft plastic pipe, which is called high-density polyethylene (HDPE) and is often used for force mains. Plastic PVC pipe is usually grey, sometimes green or blue; ABS pipe is black.

18.1.1 Vitrified Clay Pipe (VCP)

Vitrified clay is the most common material used in sanitary sewers, especially for small sizes. VCP pipes are made in standard sizes up to 900 mm in diameter. They are popular because of their resistance to corrosion and relatively smooth surface. Because they are brittle and easily broken, they are not manufactured in large sizes. For such cases, reinforced concrete pipe (RCP) coated with protective lining is commonly used. VCP pipe is available in various lengths ranging from 0.60 m to 2.0 m, depending on diameter.

18.1.2 Plastic Pipe

The plastic pipe used most frequently in sewer systems is polyvinyl chloride (PVC) or polyethylene (PE). PVC pipe is usually made in sizes 100 mm to 300 mm, with certain manufacturers making up to 750 mm diameter. It is light and usually made in 6.0 m lengths, though other lengths are also available. PVC pipe is used for building connections and branch sewers. PE pipes are used more for long pipelines often laid under adverse conditions, as in swamps or underwater crossings.

18.1.3 Fibreglass Pipe (FRP)

Fibre-reinforced polymer (FRP) and glass fibre reinforced plastic (GRP) pipes are available in sizes from 150 mm to 3000 mm and lengths up to 6.0 m. Bell and spigot O-ring compression joints are generally used to join the pipe sections.

18.1.4 Concrete Pipe (CP)

Concrete pipe (CP) is used extensively in storm sewer systems since it is available in large sizes and is resistant to abrasion. Concrete pipe is not used in small diameter sanitary sewers because of its vulnerability to crown corrosion. Corrosion is caused by the formation of hydrogen sulphide gas under anaerobic conditions. The oxidation of hydrogen sulphide to sulphuric acid collects in droplet form at the crown of sewer pipe and results in corrosion, hence the term crown corrosion.

DOI: 10.1201/9781003231264-18

Non-reinforced concrete pipe is available in sizes up to 600 mm and usually in shorter lengths. Non-reinforced concrete pipe in sizes 100 mm to 600 mm in diameter and reinforced concrete pipe in sizes 200 mm to 3.0 m in diameter are generally available for gravity sewers. A number of joint designs are available depending on the degree of water tightness required. Casketed tongue-and-groove joints can be used when infiltration is a problem. Protective linings should be used where excessive corrosion is likely to occur.

For large sizes, reinforced cement concrete (RCC) pipe is used due to its strength. Reinforced cement concrete pipe is used in large sanitary sewers such as trunk sewers where VCP sizes are not available. In such cases, the pipe is protected against corrosion by applying epoxy, providing ventilation, and laying pipes on relatively steeper grades to achieve high flushing velocities. The advantages of concrete pipe are the relative ease with which the required strength may be provided, the wide range of pipe sizes, the long laying lengths, and the rapidity with which the trench can be opened and backfilled. Concrete pipes are preferred for medium and large sewer pipes such as main and branch sewers. When specifying concrete pipe, it is necessary to give the pipe diameter, class or strength, the method of jointing, the type of protective coating and lining, if any, and any other special requirements for concrete.

18.1.5 Asbestos-Cement Pipe (ACP)

Pipe made up of asbestos fibre and cement is used in sewerage systems. It is available in sizes ranging from 100 mm to 900 mm in diameter. Jointing is accomplished by compressing rubber rings between pipe ends and sleeves. Asbestos-cement or cast iron fittings are used. The advantages of using this pipe are that it is lightweight and easy to handle at long laying lengths, forms tight joints, is quick to install, and is corrosion-resistant in most natural soil conditions. However, pipes of this type are prone to sulphide corrosion. When specifying the pipe, the diameter, class or strength, and type of joint should be specified. ACPs have a smooth surface, so they are hydraulically more efficient. They are brittle and structurally weak, so more commonly used as verticals.

18.1.6 Brick Masonry

Brick masonry had been used for large-diameter sewers in the past. Due to its high cost, lack of durability and other factors, brick is now used only in special applications.

18.1.7 Cast Iron Pipe

Cast iron (CI) pipe is available from 100 mm to 1200 mm in diameter with a variety of jointing methods. They are used in gravity sewers where tight joints are essential. The advantages of cast iron pipe are long laying lengths with tight joints, its ability to withstand high internal pressure and external load, and its corrosion resistance in most natural soils. They are very vulnerable to corrosion unless protected by lining with paint or cement concrete. They are relatively costly and used under special circumstances such as heavy external loads, protection against contamination, crossing low-level areas, the prevalence of wet conditions, and temperature variations and vibrations. When specifying cast iron pipe, it is necessary to give the pipe class, the joint type, the type of lining and the type of exterior coating.

18.1.8 Steel Pipe

Steel pipe is used for force mains and inverted siphons. Other pipe materials used for special applications in wastewater collection systems are smooth-wall and corrugated steel pipe, bituminized fibre pipe and reinforced resin pipe. The flexibility of galvanized corrugated steel permits the fabrication of a variety of conduit shapes with a choice of protective coatings. Available circular sizes are in the range of 200 mm to 2400 mm in diameter. Pipe sections are generally furnished up to 6.0 m in length in multiples of 60 cm. Coupling bands that may be single-piece, two-piece or an internal expanding type used in lining work joint the sections. To increase durability and to resist corrosion, galvanized pipes can be coated with bituminous material. The advantages of steel pipe are that it is lightweight, allows for long laying lengths, is easy to ship, adaptable to stacking methods, flexible, provides strong mechanical joints, and able to adjust to trench-loading conditions.

18.1.9 Cast-in-Place Reinforced Concrete

Sewers are constructed of cast-in-place reinforced concrete when the required size is more economical to construct in situ, when a special shape is required, and when headroom and working space are limited. Forms for concrete sewers should be unyielding and tight and should produce a smooth sewer interior. Methods for resisting corrosion are the same as those for concrete pipe.

18.2 LAYOUT AND INSTALLATION

The various steps involved in the layout and construction of sewer mains are described in the following sections.

18.2.1 Setting Out

This is the first step in the construction of any work. In sewerage work, this is carried out starting from the tail end or the outfall end and proceeding upwards, marking sewer lines along proposed routes on the ground, fixing pegs at intervals of 15 m and establishing temporary benchmarks with respect to the fixed benchmarks.

The advantage gained by starting from the tail end is the utilization of the sewers from the beginning, thus ensuring that the functioning of the sewerage scheme does not have to wait until the completion of the entire project work.

18.2.2 Alignment and Gradient

Sewers are laid to the correct alignment and gradient, setting their positions and levels to ensure a smooth gravity flow. This is done with the help of suitable boning rods and sight rails, and with accurate levelling instruments (**Figure 18.1**).

The sight rail is a horizontal wooden board secured to two vertical posts called uprights using heavy steel clamps, fixing it immovably to the correct line and level. A boning rod or traveller is a vertical wooden post suitably shod with iron and fitted with a crosshead or tee and is of length equal to the height of the sight rail above the invert line of the sewer. This can move to and fro in the trench to give the invert line on the prepared bed of the sewer. Both the boning rod and the sight rail have their centreline accurately marked with a thin saw-cut and are painted black and white to aid visibility. At least three to four sight rails are always maintained at the correct level and gradient along the line of the sewer at the sight of sewer construction.

18.2.3 Excavation of Trenches

This work is usually carried out in open cutting. Tunnelling is adopted only when large sewers are to be laid at considerable depth below the ground level. The excavation is made to have trenches of such lengths, widths and depths as enable the sewers to be properly constructed. Usually, not more

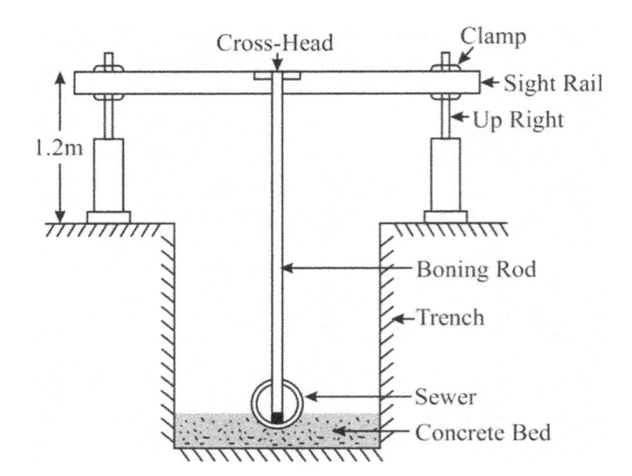

FIGURE 18.1 Setting Out Sight Rail

than about an 18 m length of any trench in advance of the end of the constructed sewer is to be left open at any time in busy streets and localities. The width is chosen based on two considerations:

1. To permit thorough ramming of the backfill material around the pipe.
2. To enable the construction of tight joints.

At least 20 cm of clear space should be left on each side of the barrel of the pipe so that the minimum clear width of the trench is equal to the external diameter of the pipe barrel plus 40 cm. For other types of sewers, the minimum clear width should be the greatest external width of the structures to be built therein. The depth of excavation should be such as to enable the sewers to be laid at their proper grades on the bed of the trenches. Except where the soil is very firm, it is usual to provide a bed of concrete on the bottom of the trench and rest the sewer thereon. Suitable recesses are left on the bed in order to accommodate the socket-end of the pipe sewer.

In countries like India, excavation is usually carried out manually with pickaxes and shovels. In most developed countries, backhoe and similar power equipment are used to excavate trenches. The broken turf, pavement, etc., is carefully stacked out for use in reinstatement. The excavated material is stacked sufficiently away from the edge of the trench to form spoil banks of ordinary size. Excavation in roads is done so as to cause minimal obstruction to traffic and to ensure public safety by erecting suitable warning signals at the site of trenches. Excavation below the water table is done after dewatering the trenches.

Timbering of Trenches

For depths exceeding 2 m, it is necessary to timber the trenches to prevent the sides from caving in, which may cause possible damage to the foundations of the adjoining property. The type of timbering depends chiefly upon the type of soil met with. It consists of vertical poling boards with horizontal whaling suitably strutted through hard wood wedging of adequate dimensions and strength. For soil that is loose or sandy, close timbering formed by additional vertical boards called runners is used (**Figure 18.2**).

Steel sheet piling is sometimes resorted to in place of timbering in case of badly waterlogged areas or in other situations where timber is not easily available. Steel sheeting is more watertight, stronger and durable. Though costlier than timber, it can be used many times without disintegrating, and hence it is more economical in works of larger scale. Timbering or sheeting is usually withdrawn

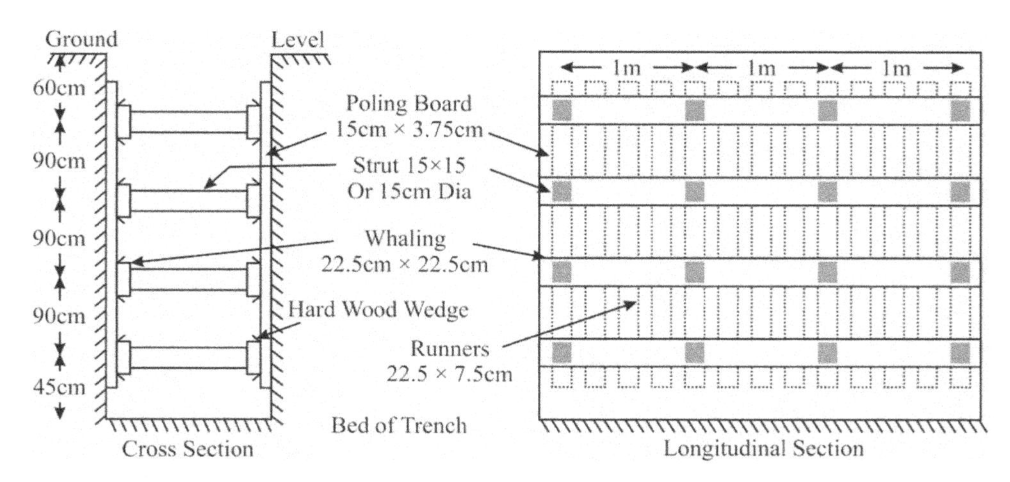

FIGURE 18.2 Timbering of Trenches

after the sewer has been laid, though sometimes it is necessary to leave it off as such, particularly in the case of wet trenches, which may otherwise be damaged.

Dewatering in Trenches

Where the subsoil water level is very near the ground level, the trench can be wet and muddy, and it becomes difficult for sewers to be constructed. Methods suggested to dewater trenches include direct drainage, drainage by an underdrain, sump pumping and wellpoint drainage (**Figure 18.3**).

Wellpoint drainage consists of driving or jetting underwater pressure wellpoints alongside the trench to a common header pipe at intervals of about 1.5 m or less. These points are connected to a common header pipe, which, in turn, is connected to a pump.

In sump pumping, water is collected in a sump made out in the trench from where water is pumped. The pump must be continuously worked day and night, otherwise water will keep flowing into the trench. This method can be used on small jobs and where the subsoil strata is not very sandy, otherwise the sides of the trench are likely to cave in due to continuous pumping. Wellpoint drainage is particularly suitable for large jobs and where the subsoil strata consists of quicksand or running sand.

18.2.4 Bedding

When installing new or repaired sections of a sewer main, careful attention must be paid to the pipe bedding. Bedding refers to the type and preparation of the material that the pipe is placed in when it is being installed and repaired. The preferred method of bedding pipe is to install it in bedding that is supported along its whole length by a bed of sand or gravel spread to an even depth. The sand or gravel should be slightly compacted. In some cases, it may be preferable to lay the pipe directly on the bottom of the trench. This is acceptable only if the trench has been smoothed out to support the pipe along its entire length and the areas under the couplings have been slightly scooped out. Couplings should never rest directly on the trench bottom because they are slightly flexible and will break if not allowed to move. Bed preparation for plastic pipes needs to be done more carefully as these pipes deform when backfilled. Bedding is classified into four types.

Class D bedding is the weakest and least desirable type and is rarely recommended for sewer construction. In this type of bedding, the bottom of the trench is left flat, and the barrel of the pipe is not fully supported because of the protruding bell ends. Backfill is placed loosely without proper compaction.

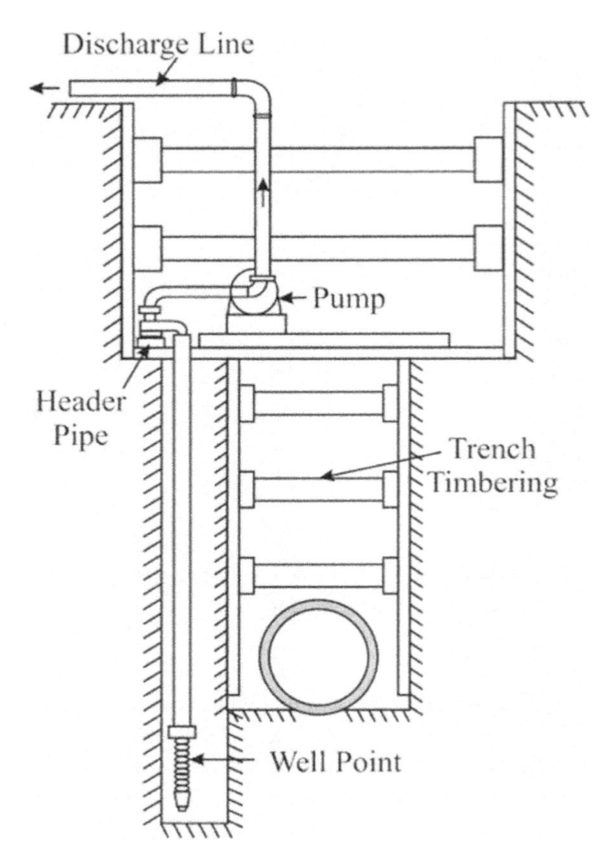

FIGURE 18.3 Wellpoint System of Drainage

Class C, called ordinary bedding, has compacted granular material placed under the pipe and partially extending up the pipe barrel. This provides reasonable support, with a load factor of 1.5. Thus, field strength is increased to 150% of the crushing strength.

Class B, or first-class bedding, has the compacted granular material extending halfway up the barrel of pipe. In addition, backfill is compacted carefully over the top of the pipe, thus achieving a load factor of 1.9.

Class A bedding is the superior type of bedding with a load factor as high as 2.8. The pipe barrel is cradled in concrete and the backfill is carefully compacted.

18.2.5 Laying

After setting the sight rails over the trench, small pegs at intervals of 3 m or so are driven such that the tops of the pegs are level with the invert-line of the pipe sewer, as shown in **Figure 18.4**.

This is achieved by adjusting the uprights of sight rails, stretching a string line bb' across the centre marks on the sight rails and moving a boning rod with its crosshead nearly touching the string line, its base shoe resting on the trench bed. If the horizontal distance between consecutive sight rails is 15 m, and the desired gradient of the sewer is 1:60, the drop in elevation must be 0.25 m or 25 cm. After transferring the centreline of the sewer to the bottom of the trench, the latter is trimmed off slightly to enable the inside of the pipe barrel to conform to the invert line. The pipe can now be laid

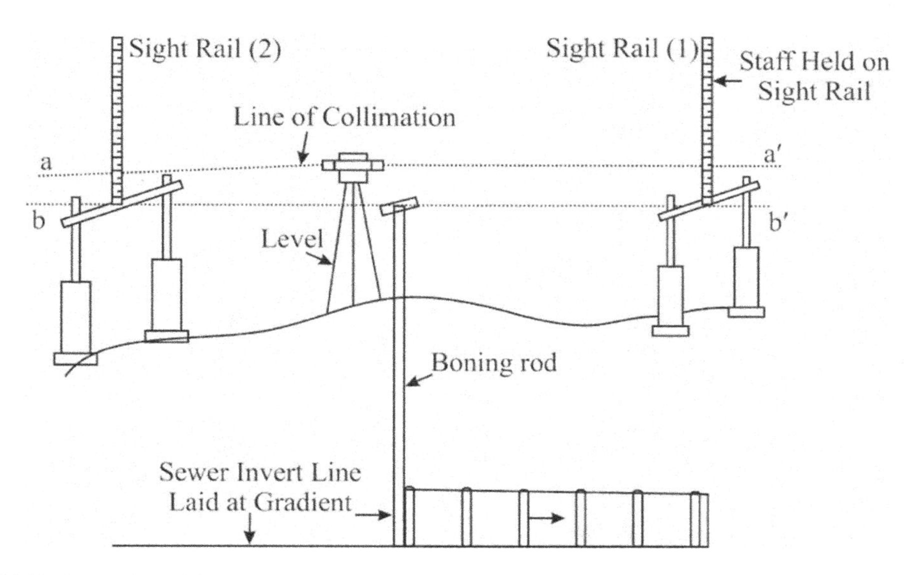

FIGURE 18.4 Laying of Sewer

on the prepared bed. To enable the pipes to be laid in straight length, the method is to measure, from the centreline already marked, a distance equal to half the external diameter of the socket. At this distance, stretch a line at half the height of the pipe so that when the pipes are laid not quite touching this line, they will be in a straight line.

The pipes are usually laid uphill with their sockets facing the direction of flow. In this way, the spigot of each pipe can be easily inserted into the socket of the pipe already laid. Once laid and pressed in position, the pipe is tested for level by passing a straight edge through it and seeing that the edge rests squarely upon the level pegs, as well as on the invert of the pipe throughout its length. Any departure from there would mean that the pipe is laid too high or too low. This should be corrected accordingly.

18.2.6 Lasers

In modern sewer construction, laser beams are used to establish the specified slope of the pipe. A laser can project an intense but narrow beam of light over a long distance. The laser is securely mounted in the manhole and the slope of the light beam is accurately set to match the required slope of the pipe. A transit mounted on the manhole is used to establish pipe alignment from the field reference points and to transfer the alignment down to the laser. Lasers can maintain accuracies of 0.01% over a distance of up to 300 m. In other words, the invert elevations can be set accurately to within 30 mm in a km length of the sewer line.

Example Problem 18.1

A 120 m reach of sewer is to be designed with a flow capacity of 100 L/s, flowing full. The street elevation at the upper manhole is 80.00 m and at the lower manhole is 77.60 m. Select the appropriate size and slope of the pipe invert elevations. Assume a minimum cover of 2.0 m above the crown of the pipe.

Given: n = 0.013
D = ?
Z_1 = 80.00 m
Z_2 =77.60 m
L = 120 m
Q = 100 L /s

SOLUTION:

$$S = \frac{(80.0 - 77.6)\ m}{120\ m} = 0.02\ m/m$$

$$D = \left(\frac{Q_F \times n}{0.312\sqrt{S}}\right)^{0.375} = \left(\frac{0.1 \times 0.013}{0.312 \times \sqrt{0.02}}\right)^{0.375} = 0.266\ m$$

$$= 270\ mm = 300\ mm\ (12\ in\ NPS)$$

$$Q_F = \frac{0.312}{n} \times D^{8/3} \times \sqrt{S} = \frac{0.312}{0.013} \times 0.30^{8/3} \times \sqrt{0.02}$$

$$= 0.30^{8/3} \times \sqrt{0.02} = 1.52\ m^3/s = \underline{150\ L/s}$$

$$\frac{Q}{Q_F} = \frac{100}{152} = 0.66 \quad Corresponding \quad \frac{v}{v_F} = 1.07 \quad \frac{d}{D} = 0.60$$

$$d = D \times 0.60 = 300\ mm \times 0.6 = 180\ mm$$

$$v_F = \frac{0.40}{n} \times D^{2/3} \times \sqrt{S} = \frac{0.40}{0.013} \times 0.30^{2/3} \times \sqrt{0.02} = 1.94 = 1.9\ m/s$$

$$Z_1 = 80.00\ m - 2.0\ m - 0.30\ m = 77.70\ m$$

$$Z_2 = Z_1 - drop = 77.70\ m - 0.20 \times 120\ m = 75.30\ m$$

18.2.7 Jointing

The characteristics of a good joint include water tightness, resistance to root penetration, resistance to corrosion, a reasonable degree of flexibility, and durability. There are several types of joints. For small concrete pipes made with bell-and-spigot ends, the joints may be similar to those used on vitrified clay pipe. A bell-and-spigot joint is shown in **Figure 18.5**. For larger concrete pipes with tongue-and-groove ends, joints can be made with mortar or bituminous compounds or with rubber gaskets. For asbestos-cement pipe, the joint consists of a collar or coupling and a pair of rubber rings. For bell-and-spigot concrete pipe, cement mortar is used to pack against the hemp, or jute caulked into the annular space of the pipe, and to butt around the pipe joint forming a 45-degree bevel.

After truly bedding the first pipe, the second pipe is laid. A ring of tarred yarn soaked in cement slurry is passed around the spigot of the second pipe so that when driven home it is supported in the spigot of the first pipe. The yarn ring is caulked tightly so as to fill about one-quarter of the depth of the socket. Caulking helps to make the joints watertight.

The two pipes can be tested for gradient by passing the straight edge along the inverts of the two pipes and onto the nearest level-peg up the gradient. The level-pegs are removed as the work proceeds. When the second pipe has been truly bedded, the socket of the first is filled with cement

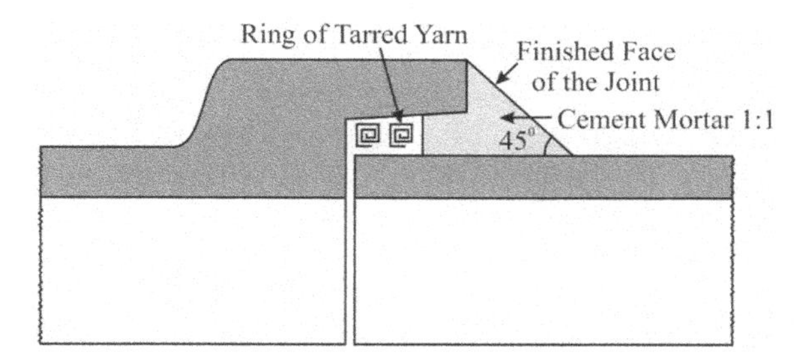

FIGURE 18.5 Bell-and-Spigot Joint

mortar of a stiff consistency (1:1), worked well and finally finished off with a splayed joint or fillet formed with a trowel making an angle of 45° with the barrel of the pipe. Any cement or other extraneous matter sticking inside the pipe or joint is removed with a scraper before the next pipe is laid.

Joining VCP

Bell-and-spigot VCPs are connected by compression joints with seals of plastic to prevent leakage, infiltration, and penetration of roots. The bell end has a polyester liner and the spigot end has a compression ring in the annular space. In the bell section of the pipe already laid, the spigot end of the new section is pushed to form a tight seal. In other cases, both bell and spigot have polyurethane elastomer seals. Before joining, seals are lubricated, and the spigot end is pushed into the bell section of the already laid pipe using a pipe puller. For plain-end pipes, a PVC sleeve or collar is attached in the factory. The seal on the plain end is lubricated and pushed into the collar to form a compression joint.

Jointing ACP

Asbestos-cement pipe has tapered ends. To join two ACPs together, an asbestos-cement collar is used that has rubber gaskets at each end. To install the collar, the gaskets must be free of all grit and dirt. The gaskets are greased with manufacturer recommended lubricant. A joint is made with two asbestos-cement pipes by carefully forcing the tapered end of each pipe into the coupling. Some makes of couplings have a wedged type gasket that must be placed in the correct position in order to make a good seal. ACP must be very carefully bedded or supported, as it has limited resistance to shearing. ACP comes in longer lengths than clay pipe and therefore requires fewer joints.

Jointing Plastic Pipe

As mentioned above, there are three types of plastic pipe, ABS, PVC and HDPE. Recently, plastic pipe has been used more frequently in sewage installation, especially for building service connections. Plastic pipe is very smooth and has excellent flow characteristics. It may be manufactured with either plain ends or a bell-and-spigot arrangement. Both types use a special jointing solvent (glue) that cements the sections together. When jointing bell and spigot pipe, the solvent is spread either on the spigot or in the bell, and the ends are quickly joined.

Plain-end pipes are joined with a collar. Solvent is applied to the end of one pipe and the collar is immediately fitted. Solvent is then applied to the end of the other pipe to be joined and it is fitted into the collar. Plastic jointing solvents have a very quick drying time. The joint must be made within seconds of the application of the solvent. Solvents are specific to each pipe. Use only ABS solvent with ABS pipe and PVC solvent with PVC pipe. High-density plastic pipe can only be jointed by a special fusion-welding machine used by the contractor at the time of installation, or using a repair clamp.

18.3 TESTING

Sewers are normally tested for leakage and straightness before they are put into service. After the installation of pipes, joint integrity and leakage need to be checked. Water or low-pressure air tests are used for this purpose. An air test is simple, but in some situations a water test is the only option.

18.3.1 WATER TEST

After sufficient time has been given for the joints to set, the section of pipe usually between two manholes is subjected to a test pressure head of 1.5 m of water in case of stoneware pipe and concrete pipes, and 9 m of water in case of cast iron pipes (**Figure 18.6**).

The test is carried out by plugging the lower end of the pipe-sewer with a rubber bag equipped with a canvas cover that is inflated by blowing air, and filling the section with water. The upper end is plugged with a connection to a hose ending in a funnel, which can be raised or lowered until the required head is maintained for observation. The tolerance of 1 to 2 L/cm. km length of sewer may be allowed for a period of 10 minutes. Any subsidence of test water may be attributed to leakage at pipe joints or in the defective lengths of pipes, which should, therefore, be cut out and made good. Subsidence may also be due to some absorption or sweating of pipes or joints.

18.3.2 AIR TESTING

Leakage and integrity of joints can also be done by low-pressure testing. To do this, plug both ends of a section of the sewer to be tested before subjecting that section of pipe to low-pressure air. The air must be maintained at a minimum pressure of 25 kPa for the specified time for each diameter. Duration times vary with changes in pipe size and length of test sections.

A maximum pressure drop of 3.5 kPa is permitted within the specified time duration. Should the pressure drop be greater than 3.5 kPa (0.5 psi) within the specified time duration, deficiencies should be repaired and retesting must be performed until a successful test is achieved. Sources of leaks may be dirt in an assembled gasketed joint, incorrectly tightened service saddles or improper plugging or capping of sewer lateral piping. If there is no leakage (i.e. zero pressure drop) after one hour of testing, the section should be passed and presumed free of defects.

If there is groundwater present at a level higher than the pipe invert during the air test, the test pressure should be increased to a value of 25 kPa greater than the water head at the bottom of the sewer pipe.

Tests for straightness and obstruction are carried out to check pipe alignment. Two common tests for this purpose are the ball test and the mirror test.

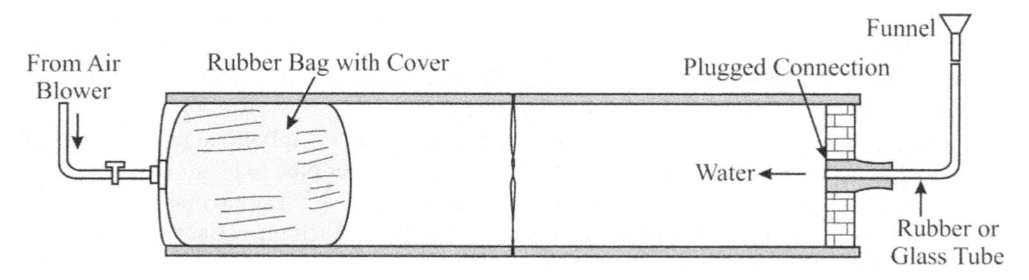

FIGURE 18.6 Water Test

Ball Test

The ball test is carried out by inserting at the high-end of the sewer or drain a smooth ball with a diameter 13 mm less than the pipe bore. In the absence of any obstruction, such as yarn or mortar projecting through the joints, the ball shall roll down the invert of the pipe and emerge at the lower end.

Mirror Test

This test is done by placing a mirror at one end of the sewer line and a lamp at the other end. If the pipeline is straight, the full circle of light can be observed. If the pipeline is not straight, this would be apparent. The mirror will also indicate any obstruction in the pipe barrel.

Smoke Test

This is carried out for drainage pipes located in buildings. Soil pipes, waste pipes, vent pipes and all other pipes when above ground should be approved gas-tight by a smoke test conducted under a pressure of 25 mm of water, maintained for 15 minutes after all trap seals have been filled with water. The smoke is produced by burning oil waste, tarpaper, or similar material in the combustion chamber of a smoke machine (chemical smokes are not considered satisfactory).

18.3.3 Backfilling

After a sewer has been constructed and tested, the trenches need to be refilled. The work involved should be carried out with due care, particularly the selection of the soil used for backfilling around the sewer, to ensure the further safety of the sewer.

The filling in the trenches and up to about 0.75 m above the crown or soffit of the sewer should be made in the finest selected material placed carefully in layers of 15 cm thickness, then watered and evenly rammed. After this, the excavated topsoil, turf, pavement or road metal are replaced as the top filling material, then rammed and satisfactorily maintained until the surface has been restored.

18.4 STRUCTURAL REQUIREMENTS

The structural design of a sewer requires that the supporting strength of the conduit as installed, divided by a suitable factor of safety, must equal or exceed the loads imposed on it by the weight of earth and any superimposed loads. The supporting strength of buried conduits is a function of installation conditions as well as the inherent strength of the pipe itself. Since installation conditions have such an important effect on both load and supporting strength, a satisfactory sewer construction project requires the attainment of design conditions in the field.

1. Rigid pipes support loads in the ground by virtue of resistance of the pipe wall as a ring in bending.
2. Flexible pipes rely on the horizontal thrust from the surrounding soil to enable them to resist vertical load without excessive deformation.
3. Intermediate pipes are those pipes that exhibit behaviour between those in (a) and (b). They are also called semi-rigid pipes.

18.4.1 Loading Conditions

The three general types of loading conditions are trench, embankment and tunnel. Out of these, trench conditions are more common.

Trench Conditions

Trench conditions exist when a conduit is laid in a relatively narrow trench or ditch of width W_d cut out of undisturbed soil and then backfilled with original soil in level with the ground surface. If the trench has sloping sides, the width W_d is taken equal to the horizontal plane tangential to the top of the sewer pipe.

Embankment Conditions

Embankment conditions are defined as those in which the conduit is covered with fill above the original ground surface or when a trench in undisturbed ground is so wide that trench wall friction does not affect the load on the pipe. This classification is further subdivided into positive projection and negative projection.

Positive projection exists when the top of the conduit is above the adjacent original ground surface. Negative projection exists when the top of the pipe is below the adjacent original ground surface in a trench that is narrow with respect to the size of pipe and depth of cover, and when the native material is of sufficient strength that the trench shape can be maintained dependably during the placing of the embankment.

Tunnel Conditions

Tunnel conditions are defined as those in which the conduit is installed in a relatively narrow trench cut in undisturbed ground and covered with earth backfill to the original ground surface.

18.5 DEAD LOADS

The vertical load on a sewer pipe is the result of the weight of the prism of soil within the trench and above the top of the pipe and the friction or shearing forces generated between the prism of soil in the trench and the sides of the trench. The backfill tends to settle in relation to the undisturbed soil in which the trench is excavated. This tendency for movement induces upward shearing forces that support a part of the weight of the backfill. Hence, the resultant load on the horizontal plane at the top of the pipe within the trench is equal to the weight of the backfill minus the upward shearing forces, as indicated in **Figure 18.7**. The width of the trench and the unit weight of the backfill soil have a direct influence on the load on the pipe. Thus, the width should be kept to an absolute minimum while providing sufficient working space at the sides of the pipe to caulk joints properly, insert and strip form, and compact backfill. The load is also influenced by the coefficient friction between the backfill and the sides of the trench and by the internal friction of the backfill soil.

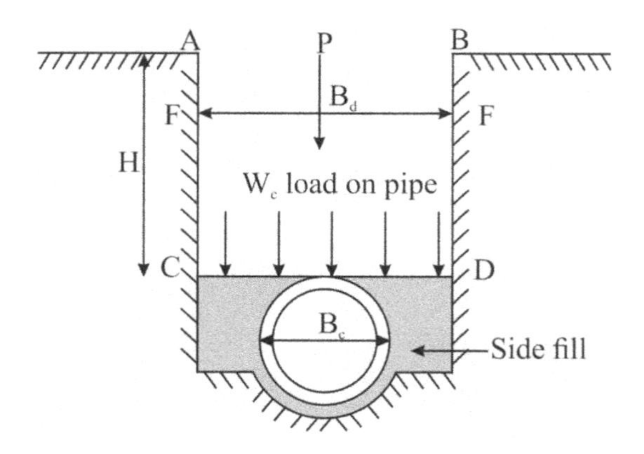

FIGURE 18.7 Load on Pipe (Trench Conditions)

Anson Marston developed methods for determining the vertical load on buried conduits due to gravity earth forces in all of the most commonly encountered construction conditions. His methods are based on a theory which states that the load on a buried conduit is equal to the weight of the prism of earth directly over the conduit, called the interior prism, plus or minus the frictional shearing forces transferred to that prism by the adjacent prisms of earth. The magnitude and direction of these frictional forces depend upon the relative settlement between the interior and adjacent earth prisms.

$$\text{Marston equation: } W = C_d \times \gamma \times B_d^2$$

W = vertical load per unit length, kN/m
γ = specific or unit weight of earth, kN/m^3
B_d = trench width at the crown of pipe
C_d = a constant that measures the effect of:

- The ratio of the height of the fill to the width of the trench.
- The shearing forces between the interior and adjacent earth prisms.
- The direction and amount of relative settlement between the interior and adjacent earth prisms for embankment conditions.

The value of C_d is read from **Figure 18.8** or computed using an empirical formula.

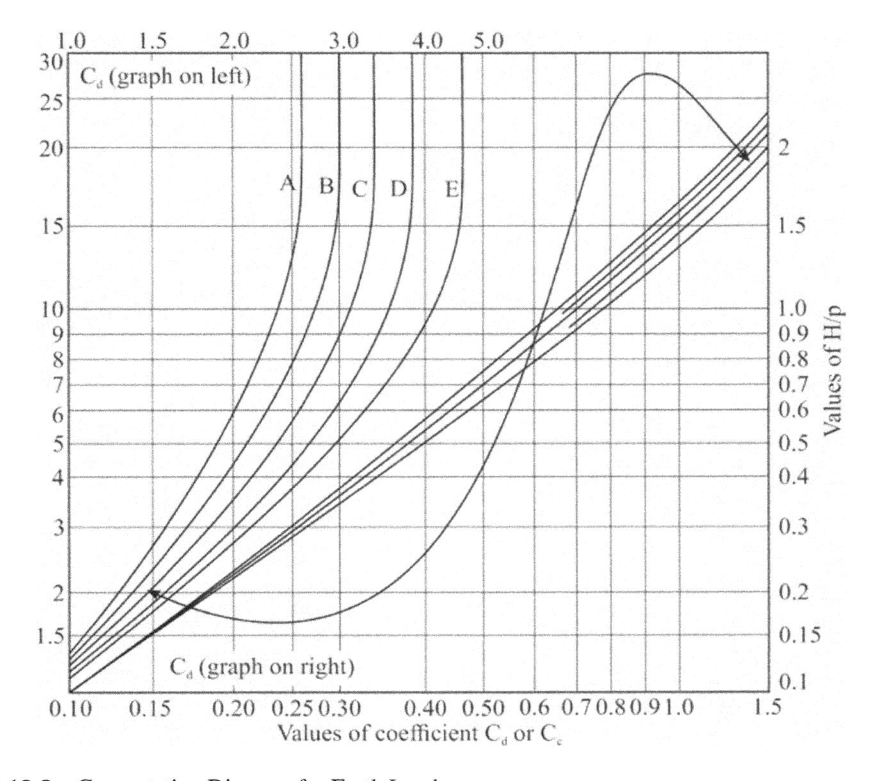

FIGURE 18.8 Computation Diagram for Earth Loads

TABLE 18.1
Factor Kμ′ for Various Fill Materials

Curve	Type of Fill	Kμ′
Curve A	Granular material	0.1924
Curve B	Sand and gravel	0.165
Curve C	Saturated topsoil	0.150
Curve D	Ordinary clay	0.130
Curve E	Saturated clay	0.110

Empirical Relationship

The value of C_d can be computed using an empirical formula. The value of the coefficient for a given backfill soil decreases exponentially as the ratio of soil cover to width increases. An empirical relationship is developed for computing the value of the coefficient.

$$C_d = \left[\frac{1-e^{-2K\mu' H/B_d}}{2K\mu'}\right]$$

Where μ' is the friction coefficient and K is the ratio of lateral unit pressure to vertical unit pressure. Maximum values of Kμ' for fill materials are shown in **Table 18.1**.

- **Curve A** represents granular material without cohesion.
- **Curve B** represents sand and gravel.
- **Curve C** represents saturated topsoil.
- **Curve D** represents ordinary clay.
- **Curve E** represents saturated clay.

Example Problem 18.2

A 300 mm diameter sewer pipe is placed in a 2.5 m deep rectangular trench that is 0.60 m wide. The trench is backfilled with clay that has a unit weight of 19 kN/m³. Compute the dead load. Assume the pipe wall thickness is 30 mm.

Given: D = 300 mm = 0.30 m
H = 2.5 m
γ = 19 kN/m³
B = 0.60 m
W = ?

SOLUTION:

$$\frac{H}{B} = \frac{2.5\ m}{0.60\ m} = 4.16 = 4.2 \quad Read\ C_d = 2.5$$

$$W = C_d \times \gamma \times B_d^2 = 2.5 \times \frac{19\ kN}{m^3} \times (0.6\ m)^2 = 17.1 = \underline{17\ kN/m}$$

Example Problem 18.3

Determine the load on a 600 mm diameter rigid pipe under 3.0 m of saturated topsoil (16 kN/m³) in trench conditions. The side fill is 30 cm wide on each side. Assume the pipe wall thickness is 50 mm.

Given: D = 600 mm = 0.60 m
H = 3.0 m
γ = 16 kN/m³
W = ?

SOLUTION:

$$B_d = 0.60\ m + 0.1 + 2 \times 0.30\ m = 1.3\ m$$

$$\frac{H}{B} = \frac{3.0\ m}{1.3\ m} = 2.31 \quad \textit{From table } K\mu' = 0.15$$

$$C_d = \left[\frac{1 - e^{-2K\mu' H/B}}{2K\mu'}\right] = \left[\frac{1 - e^{-2\times0.15\times2.31}}{2\times0.15}\right] = 1.666 = 1.67$$

$$W_c = C_d \times \gamma \times B_d^2 = 1.67 \times \frac{16\ kN}{m^3} \times (1.3\ m)^2 = 45.06 = \underline{45\ kN/m}$$

If the pipe is flexible and the soil at the sides is compacted to the extent that it will deform under vertical load the same amount as the pipe itself, the side fills may be expected to carry their proportional share of the total load. For a situation of this kind, the trench load formula is modified as follows:

$$\boxed{\textit{Flexible pipes: } W = C_d \times \gamma \times B_d \times B_c}$$

B_d = width of trench or ditch at the crown of the pipe, m
B_c = outside width of the conduit or pipe, m

Example Problem 18.4

Determine the load on a 750 mm diameter flexible conduit, with a wall thickness of 50 mm, installed in a trench 1.4 m wide at a depth of 3.4 m. The trench is cut in clay soil with a specific gravity of 1.90.

Given: D = 750 mm = 0.75 m
H = 3.4 m
B_d = 1.4 m
SG = 1.90
W = ?

SOLUTION:

$$B_c = 0.75\ m + 2 \times 0.050\ m = 0.85\ m$$

$$H = 3.4\ m - 0.85\ m = 2.55\ m \qquad \frac{H}{B_d} = \frac{2.55\ m}{1.4\ m} = 1.82\ m$$

$$C_d = \left[\frac{1-e^{-2K\mu'H/B}}{2K\mu'}\right] = \left[\frac{1-e^{-2\times0.13\times1.82}}{2\times0.13}\right] = 1.449 = 1.45$$

$$W_c = C_d \times \gamma \times B_c \times B_d = 1.449 \times \frac{1.9 \times 9.81\ kN}{m^3} \times 0.85\ m \times 1.4\ m$$

$$= 32.16 = \underline{32\ kN/m}$$

18.6 SUPERIMPOSED LOADS

Two types of superimposed loads are encountered commonly in the structural design of sewers, concentrated load and distributed load.

18.6.1 Concentrated Load

The formula for load due to superimposed concentrated load, such as a truck wheel, as shown in **Figure 18.9**, is given by the following formula:

$$\boxed{W_{sc} = C_s \times \left(\frac{P \times F}{L}\right)}$$

W_{sc} = the load on the conduit, kN/m
P = the concentrated load, kN
F = the impact factor
L = the effective length of the conduit, m
C_s = the load coefficient, which depends upon $B_c/2H$ and L/2H where
H = the height of fill from the top of the conduit to the ground surface, m
B_c = width of conduit, m

The effective length of a conduit is defined as the length over which the average load due to surface traffic units produces the same stress in the conduit wall, as does the actual load, which varies in

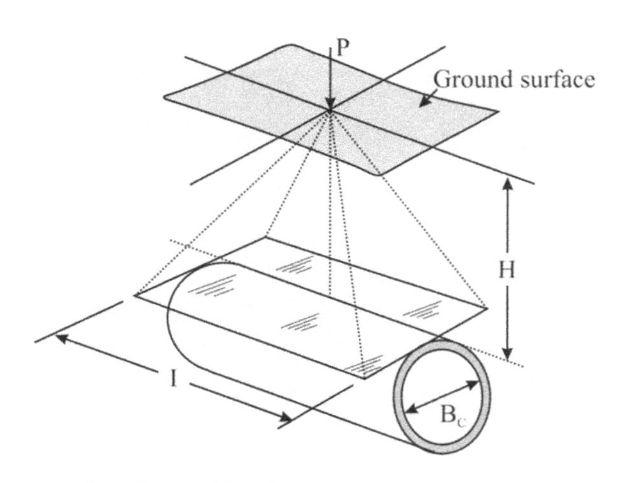

FIGURE 18.9 Concentrated Superimposed Load

TABLE 18.2
Values of Load Coefficients

$\frac{D}{2H}$ or $\frac{B_c}{2H}$	$\frac{M}{2H}$ or $\frac{L}{2H}$													
	0.1	**0.2**	**0.3**	**0.4**	**0.5**	**0.6**	**0.7**	**0.8**	**0.9**	**1.0**	**1.2**	**1.5**	**2.0**	**5.0**
0.1	0.019	0.037	0.053	0.067	0.079	0.089	0.097	0.103	0.108	0.112	0.117	0.121	0.124	0.128
0.2	0.037	0.072	0.103	0.131	0.155	0.174	0.189	0.202	0.211	0.219	0.229	0.238	0.244	0.248
0.3	0.053	0.103	0.146	0.190	0.224	0.252	0.274	0.292	0.306	0.318	0.333	0.345	0.355	0.360
0.4	0.067	0.131	0.190	0.241	0.284	0.320	0.349	0.373	0.391	0.405	0.425	0.440	0.454	0.460
0.5	0.079	0.155	0.224	0.284	0.336	0.379	0.414	0.441	0.463	0.481	0.505	0.525	0.540	0.548
0.6	0.089	0.174	0.252	0.320	0.379	0.428	0.467	0.499	0.524	0.544	0.572	0.596	0.613	0.624
0.7	0.097	0.189	0.274	0.349	0.414	0.467	0.511	0.546	0.584	0.597	0.628	0.650	0.674	0.688
0.8	0.103	0.202	0.292	0.373	0.441	0.499	0.546	0.584	0.615	0.639	0.674	0.703	0.725	0.740
0.9	0.108	0.211	0.306	0.391	0.463	0.524	0.574	0.615	0.647	0.673	0.711	0.742	0.766	0.784
1.0	0.112	0.219	0.318	0.405	0.481	5.44	0.597	0.639	0.673	0.701	0.740	0.744	0.800	0.816
1.2	0.117	0.229	0.333	0.425	0.505	0.572	0.628	0.674	0.711	0.740	0.783	0.820	0.849	0.868
1.5	0.121	0.238	0.345	0.440	0.525	0.596	0.650	0.703	0.742	0.774	0.820	0.861	0.894	0.916
2.0	0.124	0.244	0.355	0.454	0.540	0.613	0.674	0.725	0.766	0.800	0.849	0.894	0.930	0.956

intensity from point to point. It is general practice to use an effective length of 1.0 m for conduits greater than 1.0 m long and the actual length for shorter conduits.

The dynamic loads caused by traffic produce an impact and this must be considered in load computations. The following impact factors are recommended:

- Highway traffic 1.50
- Railway traffic. 1.75
- Airfield runways 1.00
- Airfield taxiways, aprons, hard stands… 1.50

Example Problem 18.5

Determine the load on a 600 mm diameter pipe under 1.0 m of cover caused by a 50 kN truck wheel applied directly above the centre of the pipe. Assume the pipe section is 0.75 m long, the wall thickness is 50 mm and the impact factor is 1.5.

Given: D = 600 mm L = 0.75 m
H = 1.0 m
W = ?

SOLUTION:

$$B_c = 0.60\ m + 2 \times 0.050\ m = 0.70\ m$$

$$\frac{B_c}{2H} = \frac{0.70\ m}{2 \times 1.0\ m} = 0.35 \quad \frac{L}{2H} = \frac{0.75\ m}{2 \times 1.0\ m} = 0.375$$

$$C_s = 0.190 + \frac{(0.241 - 0.19)}{(0.4 - 0.3)} \times 0.05 = 0.2155$$

$$W_{sc} = C_s \times \left(\frac{P \times F}{L}\right) = 0.2155 \times \left(\frac{50\ kN \times 1.5}{0.75\ m}\right) = 21.55 = 22\ kN/m$$

18.6.2 Distributed Loads

In the case of a superimposed load distributed over an area of considerable extent, the following formula is used for load computation:

$$\boxed{W_{sd} = C_s \times \left(\frac{P \times F}{B_c}\right)}$$

W_{sd} = load on conduit, kN/unit length
p = intensity of distributed load, kN/m^2
F = impact factor
B_c = width of the conduit, m
C_s = load coefficient, which depends upon D/2H and M/2H (**Table 18.2**) where
H = height from the top of the conduit to the ground surface, m
D and M = width and length, respectively, of the area over which the distributed load acts

18.7 FIELD SUPPORTING STRENGTH

Sewers must support both dead loads and live loads. Dead load refers to load due to backfill and live load refers to load due to vehicular traffic. Factors affecting dead load and live load have been discussed. Once the load on the sewer pipe is known, it is important to compare it with the load-carrying capacity of the sewer pipe.

18.7.1 Load-Carrying Capacity

The load-carrying capacity of a sewer will depend on two key factors: pipe crushing strength and class of pipe bedding. The crushing strength of a pipe is determined by a standard laboratory procedure and is specified in terms of load per unit length, for example, kN/m. The procedure is called a three-edge bearing test because the load is applied to the test sections along three edges of the pipe barrel only. The minimum crushing strengths of pipe materials and sizes are published by pipe manufacturers. For example, the typical crushing strength values of vinyl chloride pipe (VCP) are presented in **Table 18.3**.

TABLE 18.3
Strength of VCP of Various Sizes

	Strength, kN/m	
Nominal Size mm	**Standard**	**Extra**
200	20.4	32.0
250	23.2	35.0
300	26.3	37.9
380	29.2	42.3
460	32.0	48.1

18.7.2 Load Factor

Proper bedding always increases the actual or field supporting strength of the installed pipe above the crushing strength by distributing the load over the pipe circumference. The ratio of the actual field supporting strength to crushing strength is called the load factor.

$$Load\ factor,\ LF = \frac{Field\ supporting\ strength}{Crushing\ strength}$$

$$\boxed{Field\ supporting\ strength = LF \times Crushing\ strength}$$

In addition to the load factor provided by the pipe bedding, a safety factor, SF, is applied to the computations to arrive at the safe supporting strength as follows:

$$\boxed{Safe\ supporting\ strength = \frac{LF}{SF} \times Crushing\ strength}$$

A safety factor of 1.5 is commonly used for clay and plain concrete sewers to compensate for the poor quality of the material or faulty construction.

Example Problem 18.6

Assume the sewer pipe in Example Problem 18.1 is of standard strength VCP. What class of bedding should be specified for construction using a safety factor of 1.5?

Given: W = 17 kN/m
SF = 1.5
Crushing Strength = 26.3 kN/m (**Table 18.3**)
LF = ?

SOLUTION:

$$LF = \frac{Safe\ supporting\ strength \times SF}{Crushing\ strength} = \frac{17\ kN}{m} \times 1.5 \times \frac{m}{26.3\ kN} = 0.969 = 1.0$$

As per calculations, the lowest class of bedding with an LF of 1.1 should be adequate. However, since class D bedding is not recommended for good construction, class C bedding is selected.

Example Problem 18.7

A 200 mm diameter VCP sewer pipe is placed in a 0.60 m wide trench with 3.0 m of cover. The trench is backfilled with saturated clay that has a unit weight of 20.4 kN/m^3. Determine the required bedding condition for standard strength and extra-strength pipe. Use a safety factor of 1.5.

Given: D = 200 mm
H = 3.0 m
γ = 20.4 kN/m^3

B = 0.60 m
LF = ?

SOLUTION:

$$\frac{H}{B} = \frac{3.0\ m}{0.60\ m} = 5.0$$

From **Table 18.1**, the value of the factor Kμ' for saturated clay is 0.11.

$$C_d = \left[\frac{1-e^{-2K\mu' H/B}}{2K\mu'}\right] = \left[\frac{1-e^{-2\times 0.11\times 5}}{2\times 0.11}\right] = 3.03 = 3.0$$

$$W_c = C_d \times \gamma \times B_d^2 = 3.03 \times \frac{20.4\,kN}{m^3} \times (0.6\,m)^2 = 22.27 = \underline{22\,kN/m}$$

Crushing strength values for standard and extra-strength VCP pipe as read from **Table 18.3** are 20.4 and 32 kN/m, respectively.

$$LF(Standard) = \frac{Safe\,supporting\,strength \times SF}{Crushing\,strength} = \frac{22.27\,kN}{m} \times 1.5 \times \frac{m}{20.4\,kN} = 1.63$$

This would require Class B bedding, which provides LF = 1.9.

$$LF(Extra\,strength) == \frac{22.27\,kN}{m} \times 1.5 \times \frac{m}{32\,kN} = 1.04 = 1.0$$

Thus, if extra-strength pipe is used, Class C bedding will suffice.

Discussion Questions

1. Explain the steps involved in sewer construction.
2. Describe the procedure for laying sewers with correct alignment and slope with the use of lasers.
3. Why is pipe bedding important in sewer construction? Discuss the different classes of bedding.
4. Explain the tests on sewers for straightness and obstruction.
5. Why are pipe joints important in sewer construction? What are the requirements of a good pipe joint?
6. Explain the terms timbering of trenches, backfilling of trenches, and wellpoint drainage
7. Explain how the load on sewers is taken into consideration when designing sewers?
8. What are the different materials with which sewers are made? Discuss their suitability with respect to the size and type of sewer pipe.
9. List five points that should be important considerations in the selection of materials for sewer pipes.
10. Explain the procedure for jointing plastic pipes.
11. What are the Indian standard specifications for the different types of pipes used for sewer construction?

12. What is the purpose of the water test and how is it carried out in the field?
13. Describe the different types of loads on sewer pipes laid underground.
14. The preparation of bedding for flexible pipes needs extra care compared with rigid pipes. Comment.
15. Define loading factor and how it is used to select proper bedding for sewers.
16. Is it a good practice to overdesign a sewer pipeline? Why?

Practice Problems

1) Work out the fill load on a 1200 mm inside diameter concrete sewer pipe installed in a trench of 2.2 m width and 3.2 m depth. The sewer pipe is 65 mm thick and the unit weight of the fill (Kμ' = 0.15) is known to be 19 kN/m^3. (69 kN/m)
2) Determine the load on a 450 mm diameter concrete pipe with a wall thickness of 35 mm installed in a trench 0.85 m wide at a depth of 3.0 m. The trench is cut into clay soil with a specific gravity of 1.8. (26 kN/m)
3) Determine the load on a 600 mm diameter rigid pipe under 2.5 m of saturated topsoil in trench conditions. The side fill is 30 cm wide on each side. Assume the pipe wall thickness is 50 mm and the weight density of the wet fill is 17 kN/m^3. (42 kN/m)
4) Repeat the previous problem assuming the fill is saturated clay with a specific gravity of 1.9. (50 kN/m)
5) Determine the load on a 750 mm diameter flexible conduit with a wall thickness of 50 mm installed in a trench 1.5 m wide at a depth of 3.5 m. The trench is cut into topsoil with a specific gravity of 1.60. (27 kN/m)
6) Determine the load on a 600 mm diameter pipe under 1.3 m of cover caused by a 65 kN truck wheel applied directly above the centre of the pipe. Assume the pipe section is 0.75 m long, the wall thickness is 50 mm and the impact factor is 1.5. (19 kN/m).
7) A 100 m reach of sewer is to have a minimum capacity of 200 L/s. The street elevation at the upper manhole is 103.55 m and the lower manhole is 101.05 m. Select the appropriate size and slope for this reach and establish the pipe invert elevations. Assume a minimum cover of 2.0 m above the crown of the pipe. (350 mm, 103.20 m, 100.70 m)
8) A sewer reach of length 100 m is to carry 40 L/s flowing full (**Figure 18.10**). Select the required size of sewer and invert elevations. (200 mm, 347.80 m, 345.67 m)
9) The second reach of the sewer line in **Figure 18.10** is 120 m long and is required to carry wastewater at a peak rate of 80 L/s. Select the required size of sewer and invert elevations. (300 mm, 345.47 m, 344.60 m)
10) A 200 mm diameter sewer pipe is placed in a 3.0 m deep, 0.90 m wide trench and backfilled with sand (17.2 kN/m^3). Using a safety factor of 1.5, select appropriate bedding for the standard strength VCP sewer. (LF = 1.85, Class B bedding)
11) Repeat the previous problem if the VCP sewer is extra strength. (LF = 1.2, Class C bedding)
12) A sewer pipe with a diameter of 300 mm is placed in a 3.2 m deep, 0.75 m wide trench filled with saturated clay (20.4 kN/m^3). Work out the dead load on the pipe. (30 kN/m)

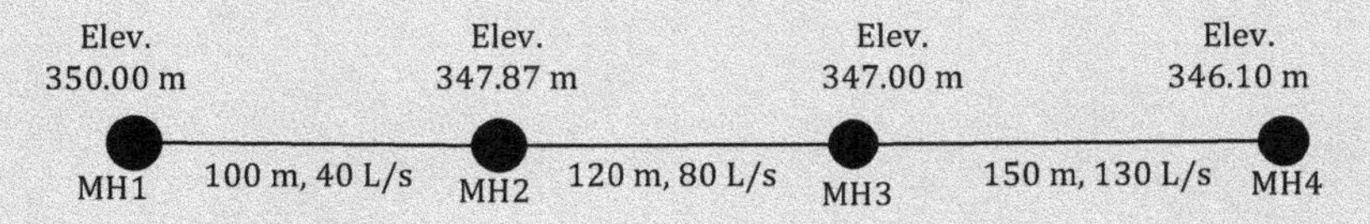

FIGURE 18.10 Sewer Line Reaches

19 Lift Stations

Wastewater pumping is in many ways different to fresh water pumping. Wastewater flow varies hourly, seasonally and yearly depending on growth and climatic conditions. The pumping system therefore should be capable of handling both peak and minimum flows. Retaining wastewater too long in a wet well can lead to septicity and odours. Therefore, the detention time in a sump should not exceed 30 minutes and wells should be properly ventilated to release gases produced by the decomposition of organics in wastewater.

A wet well pumping system usually consists of two pumps of capacity at least equal to peak flow. Two pumps ensure 100% standby capacity. In situations where flows exceed peak flows, both pumps operate. Operation is such that both pumps take turns operating as the lead pump. This keeps both the pumps in good condition and does not cause excess wearing of either one.

19.1 WET WELL LIFT STATIONS

A wet well lift station consists of a holding reservoir, which receives and holds the sewage as it pours in from the sewer mains; pumps to lift the sewage to the disposal area; electric motors to power the pumps; the necessary pipes, valves and fittings to move the sewage to the pump and then to the point of discharge; and a control system that starts and stops the pumps automatically. In a wet well station, the pumps are situated in the holding reservoir (**Figure 19.1**).

Since wastewater carries solids, the diameter of the rising main should be such as to provide a minimum flow velocity of 0.6 m/s. To minimize the premature wearing out of the pump and motor, the frequency of pump starts is usually restricted to 5/h. More frequent starts will shorten the useful life of the pump and motor. If the minimum flows are less than 20% of the average flows, odours can develop. In such situations, providing proper ventilation is very important.

19.2 DRY WELL LIFT STATIONS

Dry well lift stations consist of two pits or wells: a wet well and a dry well. The wet well receives and holds the sewage as it comes from the sewer main. The dry well houses the pumps, motors, piping, valves, fittings and electrical controls necessary to move the sewage from the wet pit to the disposal area. The dry well allows easy access to the pump, motor and electrical controls. The standby motor (if one is present) is mounted on the floor of the lift station above the dry pit and supplies power to one of the pumps in the event of a power failure. **Figure 19.2** shows a typical dry well lift station.

19.3 WASTEWATER FLOW PUMPS

One main difference compared to potable water pumps is that sewage pumps must be capable of passing solids. For this reason, non-clog pumps or pumps with large impellers are used. Screens are placed to retain large debris, and in some installations a shredder may also be used. The centrifugal pump is the most commonly used pump in lift stations.

With submersible pumps, a rail system is used to remove the pump. Self-priming sewage pumps are suitable when suction lift is less than 5.5 m. Unlike submersible pumps, access to self-priming pumps is direct. This allows for ease of maintenance, and the operation of the pump can be observed readily.

DOI: 10.1201/9781003231264-19

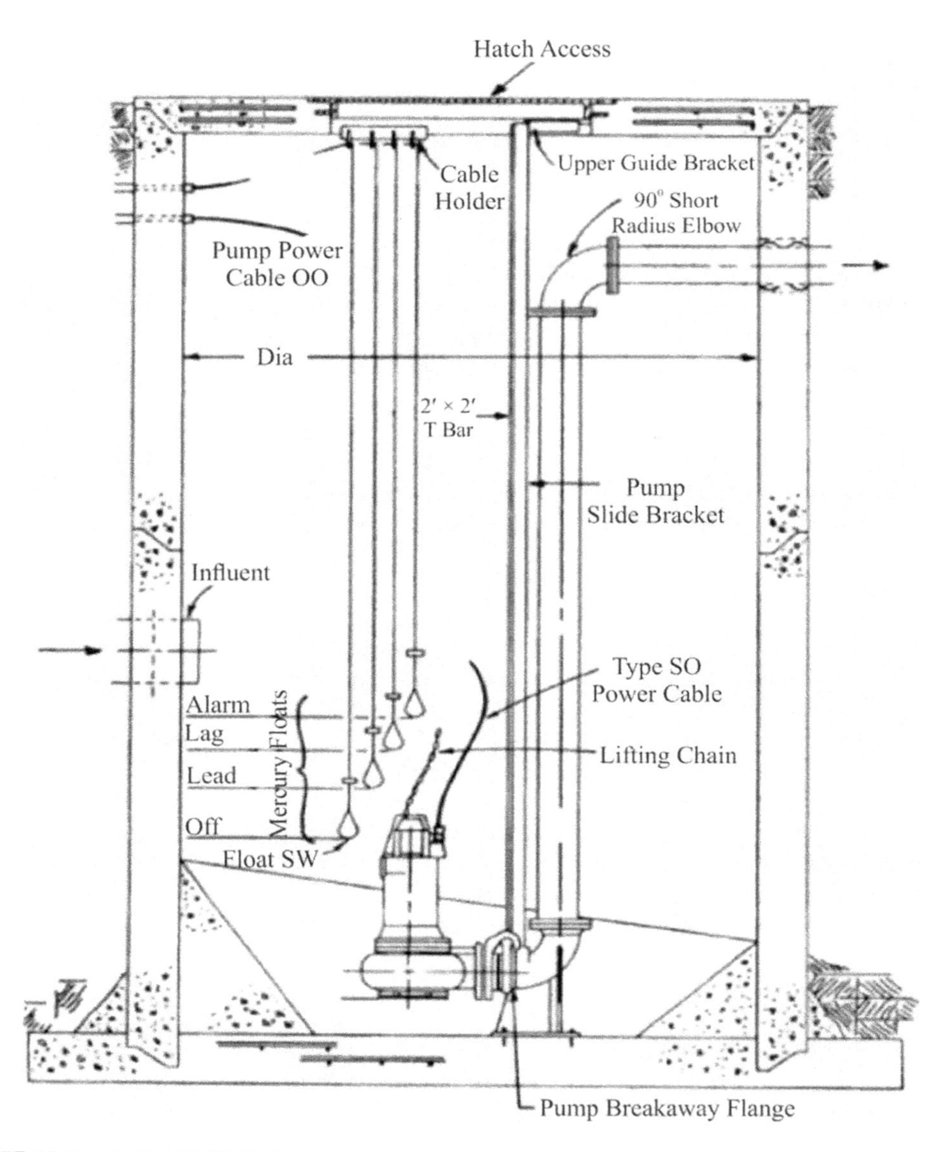

FIGURE 19.1 Wet Well Lift Station

In dry well applications, non-clog vertical mounted centrifugal pumps are used. Wastewater from the wet well side is conveyed to the pump suction by piping. Since the pump sits below the wastewater level in the dry well, it always operates under positive head conditions and hence does not require priming.

19.4 WET WELLS

The main purpose of the wet well is to provide short-term storage. This reduces stress on the pumping equipment by keeping the pump cycle time to more than 8 minutes. Frequent start time results in overheating and shortens the useful life of motors.

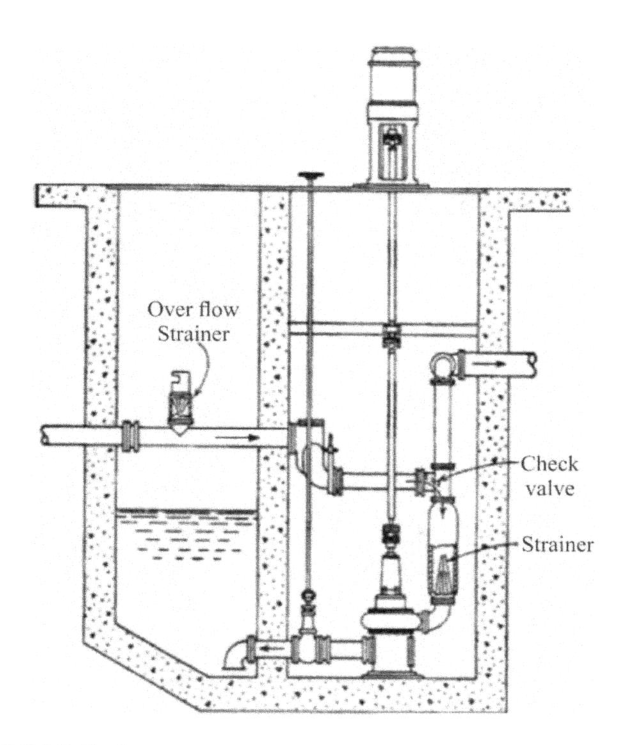

FIGURE 19.2 Dry Well Lift Station

19.5 SCREENS

As the name indicates, screens are installed to retain large debris from wastewater. To prevent large head built up and restrictions to flow, screens must be cleaned frequently. Comminutors and barminutors are used where large amounts of debris are expected.

19.6 ELECTRICAL AND CONTROLS

The electrical system in a lift station consists of the main breaker that feeds the motor control panel and auxiliary electrical systems. The motor control panel houses the control system for the motors and other electrical system in the lift station. The controls include starters, fuses, heater strips, coils, relays and switches.

The level-sensing devices used in controlling the pump operation can be floats, electrodes, bubblers, mercury tilt and sonic. Compressed air is supplied if a bubbler tube is used. The pressure required to bubble the air through the wastewater column is proportional to the height of wastewater above the tube end. In sonic units, the time it takes for a high-frequency sound to return indicates the depth of the wastewater.

19.7 LIFT STATION MAINTENANCE

In both wet pit and wet pit-dry pit lift stations, maintenance involves mainly keeping the station and the wet pits clean and keeping the pump or pumps unclogged and operating at the proper capacity. The main items requiring regular attention are screen baskets and bar screens, sand, grease or debris build-ups on the walls or floor of wet pits and sump pumps in dry pits.

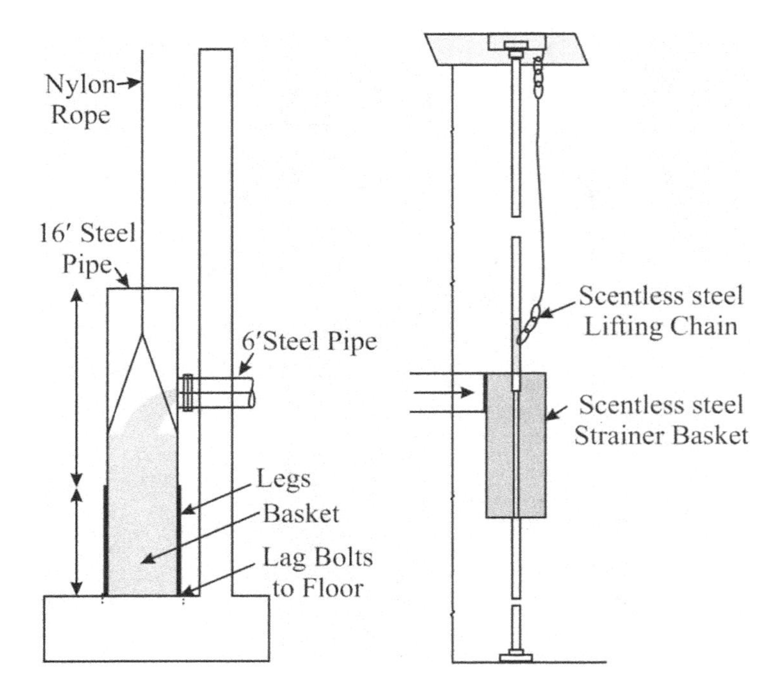

FIGURE 19.3 Screening Basket

19.7.1 Screening Baskets

To empty a screening basket (**Figure 19.3**), the basket is pulled up to the surface and the accumulated material is emptied into a bucket or similar receptacle. Material must be appropriately disposed of (e.g. burning, burying or sending to landfill). A bar screen serves the same purpose as the screening basket, collecting the larger pieces of debris in sewage before they reach the lift station pumps and clog the impellers. A bar screen is usually installed in the nearest manhole upstream from the lift station. The manhole will be no more than 30 m from the lift station and much closer in most cases. Screening baskets and bar screens must be checked regularly and cleaned out before they are full. Once the screens are full, the material starts to wash over and may go into the pumps. If this happens, they need to be dismantled and cleaned, or the lift station pumps need to be repaired.

19.7.2 Floor Maintenance

Grease, grit, sand and other material present in sewage will build up over time on the walls, floor, piping and pumps in a wet pit. If the sand and grit are allowed to build up they will work into the pumps, causing rapid wear of the impellers and impeller chambers. The wet well should be cleaned regularly to avoid excessive wear or damage to the pumps.

19.7.3 Sump Pump Operation and Maintenance

Sump pumps are located in the dry pit of a lift station. The sump pump removes water caused by the seepage of groundwater into the dry pit, excess from water-lubricated bearings, and spills left from servicing or washing out sewage pumps. The sump pit must be regularly cleaned to prevent clogging of the sump pump. The sump pit should be checked for clogging material and any found should be promptly removed and disposed of. The sump pump should also be checked to make sure that it is operable. Lifting the float that triggers the pump does this. If a hum is heard, the pump is operable. If the pump does not turn on when the float is lifted, the pump is not working and should be repaired or replaced.

19.8 PUMP OPERATING SEQUENCE

Most lift stations are controlled automatically. Level controllers in the wet well turn the pumps on and off depending on the level in the well. In a dual-pump system, the sequence of operation is described below:

1. The lead pump comes on when the water level reaches the pump start level. Below this level, no pump is running, and incoming wastewater is being stored.
2. At some stations, the role of the lead pump is automatically alternated between the two pumps with each pumping cycle. This way, as discussed before, each pump receives equal wear. At stations not equipped with an alternating device, the lead pump is alternated manually every week.
3. During peak flows, the water level will continuously rise when the lead pump is running. As the water level reaches above the start level, the second pump comes on.
4. If water flows are exceptionally high, such as during storm periods, the water level may continue to rise. In such cases, as the water level reaches the flooding level, an alarm is activated.
5. As the water level starts dropping, the first pump will stop. During low flows, when the water level drops below the lead pump start level, the lead pump stops to complete the pumping cycle.

19.8.1 LEVEL SETTING

Too many starts can damage the pumping equipment. Experience has shown that each pumping cycle should be preferably more than 10 minutes, or in the range of 6–8/h. Frequent starts will not allow the motor and starters to cool down properly and may result in overheating. The lead pump's stop level is normally set at the minimum submergence level to prevent pump cavitation. The start level is determined using the following formula.

$$h_{start} = \frac{V_S}{A} = \frac{t_C}{A} \times \frac{Q_i}{Q_p} \times \left(Q_p - Q_i\right)$$

V_s = volume of storage between the stop and start level
A = area of cross-section of wet well
t_c = desired pump cycle time
Q_p = pumping rate
Q_i = wastewater influent rate

Example Problem 19.1

Work out the pump start level in a 2.5 m diameter wet well if the desired pumping cycle time is 10 minutes (6 /h). The wastewater average inflow rate is 12 L/s and the pump flow rate is 15 L/s.

Given: D = 2.5 m
t_c = 10 min
Q_i = 12 L/s
Q_p = 15 L/s

SOLUTION:

$$h_{start} = \frac{t_C}{A} \times \frac{Q_i}{Q_p} \times \left(Q_p - Q_i\right) = \frac{10\ min}{0.785(2.5\ m)^2} \times \frac{15}{12}(15-12)\frac{L}{s} \times \frac{60\ s}{min} \times \frac{m^3}{1000\ L} = 0.46\ m$$

The pump start must be set at 50 cm above the submergence level (stop level).

19.8.2 Pumping Rate in Lift Stations

Even with good maintenance of screening baskets and lift station walls and floors, some debris may enter pumps or force mains and begin to clog them up. Clogging in pumps or force mains will often show up first as a drop in the lift station pump's capacity. If the pump is not moving sewage at the rate indicated by its capacity rating, there may be debris clogging the pump or force main. This situation must be corrected before it causes serious problems leading to major pump repairs and the complete blockage of sewer lines. The actual pumping rate of a lift station can be checked against its rated capacity by observing and timing the pump during its operation. The steps for this are as follows:

1. Find out the capacity rating for the pump as indicated by the manufacturer's information sheets or pump table.
2. Ideally, check the pumping rate for the operating head using the performance curves. A decrease in pumping rate might indicate a problem with the pump that needs to be investigated.
3. Select and tag two ladder rungs close to the bottom of the wet pit. These will be used as benchmarks for high and low wastewater levels in the wet pit.
4. Measure the distance in metres between the two rungs. This measurement will give the depth between the high and low water levels.
5. Calculate the number of litres of wastewater contained in the wet pit between these two levels.
6. Shut off the pump and wait for the wastewater level to rise over the top rung selected.
7. Restart the pump when the water has risen over the top rung. As the water level drops, time when the top rung reappears. Continue timing until the lower rung is out of the water.
8. Write down the time it took for the water level to drop from the top to the bottom ladder rung in minutes.

Example Problem 19.2

A wet well measures 3.0 m × 2.5 m. To check the pumping rate, the influent valve is closed and the drop in water level is observed. Over 5.0 minutes, the water level in the well dropped by 65 cm. Work out the pumping rate.

Given: A = 3.0 m × 2.5 m
Δd = 65 cm = 0.65 m
Δt = 5.0 min
Q_p = ?

SOLUTION:

$$Q_p = \frac{\Delta V}{\Delta t} = \frac{3.0\ m \times 2.5\ m \times 0.65\ m}{5.0\ min} \times \frac{min}{60\ s} \times \frac{1000\ L}{m^3} = 16.25 = \underline{16\ L/s}$$

Example Problem 19.3

A wet well measures 2.5 m × 3.0 m. Influent to the well measured before the test is 22 L/s. If the water level drops by 6.0 cm in 5.0 minutes, what is the pumping rate in L/s?

Given:

$A = 3.0\ m \times 2.5\ m$
$\Delta d = 6.0\ cm$
$\Delta t = 5.0\ min$
$Q_i = 22\ L/s$
$Q_p = ?$

SOLUTION:

$$\Delta Q_S = \frac{\Delta V}{\Delta t} = \frac{2.5\ m \times 3.0\ m \times 6.0\ cm}{5.0\ min} \times \frac{m}{100\ cm} \times \frac{1000\ L}{m^3}$$

$$= \frac{90\ L}{min} \times \frac{min}{60\ s} = 1.5\ L/s$$

$$Q_p = Q_i - (-\Delta Q_S) = 22\ L/s + 1.5\ L/s = 23.5 = \underline{24\ L/s}$$

The change in storage is negative since the water level is dropping. The change will be positive if the water level rises during pumping.

Example Problem 19.4

A pumping station pumps wastewater from an intercepting sewer to the treatment plant. The wet well measures 3.8 m by 3.2 m. After the influent valve is closed, the water level in the wet well drops by 0.6 m in 5.0 minutes. After the valve is opened, the water level rises by 5.0 cm in 10.0 minutes. Calculate the influent rate.

Given: Valve closed: $A = 3.8\ m \times 3.2\ m$ $\Delta d = 0.6\ m$ $\Delta t = 5.0\ min$ $Q_p = ?$
Valve open: $A = 3.8\ m \times 3.2\ m$ $\Delta d = 5.0\ cm$ $\Delta t = 10\ min$ $Q_i = ?$

SOLUTION:

$$Pump,\ Q_p = \frac{\Delta V}{\Delta t} = \frac{3.8\ m \times 3.2\ m \times 0.60\ m}{5.0\ min} \times \frac{min}{60\ s} \times \frac{1000\ L}{m^3}$$

$$= 24.3 = \underline{24\ L/s}$$

$$Change\ in\ storage,\ \Delta Q_S = \frac{3.8\ m \times 3.2\ m \times 0.05\ m}{10\ min} \times \frac{min}{60\ s} \times \frac{1000\ L}{m^3}$$

$$= 1.01 = 1.0\ L/s$$

$$Influent\ rate,\ Q_i = Q_p + \Delta Q_S = 24.3\ L/s + 1.01\ L/s = 25.4 = \underline{25\ L/s}$$

Example Problem 19.5

Design a lift station to lift the wastewater by 11 m via a 90 m long rising main. This station is to serve a population of 35 000 with a per capita water supply of 150 L/c.d. Make appropriate assumptions as necessary.

Given: Population = 35 000 people at 150 L/c.d

ΔZ = 11 m

L = 90 m

Peaking factor = 3 ×

Assume: f = 0.03 and detention time = 15 minutes during peak flow.

SOLUTION:

$$Q_p = 3 \times 35000 p \times 0.80 \times \frac{150\ L}{p.d} \times \frac{d}{1440 \times 60\ s} \times \frac{m^3}{1000\ L}$$

$$= 0.145 = \underline{0.15\ m^3/s}$$

Assuming two pumps are running during the peak flow period, the discharge rate of each pump would be half, say 0.075 m^3/s. The rising main must have a flow velocity > 1.0 m/s to avoid any clogging. Assuming a flow velocity of 1.2 m/s, the required size of the rising main can be found.

$$D = \sqrt{\frac{1.27Q}{v}} = \sqrt{1.27 \times \frac{0.15\ m^3}{s} \times \frac{s}{1.2\ m}} = 0.398\ m = 400\ mm\ say$$

$$v_{actual} = \frac{1.27Q}{D^2} = 1.27 \times \frac{0.15\ m^3}{s} \times \frac{1}{(0.40\ m)^2} = 1.19 = \underline{1.2\ m/s}$$

The capacity of a sump well or wet well is worked out based on a detention time of 15 minutes during peak flow and additional capacity equivalent to that of the rising main.

$$Capacity\ V = V_I + V_{II} = Q_p \times t + \frac{\pi D^2}{4} \times L$$

$$= \frac{0.15\ m^3}{s} \times 15\ min \times \frac{60\ s}{min} + \frac{\pi}{4} \times (0.4\ m)^2 \times 90\ m$$

$$= 146.3 = 146\ m^3$$

Assuming an effective depth, d, of 3.0 m, the required diameter of each sump well (#=2) is:

$$D_{sump} = \sqrt{\frac{1.27V}{d}} = \sqrt{1.27 \times \frac{73\ m^3}{3.0\ m}} = 5.55 = \underline{5.6\ m}$$

Head loss in the rising main is the sum of frictional losses and minor losses. Assuming a friction factor of 0.03 and a 10 m equivalent length for minor losses, the total head loss in the rising main can be calculated:

$$Head\ loss,\ h_l = f \times \frac{L}{D} \times \frac{v^2}{2g} = 0.03 \times \frac{100\ m}{0.4\ m} \times \left(\frac{1.2\,m}{s}\right)^2 \times \frac{s^2}{2 \times 9.81\ m} = 0.550 = 0.55\ m$$

$$Pumping\ head,\ h_a = lift + h_l = 11\ m + 0.55\ m = 11.55 = 11.6\ m$$

Capacity of each pump is such as to empty half of sump water in 15 min

$$Water\ power, P_a = Q \times \gamma \times h_a = \frac{3\ m^3}{900\,s} \times \frac{9.81kN}{m^3} \times 11.55m$$

$$= 9.19 = 9.2\ kW$$

Assuming a pump efficiency of 65%, the pump power can be worked out:

$$Pump\ power,\ P_p = \frac{P_a}{E_p} = \frac{9.19\ kW}{65\%} \times 100\% = 14.1 = \underline{14\ kW}$$

Discussion Questions

1) How is pumping wastewater different from pumping potable water?
2) During the inspection of pumping stations, what important tasks should be performed?
3) Describe the procedure for verifying the pumping rate at a wastewater pumping station.
4) Describe the pump-operating sequence at a lift station.
5) Compare a dry well pumping station with a wet well pumping station.
6) Discuss the type of pumps used for wastewater pumping.
7) What are the main considerations when selecting the pumping capacity?
8) With the help of a sketch, explain the main components of a lift station with a separate dry well and wet well.
9) Modify the general energy equation for working out the pumping head for pumps at a wastewater lift station.
10) Explain the importance of preliminary treatment before water enters a wet well. What devices are commonly employed for this purpose?

Practice Problems

1. Design a pumping station to lift the wastewater by 12 m via a 190 m long rising main. This station is to serve a population of 30 000 with a per capita water supply of 130 L/c.d. Assume peak flow is three times the average flow, the sump water depth is 3.0 m, and the desired minimum flow velocity in the rising main is 1.0 m/s. Minor losses are

equal to 5 times the velocity head and friction factor of 0.03. Make other appropriate assumptions as necessary.
   a. Sump capacity = 115 m^3, sump diameter = 5.0 m).
   b. Rising main diameter = 350 mm, actual flow velocity = 1.1 m/s.
   c. Head added = 13.5 m.
   d. Waterpower = 8.4 kW, pump power = 14 kW at 60% efficiency.
2. Select pumping units to pump an average wastewater flow of 1550 m^3/d assuming peak flow is 2.5 times the average flow and both the pumps operate during peak flow period. The rising main is 550 m long and the wastewater has to be lifted through a height of 11 m.
   a. Rising main diameter = 250 mm, actual flow velocity = 0.91 m/s.
   b. Head loss = 3.00 m (f = 0.03, 50 m minor losses length), head = 14 m.
   c. Waterpower = 3.1 kW, pump power = 5.1 kW at 60% efficiency.
3. Work out the pump start level in a 3.0 m diameter wet well if the desired pumping cycle should not exceed 5/h. The average wastewater inflow rate is 8.5 L/s and the pump rate is 13 L/s. (2.70 m)
4. A wastewater pump has 300 mm delivery and 350 mm diameter suction. The reading on the delivery gauge located at the centreline of the pump is 150 kPa, while the suction gauge located 50 cm below the pump reads 20 kPa. Knowing the discharge rate is 650 L/s, determine:
   a. The total head on the pump. (16 m)
   b. The power supplied assuming a pump efficiency of 82% and a motor efficiency of 90%. (140 kW)
5. A wastewater pump has 300 mm delivery and 350 mm diameter suction. The reading on the discharge pressure gauge located at the centreline of the pump is 100 kPa, while the suction gauge located 75 cm below the pump reads 20 kPa. If the total head on the pump is 10 m, determine:
   a. The discharge rate of the wastewater. (480 L/s)
   b. The power input assuming the pumping unit is 70% efficient. (67 kW)

20 Natural Purification

Wastewater needs to be treated when the pollutant loading exceeds the assimilative capacity of the receiving water bodies. In the past, in the majority of cases, wastewater was directly discharged into receiving water bodies. It was assumed that the wastewater was diluted by a factor of hundreds so that the natural purification was good enough to maintain the quality of water in the receiving water body. The report of the Royal Commission on Sewage Disposal lays down certain standards, as indicated in **Table 20.1**.

Perennial streams and rivers are the best types of water bodies for the disposal of sewage. The most stressful period for a receiving water body is the summer when flows are at their minimum and the biochemical oxygen demand (BOD) reaction rate is accelerated due to a rise in temperature.

20.1 BIOCHEMICAL OXYGEN DEMAND

Biochemical oxygen demand (BOD) indicates the strength of wastewater. In water bodies, it indirectly describes the amount of biodegradable organic matter. The oxidation of organic matter can cause dissolved oxygen (DO) stress due to the depletion of oxygen, impairing the quality of water.

20.1.1 BOD Test

A BOD test is carried out by observing the depletion of dissolved oxygen in a sample or diluted sample after incubation over a given period at 20°C. Dilution becomes necessary when expected BOD exceeds 4.0 mg/L. In most cases, a portion of a sample (aliquot) is diluted with water saturated with oxygen and containing other chemicals/nutrients necessary for the growth of microorganisms. A sample dilution factor is chosen based on the expected BOD since water can only have a limited amount of dissolved oxygen at a given temperature. As a rule of thumb, the sample dilution factor can be chosen based on the following relationship:

$$Dilution\ factor,\ DF = \frac{Diluted\ sample}{Sample\ portion} = \frac{Expected\ BOD}{4}$$

To be on the safe side, more than one dilution is done. Only BOD dilutions with depletions in the range of 40–70% are considered valid. Results are also not valid if the final DO falls below 2.0 mg/L.

$$BOD = DF \times \left(D_i - D_f \right)$$

For industrial wastewater samples, seeds or microorganisms are added to complete a BOD test.

$$BOD = DF \times \left[\left(D_i - D_f \right) - f\left(B_i - B_f \right) \right]$$

$D_{i,}\ D_f$ = initial and final DO readings of the diluted sample
$B_{i,}\ B_f$ = initial and final DO readings of the seeded blank
f = volume of seed in dilution/volume of seed in the blank

DOI: 10.1201/9781003231264-20

TABLE 20.1
Royal Commission on Sewage Disposal Report

Dilution Factor	Standards of Purification Required
>500	No treatment required
300–500	Primary treatment required, plant effluent SS < 150 mg/L
150–300	Chemical precipitation is also required so that the effluent suspended solid content is < 50 mg/L
<150	Secondary stage of treatment is needed to produce effluent with SS < 30 mg/L and BOD_5 < 20 mg/L

Example Problem 20.1

A BOD test was done on a wastewater sample by making a 1% dilution. Initial and final DO readings in the diluted samples are 7.95 and 2.15 mg/L, respectively, and for the seeded blanks are 8.15 mg/L and 7.95 mg/L, respectively. Calculate the BOD of the sample.

Given: DF = 100%/1% = 100
f = 100%-1% = 99%
BOD = ?

SOLUTION:

$$BOD = DF \times \left[\left(D_i - D_f\right) - f\left(B_i - B_f\right)\right]$$

$$= 100 \times \left[(7.95 - 2.15)\frac{mg}{L} - \frac{99\%}{100\%}(8.95 - 7.95)\frac{mg}{L}\right]$$

$$= 480.2 = \underline{480\ mg/L}$$

20.1.2 BOD Reaction

A BOD reaction is a first-order reaction at a given temperature. The rate at BOD exerted (BOD_e) is directly proportional to the BOD remaining (BOD_r). The BOD exerted is BOD ultimate, BOD_u minus BOD remaining, BOD_r.

$$First\ order\ reaction,\ \frac{dBOD_r}{dt} = -kBOD_r$$

On integration, the BOD remaining after duration, t, is given by the following expression:

$$BOD_r = BOD_u e^{-kt} \quad or \quad BOD_t = BOD_u\left(1 - e^{-kt}\right)$$

The value of the BOD rate constant k depends on the temperature and biodegradability of the wastewater. The typical value of k at 20°C is 0.23/d. The rate constant k for other temperatures can be modified using the following empirical equation:

$$k_T = k_{20}\theta^{(T-20)} = k_{20} \times 1.047^{(T-20)}$$

For temperatures less than 20°C, the value of θ is 1.35.

20.1.3 Determination of Rate Constant

The rate constant of a BOD reaction indicates the biodegradability of a given wastewater. It can be determined based on BOD test data. The exerted BOD equation can be modified as follows:

$$BOD_{t+1} = BOD_u\left(1 - e^{-k(t+1)}\right) \; or$$

$$BOD_{t+1} = BOD_u - BOD_u e^{-k} + e^{-k} BOD_t \; or$$

$$BOD_{t+1} = BOD_u\left(1 - e^{-k}\right) + e^{-k} BOD_t$$

The above equation shows that the relationship between BOD_{t+1} and BOD_t is linear. That is to say, the plot of BOD_{t+1} versus BOD_t fits a straight line of the form y = Ax + B.
Comparing, we get

$$A = e^{-k} \quad or \; k = -\ln A$$

$$B = BOD_u\left(1 - e^{-k}\right) = BOD_u(1 - A) \quad or \;\; BOD_u = \frac{B}{(1 - A)}$$

By fitting an equation or running linear regression, the values of constants A and B can be found and hence both k and BOD_u can be determined.

Example Problem 20.2

Assuming a BOD reaction constant value of 0.23/d, figure out what percentage of ultimate BOD is exerted after a duration of five days.

Given: k = 0.23/d BOD/BOD_u = ?

SOLUTION:

$$\frac{BOD_t}{BOD_u} = \left(1 - e^{-kt}\right) = \left(1 - e^{\frac{-0.23}{d} \times 5.0d}\right) = \left(1 - e^{-1.15}\right) \times 100\%$$

$$= 68.33 = \underline{68\%}$$

Example Problem 20.3

A 5 day BOD for a given wastewater is 220 g/m³. Assuming BOD_5 is 70% of the BOD_u, find the 3 day BOD.

Given: k = ?
$BOD_5 = 220\ g/m^3$
$BOD_3 = ?$

SOLUTION:

$$k = -\frac{1}{t}\ln\left(1-\frac{BOD_t}{BOD_u}\right) = -\frac{\ln(1-0.70)}{5d}$$

$$= 0.2407 = 0.24\,/\,d$$

$$BOD_3 = BOD_u\left(1-e^{-kt}\right) = \frac{1}{0.70}\times\frac{220\ g}{m^3}\times\left(1-e^{-\frac{0.24}{d}\times 3d}\right)$$

$$= 161.3 = \underline{160\ g\,/\,m^3}$$

Example Problem 20.4

Given that the 5 day BOD is 250 g/m³, find the ultimate BOD. Assume a BOD reaction constant of 0.23/d. What would be the 5 day BOD at a temperature of 25°C?

Given: k = 0.23/d
$BOD_5 = 250\ g/m^3$
$BOD_u = ?$
$BOD_5\ (25°C) = ?$

SOLUTION:

$$BOD_u = \frac{BOD_5}{\left(1-e^{-5k}\right)} = \frac{250g}{m^3}\times\left(1-e^{\frac{-0.23}{d}\times 5.0d}\right)^{-1} = 365.83 = \underline{366\ g\,/\,m^3}$$

$$k_T = k_{20}\times 1.047^{(T-20)} \qquad k_{25} = k_{20}\times 1.047^{(25-20)} = 0.289 = 0.29\,/\,d$$

$$BOD_5 = \frac{365.8\ mg}{L}\times\left(1-e^{-\frac{0.289}{d}\times 5d}\right) = 279.5 = \underline{280\ mg\,/\,L}$$

20.2 NATURAL PROCESS

When wastewater is discharged into a natural stream, organic matter is biodegraded aerobically. As a result of this aerobic reaction, ammonia, nitrates, sulphates, carbon dioxide, water and new cells are produced. The main factors influencing this process are dilution, sedimentation, oxidation, reduction, temperature and sunlight.

High dilution ensures aerobic conditions and the completion of biodegradation. During the summer months, stream flows are minimal and biological activity is accelerated due to an increase in water temperature. This represents the worst-case scenario from a pollution point of view. The DO content of the stream water is depleted. As the water moves downstream, its DO content continues

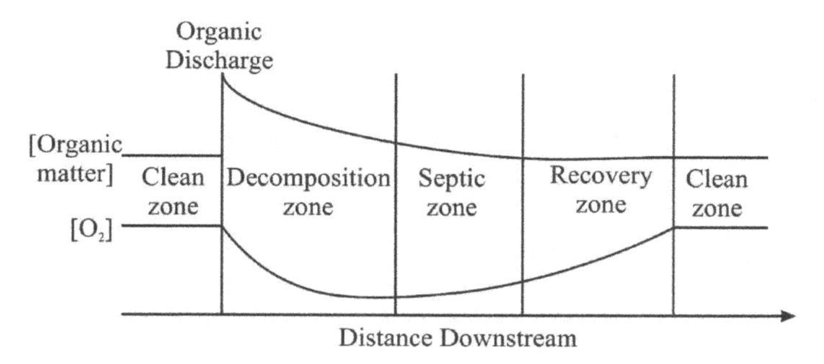

FIGURE 20.1 Zones of Natural Process

to deplete until it reaches a minimum level called the critical point. Whereas decomposition causes an oxygen deficit, re-aeration or natural oxygenation adds oxygen into stream water. At the critical point, both the rates become equal. After the critical point, oxygenation exceeds the depletion rate and the recovery of DO begins. The self-purification process can thus be divided into four distinct stages or zones, described briefly in the following sections.

20.2.1 Zone of Degradation

As wastewater is discharged into a water body, it goes through various phases of degradation, as shown in **Figure 20.1**. The water in the zone of degradation or decomposition is turbid, and sludge forms at the bottom of the stream. The DO level is suddenly reduced due to mixing, and the deoxygenation rate exceeds the re-aeration rate. The decomposition of solid matter takes place in this zone and anaerobic conditions prevail.

20.2.2 Zone of Active Decomposition

This zone is under the most stress due to pollution. As a result, the DO level falls to its minimum and anaerobic conditions set in. Hence, it is also called the septic zone. Sludge at the bottom goes through anaerobic biodegradation producing methane and hydrogen sulphide. Towards the end of this zone, the oxygenation rate becomes greater than the deoxygenation rate and DO starts picking up.

20.2.3 Zone of Recovery

As the name indicates, here, the process of recovery starts, and stream water starts returning to its former condition. The aerobic stabilization of soluble BOD takes place in this zone, thus BOD falls and the DO level goes up. Near the end of this zone, microscopic life reappears, fungi levels decrease, and algae start establishing.

20.2.4 Clear Water Zone

Here, the stream returns to its former condition and the water becomes clear, and recovery is assumed to be complete. The oxygen level reaches close to saturation.

20.3 FACTORS AFFECTING SELF-PURIFICATION

Some of the key factors that affect the self-purification process are as follows.

Dilution

When sufficiently diluted water is available in the receiving water body, the DO level in the receiving stream may not be stressed to critical levels due to the availability of sufficient DO. This is because before the discharge of the wastewater, the receiving water has DO close to saturation.

Current

When a strong water current is available, the discharged wastewater will be thoroughly mixed with stream water, preventing the deposition of solids. In weak currents, the solid matter from the wastewater will get deposited at the bed following decomposition and a reduction in DO level.

Temperature

The capacity of water to retain DO is directly related to temperature. The colder the temperature, the higher the DO saturation level. The quantity of DO available in stream water is higher in cold temperatures than in hot temperatures. Biological activity increases with an increase in temperature; hence, self-purification will take less time in warmer climates than cold weather.

Sunlight

Algae produce oxygen in the presence of sunlight due to photosynthesis. Therefore, sunlight helps in the purification of a stream by adding oxygen through photosynthesis.

Rate of Oxidation

DO depletion occurs due to the oxidation of organic matter discharged into the river. The rate at which this occurs is faster at higher temperatures and slower at lower temperatures. The rate of oxidation of organic matter depends on the biodegradability of the organic matter and the type of fauna and flora present.

Rate of Re-Aeration

The rate of re-aeration is dependent on the oxygen deficit, rate of turbulence, flow velocity and wind aeration. The re-aeration constant is higher for swift streams as compared to sluggish streams.

20.4 OXYGEN SAG CURVE

As waste is discharged into the stream, it lowers the DO level in the water. As mixed water moves downstream, the oxygen deficit increases as BOD is exerted. The plot of DO versus distance or time from the point of mixing downstream is called the oxygen sag curve, as shown in **Figure 20.2**.

Immediately after mixing, the oxygen deficit is called the initial deficit, D_0. As water moves downstream, this deficit increases to a maximum, D_c, also called the critical point. At the critical point, the DO_c level reaches a minimum, and oxygenation and deoxygenation rates become equal. Beyond the critical point, the oxygenation rate is greater than the deoxygenation rate, and the DO in the stream water starts going up and the stream starts recovering. The critical point is of great interest since it allows for the determination of the amount of waste that can be naturally purified while maintaining a minimum DO level.

Re-aeration and deoxygenation are both assumed to be first-order reactions. Based on this assumption, the oxygen deficit is described by the first-order differential equation suggested by Streeter and Phelps.

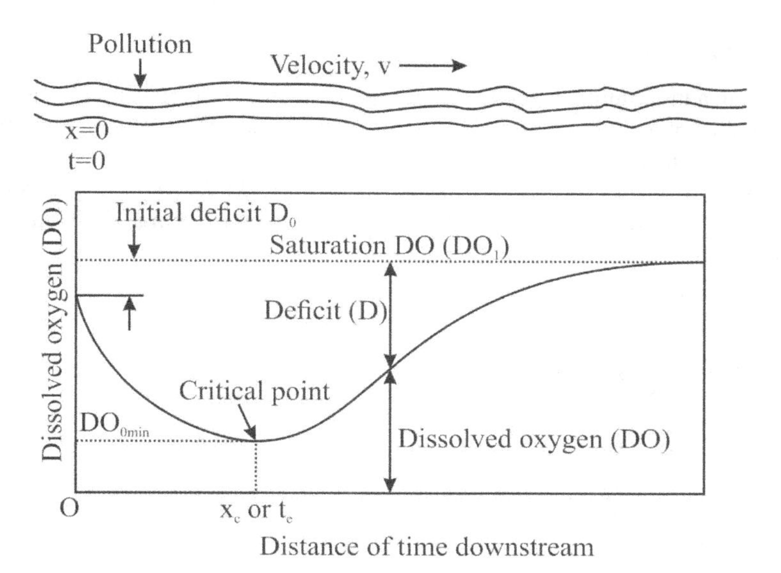

FIGURE 20.2 Oxygen Sag Curve

$$\frac{dD_t}{dt} = K_1 BOD_r - K_2 D_t \quad Solving\ this,\ D_t$$

$$D_t \quad Solving\ this,\ D_t = \frac{K_1 BOD_u}{K_2 - K_1}\left[e^{-K_1 t} - e^{-K_2 t}\right] + D_0 e^{-K_2 t}$$

Where, D_t = DO deficit at time t
BOD_r = BOD remaining at time t
K = reaction rate constant
Sub 1 = deoxygenation
Sub 2 = Oxygenation or re-aeration

Knowing the deficit D_t, dissolved oxygen DO can be found:
$DO_t = DO_{sat} - D_t$

The determination of the critical point is of great engineering significance and can be found by equating the first differential to zero.

$$t_c = \frac{1}{K_2 - K_1} ln\left[\frac{K_2}{K_1}\left(1 - \frac{D_0 \left(K_2 - K_1\right)}{K_1 BOD_u}\right)\right]$$

$$Max.\ deficit,\ D_c = \frac{K_1 BOD_u}{K_2} e^{-K_1 t_c}$$

As mentioned earlier, time and distance are related by the flow velocity, u. Thus the critical point can be expressed in time or distance travelled. By knowing the time to the critical point, some numerical examples are presented to further illustrate oxygen sag analysis.

Example Problem 20.5

In a stream with a minimum flow of 1.6 m^3/s containing on average 5.0 mg/L of BOD, the DO is 8.0 mg/L with a re-aeration constant of 0.60/d. A wastewater plant effluent of 0.10 m^3/s with a BOD of 20 mg/L and DO of 3.0 mg/L is discharged into the stream. The BOD rate constant is 0.25/d and the average flow velocity in the stream is 0.20 m/s. Assuming a DO saturation of 9.2 mg/L, calculate the DO in the stream water at locations 25 km and 50 km downstream of the discharge point.

Given: $K_1 = 0.25/d$
$K_2 = 0.60/d$
$BOD_w = 20\ g/m^3$
$Q_w = 0.10\ m^3/s$
$v = 0.2\ m/s$
$DO_w = 3.0\ g/m^3$
$DO_{st} = 8.0\ g/m^3$
$BOD_{st} = 5.0\ mg/L$
$Q_{st} = 1.60\ m^3/s$
$DO_t = ?$

SOLUTION:

Note: subscript m is used for mixed stream.

$$BOD_m = \frac{BOD_w \times Q_w + BOD_s \times Q_s}{Q_w + Q_s} = \frac{20 \times 0.1 + 5 \times 1.6}{0.1 + 1.6}$$

$$= 5.88\ g/m^3$$

$$BOD_u = \frac{BOD_5}{\left(1 - e^{-5k}\right)} = \frac{5.88\ g}{m^3} \times \frac{1}{\left(1 - e^{-\frac{0.23}{d} \times 5d}\right)} = 8.24 = 8.2\ g/m^3$$

$$DO_m = \frac{DO_w \times Q_w + DO_s \times Q_s}{Q_w + Q_s} = \frac{3.0 \times 0.1 + 8.0 \times 1.6}{0.1 + 1.6}$$

$$= 7.705 = 7.7\ g/m^3$$

$$Deficit,\ D_m = DO_{sat} - DO_m = 9.20 - 7.7 = 1.5\ g/m^3$$

$$Travel\ time,\ t = \frac{s}{v} = 25\ km \times \frac{s}{0.2\ m} \times \frac{1000\ m}{km} \times \frac{h}{3600\ s} \times \frac{d}{24\ h}$$

$$= 1.446 = 1.4\ d$$

$$D_t = \frac{K_1 BOD_u}{K_2 - K_1}\left[e^{-K_1 t} - e^{-K_2 t}\right] + D_0 e^{-K_2 t}$$

$$= \frac{0.25}{0.60 - 0.25} \times \frac{8.2g}{m^3} \times \left[e^{-0.25 \times 1.4} - e^{-0.6 \times 1.4}\right] + \frac{1.5\ g}{m^3} e^{-0.6 \times 1.4}$$

$$= 2.249 = 2.25 \text{ mg/L}$$

$$DO_t = DO_{sat} - D_t = (9.0 - 2.25) = 5.75 = 5.8 \; g/m^3$$

Oxygen deficit at a location 50 km (t = 2.9 d) downstream.

$$D_t = \frac{K_1 BOD_u}{K_2 - K_1}\left[e^{-K_1 t} - e^{-K_2 t}\right] + D_0 e^{-K_2 t}$$

$$= \frac{0.25}{0.60 - 0.25} \times \frac{8.2 g}{m^3} \times \left[e^{-0.25 \times 2.9} - e^{-0.6 \times 2.9}\right] + \frac{1.5 \; g}{m^3} e^{-0.6 \times 2.9}$$

$$= 2.07 = 2.0 \; mg/L$$

$$DO_t = DO_{sat} - D_t = (9.0 - 2.07) = 6.93 = 6.9 \; g/m^3$$

$$t_c = \frac{1}{K_2 - K_1} ln\left[\frac{K_2}{K_1}\left(1 - \frac{D_0 (K_2 - K_1)}{K_1 BOD_u}\right)\right]$$

$$= \frac{d}{0.60 - 0.25} \ln\left(\frac{0.60}{0.25}\right)\left(1 - \frac{1.5(0.60 - 0.25)}{0.25 \times 8.2}\right) = 1.656$$

$$= 1.7 \; d \; (29 \; km)$$

This further confirms that recovery starts after 1.7d, hence DO starts picking up as shown by the above calculations.

Example Problem 20.6

A town discharges sewage containing BOD_5 of 220 mg/L at a rate of 5.0 m^3/s into a stream with a minimum flow of 110 m^3/s. Calculate the critical DO in the water downstream. Assume a BOD rate constant of 0.23/d, an oxygenation rate constant of 0.81/d and the saturated DO is 9.0 mg/L.

Given: K_1 = 0.23/d
K_2 = 0.81/d
BOD_5 = 220 g/m^3
Q_w = 5.0 m^3/s
Q_{st} = 110 m^3/s
DO_c = ?

SOLUTION:

$$BOD_m = \frac{BOD_w \times Q_w + BOD_s \times Q_s}{Q_w + Q_s} = \frac{220 \times 5 + 0 \times 110}{5 + 110} = 9.56 \; g/m^3$$

$$BOD_u = \frac{BOD_5}{\left(1-e^{-5k}\right)} = \frac{9.56\ g}{m^3} \times \frac{1}{\left(1-e^{-\frac{0.23}{d}\times 5d}\right)} = 13.98 = 14\ g/m^3$$

$$DO_m = \frac{DO_w \times Q_w + DO_s \times Q_s}{Q_w + Q_s} = \frac{0\times 5 + 9\times 110}{5+110} = 8.60\ g/m^3$$

$$D_m = DO_{sat} - DO_m = 9.0 - 8.60 = 0.40\ g/m^3$$

$$t_c = \frac{1}{K_2 - K_1} ln\left[\frac{K_2}{K_1}\left(1 - \frac{D_0\left(K_2 - K_1\right)}{K_1 BOD_u}\right)\right]$$

$$= \frac{d}{(0.81-0.23)} \ln\frac{0.81}{0.23}\left[1 - \frac{0.40(0.81-0.23)}{0.23\times 14}\right] = 1.3\ d$$

$$D_c = \frac{K_1 BOD_u}{K_2} e^{-K_1 t_c} = \frac{0.23}{0.81} \times \frac{14\ g}{m^3} \times e^{-\frac{0.23}{d}\times 1.33d} = 2.92 = 2.9\ g/m^3$$

$$DO_c = DO_{sat} - D_c = (9.0 - 2.92) = 6.07 = 6.1\ g/m^3$$

Example Problem 20.7

A town discharges secondary effluent containing BOD_5 of 25 mg/L at a rate of 5.0 m^3/s into a stream with a minimum flow of 65 m^3/s. Calculate the critical DO in the water downstream. Assume a BOD rate constant of 0.22/d, an oxygenation rate constant of 0.75/d and the saturated DO is 9.0 mg/L. Assume the DO of the effluent is 1.0 mg/L.

Given: $K_1 = 0.22/d$
$K_2 = 0.75/d$
$BOD_5 = 25\ g/m^3$
$Q_w = 5.0\ m^3/s$
$Q_{st} = 65\ m^3/s$
$DO_c = ?$

SOLUTION:

$$BOD_m = \frac{BOD_w \times Q_w + BOD_{st} \times Q_{st}}{Q_w + Q_{st}} = \frac{25\times 5 + 0\times 5}{5+65} = 1.78\ g/m^3$$

$$BOD_u = \frac{BOD_5}{\left(1-e^{-5k}\right)} = \frac{1.78\ g}{m^3} \times \left(1 - e^{-\frac{0.22}{d}\times 5d}\right)^{-1} = 2.676 = 2.7\ g/m^3$$

$$DO_m = \frac{DO_w \times Q_w + DO_s \times Q_s}{Q_w + Q_s} = \frac{1\times 5 + 9\times 65}{5+65} = 8.428 = 8.43\ g/m^3$$

$$D_m = DO_{sat} - DO_m = 9.0 - 8.43 = 0.57\ g/m^3$$

$$t_c = \frac{1}{K_2 - K_1} ln\left[\frac{K_2}{K_1}\left(1 - \frac{D_0(K_2 - K_1)}{K_1 BOD_u}\right)\right]$$

$$= \frac{d}{(0.75-0.22)} \ln\frac{0.75}{0.22}\left[1 - \frac{0.57(0.75-0.22)}{0.22\times2.68}\right]$$

$$= 0.958 = 0.96\ d$$

$$D_c = \frac{K_1 BOD_u}{K_2} e^{-K_1 t_c} = \frac{0.22}{0.75} \times \frac{2.68\ g}{m^3} \times e^{-\frac{0.22}{d}\times 0.96 d} = 0.636 = 0.64\ g/m^3$$

$$DO_c = DO_{sat} - D_c = (9.0 - 0.64) = 8.36 = 8.4\ g/m^3$$

20.5 DILUTION INTO SEA

Because saturated DO in water decreases with increasing solid content, the saturated DO of the sea water is about 80% that of fresh water. Since sea water is denser than sewage water, which usually has a higher temperature, sewage discharged into the sea spreads on the surface in a thin layer called sleek. Moreover, sea water contains a large quantity of solids that react with some of the sewage solids to precipitate, resulting in the formation of sludge banks. As more waste is dumped, the capacity of the sea water to absorb sewage is reduced and results in undesirable conditions. However, since the sea contains a large volume of water, such problems can be overcome if the sewage is discharged deep into the sea and further away from the coastline. To achieve this, the following points should be kept in mind.

1. The wastewater should be discharged at least 1 km away from the shoreline.
2. The outfall should be designed to ensure proper dilution and adequate mixing. This is achieved by providing a multipoint diffuser.
3. The minimum depth of the water at the outfall point should be 3 to 5 m.
4. Wastewater should be discharged only during low tide. Provisions must be made to hold wastewater during high tide.
5. While deciding the position of the outfall, the direction of the wind and ocean currents should be taken into consideration.

20.6 DISPOSAL BY LAND TREATMENT

Land treatment refers to the disposal of treated or raw wastewater spread over the land surface. Some of the wastewater evaporates, and some enters the soil by infiltration. Wastewater percolating through soil pores leaves behind organic solids that are oxidized aerobically and increase the fertility of the soil. Disposal by land treatment is also called sewage farming.

20.6.1 SEWAGE FARMING

In sewage farming, in addition to the disposal of partly treated wastewater, crops are raised using nutrients and other microelements from wastewater. The most common methods of sewage farming and treatment of wastewater are irrigation, rapid infiltration and overland run-off.

Irrigation and infiltration are suitable for lands with coarse-textured soils with relatively high percolation rates. Raw sewage applied in large amounts can result in plugging of the soil surface, significantly reducing infiltration rates. Thus, partially treated wastewater can greatly reduce plugging

TABLE 20.2
Recommended Loading Rate

Soil Type	Application Rate (mm/d)
Sandy	22–25
Sandy loam	15–20
Loam	10–15
Clayey loam	5–10
Clay	3–5

and allow voids to remain open. Rapid infiltration is practised where the percolation rate is in the range of 6–25 mm/min. Irrigation is suitable when percolation rates are 2 to 6 mm/min, and overland run-off is the best choice when the rate is below 2.0 mm/min.

With the exception of overland run-off, straining and biological filtration reduce the organic content of wastewater. In sewage farming, as wastewater percolates, soil holds moisture, nutrients and organic matter to sustain crops. Rapid infiltration is suitable for recharging groundwater. Since wastewater carries pathogens, the safety of workers must be ensured. In addition, leafy vegetables and root crops like potatoes are not recommended for sewage farming. The public must be warned about the risks of using produce from sewage farms.

In sewage farming, wastewater effluents from primary or secondary treatment plants can be applied to the land by surface irrigation methods such as border or furrow irrigation. Sprinkler irrigation is more common in western countries. Subsurface irrigation such as drip irrigation is recommended for highly treated wastewater only. The land disposal method is preferred in the following situations:

1. A large area of land, preferably with permeable soils.
2. Hot climatic conditions, due to high evaporation.
3. Low rainfall and scarcity of irrigation water.
4. When natural water bodies are not available in the vicinity.

20.6.2 Broad Irrigation

Broad irrigation is similar to sewage farming except that its main purpose is the disposal of wastewater and not the growing of crops, as in sewage farming. For this reason, it is also called effluent irrigation. Another key difference is that in broad irrigation, both raw and treated effluent can be applied, whereas in sewage farming, only treated wastewater is applied. In broad irrigation, raw or treated wastewater is applied on vacant land with coarse-textured soil and moving sewage retained in voids is treated aerobically.

20.6.3 Sewage Sickness

Sewage sickness is the term applied to land when sewage is continuously applied, and voids are filled and collaged to prevent the entry of air. To avoid sewage sickness, the following preventive measures should be taken.

1. Wastewater should be pre-treated to reduce suspended solids and BOD content.
2. Primary treatment should reduce solid content by 50%, thus allowing the voids to remain partially open and provide aeration.
3. To let the sick land rest, there should be provision for extra land.

4. Subsurface drainage should be provided to allow for the collection of percolating effluent.
5. Soils in the land chosen for sewage treatment should be coarse-textured.
6. Crops should be rotated.

Example Problem 20.8

A town with a population of 50 000 plans to dispose of its sewage through land application. Assuming a per capita sewage production of 150 L/c·d, determine the land area required assuming a soil capacity of 75 mm/d. In another scenario, determine the percolation rate if the sewage is pre-treated to reduce solids content, and the per capita wastewater production increases to 180 L/c·d due to more lavish lifestyles.

Given: Population = 50 000 at 150 L/c·d
rate = 75 mm/d

SOLUTION:

$$Q = \frac{150\ L}{p.d} \times 50000\ p \times \frac{m^3}{1000\ L} = 75000\ m^3/d$$

$$A_{min} = \frac{75000\ m^3}{d} \times \frac{d}{75\ mm} \times \frac{1000\ mm}{m} \times \frac{ha}{10000\ m^2} = 100\ ha$$

Providing 50% for resting and crop rotation, $A = 1.5 \times 100\,ha = \underline{150\,ha}$

$$I = \frac{Q}{A} = \frac{180\ L}{p.d} \times 50000\ p \times \frac{1}{100\ ha} \times \frac{ha}{10000\ m^2}$$

$$= \frac{90\ L}{m^2.d} \times \frac{m^3}{1000\ L} \times \frac{1000\ mm}{m} = \underline{90\ mm/d}$$

20.7 COMPARISON OF DISPOSAL METHODS

The dilution method requires that large water bodies assimilate water, whereas the land treatment disposal method requires a large area of pervious land. Thus, the following points can be made when comparing disposal methods.

- Since in urban areas land is relatively expensive, the dilution method is more suitable and preferred.
- Conversely, disposal through land application is preferred in rural areas.
- In the dilution method, the assimilative capacity of the receiving water must be considered, and the quality of effluent must be good as not to cause any pollution and keep the water body healthy.
- In the broad irrigation method of wastewater disposal, raw wastewater can be applied. However, care must be taken to avoid sewage sickness in the soil.
- In disposal by land application, sewage can be applied by flooding the land surface. The topographic features of the land may be such that the pumping of waste is required. However, this is not usual in dilution methods.

- In the dilution method, when wastewater effluent is discharged to a relatively small stream, it is important to check for adverse conditions that occur in summer due to low temperature and low flow.
- In sewage farming, good management is necessary and safeguards must be in place to prevent health risks to workers and the public.
- In the broad irrigation method, disposal must be managed to prevent any risks to groundwater, such as contamination.

Discussion Questions

1. Assuming BOD is a first-order reaction, develop the relationship between ultimate BOD and BOD exerted at any time t.
2. Describe the various stages of self-purification after waste is discharged.
3. Dilution is not a solution. Comment
4. What is land treatment? Discuss the conditions under which it is suitable.
5. Discuss the factors that affect the self-purification capacity of a river.
6. Draw a typical oxygen sag curve and label it with the various stages of self-purification.
7. Write a note on the disposal of sewage into the sea.
8. Compare the deoxygenation rate constant with the oxygenation rate constant.
9. Explain the terms sewage farming and soil sickness.
10. Prove that the relationship between BOD_{t+1} and BOD_t is linear. Derive the relationship between the rate constant and the slope of the straight line.
11. Describe the steps to carry out a BOD test on a sample of wastewater.
12. Explain why dilution of a wastewater sample is required to carry out and successful BOD test. What is the rule of thumb when deciding the dilution factor?
13. Define the terms oxygenation and deoxygenation as applied to natural purification in streams. How does it affect the shape of the oxygen sag curve?
14. Explain the difference between the dilution of sewage effluents discharged into a river and those discharged into sea water.
15. In the majority of the cases, especially in developing countries, disposal of sewage by treatment is better than disposal by dilution. Discuss.

Practice Problems

1. Assuming a BOD reaction constant value of 0.20/d, what percentage of ultimate BOD is exerted over a 5.0 day period? (63%)
2. A BOD test is performed on a river sample and the 5 day BOD is found to be 35.0 mg/L. Assuming a K of 0.23/d, determine the ultimate BOD. (51 mg/L)
3. For making a 2.0% dilution, 20 mL of sample aliquot was poured in a graduated cylinder. Diluting water was added to make it to 1000 mL. After mixing, the diluted sample was transferred to BOD bottles. The initial DO reading was 8.32 mg/L and the average final DO reading of three BOD bottles was observed to be 4.20 mg/L. What is the BOD of the sample? (210 mg/L)
4. The BOD of sewage after three days of incubation at 27°C was found to be 110 mg/L. Determine the BOD_5 at 20°C assuming rate constant $K_{20} = 0.23$/d and temperature coefficient $\theta = 1.047$. ($BOD_u = 180$ mg/L, $BOD_5 = 120$ mg/L)
5. A sewage treatment plant discharges treated wastewater at a rate of 1.5 m^3/s with a BOD of 35 mg/L and a DO of 2.0 mg/L. The minimum flow in the river is 6.0 m^3/s

containing a BOD of 5.0 mg/L and a DO is at 85% of the saturation level. The temperature of both the discharge and river water is at 20°C. Assume an oxygenation rate constant of 0.20/d and deoxygenation rate constant of 0.62/d. Calculate the minimum critical level of DO in the river water. (6.6 mg/L)

6. A city discharges untreated sewage at 13 m^3/s with a BOD of 260 mg/L in a river with a minimum flow of 160 m^3/s and a DO of 9.0 mg/L fully saturated. Assume a BOD rate constant of 0.22/d and an oxygenation constant of 0.92/d. Find out the critical DO in the stream. (4.6 mg/L after 2.2 d)
7. A river flowing at a rate of 22 m^3/s receives wastewater discharge at a rate of 0.5 m^3/s. The initial DO of the river water is 6.3 mg/L and the DO content in the wastewater is 0.6 mg/L. Lab tests have indicated that the BOD_5 of river water is 3.0 mg/L and the wastewater discharged into the river has a BOD_5 of 130 mg/L. Consider a saturation DO of 8.22 mg/L and deoxygenation and re-aeration rate constant values of 0.10/d and 0.30/d (log10 based), respectively. Find the critical DO deficit and DO in the river. (D_0 =2.1 mg/L BOD_u = 8.5 mg/L, t_c = 0.96 d, D_c = 5.7 mg/L, DO_c = 2.5 mg/L)
8. A municipal wastewater treatment plant discharges secondary effluent to a river. The worst condition occurs in the summer when the treated wastewater is found to have a maximum flow rate of 10 ML/d containing a BOD of 30 mg/L, a DO of 1.5 mg/L and a temperature of 25°C. Upstream of the waste discharge point, the minimum flow in the stream is 0.65 m^3/s with a BOD of 3.0 mg/L, a DO of 7.0 mg/L and a temperature of 22°C. The re-aeration constant is estimated to be 0.83/d at 20°C and assume θ = 1.016 for temperature correction. Calculate the DO deficit for t = 1 to 6 days after wastewater discharge.

t, d	1	2	3	4	5	6
D, mg/L	6.3	6.7	5.9	4.9	3.9	3.1

(critical DO deficit = 2.48 mg/L and the distance at which it will occur = 34 km)

9. The 5 day BOD of a certain wastewater has been found to be 600 mg/L. Assuming a reaction rate constant of 0.25/d, what is the ultimate BOD of this wastewater? What fraction of this waste would remain unoxidized after a period of 10, 15 and 20 d? (860 mg/L, 9.1%, 2.7%, 1%)
10. The 5 day BOD of a wastewater sample at 20°C has been found to be 250 mg/L. Assuming the reaction rate constant is 0.28/d at 20°C, what is the 15 day BOD of this wastewater at 25°C? (330 mg/L)
11. A wastewater sample is incubated for 1 day at 30°C with an indicated 1 day BOD of 100 mg/L. Based on this, determine the 5 day BOD at 20°C? (220 mg/L)
12. A treated wastewater effluent at a temperature of 25°C is discharged at a rate of 1.5 m^3/s into a river. A minimum river flow of 5.0 m^3/s is observed during the summer months when the river water temperature reaches 25°C. The raw wastewater 5 day BOD at 25°C is known to be 200 mg/L and that of the river water upstream of the outfall is 1.0 mg/L.
    a. What is the ultimate BOD of the river water after mixing if raw wastewater is discharged directly into the river during summer? (61 mg/L)
    b. After how long will the river water reach minimum DO? Assume the saturated DO is 8.3 mg/L, the DO of the river water upstream is 7.5 mg/L and the DO of the raw sewage is zero. Further assume the re-aeration constant is 0.65/d. (2.5 d)
    c. What is the minimum DO level at the critical point? (Zero)
13. Repeat problem 12 above if secondary effluent containing 15% of raw BOD and a DO of 2.5 mg/L is discharged. (10 mg/L, 3.5 d, 6.6 mg/L)

14. In a BOD test, 4.0 mL of a raw wastewater sample was added to a 300 mL BOD bottle and filled by adding aerated dilution water. The initial DO of the diluted sample was observed to be 8.1 mg/L. After incubation for a period of five days at 20°C, the final DO was read to be 4.5 mg/L. What is the BOD of the sample tested? (270 mg/L)
15. Determine the theoretical BOD of a 150 mg/L C_4H_7OH solution. (370 mg/L)
16. The data of BOD_{t+1} versus BOD_t of a given wastewater sample is fitted to a linear equation: $BOD_{t+1} = 0.663BOD_t + 66.0$.
    a. Find the BOD rate constant and ultimate BOD. (0.41/d, 200 mg/L)
    b. Based on the values found in a, determine the 8 day BOD. (192 mg/L)
17. The ultimate BOD of a wastewater sample is known to be 150% of 5 day BOD.
    a. What is the BOD rate constant at 20°C? (0.22/d)
    b. What is the BOD rate constant at 30°C? (0.35/d)
18. For previous Practice Problem 17, the 5 day BOD is known to be 220 mg/L.
    a. What will the 4 day BOD be at 30°C? (250 mg/L)
    b. What fraction of a 5 day BOD is that of the ultimate BOD at 30°C? (82%)
19. Biodegradable matter in a wastewater sample is estimated to be 300 mg/L with a rate constant of 0.30/d. How much matter remains after three days? (120 mg/L)
20. Multiple BOD tests were run on a sample of river water and the results are shown in the table below. Regress a linear equation for the data of $BODt_{+1}$ versus BOD_t ($BOD_{t+1} = 0.781BOD_t + 15.7$) and find the values of the rate constant and BODu. (0.25/d, 72 mg/L)

Time, d	0	1	2	3	4	5	6	7
BOD, mg/L	0	18	28	36	43	48	55	60

21 Wastewater Characteristics

The term wastewater refers to water carrying waste. Wastewater from domestic, commercial, institutional and infiltration sources combine in the sewer system and arrive at the sewage treatment plant to be cleaned before being sent back to the environment. The combined flow is called municipal wastewater. In many situations, some industrial flows also form part of municipal wastewater. However, the nature of industrial wastewater has to be compatible with the treatment process at the municipal plant. The purpose of the treatment plant is to remove solids, biochemical oxygen demand (BOD), nutrients and pathogens. If contributions from industrial sources are significant, the quality of municipal wastewater may be directly affected, and hence the operation of the wastewater treatment plant.

21.1 TREATMENT FACILITY

In a wastewater treatment plant, the solids, BOD and pathogens are removed from incoming wastewater as water passes through various unit processes and operations. In a way, a wastewater treatment plant can be thought of as a place for unloading sewage water. Based on the level of treatment, a wastewater treatment facility can be primary, secondary or tertiary, as shown in **Figure 21.1**.

Preliminary treatment includes screening, shredding and grit removal. These processes remove the coarse material and grit before the water flows to the primary treatment area. Irrespective of the stages of treatment, preliminary treatment is always there.

Primary treatment removes solids from incoming wastewater by gravity settling and floatation. Primary treatment typically removes 50% of suspended solids (SS) and 35% of BOD.

Secondary treatment processes usually follow primary treatment and commonly consist of biological processes. Colloidal and dissolved organics not removed by the preceding treatment are converted into biomass, energy, carbon dioxide and water. The term biomass is used to indicate microorganisms that break down organics and convert them to a more stable form. A fraction of raw solids are converted to carbon dioxide and water, and the remaining fraction to new cells and energy. Many plant operators visualize microorganisms as workers that perform the job of stabilizing organic waste. Typically, the effluent from a secondary treatment plant (secondary effluent) contains less than 15 mg/L each of suspended solids and BOD. In most cases, secondary treatment of municipal wastewater is sufficient to meet effluent standards. However, depending on the use of the receiving waters, further treatment may be required. Secondary treatment has become more or less the norm of the day. It is for this reason that many of the existing primary plants are being upgraded.

Tertiary treatment is for the further removal of solids and nutrients. Filtration, nitrification–denitrification, and further removal of BOD by polishing lagoons are examples of tertiary treatment. In cases where the water downstream is used as a source for drinking water supplies, tertiary treatment in different forms is used to meet more stringent discharge criteria. Tertiary treatment is expensive, but there is a higher degree of flexibility and control.

Advanced treatment is an add-on process to primary and secondary treatment. Phosphorus removal by chemical precipitation carried out as part of primary or secondary treatment is called advanced wastewater treatment. However, phosphorus removal following secondary treatment is called tertiary treatment.

DOI: 10.1201/9781003231264-21

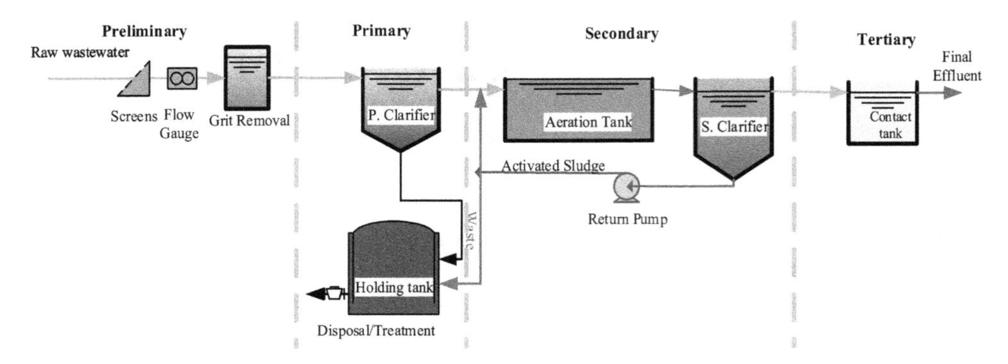

Treatment	Biochemical Oxygen Demand, BOD		Suspended Solids, SS	
	Percent Removal %	Effluent mg/L	Percent Removal %	Effluent mg/L
Primary	30-40	90-150	40-60	100-150
Secondary	95	15	90-95	15
Tertiary	98	5	98	5

FIGURE 21.1 Flow Schematic of a Wastewater Treatment Plant

21.2 DOMESTIC WASTEWATER

Domestic wastewater originates in the kitchen, bathroom and laundry room. Wastewater generated from toilets is also called black water, and from kitchen and baths is called grey water. Grey water, in some cases, may be recycled. Sanitary water is the term applied to domestic water plus water from commercial areas and some industrial water. The term municipal water applies to sanitary water and infiltration and inflow. It may contain large amounts of industrial wastewater. However, the industrial wastewater that is allowed must be treatable by conventional wastewater treatment processes. Typical municipal wastewater contains only 0.1% solids while the remaining 99.9% is made up of water. The average daily volume of wastewater production in North America is typically 450 L/person. In India, per capita wastewater production is much less and can be considered in the range of 120–200 L/c·d with 150 L/c·d as an average value.

21.3 PHYSICAL CHARACTERISTICS

Raw sewage is highly turbid. Normal fresh sewage smells musty but not unpleasant. Obnoxious smells, if present, indicate old septic sewage. Septic or partially decomposed sewage is dark, sometimes black with a sulphurous odour due to the presence of hydrogen sulphide. Hydrogen sulphide is toxic at low levels and causes corrosion. Another gas, methane, produced due to anaerobic conditions, is odourless but very explosive. For a comparison of the characteristics of three flow streams – raw, primary effluent and secondary effluent – refer to **Table 21.1**.

Wastewater temperature is typically higher than the tap water that passes through various dwellings. The temperature of wastewater affects the biological reaction rates in secondary treatment. As a rule of thumb, every 10°C increase in temperature doubles the rate of biological activity.

Wastewater becomes less dense at higher temperatures, which affects the settling characteristics of solids. A sudden increase in temperature may be an indication of industrial discharge. On the other hand, the intrusion of storm water can cause a sudden drop in wastewater temperature.

Turbidity is related to the presence of particulate matter in a flow stream. It is useful in assessing the quality of secondary effluents. The colour of fresh sewage is grey and becomes darker as wastewater gets septic. Any other type of colour again may be due to industrial discharges.

TABLE 21.1
General Characteristics of Flow Streams

Parameter	Raw Sewage	Primary Effluent	Secondary Effluent
Temperature	Generally warm	Lower temp.	Lower temp.
Turbidity	High in solids	Non-settling	No visible solids
Colour	Milky grey to black	Greyish to colourless	Clear colourless
Odour	Musty to sulphurous	Musty to sulphurous	Fresh

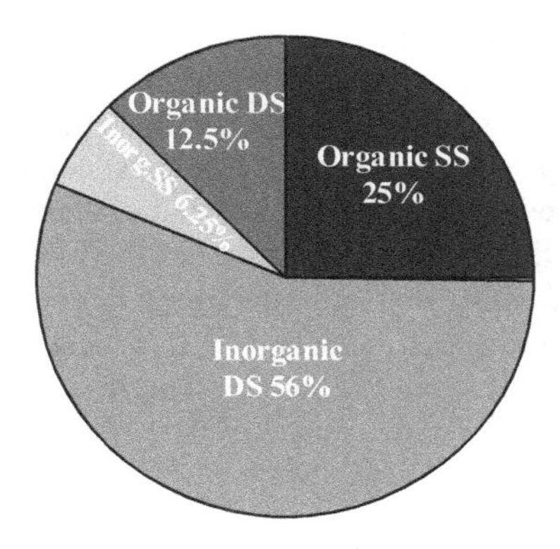

FIGURE 21.2 Solids

Normal municipal wastewater contains only 0.1% of solids – 99.9% is water. The solids in wastewater are classified as settleable and non-settleable (dissolved, colloidal and floatable). Settleable solids can be determined by running a one-hour settling test using an Imhoff cone. This test on raw wastewater can be used to estimate the removal of solids during primary treatment. Typically, the measurements fall in the range of 10–20 mL/L.

21.4 CHEMICAL CHARACTERISTICS

Suspended solids are those that are visible. The term suspended solids is usually applied to solids retained on a filter paper with openings 0.45µm in size. They may be settleable or in suspension and can be removed by the process of settling or filtration.

Dissolved solids cannot be seen as they are in solution or liquid form. The total of suspended and dissolved solids is termed total solids. The volatile solids test is where organics are burned off at 550°C, leaving only inorganic or inert material. **Figure 21.2** shows the relationship of the above terms. Solids can also be categorized as inorganic and organic. Organic solids can be broken down by biological processes while inorganic solids cannot. Organic solids are measured using the volatile portion of the solids.

21.4.1 Dissolved Gases

In addition to dissolved oxygen, sewage water may contain other gases such as carbon dioxide, ammonia and hydrogen sulphide as a result of decomposition, as well as nitrogen dissolved from the atmosphere. The presence of gases such as hydrogen sulphide can indicate the septicity of water. This is especially true for large cities with a complex collection system. Sewage takes more time to reach the plant and develop anaerobic conditions.

21.4.2 Alkalinity and pH

Whereas alkalinity indicates the capacity of wastewater to neutralize acid, pH indicates the degree of acidity or alkalinity. Both properties define the chemical environment and hence the rate of chemical and biological reactions. Raw wastewater is typically of neutral pH and falls in the range of 6–8. At a pH of below 6, biological activity is almost non-existent. When wastewater gets septic due to anaerobic conditions, it becomes darker and lowers the pH. If the oxidation of ammonia or nitrification is significant in secondary treatment, it will consume alkalinity.

21.4.3 Biochemical Oxygen Demand

The strength of sewage is measured by determining the biochemical oxygen demand (BOD). This parameter measures the quantity of oxygen utilized in the decomposition of organic matter in a sample of wastewater over a specific period of time and at a specific temperature. If not indicated, the BOD values refer to five days' incubation time. The most common tests performed for process control and regulatory requirements are the BOD test and the solids test. BOD is an indirect measure of organic matter. Since the BOD test takes a minimum of five days to complete, other tests such as the chemical oxygen demand (COD) test are used for process control. Most common regulatory requirements require 85% BOD removal in secondary treatment plants and a monthly average of suspended solids in secondary effluent not exceeding 30 mg/L.

21.4.4 Chemical Oxygen Demand

Chemical oxygen demand (COD) is commonly used to describe the strength of industrial wastewater. It takes a minimum of five days to perform a BOD test. Hence a BOD test is not suitable for process control purposes. The COD refers to the total oxygen demand and thus is usually greater than the BOD value. For typical municipal wastewater, the BOD is about 60% of the COD. Some people use total organic carbon (TOC) as a measure of the strength of a given wastewater. As a rough estimate, BOD can be assumed to be 250% of the TOC values.

21.4.5 Nutrients

Nutrients, including nitrogen and phosphorous, are essential for the growth of organisms. Lack of nutrients may hinder their ability to decompose organic matter properly. On the other hand, nutrients can also cause havoc in receiving streams by fertilizing algae and other aquatic weeds, which can lead to eutrophication, the premature ageing of natural water systems. If adequate treatment is not obtained, large amounts of solids may settle on the bottom, covering spawning beds. These solids may also decrease sunlight penetration to plants due to the cloudiness created from suspended material. The organic material can also put a demand on the receiving waters oxygen supply, which can kill fish due to the lack of oxygen. The principal nitrogenous compounds in domestic sewage are proteins, amines, amino acids and urea. Phosphorus is contributed by food residue containing phosphorus and phosphorus-based synthetic detergents. As shown in **Table 21.2**, a secondary treatment plant can remove about 25% of the nitrogen and about 15% of the phosphorus content of wastewater.

Nitrogen and phosphorus are essential for the growth of organisms in the biological process. The generally accepted value of the biochemical oxygen demand/nitrogen/phosphorus mass ratio required for biological treatment is 100:5:1.

21.4.6 Toxins

Conventional wastewater treatment is not designed to remove heavy metals and pesticides, and similar chemicals. On the other hand, these toxins can run havoc on biological treatments. Through sewer use control programs, every effort should be made to prevent discharges containing toxins.

TABLE 21.2
Nutrient Removal

	N, mg/L				P. mg/L	
Flow Stream	**NH_4**	**Organic**	**Nitrite**	**Nitrate**	**Total**	**Soluble**
Raw sewage	15–50	25–85	< 0.1	< 0.5	6–12	4–6
Primary effluent	15–50	25–85	< 0.1	< 0.5	4–8	4–6
Secondary effluent	0–1	5–20	< 5.0	> 10	3–6	2–5

21.5 BIOLOGICAL CHARACTERISTICS

Bacteriological testing is done to determine the efficiency of the disinfection process. This is usually done by testing for indicator organisms, such as total coliform and faecal coliform in a given flow stream. These results are usually reported as number of colonies/100 mL of the sample by performing a membrane filtration test.

In secondary treatment, a microscopic examination of the activated sludge can be used to study the health of the activated sludge process. An important part of this examination is to determine the distribution of various organisms, particularly protozoa, and the presence of filamentous bacteria.

21.6 PERCENT REMOVAL

The efficiency of a treatment process is its effectiveness in removing various contaminants from the wastewater. Removal efficiency is defined as the fraction of the constituents removed from the incoming (influent) flow to the treatment process. In terms of influent and effluent concentrations of the pollutant removed, percent removal, PR, can be expressed as:

$$PR = \frac{(C_i - C_e)}{C_i} \times 100\%$$

$$C_e = C_i\left(1 - \frac{PR}{100}\right)$$

Subscripts i and e refer to the influent and effluent flow streams, respectively.

In wastewater treatment, the efficiency of a process is more commonly defined in terms of BOD or SS removal. The above equation for efficiency can be rearranged to calculate the concentration of the effluent (C_e) or influent (C_i) in a given treatment process. The same equations will also apply to a treatment system consisting of more than one treatment process.

Example Problem 21.1

A paper mill effluent with a SS content of 450 mg/L is fed to a clarifier. If the clarifier's effluent has a SS content of 50 mg/L, what is the SS removal efficiency?

Given: SS_i = 450 mg/L
SS_f = 50 mg/L
PR = ?

SOLUTION:

$$PR = \frac{(SS_i - SS_e)}{SS_i} \times 100\% = \frac{(450 - 50)mg/L}{450mg/L} \times 100\% = 88.8 = \underline{89\%}$$

Example Problem 21.2

Raw municipal wastewater entering a trickling filter plant contains 220 mg/L of BOD. If the average primary removal is 35%, what is the minimum secondary removal required to produce plant effluent not exceeding 30 mg/L?

Given: SS_i = 220 mg/L
PR_I = 35%
PR_{II} = ?
BOD_e = 30 mg/L

SOLUTION:

$$BOD_e = BOD_i\left(1 - \frac{PR}{100}\right) = \frac{220\ mg}{L} \times \left(1 - \frac{35\%}{100\%}\right) = 143.0 = \underline{140\ mg/L}$$

$$PR = \frac{(BOD_i - BOD_e)}{BOD_i} \times 100\% = \frac{(143 - 30)mg/L}{450mg/L} \times 100\%$$

$$= 0.790 = \underline{79\%}$$

21.7 INDUSTRIAL WASTEWATER

The characteristics of industrial waste depend on the industrial processes and raw materials used. Since characteristics vary widely from industry to industry, each industry must customize its pre-treatment system through chemical, biological or physical processes.

Pre-treatment should only be used after all industrial plant controls such as recycling, equipment changes and process modifications have been applied. Modern industrial plant design dictates the segregation of wastes for individual pre-treatment, controlled mixing, or separate disposal. Municipal sewer use by-laws control sewer uses or joint treatment agreements between industry and municipal plants. Such agreements are called extra strength surcharge agreements, and these charges are assessed for BOD and SS over the average municipal limits.

21.7.1 Equivalent Population

For interpretation purposes, industrial waste strength is related to the number of people using the system. This is called population equivalence and is often calculated for BOD and flow. The population equivalence is used to compare the strength of industrial wastes as compared to municipal waste.

$$Per\,Capita\,BOD = \frac{200\ L}{c.d} \times \frac{220\ mg}{L} \times \frac{g}{1000\ mg} = 44g/c.d$$

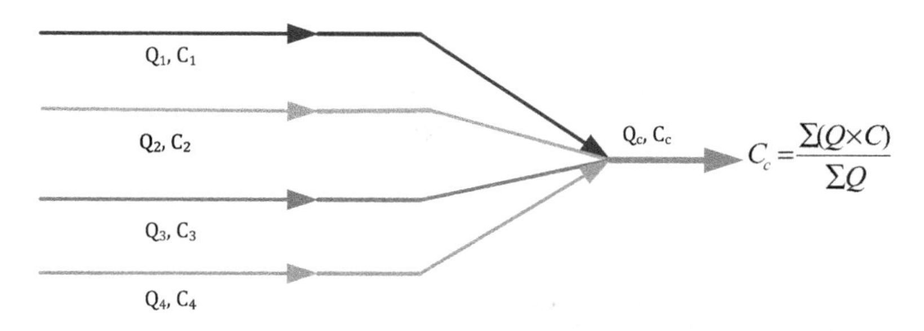

FIGURE 21.3 Composite Concentration

In India, the hydraulic equivalent population is based on a per capita contribution of flow that is 200 L/c. d. The BOD equivalent population is based on the BOD load of 45 g/c.d. In America, the hydraulic equivalent population is based on a per capita contribution of flow that is 450 L/c·d and the BOD equivalent is 90 g/c.d.

21.7.2 Composite Concentration

When the industrial contribution is significant, it may seriously affect organic and hydraulic loading. Since industrial discharges are usually of high strength, even small flows can cause significant changes in BOD loading. When estimating the composite concentration of municipal wastewater, the total BOD load is divided by the total flow, as shown in **Figure 21.3**.

$$\text{Composite concentration, } \bar{C} = \frac{\Sigma Q_i C_i}{\Sigma Q_i} = \frac{Q_1 \times C_1 + Q_2 \times C_1 + \ldots}{Q_1 + Q_2 + \ldots}$$

It is very important to keep in mind that the composite concentration of municipal wastewater is not simply the average of all the flow streams. Rather, it is the weighted average of the various discharges contributing to municipal flows. The following example illustrates the computation of the composite concentration of municipal wastewater received at the municipal facility.

Example Problem 21.3

Municipal wastewater from a small community of 8200 people, in addition to sanitary sewage, consists of food processing wastewater at a flow of 120 m³/d containing BOD of 950 mg/L. A dairy flow of 550 m³/d and a BOD concentration of 1200 mg/L also discharge into the sanitary system. Assuming typical values for sanitary discharge, calculate the municipal wastewater strength and flow.

Contributor	Q, m³/d	BOD, g/m³	BOD, kg/d
Sanitary	3690	200	738
Food processing	120	950	114
Dairy	550	1200	660
Total	4360	350	1512

SOLUTION:

$$Q = \frac{450\ L}{p.d} \times 8200\ p \times \frac{m^3}{1000\ L} = 3690\ m^3 / d$$

$$M_{BOD}(Sanitary) = \frac{3690\ m^3}{d} \times \frac{200\ g}{m^3} \times \frac{kg}{1000\ g} = 738\ kg / d$$

$$M_{BOD}(Food\ P.) = \frac{120\ m^3}{d} \times \frac{950\ g}{m^3} \times \frac{kg}{1000\ g} = 114\ kg / d$$

$$M_{BOD}(Dairy) = \frac{550\ m^3}{d} \times \frac{1200\ g}{m^3} \times \frac{kg}{1000\ g} = 660\ kg / d$$

$$\bar{C} = \frac{\Sigma Q_i C_i}{\Sigma Q_i} = \frac{1512\ kg}{d} \times \frac{d}{4360\ m^3} \times \frac{1000\ g}{kg} = 346.7 = \underline{350\ g / m^3}$$

21.8 INFILTRATION AND INFLOW

Infiltration is defined as run-off water and groundwater leaking into the sewer system through joints, porous walls, manholes or breaks. The extraneous flows that enter a sanitary sewer from sources other than infiltration, such as basement drains, roof leaders and other illegal clean water connections, are referred to as inflow.

The infiltration flows for a given facility can be estimated by comparing the dry weather flow and wet weather flow hydrograph. The difference between the two is due to the excess water. Though infiltration is mostly experienced only during a high storm run-off period, the flow due to infiltration can be as much as twice or greater than that of the dry weather flow. The excessive flows create problems including wash-out of the tanks, poor removals, high chemical costs, flow bypass and flooding of streets and basements. Acceptable values of infiltration and inflow are about 10% of the average daily flow. However, in modern wastewater collection systems, due to better pipe joints and more stringent sewer use by-laws, infiltration and inflow is much less than 10%.

21.9 MUNICIPAL WASTEWATER

Municipal wastewater, consisting of domestic waste, infiltration–inflow and industrial discharges, enters into the sanitary sewer system at one point or the other. Collectors must be designed for specific uses whether the flows are domestic, industrial, infiltration or a combination of all three. The design must be able to handle peak hourly flows. The population on the collectors, as well as types of industrial waste and their characteristics, can be used to calculate expected BOD and hydraulic loadings. In new systems, storm flows are carried by separate sewer systems called storm sewers. Though the collection system is designed based on the peak flow rates, the wastewater treatment plant is designed based on the maximum daily average flow rates.

21.9.1 Hydraulic and Organic Loading

Wastewater reaching a sewage treatment plant varies from hour to hour in terms of quality (BOD, SS) and quantity (flow Q). A typical discharge pattern is shown in **Figure 21.4**. A typical flow pattern or flow hydrograph is diurnal in nature, thereby hydraulic loading and organic loading to a given treatment process vary depending on the time of day.

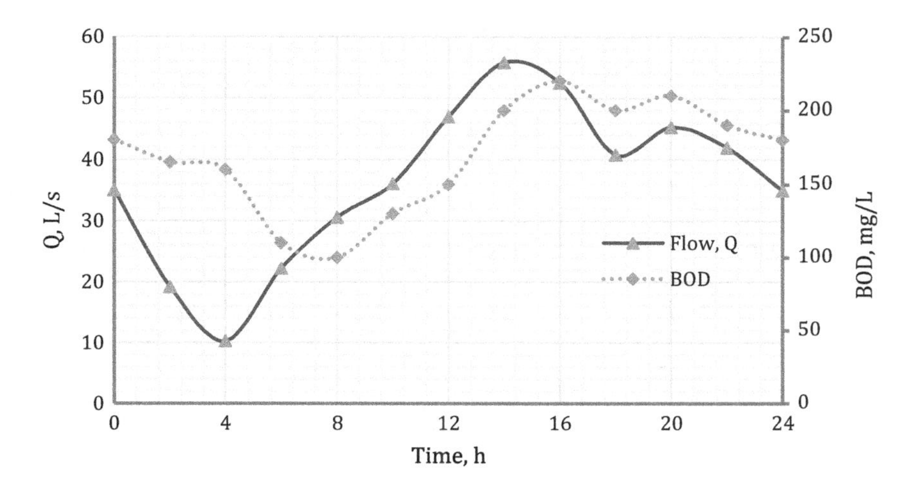

FIGURE 21.4 Flow and BOD Hydrograph

21.9.2 Main Points

The following points should be noted when studying the flow variation at municipal sewage treatment plants:

- Hourly flow rates range from 0.2 to 2.5 times the daily average flow.
- In larger communities, peaks tend to level out, therefore hourly flow ranges from a minimum of 50% to a maximum of 200% of the daily average flow.
- Lowest flow occurs in the early morning at approximately 5 am and maximum flow occurs at approximately noon. In some communities, a smaller second peak flow occurs in the early evening hours (diurnal variation).
- The strength of peak wastewater in terms of BOD follows a similar pattern as flow.
- Summer flows frequently exceed winter flows.
- The flow hydrographs with and without significant infiltration and inflow are respectively called wet weather and dry weather flow hydrographs.
- Large volumes of industrial wastes can distort the typical shape of a flow hydrograph. If the discharge takes place during daytime hours, it can accentuate the peak flows.
- Knowledge of hydraulic and organic (BOD) loading are essential in evaluating the operation of a treatment plant.
- The difference between wet and dry weather hydrograph represents infiltration and inflow contribution.
- A flow hydrograph can be converted to a pollutograph by knowing the concentration of the pollutant.

21.10 EVALUATION OF WASTEWATER

For the accurate evaluation of wastewater to be treated, it is important that the sampling techniques be suited to the task at hand. The samples must also provide an accurate representation of the characteristics to be analyzed. The results from the sample testing are only as accurate as the techniques used and the quality of the sample. Samples can be collected manually or using automatic samplers. Irrespective of the technique used, the sample must be representative. Sampling location should be selected where the flow stream is smooth and well mixed. Scoping the sample from the surface or the bottom would be a biased sample. Programmable automatic sampling is becoming very common. Manual sampling is usually done for spot-checking and similar uses.

The purpose of sampling is to measure loadings on plant processes, removal efficiencies and plant compliance. A sampling program is designed based on the characteristics of the parameter to be analyzed. To provide accurate and representative analysis, proper sampling must be achieved. There are two basic types of samples: grab and composite.

A grab sample is a sample taken at a given place and time. Discrete samples are usually taken for measurements of parameters that are constantly changing or unstable, such as chlorine residuals and dissolved oxygen. Grab samples are also used on parameters that do not change significantly, such as sludge and mixed liquor suspended solids (MLSS).

A composite sample is a combination of individual grab samples. A simple composite sample is prepared by combining equal volumes of individual samples.

A flow proportional sample is made up of individual samples of volume proportional to flow at the time of collection. Composite samples are required for a representative analysis where there is variation in the flow or concentration. A simple composite sample is acceptable when variations are less than 15%. However, a flow proportional sample is the most representative.

21.10.1 Automatic Sampling

Automatic sampling has become very common for monitoring the quality of flow streams both in sewers and at the wastewater treatment plant. Automatic sampling is becoming more common as the cost of purchasing new equipment is getting lower. If correctly programmed, an automatic sampler can collect discrete, composite and flow composite samples. This method eliminates human error and the drudgery of sampling when hourly or 24-hour samples need to be collected for regulatory purposes. Most automatic samplers are equipped with a peristaltic pump to collect the sample. In wastewater collection systems, the information provided by industrial dischargers needs to be verified. The automatic sampler and the flow-measuring device are usually housed in the sampling station through a probe submerged in the flow stream. The composite sample is collected in a single bottle; multiplex bottles collect discrete or composite samples. Timing, frequency of sampling and multiplexing can be achieved by programming the sampler accordingly.

21.10.2 Automatic Compositing

Flow proportional composite samples are slightly more difficult. A sampling device can be connected to a flow meter, which can increase or decrease the number of samples or the sample volume size with an increase or decrease in flow. As mentioned earlier, if a flow meter is not directly connected, the sampler can be programmed to take a series of grab or discrete samples that are mixed manually later in proportion to flow chart readings.

21.10.3 Manual Compositing

To determine aliquot volume for flow proportional samples, hourly flow readings are required. The volume of aliquot required from a given hourly sample is calculated as follows:

$$\boxed{Aliquot,\ A_i = Factor \times Q_i \quad Factor = \frac{V_C}{\Sigma Q_i} = \frac{V_C}{\bar{Q} \times N}}$$

V_C = volume of the composite sample
N = number of individual samples
QQ_i = flow rate in the ith discrete sample
ΣQ = total flow

To make the calculations easy, calculate the constant factor first. To determine the portion of sample for the ith sampling event, simply multiply by the flow for that event. When conditions do not allow for the immediate testing of samples, such as in the case of a composite sample, samples should be properly preserved to maintain the integrity of the sample. Peak flow or maximum daily flows usually occur around noon. The lowest flows occur in the early morning hours.

Example Problem 21.4

A 2.0 L composite sample is to be prepared manually. An automatic sampler was programmed to collect individual samples every hour. Based on the flow data given below, find the aliquot volume to be taken from each hourly sample to make the composite sample.

Time t, h	1	2	3	4	5	6	7	8
Flow Q, ML /d	11	12	15	22	20	15	12	10

SOLUTION:

Multiplication factor

$$F = \frac{V_C}{\Sigma Q_i} = \frac{2.0\,L}{117 ML/d} \times \frac{1000\,mL}{L} = 17.0 = \underline{17 mL/MLD}$$

SAMPLE OF CALCULATIONS (NOON HOUR):

$$A_{12} = F \times Q_{12} = \frac{17\ mL}{ML/d} \times 20 ML/d = 340.0 = \underline{340\ mL}$$

TABLE 21.3
Table of Computations

#	Clock hour	Q, ML/d	Ai, mL
1	8:00	11	190
2	9:00	12	200
3	10:00	15	260
4	11:00	22	370
5	12:00	20	340
6	13:00	15	260
7	14:00	12	200
8	15:00	10	170
	Σ	117	1990

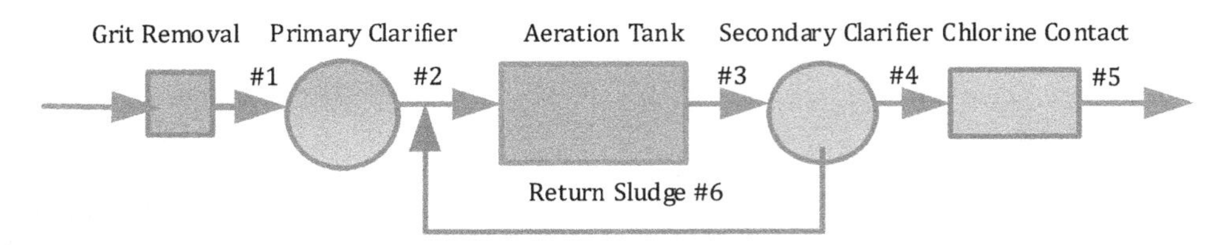

FIGURE 21.5 Standard Sampling Locations

21.10.4 Sample Locations

Accurate sampling and testing are important in plant operation and control. In addition, regulatory samples are collected for compliance. The most common and standard sampling locations are shown in **Figure 21.5**.

1. Raw sewage is sometimes taken after screening if plugging of the sampler is a problem. This composite sample is used to determine the loading to the plant.
2. The second sample location is located after the primary clarifier. This composite sample is used to measure removal by primary clarification and loading on the aeration tank.
3. The third clarification sampling location measures the MLSS. This grab sample is used in the 30-minute settling test, suspended solids concentration test, oxygen uptake test and microscopic survey test. These tests, along with visual observations, are used to operate and control the biological treatment process.
4. The fourth sample location is preferably taken before disinfection but after the secondary clarifier. This composite sample is used to determine the efficiency of the plant operation and the load on the receiving stream of water.
5. The fifth location uses a grab sample to monitor the chlorine residual.
6. The sixth sample location is for sampling the return sludge. A grab sample is required to determine the solids concentration in the activated sludge.

Discussion Questions

1. What are common tests carried out in the laboratory of a wastewater treatment facility? Explain their significance.
2. Distinguish between volatile and non-volatile solids. What is the importance of volatile solids in wastewater treatment?
3. Explain the relationship between BOD and COD. Name another parameter that can be used to estimate BOD other than COD.
4. Explain why inorganic compounds such as $MgCO_3$, which are unstable when exposed to heat, can induce an error in the measurement of volatile solids.
5. Give the characteristics and composition of raw sewage.
6. Name the most important parameters used to characterize sewage and describe their significance.
7. State the general requirements to be observed in ensuring correct sampling.
8. Explain the term population equivalent. In what cases will the BOD population equivalent not be equal to the hydraulic population equivalent?
9. What are the advantages of automatic sampling?
10. Distinguish grab, simple composite and flow proportioned composite types of samples and describe their suitability.

11. As part of wastewater plant operation and control, what are the standard sampling locations in a secondary treatment plant?
12. Sketch a typical wastewater flow hydrograph and explain diurnal variation.
13. Search the internet and find the typical characteristics of raw sewage in India and North America.
14. Compare raw sewage with secondary effluent in terms of temperature, odour, colour and turbidity.
15. Explain why excessive infiltration and inflows are undesirable.
16. What types of industrial wastewater should be allowed to discharge into a municipal sewer system?
17. In developed countries, industrial dischargers with strong wastes are surcharged. What are the typical limits of SS and BOD above which fees are applicable?
18. Define the term nutrients and their role in biological treatment. Does typical municipal wastewater have sufficient or excessive nutrients?

Practice Problems

1. A large dairy discharges 2.6 ML/d with a BOD content of 850 mg/L. Determine the equivalent BOD population at 45 g/c·d? (49 000)
2. The average wastewater flow from a city is 80 ML/d with a BOD content of 280 mg/L. What is the BOD equivalent population at 80 g/c.d? (280 000)
3. Calculate the concentration of total solids in a 50 mL sample from the following data: Empty dish = 64.50 g; Dish and dry sludge solids = 65.07 g (1.1%)
4. A solid test was run on a waste sludge sample and the following observations were made:
   Empty dish = 64.530 g; Dish and wet sample = 144.710 g
   Dish and dried sample = 68.950 g; Dish and ash = 65.735 g
   Determine the concentration of total solids and total volatile solids.
   (5.5%, 4.0%)
5. Based on a suspended solids removal efficiency of 40% at the primary clarifier, determine the expected concentration of SS in the primary effluent. The suspended solid in the raw wastewater is 235 mg/L. What must the minimum removal be to achieve a primary effluent concentration not exceeding 150 mg/L? (140 mg/L, 36%)
6. Determine the concentration of total suspended solids and the volatile fraction in a 25 mL sample from a meat processing plant wastewater based on the following data:
   Filter disc = 0.2170 g; Filter and dry solids = 0.2386 g; Filter and ash = 0.2280 g
   (860 mg/L, 49%)
7. A secondary treatment plant processes an effluent with an SS concentration of 10 mg/L. The average concentration of SS in the raw wastewater composite sample is 195 mg/L. What is the SS removal efficiency of the plant? Assuming 85% to be the solids removal efficiency of the secondary unit, determine the removal efficiency of the primary unit. (95%, 66%)
8. A secondary treatment plant receives a domestic flow of 35 ML/d with a BOD content of 220 mg/L. Food processing and dairy industries contribute 8.0 ML/d of wastewater with a BOD load of 7800 kg. What is the BOD of the municipal wastewater? (360 mg/L)

9. If the contribution of suspended solids and BOD is 100 g and 110 g per capita per day, estimate the population equivalent of 50 kL of industrial wastewater containing 1000 mg/L of suspended solids and 3300 mg/L of BOD. (500, 1500)
10. A paper mill plant effluent contains 50 mg/L of SS.
    a. If, on average, the mill discharges 5000 m^3/day of water into the river, calculate the SS loading into the receiving stream. (250 kg/d)
    b. If the legislative requirement limits the loading to 75 kg/d, what is the maximum content of SS allowed in the effluent? (15 mg/L)
11. A dairy discharges effluent into a city sewer system at a rate of 6.5 ML/d carrying a BOD of 880 mg/L.
    a. What is the equivalent population based on a per capita BOD of 75 g/c·d? (76,000)
    b. If the daily production of wastewater is 55 ML/d with a BOD content of 250 mg/L, what would the average BOD of the wastewater reaching the municipal plant be? (320 mg/L)
12. As shown in **Table 21.4** below, the hourly flow readings and samples of raw wastewater entering the plant were collected.
    a. What percentage of the average flow is low flow and what percentage is peak flow? (34%, 170%)
    b. What sample aliquot is required for the low and peak flow hour to make a 3.0 L flow proportioned composite sample? (42 mL, 220 mL)
13. Suspended solids analyses were conducted on a 25 mL sample of mixed liquor in triplicate. The average values of the three trials are as follows:
    Filter disc = 0.2165 g; Filter and dry solids = 0.2695 g; Filter and ignited solids = 0.2285 g
    a. What is the concentration of suspended solids? (2120 mg/L)
    b. What percentage of solids is volatile (volatile fraction, VF)? (77%)
14. For Practice Problem 13, a sample of mixed liquor was obtained from an activated sludge plant with a total aeration tank capacity of 2.5 ML.
    a. What mass of dry solids is kept in the aeration tanks? (5300 kg)
    b. If volatile solids represent biomass, find biomass. (4080 kg)
    c. What is the mass of inert solids in the aeration? (1220 kg)
15. An automatic sampler needs to be programmed to collect a simple composite sample of 3.0 L. What should the setting for an hourly sample aliquot be if:
    a. The desired composite sample is for a 24-hour period? (125 mL)
    b. The desired composite sample is for an eight-hour period? (375 mL)

TABLE 21.4
Flow Hydrograph (Pr. Prob. 12)

Time	ML/d	Time	ML/d	Time	ML/d
0800	18.5	1600	17.5	2400	13.1
0900	21.8	1700	16.0	0100	10.2
1000	24.7	1800	15.8	0200	8.2
1100	26.9	1900	16.7	0300	7.3
1200	28.1	2000	17.5	0400	5.8
1300	25.5	2100	16.7	0500	5.5
1400	23.3	2200	16.0	0600	7.5
1500	20.4	2300	14.6	0700	11.0

c. If the sample is collected every 15 minutes, what aliquot setting is required to make an eight-hour composite? (94 mL)

16. A certificate of approval of a wastewater treatment facility limits the phosphorus loading to 60 kg/d. The average daily flow of this plant is 75 ML/d.
   a. What is the maximum allowable concentration of phosphorus in the plant effluent? (0.80 mg/L)
   b. To be on the safe side, an operating engineer decides to have a target of 0.50 mg/L. If the raw wastewater contains 7.0 mg/L of phosphorus and primary removal is only 32%, what removal is required by the secondary treatment? (89%)
   c. The secondary unit of the plant is capable of achieving 75% removal. It is suggested to enhance phosphorus removal in the primary unit by chemical precipitation. What is the minimum removal required in the advanced primary treatment? (71%)

22 Preliminary Treatment

As the name indicates, preliminary treatment occurs at the head end of the plant and removes materials that might impair or harm head works or the operation of downstream processes, as shown in **Figure 22.1**. These materials usually consist of wood, rags, plastic and grit. Grit refers to inorganic material, like sand, that is not biodegradable.

Preliminary treatment devices are designed to remove or reduce large solids, grease, scum and grit before any further treatment of sewage. The removal of these materials protects pumps and other treatment devices from possible damage. If the preliminary treatment devices do not function as intended, maintenance costs for pump repairs, digester and clarifier clean-outs, etc., will increase. In addition, the effective capacity of treatment would be reduced due to the space occupied by the inert material. This would result in the lowering of hydraulic efficiency. If grit and other hard and heavy stuff are not removed in the earlier stages of treatment, they can pose serious problems, such as:

- Wearing out pumps, pipes and other equipment faster.
- Mixing with sewage sludge and interfering with their digestion.
- Adulterating the sludge and reducing its manurial and fuel value.
- Occupying a large volume in digestion and air tanks, reducing their effective capacity and blocking passages in channels and sludge pipes.

The devices usually associated with preliminary treatment are briefly discussed.

22.1 SCREENS

Screens are used to remove materials, debris-like rocks, cans, rags, rubber goods, toys, bits of wood, etc. If this material is not screened, it may damage equipment, interfere with the process or result in aesthetically undesirable effluent. Thus, this is the first step in the treatment of wastewater. The two basic types available are coarse screens and fine screens.

22.1.1 Coarse Screens

Coarse screens, commonly called trash rack or bar screens, generally have bars spaced from 2 cm to 15 cm. The screens are usually installed at an angle to facilitate manual cleaning, but some units are available that can be mechanically cleaned. Trash racks are normally installed at pumping stations. In most plants, bar screens with a spacing of 50 mm are used.

22.1.2 Fine Screens

Fine screens were originally used in place of sedimentation tanks. They are not commonly used in sewage treatment because the mesh will accumulate material and plug very quickly, causing what is called head loss in the system. There are other operating and economic problems as well.

DOI: 10.1201/9781003231264-22

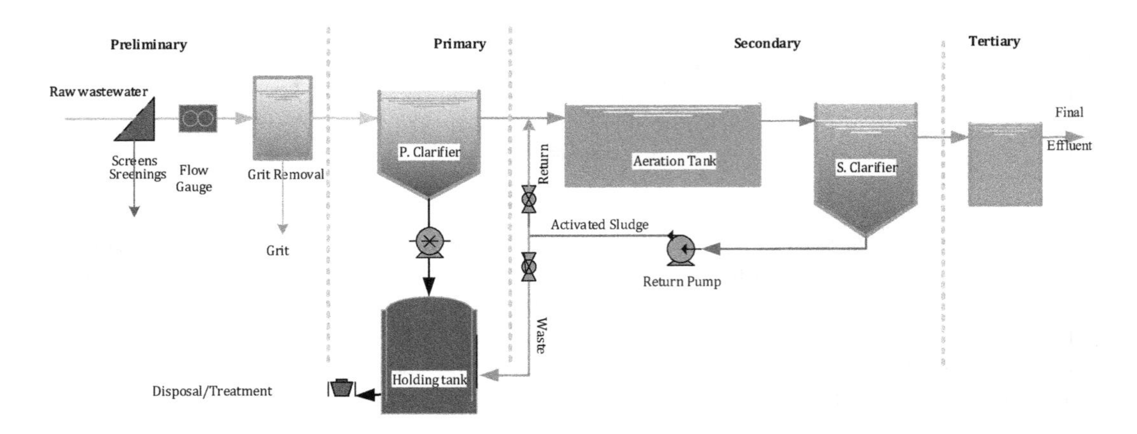

FIGURE 22.1 A Typical Wastewater Treatment Plant

22.1.3 Mechanically Cleaned Screens

A mechanical rake cleans vertical or inclined bar screens. The accumulated material on the screen is pulled up the screen and wiped off into a hopper. Screenings are regularly removed from the hopper to prevent nuisance odours and ensure adequate capacity for incoming screenings.

22.1.4 Cleaning Screens

During dry weather periods, coarse trash racks should be cleaned daily. During storm periods, they should be cleaned two to five times per day to maintain a free flow of sewage through the process. Failure to clean the screens can result in one or more of the following:

- Septic action upstream of the sewer.
- Blockage of the sewer upstream.
- Surcharge of the sewer.
- Shock load on sewage units when the screens are finally cleaned.

Coarse screens, when mechanically cleaned, offer the following advantages:

- Reduced labour costs.
- Better flow conditions in the process.

22.1.5 Volume of Screenings

The quantity of screenings will depend on the characteristics of wastewater and the screen opening. More screenings are required during storm periods and spring run-off. The production of screenings varies from plant to plant. The volume of material or screenings removed is difficult to estimate accurately. Experience has shown that screens with openings of 25 mm to 60 mm will collect between 10 L and 100 L of screenings per million litres (ML) of sewage.

22.1.6 Disposal of Screenings

Screenings may be disposed of by burial, incineration, grinding or digestion. Burying and incinerating are the usual methods of disposal because they are the most economical. Most municipalities use one of these methods for disposal. Screenings are removed in covered containers. When burying a screening, odour may be prevented by sprinkling powdered lime or other odour-control chemicals on the material. An earth cover of 30–60 cm will usually give the best results for bacterial activity. Grinding devices have been used in the past. Ground screenings are redirected to the influent flow for treatment in the process. This method has proven unsatisfactory, however, as it can create digester problems. Screenings received from grinders can cause digester foaming and excessive scum blankets.

22.1.7 Flow Through Screens

To minimize head loss as wastewater passes through screens, it is recommended to keep the screen flow velocity < 1.0 m/s, typically 0.80 m/s. High head losses will allow the screenings to pass through. Head loss is usually not allowed to exceed 15 cm and can be worked out by knowing the approach velocity, v_a, and screen flow velocity, v, as follows:

$$\text{Head loss, } h_L = 1.43\left(\frac{v^2 - v_a^2}{2g}\right)$$

In addition, screens are usually placed at an angle ranging from 30–60 degrees in the direction of flow. By doing this, flow screen area is increased and flow velocity is reduced.

Example Problem 22.1

A screen is designed to provide a flow velocity of 0.80 m/s. The screen bars are 10 mm each with a spacing of 50 mm. Find the head loss when a) the screen is clean and b) the screen is half clogged.

Given: v = 0.80 m/s
gross/open = (5+1)/5 = 6/5
h_L = ?

SOLUTION:

$$v_a = \frac{0.80\ m}{s} \times \frac{5}{6} = 0.66\ m/s$$

$$h_L(clean) = 1.43\left(\frac{v^2 - v_a^2}{2g}\right) = 1.43 \times \left(0.8^2 - 0.667^2\right)\frac{m^2}{s^2} \times \frac{s^2}{2 \times 9.81m} = 0.0134 = \underline{0.013\ m}$$

When the screen is half clogged, the screen flow velocity will be doubled.

$$h_L(clogged) = 1.43\left(\frac{v^2 - v_a^2}{2g}\right) = 1.43 \times \left(1.6^2 - 0.667^2\right)\frac{m^2}{s^2} \times \frac{s^2}{2 \times 9.81m} = 0.153 = \underline{0.15\ m}$$

Under half-clogged conditions, head loss through the screen reaches the limit, so it must be cleaned more frequently.

22.2 COMMINUTION OF SEWAGE

Comminutors, barminutors, or rotogrators are trade names used by different manufacturers to identify their shredding devices. By shredding the screenings rather than removing them, material is put back in the wastewater stream. This may seem attractive at first sight but keep in mind that the material has become part of the wastewater and can interfere with other processes following the screens. One of the main problems associated with comminutors is high maintenance, as the cutting teeth wear out quickly.

A barminutor is a combination of a bar screen and a rotating drum with teeth. The rotating drum travels up and down the bar screen. This piece of equipment is used to shred and grind material small enough to pass through the screens of the grinding unit. Shredders should be installed with a bypass equipped with a bar screen to facilitate the removal of settled material and allow for the inspection of equipment components, such as the cutting edges. Comminuting devices are normally operated continuously and are usually located ahead of the grit removal units.

22.3 FLOW MEASUREMENT

Even though flow measurement is not a wastewater treatment process, its importance can be overemphasized. This information is required to adjust chemical feed rates, the frequency of sludge withdrawal, recirculation rates, airflow rates, plant operation, plant capacity and future planning. Some of the common wastewater flow measuring devices are discussed below.

22.3.1 PARSHALL FLUME

The best equipment for flow measurement in wastewater treatment plants is the Parshall flume (**Figure 22.2**), equipped with an automatic flow recorder and a totalizer. Flow moving freely through the flume can be calculated by measuring flow depth upstream.

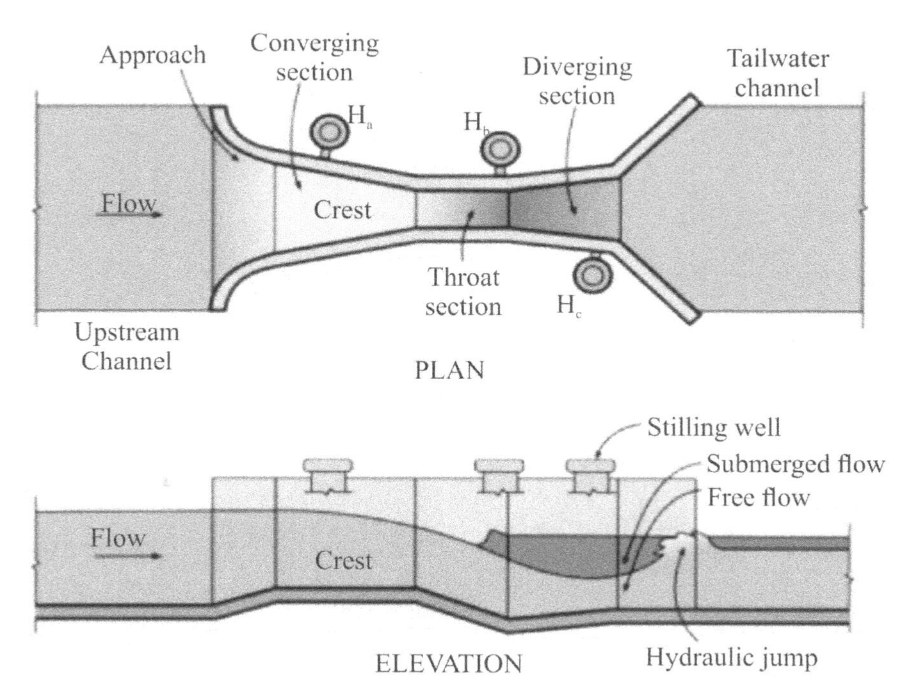

FIGURE 22.2 Parshall Flume

A stilling well is usually provided to hold a float, bubbler tube and other depth sensors such as an ultrasonic sensor, which is connected to a transmitter and a flow recorder. As shown in **Figure 22.2**, the Parshall flume has a converging section, a throat section and a diverging section. The Parshall flume is preferred since it is self-cleaning and causes minimum head loss. The size of the Parshall flume is indicated by the width of the throat section, and its flow equation is written in terms of width of throat and head of water upstream. For free flow conditions, the flow equation has the general form of $Q = KH^n$. However, the values of the coefficient are dependent on the width of the throat section. For flumes with a 30 cm throat, the flow equation is $Q = 0.61H^{1.522} - SI$.

22.3.2 Palmer-Bowlus Flume

Palmer-Bowlus flume is more commonly employed to measure the flow in a sewer pipe since it can fit into half sewer sections. The Palmer-Bowlus flume is usually not available in standard sizes. Its rating equation is $Q = 0.5(H + 0.00177)^{1.9573} \quad - SI$.

22.3.3 Weirs

Weirs or notches can also be employed for flow measurement. The most common shapes are rectangular, triangular (V-notch) and trapezoidal (Cipolletti). Weirs are not preferred for the measurement of raw wastewater since they are prone to solid build-up behind the weir, which affects the flow profile. However, weirs can be used to measure plant effluent flows, which are required by regulation in many cases. A triangular or V-notch weir is more accurate for measuring small flows. For standard sharp-crested triangular weirs, the general flow formula is:

$$Q = \frac{8}{15} \times C_d \times tan\left(\frac{\theta}{2}\right) \times \sqrt{2gH^5} \quad \textit{Any consistent units}$$

Where θ is the notch angle and C_d is the discharge coefficient. The minimum value of the discharge coefficient is 0.58 and the typical value is 0.60.

22.4 GRIT REMOVAL UNITS

Grit, such as sand, stones and gravel, may find its way into a sewer system and be carried by the sewage to the treatment plant. When discussing grit removal, the following points about grit should be kept in mind.

- Grit is inorganic, inert or non-biodegradable.
- Grit is much denser (SG = 2–2.7) than organics (SG = 1.1–1.2).
- Grit is quick settling because of its denseness.
- When combined with other materials like grease, grit can solidify.

Grit removal units are installed after screening equipment in the process to protect mechanical equipment from abrasion, avoid pipe-clogging, and reduce sedimentation load on the primary clarifier and digesters. Grit settling in aeration tanks and digesters occupies dead space and reduces their effective capacity.

Depending on the characteristics of the wastewater and plant operation, some organic matter will always settle out with grit. Since it is more important to remove all the grit at the beginning, some organic matter will always be expected. Due to the presence of organic matter, the grit containers and disposal is similar to that of screenings.

22.4.1 Settling Velocity

Being inorganic, grit follows discrete settling. Since particles are usually >0.2 mm, transitional flow conditions prevail and Stokes' law becomes inapplicable. For particles falling in the range of 0.1–1.0 mm, as in the case of grit particles, settling velocity is given by Hazen's equation.

$$\textit{Settling velocity, } v_s = 418D(G_s - 1)\left[\frac{3T+70}{100}\right]$$

D = diameter of the particle in m
G_s = relative density of grit material
T = temperature in °C

In theory, the overflow rate should be set equal to the settling velocity of the grit particles. Since there is always some short-circuiting, the overflow rate chosen is lower than the settling velocity. Actual settling velocity will be less than as given by these equations since the real particles are not spherical.

$$\textit{Modified Hazen equation}; v_s = 61D(G_s - 1)\left[\frac{3T+70}{100}\right]$$

Example Problem 22.3

A grit chamber is to be designed to handle a peak flow of 65 ML/d. Assume a grit particle size of 0.20 mm and a relative density of 2.5. Assume a temperature of 20°C. Select the size of the grit chamber.

Given: Q = 65 ML/d
D = 0.20 mm
G_s = 2.5
T = 20°C

SOLUTION:

$$v_s = 61D(G_s - 1)\left[\frac{3T+70}{100}\right] = 61 \times 0.0002\ m \times 1.5 \times \left[\frac{60+70}{100}\right] = 0.0292 = \underline{0.029\ m/s}$$

Since the grit chamber is not 100% efficient, the overflow rate, v_o, is chosen as 60% of v_s.

$$A_s = \frac{Q}{v_o} = \frac{65\ ML}{d} \times \frac{1000\ m^3}{ML} \times \frac{s}{0.60 \times 0.029\ m} \times \frac{d}{24\ h} \times \frac{h}{3600\ s} = 43.2 = \underline{43\ m^2}$$

Select two channels for flexibility. Assuming an effective depth of 1.0 m and maximum flow velocity of 0.30 m/s, the desired cross-section of the grit channel is:

$$A_X = \frac{Q}{2 \times v_H} = \frac{65000\ m^3}{d} \times \frac{s}{2 \times 0.30\ m} \times \frac{d}{24 \times 3600 s} = 1.25\ m^2$$

$$Width, B = \frac{A_X}{d} = \frac{1.25\ m^2}{1.0\ \ m} = 1.25 = \underline{1.3\ m}$$

$$L = \frac{A_S}{B} = \frac{43\ m^2}{2 \times 1.3\ \ m} = 16.53 = \underline{17\ m}$$

$$H = 1.0\ m + 0.25\ m\,(f.b.) + 0.25\ m(storage) = \underline{1.5\ m}$$

22.4.2 Grit Channels

Grit particles will settle faster than organic putrescible solids because they are heavier. Grit channels are usually designed to maintain a flow velocity of 0.3 m/s at design flow, which is usually sufficient to keep the organic matter in suspension while allowing the heavier particles to settle. Grit channels are usually rectangular and velocity control is achieved by installing a proportional weir at the effluent end of the channel. In some cases, velocity control can be achieved by changing the shape of the channel. However, in such cases, a flow control device like a Parshall flume is installed at the end. There is usually more than one channel to accommodate peak flows and for cleaning purposes.

22.4.3 Aerated Grit Chamber

Grit chambers using air to separate the lighter materials from the heavier ones are called aerated grit chambers. Sewage flows into the aerated grit chamber and the heavier particles settle to the bottom as the sewage rolls in a spiral motion from entrance to exit. The lighter organic particles eventually roll out of the tank. The grit at the bottom of the tank is directed to a grit hopper where it is removed by a clamshell bucket or air lift unit.

Aerated grit chambers are designed to provide 3 to 5 minutes of detention time based on the peak flow. Aeration keeps the organics in suspension and helps to freshen up the wastewater. The velocity of the rolling action in the chamber dictates the removal efficiency. If too slow, organics will settle out. If the rolling velocity is properly adjusted, the settled material is usually very clean of organics and does not need any washing.

22.4.4 Detritus Tank

Short-period sedimentation in a tank that operates at substantially constant levels produces a mixture of grit and organic solids called detritus. A detritus tank is basically a grit clarifier. The lighter organic solids are subsequently removed from or washed out of the mixture. Several manufacturers specializing in sewage disposal equipment have perfected this type of equipment. For example, one such unit not only removes the grit but also washes it.

The grit-collecting mechanism is installed in a square, shallow, concrete tank with filled-in sloping corners. Sewage enters along one side of the tank through adjustable vertical gates, which are set to provide a uniform influent velocity across the entire width of the unit. The sewage then flows in straight lines across the tank and overflows at a weir constructed along the outlet side of the tank.

The collecting mechanism consists of two structural steel arms attached to a vertical shaft and fitted with outward raking blades with scoops on the ends. As the rakes revolve, settled grit is ploughed outward to the radius where the end scoops collect and discharge it to a hopper at one side of the tank.

22.4.5 Cyclone Separators

Grit removal is possible by mechanical means such as a centrifugal unit. Centrifugal units are usually liquid cyclones. The wastewater is introduced tangentially into a cylindrical, conical housing. The heavier, larger particles of grit are thrown to the outside wall and collected for disposal. Cyclone separators are very compact and are inexpensive compared to channels and grit chambers. One of the main disadvantages is their flexibility of operation. It is difficult to regulate the quality of the grit that is removed by the unit.

22.4.6 Grit Disposal

Average figures indicate that from 20 L to 50 L of grit can be collected per million litres of sewage. The grit must be removed before it is carried by the stream flow into the primary clarifier, digester or chlorine contact chamber. It is good practice to periodically check that the grit is not carried to the clarifier, digester or chlorine contact chamber, as there it would still have to be removed but with much more difficulty and expense. The grit removal facility can reduce unnecessary maintenance costs more than any other unit. If these facilities malfunction because of problems or improper operation, the result will be plugged lines, abraded impellers and grit-filled treatment tanks.

The disposal of grit is usually done by burial or dumping at the municipal dump or in a landfill site. If the grit is adequately washed (having less than 3% volatile solids remaining as determined by lab tests), it may be used as fill around the plant or to re-sand sludge drying beds.

22.5 PRE-AERATION

Aeration basins may precede or follow screens and grit chambers. In general, pre-aeration tanks are designed for detentions of 5 to 15 minutes for grease removal, using 0.1–1.0 L of air per cubic metre of sewage treated. If flocculation of the fine suspended solids in the raw sewage is also attempted, the detention period must usually be extended to at least 15 minutes up to as much as 60 minutes, the average time being about 30 minutes.

Pre-aeration of raw wastewater is becoming common, especially in cities and towns where the sewage gets septic before it reaches the plant. Raw sewage is aerated for one or more of the following purposes:

1. To remove gases from the sewage, especially hydrogen sulphide, which creates odour problems and increases the chlorine demand of sewage.
2. To promote floatation of excessive grease, which can then be removed from the raw sewage early in its treatment.
3. To aid in the coagulation of colloids (finely divided suspended solids) in the raw sewage to obtain a higher removal of suspended solids by primary settling.
4. To freshen the wastewater by increasing the oxygen content of the wastewater and by eliminating odours.
5. To enhance settling. The addition of air in water reduces its density and thus accelerates gravity settling.
6. Pre-aeration is more economical in the activated sludge process since aeration system is in place for that process.

22.6 PROCESS CALCULATIONS

In addition to detention time and overflow rate, horizontal flow velocity calculations are made to control the operation of grit removal units, especially grit channels. The relationship between

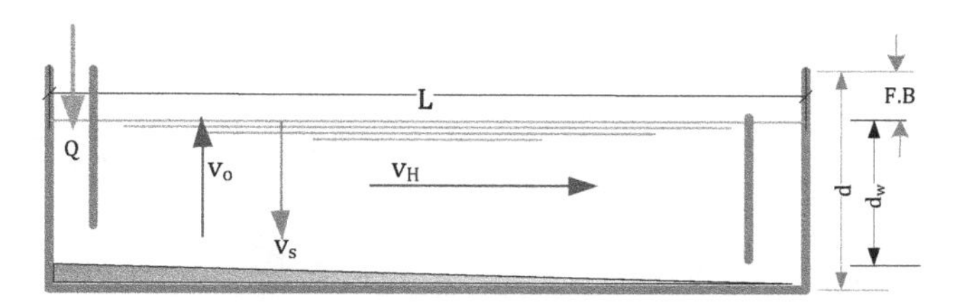

FIGURE 22.3 Flow Through a Grit Channel

settling velocity and flow velocity can be used to determine what size particles will be removed (**Figure 22.3**).

$$\text{Detention time, } t_d = \frac{L}{v_H} = \frac{d}{v_o} = \frac{V}{Q} \quad \text{and} \quad v_H = \frac{Q}{A_X} = \frac{L}{t_d}$$

v_H = horizontal flow velocity
v_o = overflow rate
L = length of the chamber/channel
W = width
d = depth
H = height = d + f.b. + storage
A_x = B × d = area of cross section
A_S = L × B = surface area

Example Problem 22.4

A grit chamber is designed to remove grit material with an average particle size of 0.20 mm and a settling velocity ranging from 16–22 mm/s. By employing a proportional weir at the exit end, a flow velocity of 0.30 m/s will be maintained. Determine the dimensions of the grit channel to carry a peak flow of 9.5 ML/d.

Given: Q = 9. 5 ML/d d = 1.0 m assumed
v_s = 0.02 m/s
v_H = 0.30 m/s
L, B = ?

SOLUTION:

$$\text{Sectional } A_x = \frac{Q}{v_H} = \frac{10\ ML}{d} \times \frac{s}{0.30\ m} \times \frac{1000\ m^3}{ML} \times \frac{d}{24\ h} \times \frac{h}{3600} = 0.385\ m^2$$

$$\text{Width, } B = \frac{A_x}{d} = \frac{0.385\ m^2}{1.0\ m} = 0.385\ m = \underline{0.40\ m}\ \text{say}$$

$$\text{Detention time, } t_d = \frac{d}{v_s} = 1.0\ m \times \frac{s}{0.02\ m} = 50.0 = \underline{50\ s}$$

$$Length,\ L = v_H \times t_d = \frac{0.30\ m}{s} \times 50.0\ s = 15.0 = \underline{15\ m}$$

A grit channel measuring 15 m × 0.40 m × 1.0 m + f.b. will do the job.

Discussion Questions

1. Differentiate between unit operations and unit processes.
2. Explain why modern treatment plants are at least secondary. What biological treatments are commonly employed in the treatment of municipal wastewater?
3. Which parameter is used to describe surface loading, and how is it related to detention time?
4. What is the first unit operation in the treatment of wastewater? Discuss it.
5. Discuss the importance of grit removal at the beginning of wastewater treatment.
6. What are the major sources of grit in municipal wastewaters? What treatment methods are commonly used to remove grit?
7. Grit settling is discrete particle settling. However, Stokes' law is not applicable. Explain why.
8. What is a velocity control device? What is the most common device used for this purpose?
9. In what situations should pre-aeration grit removal chambers be recommended?
10. Discuss the importance of flow measurement as part of preliminary treatment.
11. What flow measuring devices are commonly employed in wastewater treatment plants?
12. The Parshall flume flow formula in USC units is $Q = 4WH^{1.522W^{0.026}}$. Modify this formula for SI units when the width of the flume is 2.0 ft.
13. Modify the triangular weir general equation assuming a C_d of 0.60 and the angle of the weir is 90°.
14. List the advantages of pre-aeration.

Practice Problems

1. A channel-type grit chamber has a flow velocity of 0.30 m/s, a length of 10 m and a depth of 0.90 m. At what settling velocity can grit particles be completely removed? (0.03 m/s)
2. A grit channel has a flow velocity of 0.28 m/s, a length of 10 m and a depth of 1.0 m. For grit with a specific gravity of 2.5, determine the smallest diameter grit particles that can have 100% removal? (0.028 m/s, 0.24 mm)
3. A channel-type grit chamber is to be installed in a water pollution control plant with a daily flow of 9500 m^3/d. The flow velocity is controlled by an effluent weir at a rate of 0.32 m/s. What should the channel section be for a depth-to-width ratio of 1:1.5? (0.75 m × 0.50 m + f.b)
4. A grit chamber is designed to remove grit particles with an average size of 0.2 mm and a specific gravity of 2.65 when the average sewage temperature is 20°C. A flow-through velocity of 25 cm/s will be maintained by installing a proportional weir at the exit end. What size channel is needed for a carrying peak flow of 12 ML/d? Assume the design overflow rate to be 60% of settling velocity and flow depth of 1.0 m. (v_s = 2.6 cm/s, v_o = 1.6 cm/s, L = 16 m, W = 0.6 m, d = 1.0 m + f.b.)

5. Design a grit chamber to remove grit particles with a settling velocity of 20 mm/s, typical of inorganic particles of size 200 µm and an SG of 2.65. Assume a parabolic flume at the effluent end will maintain a flow velocity of 0.30 m/s. Design the channel to treat a maximum flow of 12 ML/d. Assume a water depth of 1.1 m in the channel. Draw the plan view and cross-section of the designed channel. (L = 13 m, W = 0.4 m)
6. For a 90° notch, what is the head achieved for a flow of 150 L/s passing through it? (41 cm)
7. The flow equation for a Parshall flume in SI units is $Q = 2WH^{1.57W^{0.026}}$. Compute the flow rate when head is 0.24 m and width is 60 cm. (0.13 m^3/s)
8. A bar screen is fitted with 10 mm thick bars at a spacing of 40 mm and designed to maintain a screen velocity of 0.80 m/s when clean. What would the head loss through the screen be when the screen is 1/3rd clogged? (7.5 cm)

23 Primary Treatment

The main purpose of primary treatment is to separate and remove settleable and floatable solids and allow for the thickening of sludge. The principle of primary settling is to slow the flow of sewage in the clarifier to below 10 mm/s so that solids can settle. Since this is primarily a physical process it is known as operation. Settling under gravity can result in the removal of 40% to 60% of suspended solids and 30% to 40% of BOD.

Chemical precipitation addition with alum or other coagulants can remove much of the colloidal material, bringing the suspended solids removed to 80% or 90%. Other benefits of primary settling include equalization of side stream flows and removal of BOD as part of settleable solids. However, the quantity of sludge will be higher and the chemical sludge may be hard to dewater. The basic components of sedimentation tanks include baffles to slow down influent flow while overflow weirs even out effluent flow. Mechanical skimmers collect scum from the surface and deposit it in scum pits. Collector arms or skimmers then scrape the settled sludge into a hopper. The two main designs of primary clarifiers are shown in **Figure 23.1**.

23.1 FACTORS AFFECTING SETTLING

Many factors will influence settling characteristics in a particular clarifier. The most important ones include temperature, short-circuiting, detention time, surface loading (overflow rate), solids loading and weir loading.

23.1.1 Temperature

In general, settling improves with an increase in temperature. This is because water becomes denser at a lower temperature. The density of water reaches its maximum at 4°C. Another factor that contributes to poor settling in cold water is an increase in the viscosity of the water. Additional detention time compensates for a low settling rate. Due to low temperatures in winter, wastewater and sludge can be held for longer without going septic.

23.1.2 Short-Circuiting

When the flow velocities are non-uniform, short-circuiting occurs. A high flow velocity in a given section may decrease the detention time. Short-circuiting can easily begin at the inlet end of the settling tank. This is usually prevented using weir plates, baffles and a properly designed inlet channel. Short-circuiting may also be caused by density currents or stratification due to temperature or density. When warm influent flows across the top of cold water in a clarifier, short-circuiting will occur.

23.1.3 Detention Time

Water should be held in the settling tank long enough to allow the settling of solid particles. Hydraulic overloading in a tank will result in poor quality effluent due to the carryover of solid particles. However, too much detention time in primary clarification can cause septicity. Most

DOI: 10.1201/9781003231264-23

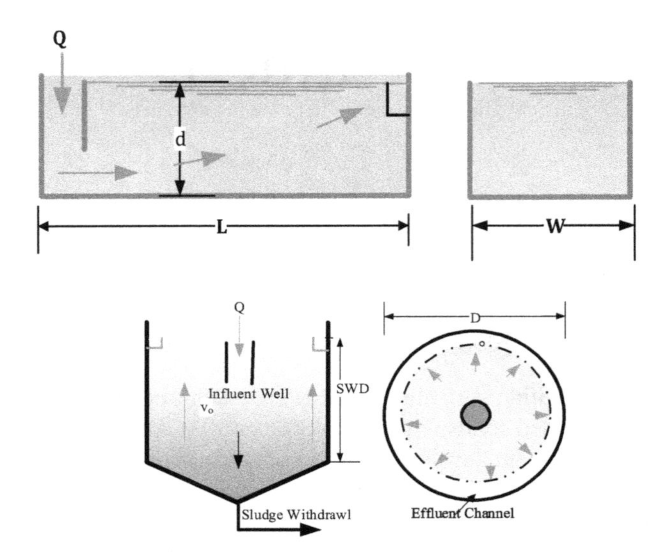

FIGURE 23.1 Rectangular and Circular Settling Tank (Clarifier)

engineers design clarifiers to provide a detention time of one to two hours. The actual value found by a dye tracer study is less than the calculated value. In summary:

- Detention time during peak flow hours is reduced.
- Shock hydraulic loading due to excessive storm flows can wash out the primary tanks.
- A longer detention time does not improve removal and might impair settling due to septic conditions and the generation of gases, and is undesirable for biological processes.

23.1.4 Surface Settling or Overflow Rate

Overflow rate indicates the rate at which water will rise in the tank before it exits and is defined as flow per unit surface area of the tank. The settling rate or velocity should be greater than the overflow rate. Overflow rates vary in the range of 16 to 32 $m^3/m^2{\cdot}d$, with 24 $m^3/m^2{\cdot}d$ being the typical value.

Although the settling of granular or discrete solids depends primarily on the surface loading, the settling of flocculent solids, as in activated sludge and chemical sludge, may also depend on tank volume and depth. Due to flocculated solids, a thicker sludge blanket is usually maintained. This typically requires the secondary tanks to be designed relatively deeper.

For a given clarifier, hydraulic detention time, $t_{d,}$ and overflow rate, v_O, are related terms. For design purposes, only one is used. In terms of overflow rate and side water depth, d, hydraulic detention time can be found as follows:

$$\text{Hydraulic detention time, } t_d = \frac{V}{Q} = \frac{d}{v_O} = \frac{L}{v_H}$$

V = volume
Q = flow rate
d = side water depth
v_o = overflow rate

TABLE 23.1
Design Parameters for Settling Tank

Types of Settling	Overflow Rate m3/m2·d		Solids Loading kg/m2.day		Depth, m	Detention time, h
	Average	Peak	Average	Peak		
Primary settling only	25–30	50–60	-	-	2.5–3.5	2.0–2.5
Primary settling with secondary treatment	35–50	60–120	-	-	2.5–3.5	
Primary settling with activated sludge return	25–35	50–60	-	-	3.5–4.5	-
Secondary settling for trickling filters	15–25	40–50	70–120	190	2.5–3.5	1.5–2.0
Secondary settling for activated sludge	15–35	40–50	70–140	210	3.5–4.5	-
Secondary settling for extended aeration	8–15	25–35	25–120	170	3.5–4.5	-

23.1.5 Weir Loading

Wastewater leaves the clarifier by flowing over weirs and into an effluent trough or channel, as shown in **Figure 23.1**. The length of the weir over which the wastewater flows should be long enough to prevent short-circuiting and the carryover of solids due to excessive velocities.

Weir loading is expressed in terms of flow per linear metre of the weir. Most designers recommend 125 $m^3/m{\cdot}d$ to 250 $m^3/m{\cdot}d$. Secondary clarifiers have a high effluent quality requirement and generally need lower weir overflow rates. In some situations, additional weir length is achieved by providing an inboard channel. These will almost double the weir length for a given circular clarifier. The inboard effluent channel is commonly used in secondary clarifiers where low weir loading is desired.

Example Problem 23.1

A wastewater flow of 3500 m^3/d enters a plant with two rectangular clarifiers. Each clarifier is 5.0 m wide, 9.0 m long and 3.0 m deep. Calculate the surface settling rate and hydraulic detention time. What is the minimum length of the weir needed not to exceed a weir overflow rate of 125 $m^3/m{\cdot}d$?

Given: Q = 3500 m^3/d (# = 2) = 3500/2 = 1750 m^3/d
L = 9.0 m
W = 5.0 m
d = 3.0 m

SOLUTION:

$$v_O = \frac{Q}{A_S} = \frac{1750\ m^3}{d} \times \frac{1}{(9.0\ m \times 5.0\ m)} = 38.8 = \underline{39\ m^3/m^2.d}$$

$$t_d = \frac{d}{v_O} = 3.0\ m \times \frac{d}{38.8\ m} \times \frac{24\ h}{d} = 1.85 = \underline{1.9\ h}$$

$$L_W = \frac{Q}{Loading\ rate} = \frac{1750\ m^3}{d} \times \frac{m.d}{125\ m^3} = 14.0 = \underline{14.0\ m}$$

Example Problem 23.2

Design a suitable rectangular clarifier for treating a wastewater flow of 10 ML/d to provide a detention time of 2.0 hours with a flow velocity not exceeding 5.0 mm/s. Make appropriate assumptions.

Given: Q = 10. ML/d
t_d = 2.0 h
v_H = 5.0 mm/s
L, B, d = ?

SOLUTION:

$$Capacity,\ V = Q \times t_d = \frac{10\ ML}{d} \times \frac{1000\ m^3}{ML} \times 2.0\ h \times \frac{d}{24\ h} = 833.3\ m^3$$

$$L = v_H \times t_H = \frac{5.0\ mm}{s} \times \frac{m}{1000\ mm} 2.0\ h \times \frac{3600\ s}{h} = 36.0 = \underline{36\ m}$$

Assume the sidewall depth to be 3.0 m. To provide space for sludge accumulation at the bottom of the tank, the overall depth of the tank will be 3.5 m.

$$B = \frac{A_s}{L} = \frac{V}{d \times L} = \frac{833.3\ m^3}{3.0\ m \times 36\ m} = 7.715 = 7.7\ m = \underline{8.0\ m\,(say)}$$

A rectangular tank with overall dimensions of 36 m × 8.0 m × 3.5 m will meet the requirements.

Example Problem 23.3

Design a primary circular clarifier to treat a flow of 3.0 ML/d such that the overflow rate does not exceed 30 $m^3/m^2 \cdot d$. Work out the detention time achieved assuming the effective depth of the tank is 3.0 m, and the weir loading rate assuming there is an effluent weir along the periphery of the clarifier.

Given: Q = 3.0 ML/d
t_d = ?
v_o = 30 m/d
D = ?

SOLUTION:

$$Surface,\ A_s = \frac{Q}{v_o} = \frac{3.0\ ML}{d} \times \frac{1000\ m^3}{ML} \frac{d}{30\ m} \times \frac{24h}{d} = 2400.0\ m^2$$

$$D = \sqrt{\frac{4A_s}{\pi}} = \sqrt{\frac{4 \times 2400\ m^2}{\pi}} = 55.27 = \underline{55\ m}$$

$$t_d = \frac{d}{v_o} = 3.0\ m \times \frac{d}{30\ m} \times \frac{24\ h}{d} = 2.40 = \underline{2.4\ h}$$

$$WLR = \frac{Q}{L_W} = \frac{3.0\ ML}{d} \times \frac{1}{\pi \times 55\ m} \times \frac{1000\ m^3}{ML} = 17.16 = \underline{17\ m^3 / m.d}$$

23.1.6 Settling Characteristics of Solids

In municipalities where there is significant industrial discharge contributing to municipal wastewater, the concentration and nature of the solids may be quite different. If the industrial effluents contain heavy solids, it may help to improve the efficiency of primary clarification. Chemical precipitation of phosphorus as part of primary treatment can also enhance solids removal. However, this results in increased sludge volume and chemical dosage rates, and would affect the dewatering characteristics of the sludge.

23.1.7 Solids Loading

Solids loading rate is more commonly used for secondary clarifiers, where the settling of biological floc is carried out. Similar to overflow rate, solid loading rate is expressed as mass of solids per unit surface area. One important difference in secondary clarifiers is increased solids due to recirculated flows. However, recirculation rates can be controlled. Solids flux or loading rate, SLR can be found as follows:

$$\boxed{SLR = (Q + Q_R) \times \frac{MLSS}{A_S}}$$

MLSS = concentration of solids in the mixed liquor
Q_R = recirculation or return rate

23.2 RECTANGULAR CLARIFIER

Rectangular clarifiers are more commonly used for primary treatment. In the past, the sludge collection mechanism was a chain and sprocket arrangement that required high maintenance. These days a travelling bridge is more common. Sludge is pushed towards the head end and clarified water flows over the weirs at the exit end. The main components of a rectangular clarifier are described in **Table 23.2**.

23.3 CIRCULAR CLARIFIER

Circular tanks can be either centrally fed or peripherally fed. In general, sludge collection mechanisms in circular tanks are operated over longer periods than collectors in rectangular tanks. Collectors

TABLE 23.2
Components of a Rectangular Clarifier

Component	Function
Tank structure	Usually made of concrete, the floor is horizontal or slopes towards the head end to allow for the collection of sludge into the hopper at the head end side.
Influent end	Pipes or channels distribute water evenly along the shorter side of the clarifier as the flow is along the length. There is always some sort of flow control such as gates, valves or weirs.
Inlet baffle	Plays an important role in distributing the water evenly and preventing short-circuiting. Small ports near the surface allow floating material to pass through.
Sludge collector	Collectors operating in rectangular tanks consist of two endless chains operating on sprocket wheels and supporting wood crossbars or flights. The flights push the sludge to a hopper at the end as they move slowly along the bottom. The system is designed so that flights move across the surface on the return trip and move floating solids to the inlet end.
	In modern plants, sludge is moved to the hopper by a scraper attached to a moving bridge on rails. This reduces the number of moving parts and hence maintenance problems as experienced in earlier systems.
Sludge hopper	Receives sludge scraped from the floor and is located at the head end. To provide room for sludge, the head end is deeper.
Scum trough	Scum is collected in a trough that extends along the width of the tank.
Sludge line	A line fitted with a valve is provided to pump the sludge out.
Effluent weir	Effluent from the clarifier flows over the weirs into the effluent channel. The role of the weir is to even out the flow to reduce flow velocity and prevent the carryover of solids.
	There are different kinds of weirs, including flat, fingers and v-notch shaped. They should be kept clean so that flow remains even and short-circuiting is prevented.
Launder	This is the effluent trough into which weirs overflow. Flow from the trough goes to the next process.

should be run often enough to prevent a build-up of solids in the tank from causing an undue load on the mechanism at start-up and damaging the equipment (**Table 23.3**).

23.4 SCUM REMOVAL

Foreign matter that rises to the surface forms scum and should be pumped out of the tank before pumping the sludge, if possible. By doing this, any grease remaining in the pipes will be scoured by the sludge when it is removed. The removal of scum, floating garbage and grease is essential for the efficient operation of settling tanks.

A scum barrier or baffle is generally provided in the flow path between the centre of the tank and the effluent weir. Excessive skimming will result in too much water being carried out with the scum, while insufficient skimming will permit scum to flow around or under the baffle and escape with the tank effluent. Scum must be removed daily, and ideally small amounts should be removed continually rather than a large batch at one time.

23.5 SECONDARY CLARIFIER

In an activated sludge flow process scheme, activated sludge is continuously shifting between the aeration tank and the secondary clarifier. The activated sludge is less dense; hence the secondary

TABLE 23.3
Main Components of a Circular Clarifier

Component	Function
Tank structure	A cylindrical tank with a conical bottom. The bottom slope is about 8%. In the centre at the bottom is the hopper.
Influent line	Usually runs under the tank and rises vertically in its centre to feed it. There is always some sort of flow control such as gates or a valve.
Influent well	A cylindrical baffle forces the water to flow down and then in the radial direction towards the effluent weir. Its main function is distributing the water evenly and preventing short-circuiting. Small ports near the surface allow floating material to pass through.
Sludge collector	A continuously moving scraper arm at the bottom scrapes the sludge towards the centre and into the hopper. A motor and reduction gear at the centre usually drive the sludge collector.
Access bridge	This bridge lies over the weir to provide access to the centre for maintenance and operation.
Skimmer arms	Skimmer arms rotating at the top surface move the floating material to the periphery of the tank. Scum is collected in a trough.
Sludge line	A line fitted with a valve is provided to pump the sludge out.
Effluent weir	Weirs in circular clarifiers are along the periphery of the clarifier. Weirs can be adjusted to maintain even flow.
Launder	This is the effluent channel trough into which weirs overflow. Flow from the effluent channel goes to the next process.

clarifiers are more sensitive to loading than primary clarifiers. The sludge blanket is typically about one-third of the depth of the clarifier. For this reason, engineers are more conservative in the design of secondary clarifiers.

Sludge removal mechanisms in secondary clarifiers tend to be different from most primary clarifier mechanisms, especially those in circular clarifiers. These secondary clarifiers are designed for continuous sludge removal by a hydrostatic system. The hydrostatic system creates suction to capture settling sludge faster. Remember, we need to pump the activated sludge back to the aeration tank as fast as possible. Too much retention time can turn the aerobic process into an anaerobic process.

Whereas primary clarification is mainly flocculent settling, secondary clarification is mainly zone settling since the influent to the secondary clarifier is mixed liquor that comprises high concentrations of biological floc. For this reason, the sludge blanket is relatively high and the whole thing settles as one mass unit, hence the name zone settling. At the bottom of the clarifier, settling changes to compression due to the mass load of overlying layers and can squeeze water out of the pores of floc particles. Since biological solids are lighter, rising gases due to denitrification can easily buoy them up to float on the surface. If there is a dominance of filamentous microorganisms, there is sludge bulking which expands the sludge blanket. This results in poor settleability and increases the concentration of solids in the secondary effluent. As mentioned aboveearlier, solids loading rate must be considered in the operation and design of secondary clarifiers.

23.6 CHEMICAL PRECIPITATION

Chemical precipitation is a modified sedimentation process in which a coagulant is employed to improve the efficiency of settling. Chemicals used include alum, ferric chloride and lime. Proper mixing of these is essential, and the dosage will vary according to the characteristics of the sewage being treated. Efficiencies of 80% to 90% removal of suspended solids and 50% to 55% removal of

BOD are common when using chemical precipitation. However, chemical precipitation is expensive and produces a high volume of sludge that may be difficult to dispose of, further increasing operating costs. When using chemical precipitations, as more sludge is produced, sludge withdrawal rates will be affected. If residual coagulant is significant, it will affect secondary treatment too.

23.7 SLUDGE HANDLING

Sludge handling operations are governed by quality in terms of solids concentration or thickness of the sludge, and quantity in terms of volume of sludge pumped. The volume of sludge produced indirectly depends on the thickness achieved and the solids removal efficiency of the primary clarifier. A thin sludge with solids concentrations as low as 0.5% is acceptable if sent to a thickener. However, this may not be acceptable if the downstream process is a digester or a dewatering unit. Under proper operating conditions, an operator should be able to obtain a solids concentration of 4–6%. The process control requires finding a sludge blanket level that will produce sludge of the desired consistency without adversely affecting removal, overloading the collector arm, and causing septicity.

As a general rule of thumb, sludge should be pumped more frequently for smaller durations as this allows for higher consistency and prevents septicity. Estimation of sludge production is illustrated in Example Problem 23.4.

Example Problem 23.4

In a wastewater treatment plant, the average daily flow to each primary unit is 4.5 ML/d containing 250 mg/L of solids. Assuming a solids removal of 45% by settling, work out the volume of primary sludge produced assuming the consistency of the raw sludge is 3.5%.

Given: Q = 4500 m^3/d
PR = 45%
SS = 250 mg/L
SS_{sl} = 3.5% = 35 kg/m^3

SOLUTION:

$$\text{Solids removed, } M_{SS} = Q \times SS_i \times PR = \frac{4.5\ ML}{d} \times \frac{250\ kg}{ML} \times \frac{45\%}{100\%} = 506.2 = 510\ kg/d$$

$$\text{Sludge volume} = \frac{M_{SS}}{SS_{sl}} = \frac{506.2\ kg}{d} \times \frac{m^3}{35\ kg} = 14.4 = \underline{14\ m^3/d}$$

Example Problem 23.5

Design a secondary circular clarifier to treat an average daily flow of 45 ML/d such that the overflow rate does not exceed 24 $m^3/m^2{\cdot}d$. Assume an MLSS of 2500 mg/L and a return sludge ratio of 25% and permissible loading of 100 $kg/m^2{\cdot}d$.

Given: Q = 45 ML/d
$Q_R = 0.25Q$
$v_o = 24\ m^3/m^2{\cdot}d$
SLR = 100 kg/m²·d
D = ?

SOLUTION:

$$A_s = \frac{Q}{v_o} = \frac{45\ ML}{d} \times \frac{1000\ m^3}{ML} \frac{d}{24\ m} = 1875 = 1880\ m^2$$

$$A_s = \frac{(Q+Q_R)\times MLSS}{SLR} = \frac{1.25\times 45\ ML}{d} \times \frac{2500\ kg}{ML} \times \frac{m^2.d}{100\ kg} = 1406 = 1410 < 1880\ m^2$$

$$D = \sqrt{\frac{4A_s}{\pi}} = \sqrt{\frac{4\times 1875\ m^2}{\pi}} = 48.79 = \underline{49\ m}$$

Two identical clarifiers are suggested.

$$D = \frac{D}{\sqrt{2}} = \frac{48.79\ m}{\sqrt{2}} = 34.5 = 35\ m$$

$$WLR = \frac{Q}{L_W} = \frac{0.5\times 45\ ML}{d} \times \frac{1}{\pi\times 35\ m} \times \frac{1000\ m^3}{ML} = 204 = \underline{210\ m^3/m.d}$$

Weir loading is on the higher side, so an inboard channel with a weir on both sides is suggested.

Discussion Questions

1. Explain why it is not necessary to define both detention time and overflow rate in the design of sedimentation tanks.
2. Compare circular tanks and rectangular tanks for primary treatment.
3. Give the advantages and disadvantages of chemical precipitation.
4. Describe the various operating parameters for primary clarifiers and the normal range of values.
5. Describe the dominant types of settling in primary clarifiers versus secondary clarifiers.
6. Describe the additional parameters that are important in the design and operation of secondary clarifiers in the activated sludge process.
7. Compare primary clarifiers with grit clarifiers.
8. Explain why primary treatment should be followed by secondary treatment.
9. Describe the main components of a rectangular clarifier and a circular clarifier.
10. What chemicals are commonly used for chemical precipitation?
11. For a rectangular sedimentation tank, derive the following relationship:

$$\frac{d}{v_O} = \frac{L}{v_H}$$

Practice Problems

1. Design a circular clarifier for a town with a population of 35 000. The average demand for water is 150 L/c·d, 75% of which is discharged as wastewater. The maximum allowable surface loading is 24 m^3/m^2·d. (15 m)
2. Size a rectangular primary clarifier to serve a population of 40 000. The average demand for water is 180 L/c·d, 75% of which is discharged as wastewater. The peaking factor is 2.5 and the allowable surface loading is 20 m^3/m^2·d based on average daily flow. Assume the length of the tank is four times the width and the effective depth is 2.5 m. Check for horizontal flow velocity. (30 m × 7.5 m × 4.0 m, 0.20 m/min $<$ 0.3 m/min)
3. A city sewage treatment plant treats an average flow of 4.5 ML/d with a peaking factor of 2.2. For appropriate removal, it is recommended that the overflow rate not to exceed 30 m^3/m^2·d. What is the required diameter of the clarifier? Check for average detention time assuming a depth of 3.0 m. Check loading during the peak hour and comment on the removal. (14 m, 2.5 h, 64 m^3/m^2·d.)
4. Determine the recommended size of two new circular clarifiers for an activated sludge system with a design flow of 20 ML/d and a peak hourly flow of 32 ML/d. Use a 30 m^3/m^2·d overflow rate for design flow and 50 m^3/m^2·d for peak hourly flow. (21 m diameter, 48 $<$ 50 m^3/m^2·d)
5. A sewage treatment plant operates with a couple of 15 m diameter clarifiers at an average flow of 13 ML/d and a peak flow of 21 ML/d. Based on the recommended overflow rates, are these adequate? Can one unit be removed from service during average flow conditions? (37 m^3/m^2·d $>$ 35 m^3/m^2·d and 59 m^3/m^2·d $>$ 60 m^3/m^2·d, not adequate)
6. A wastewater treatment plant has three primary circular units, each with a diameter of 20 m and a side water depth of 2.8 m. The inboard weirs are set at 18.5 m. The average daily flow is 18 ML/d and the peak flow is 35 ML/d. Calculate the overflow rates and weir loading, and comment on the adequacy. Is it possible to take one unit out of service? (19 m^3/m^2·d, 53 m^3/m·d, two clarifiers are adequate)
7. A secondary treatment plant is to have twin primary clarifiers. What is the minimum diameter required if it is not to exceed an overflow rate of 40 m^3/m^2·d for a design flow of 12 ML/d? (20 m)
8. Design a pair of circular primary clarifiers to handle a raw wastewater flow of 12 ML/d based on a surface overflow rate of 24 m^3/m^2·d.
   a. Find the detention time assuming a side water depth of 3.0 m. (3.0 hours)
   b. If the shape chosen is rectangular, calculate its dimensions assuming its length is four times its width (32 m × 8 m × 4.0 m)
   c. What minimum diameter circular clarifier would be needed? (18 m)
   d. Assuming there is a weir along the periphery of the circular clarifier, calculate the weir loading. (110 m^3/m·d)
   e. Estimate the daily production of sludge from each clarifier assuming solids are removed at 100 mg/L and the solids concentration of the wet sludge is 3.0%. (20 m^3/d)
9. Design a circular primary clarifier to handle raw wastewater at 7.0 ML/d to provide a minimum detention time of 2.0 hours with weir loading not exceeding 200 m^3/m·d. Calculate the surface flow rate for the designed clarifier. Assume a side water depth of 2.7 m. (17 m, 31 m^3/m^2·d)

10. Design primary clarifiers for a wastewater treatment facility in a community of 60 000 people with a wastewater contribution of 180 L/c.d. The maximum allowable overflow rate is 30 $m^3/m^2 \cdot d$. Assume a side water depth of 2.5 m.
    a. Design a rectangular tank with a length four times its width
    (40 + 4 = 44 m × 10 m × (2.5 + 0.50 + 1.0) = 4.0 m)
    b. Design a circular tank. (22 m)
    c. What is the detention time? (2.0 h)
11. A secondary clarifier is built with an inboard channel with an effluent weir on both sides. If the average daily flow to the plant is 7.5 ML/d, what must the minimum diameter of the inboard channel be so that a weir loading of 100 $m^3/m \cdot d$ is not exceeded? (12 m)

24 Activated Sludge Process (ASP)

Primary effluent contains 60% to 70% of its original organic contaminants in the form of very fine (colloidal) or dissolved organic materials not readily removed by normal mechanical or physical methods. In wastewater terminology, the soluble biochemical oxygen demand (BOD) content indicates biodegradable material. If left untreated, these will cause odours and, eventually, pollution in the receiving water bodies.

24.1 BIOLOGICAL TREATMENT

Naturally occurring bacteria in the presence of oxygen can break down colloidal and soluble BOD left over in primary effluent. This process is known as biochemical oxidation and is usually aerobic since microorganisms, primarily bacteria, need molecular oxygen to survive. Organics are broken into nitrates, phosphates, carbon dioxide, and water and new cells and energy. This is the principle used in the biological or secondary treatment of wastewater.

Although organics can be oxidized with the direct use of chemicals, this is expensive, and the end products are often toxic. The biological process, on the other hand, involves natural purification. In engineered systems, biological processes are not allowed to proceed naturally but are controlled to achieve greater removal at a faster rate. Biological treatment can be broadly classified into two main categories: suspended growth systems and fixed growth systems.

24.1.1 Suspended Growth Systems

In suspended growth systems, microorganisms are suspended in wastewater either as single cells or as a cluster of cells called biological floc. They are surrounded by wastewater containing food or BOD and molecular or dissolved oxygen (DO) for their growth. This suspension of wastewater containing BOD and microorganisms in the form of activated sludge is known as mixed liquor. The most common biological process falling into this category is the activated sludge process, as shown in **Figure 24.1**. In addition to the activated sludge process, stabilization ponds or lagoon systems are also suspended growth systems.

24.1.2 Fixed Growth Systems

Fixed growth systems, also called attached culture systems, consist of biomass adhered to inert surfaces with wastewater passing over the microbial layer. Trickling filters and rotating biological contactors (RBCs) are good examples of fixed growth systems. In both types of system, the biological process is aerobic. Whereas in an activated sludge process, oxygen is added by pumping air into mixed liquor, in a fixed growth system, wastewater passing over the microbial layer dissolves the needed oxygen from the atmosphere. In this unit, we will discuss the activated sludge process.

24.2 PRINCIPLE OF ASP

The activated sludge process (ASP) was first used in Manchester, England. Today, this process is the most widely used process for the secondary treatment of wastewater. The term activated sludge

DOI: 10.1201/9781003231264-24

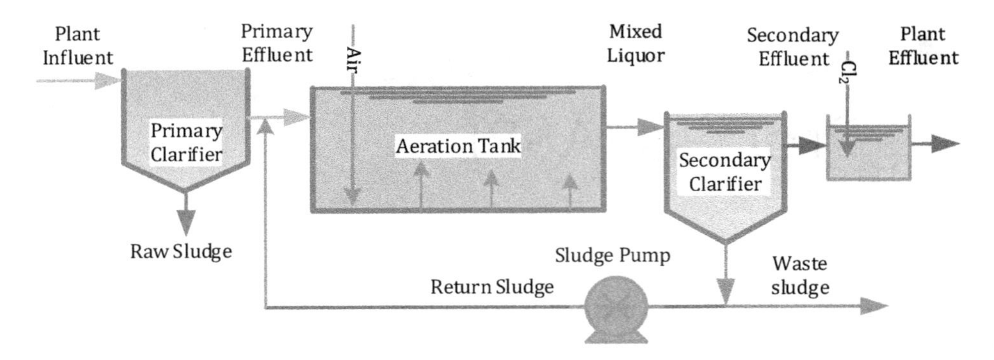

FIGURE 24.1 A Conventional Activated Sludge Plant

indicates the biomass that causes the breakdown of organics. The soluble and colloidal BOD is converted to new growth – or activated sludge – in three steps: transfer, conversion and flocculation.

24.2.1 Transfer

Transfer occurs when organic food matter meets microorganisms. In this step, food material is transferred from the water to bugs (microorganisms) through adsorption and absorption.

Organic matter in colloidal form is first adsorbed on the cell membrane. Adsorbed organics must be broken into a simple soluble form before they can be absorbed into the cell.

24.2.2 Conversion

The conversion of food matter into cell matter occurs after all the food enters the cell. The same thing happens when food is digested in our bodies. This conversion is called metabolism. During conversion, only part of the BOD is converted to new cell growth. More than half the BOD is converted to carbon dioxide, water and energy for the microorganisms.

24.2.3 Flocculation

Biological flocculation occurs when cells combine to form clusters called biological floc. This is an important step as the separation of water from sludge is determined by the settlement efficiency of biological floc. It is important to realize that these steps occur continuously and simultaneously within the process. The transfer and conversion steps are completed in the aeration tank, and the flocculation and separation of solids mainly occur in the secondary clarifier.

The activated sludge process is aerobic and, as such, must be supplied with oxygen at all times. Without oxygen, the bacteria will die, the oxidation process will come to a halt and a foul-smelling black sludge will be left. In this state, the sludge is said to be anaerobic (lacking oxygen). To dissolve oxygen into the waste, the activated sludge, the microorganisms and the primary effluent are aerated and mixed in an aeration tank. These organisms are then separated from the now treated wastes and settled out in the final clarifier to be recycled and used again. The overflow from this clarifier is a clear liquid, which, after disinfection by chlorination, is discharged into receiving water.

24.3 COMPONENTS OF ASP

The principal elements of the activated sludge process are shown in **Figure 24.2**. Its main components are discussed below.

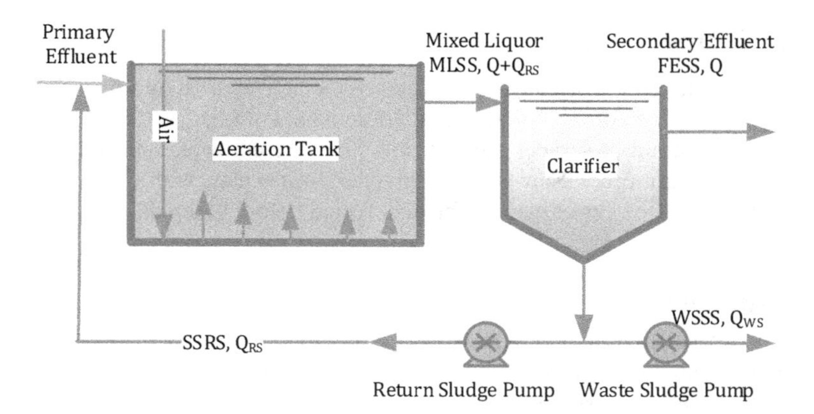

FIGURE 24.2 Schematic of an Activated Sludge Process

24.3.1 Aeration Tanks

The aeration tank is the bioreactor in which colloidal and soluble matter contributing to BOD are oxidized under aerobic conditions. The aeration tank is the heart of the activated sludge system. The breakdown of organic materials in wastewater takes place in the aeration tank. This is achieved by bringing the organic materials into contact with the bacteria in the presence of dissolved oxygen long enough to allow the breakdown to occur. The mixture of activated sludge (microorganisms) and wastewater (food) in the aeration tank is called mixed liquor.

Aeration tanks can be square, rectangular or circular, and are generally 3 to 5 m deep. They are relatively deep to allow for proper mixing and oxygen transfer. Tank size depends on the volume of sewage to be treated and its ability to hold the incoming sewage for a period of four to eight hours, called the aeration period.

Aeration tanks are generally made of concrete or steel. Oxygen is dissolved into the wastewater in tanks either by diffused aeration or surface aeration. It is essential that adequate mixing is provided so that the activated sludge is maintained in suspension. To keep the contents aerobic, a dissolved oxygen (DO) level of 1–2 mg/L is maintained. Higher levels of dissolved oxygen do not necessarily help the biological process. However, DO levels below 1.0 mg/L may encourage the growth of filamentous organisms that hinder settling.

24.3.2 Diffused Aeration

In this type of aeration system, air is blown from compressors through various devices located at the bottom of the aeration tanks, generally on one or both side walls of the tank. While oxygen is being dissolved into the liquid, a rolling action is generated to ensure thorough mixing and suspension of the activated sludge.

24.3.3 Mechanical Aeration

This technique uses blades of various designs that rotate partially submerged at the surface of the liquid with dissolved oxygen from the atmosphere. These devices splash large volumes of liquid over the surface of the tank, entraining and dissolving atmospheric oxygen into the tank contents. This also generates a pumping action for the necessary mixing. The amount of oxygen that can be dissolved varies with the speed of the device, its diameter, submergence, and the power of the drive unit. The drive motor ranges from 5 to 100 kW and the device can be as big as 3 m across.

24.3.4 Final Settling Tanks

The secondary clarifier, or final settling tank, receives the activated sludge from the aeration tank. The microorganisms in the form of sludge (called activated sludge) settle to the bottom of this clarifier where, with the aid of scraper mechanisms, they are collected and returned (recycled) to the aeration tank to treat more wastewater. The treated wastewater, with only 10% of its original contaminants remaining, flows over weirs to be disinfected before being discharged to the receiving rivers or lakes.

Secondary Clarification

Any solids that escape separation will reduce the quality of the final effluent. Thus, the clarifier must be operated so as to remove the maximum possible solids as sludge. Since activated sludge is relatively light, overflow rates are usually lower.

Settling

This function is best performed when the settling properties of the mixed liquor solids are such that surface tension is broken, quickly allowing the solids to settle. Old sludge, being heavier, may settle too quickly and reduce the particle collision necessary for capturing the fine solids (pin floc) in the upper regions of the tank.

Thickening

Another important function is thickening or allowing the solids to compact to become denser or concentrated. Simply put, thickening makes the sludge more concentrated. In addition to the hydraulic conditions under which a clarifier is operated, the concentration of solids in the return sludge will depend on the rate at which they are removed.

24.3.5 Sludge Recirculation and Wasting

Variable speed pumps take their suction from the draw-off and return the sludge to the aerator. A plant operator controls the return rate of the sludge to maintain a healthy biomass. A certain volume of sludge is occasionally wasted (Q_{WS}) from the return sludge lines, as shown in **Figure 24.2**. The sludge wastage is necessary, otherwise solids will accumulate in the system daily as a result of new growth. The wastage rate is adjusted to maintain steady-state conditions.

24.4 PROCESS LOADING PARAMETERS

The conversion of BOD into biosolids or activated sludge is carried out in the aeration tank followed by separation in the final clarifier. Both steps must complement each other in order to achieve maximum removal. Moreover, what happens in the aeration tank has a large impact on the performance of the final clarifier. The loadings affecting the activated sludge process are discussed in the following paragraphs.

24.4.1 Aeration Period

The aeration time or period indicates the hydraulic loading on the aeration tank. It is essentially the hydraulic detention time in the aeration tank based on the daily average wastewater flow. The recirculation flow is not considered when calculating the aeration period. The aeration period represents the average time for which bio-reactions take place or the microorganisms get the opportunity to break down the BOD. It is usually expressed in hours and ranges from four to 30 hours.

$$AP = \frac{Volume\ under\ aeration}{Daily\ average\ flow\ rate} = \frac{V_A}{Q}$$

24.4.2 Volumetric BOD Loading

Volumetric BOD loading on the aerator is usually expressed in terms of grams of BOD applied per day per cubic metre of aeration volume, $g/m^3{\cdot}d$. Depending on the type of activated sludge process, BOD loading can vary from 100–3000 $g/m^3{\cdot}d$. For example, BOD loading can be as small as 150 $g/m^3{\cdot}d$ in the extended aeration system and as large as 1500 + $g/m^3{\cdot}d$ in high-rate systems. BOD loadings increase with an increase in incoming BOD and decrease with an increase in aeration period, as shown below.

$$BODLR = \frac{M_{BOD}}{V_A} = \frac{Q \times BOD_{PE}}{V_A} = \frac{BOD_{PE}}{AP}$$

BOD loading and aeration period are interrelated parameters depending on the concentration of BOD in the secondary influent. BOD loading is directly proportional to BOD concentration and inversely proportional to the aeration period. Thus, for the same quality of water entering the aeration tank, if the aeration period is reduced by a factor of two, BOD loading will be doubled.

Example Problem 24.1

An aeration tank receives a primary effluent of 20000 m^3/d with a BOD concentration of 150 g/m^3. The capacity of the aeration tank is 5000 m^3. What is the current BOD loading?

Given: Q = 20 000 m^3/d
BOD = 150 g/m^3
V_A = 5000 m^3

SOLUTION:

$$BODLR = \frac{M_{BOD}}{V} = \frac{20000\ m^3}{d} \times \frac{150\ g}{m^3} \times \frac{1}{5000\ m^3} = 600.0 = \underline{600\ g/m^3.d}$$

Based on organic loading and operational control, an aeration process has the following commonly used variations: extended aeration, conventional aeration, contact stabilization, high rate, and high-purity oxygen.

24.4.3 Food to Microorganism Ratio (F/M)

The F/M ratio expresses BOD loading in relation to biomass in the system. The microbial mass is indicated by the mass of mixed liquor solids in the aeration tank. The calculation for the F/M ratio can be made using any of the following expressions.

$$\frac{F}{M} = \frac{M_{BOD}}{m_{MLSS}} = \frac{Q \times BOD_{PE}}{MLSS \times V_A} = \frac{BOD_{PE}}{MLSS} \times \frac{1}{AP}$$

TABLE 24.1
F/M Ratio and Metabolic Stages

Property	Extended	Conventional	High rate
Growth phase	Endogenous	Declining	Accelerated
Quality, BOD	Excellent	Good	Poor
Cost	High	Medium	Low
kg of BOD/ kg air	Low	Medium	High
Sludge settling	Excellent	Good	Fair

Because this ratio is of mass rate to mass, the units of F/M are that of per day. The values typically range from 0.1/d to 1.0/d. Most municipal plants are operated at an F/M ratio of 0.1 to 0.3/d. Many people express F/M ratio in terms of mass of mixed liquor volatile suspended solids (MLVSS) rather than MLSS. This is because the volatile solids more accurately represent the biomass. The volatile fraction is around 75% of the total mixed liquor solids. **Table 24.1** shows the relationship between F/M ratio and the metabolic stages of microorganisms.

The F/M ratio is an important activated sludge process control parameter used to maintain the proper balance between food supply and biomass. The F/M ratio maintained in the aeration tank defines the operation of the activated sludge. If the F/M ratio is high, there is an abundance of food and the microorganisms are in an exponential or logarithmic growth phase. When operating in this manner, the biomass is not limited by food and rapidly multiplies. Cell growth in this phase is similar to cell growth in children. The new growth is young and highly active. Because of a high activity level, the cells do not easily floc together to become heavy enough to settle. A system operated with a high F/M ratio will have depressed DO in the aerator, increased oxygen uptake rate, low BOD removal and a rapid increase in MLSS build-up.

In a process operated with a low F/M ratio (old sludge), the activity of microorganisms becomes comparable to that of older people. For example, when full growth is reached, the growth rate reduces and continues to reduce until death occurs. The growth phase becomes endogenous, or starvation occurs. Because of the lack of food during the starving phase, there would continue to be a loss in body weight as more and more cellular material is converted into energy. When the system is operated with a low F/M ratio, it is called extended aeration. A process operated by maintaining a low F/M ratio will have high DO in the aeration tank, low oxygen uptake rate (OUR), quick settling sludge, high BOD removal and high total oxygen requirements. The conventional activated sludge process is operated with an F/M ratio in the middle range of 0.2/d to 0.5/d. This creates sludge that is neither old nor young. Hence, conventional activated sludge is a good compromise between quality and quantity.

The F/M ratio is an expression of BOD loading indicating the metabolic state of the biological system. The advantage of this expression is that it defines an activated sludge process without reference to an aeration period or strength of wastewater. Two systems that are quite different may operate at the same F/M ratio. If one system has a shorter aeration period than the other, the level of MLSS can be increased to compensate for the reduction, thus operating the two systems at the same F/M ratio. The calculation of F/M ratio is illustrated in the following example problem.

Example Problem 24.2

An aeration tank receives a primary effluent of 20000 m^3/d with a BOD concentration of 150 g/m^3. The mixed liquor suspended solids concentration is 2500 g/m^3 and the aeration volume is 5000 m^3. Calculate the F/M ratio.

Given: Q = 20 000 m³/d
BOD = 150 g/m³
V_A = 5000 m³

SOLUTION:

$$\frac{F}{M} = \frac{Q \times BOD_{PE}}{MLSS \times V_A} = \frac{20000\ m^3}{d} \times \frac{1}{5000\ m^3} \times \frac{150 g/m^3}{2500 g/m^3} = 0.240 = \underline{0.24/d}$$

24.4.4 Sludge Age/Solids Retention Time

Sludge age, also called mean cell residence time (MCRT) or solids retention time (SRT), is an operational parameter related to the F/M ratio. Because biomass is recycled from the clarifier back to the aeration tank, the biosolids in activated sludge have more than one pass through the system. The only way activated sludge can exit is as suspended solids in the two flow streams: waste sludge and final effluent. Although water has only one pass through the system, activated sludge has repeated passes before it exits the system. Whereas the aeration period varies from three to 30 hours, the SRT is much greater and is measured in terms of days. It may be helpful to think of sludge age as the average length of time microorganisms stay in the system.

The SRT is calculated based on the MLSS in the aeration tank related to the total mass of biosolids leaving via the waste sludge stream and final effluent stream. Referring to **Figure 24.3**, the sludge age can be calculated as follows:

$$\text{Solids retention time, } SRT = \frac{(V_A + V_C) \times MLSS}{WSSS \times Q_{WS} + FESS \times Q_{FE}}$$

Some authors ignore the solids exiting as part of the final effluent and the volume of mixed liquor in the final clarifier. This is valid when the solids concentration in the final effluent is relatively small.

$$SRT(\text{short form}) = \frac{V_A \times MLSS}{WSSS \times Q_{WS}} = \frac{MLSS}{WSSS} \times \frac{V_A}{Q_{WS}}$$

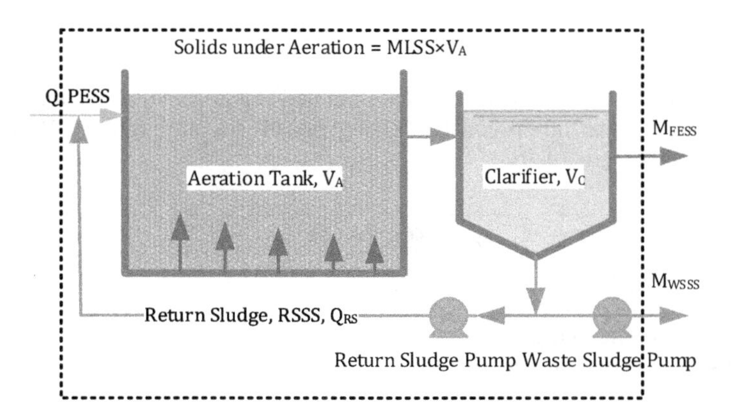

FIGURE 24.3 Mass Rate of Solids Entering and Exiting

It is logical that the short form will yield higher values of SRT. It does not matter which form of the expression is used, the important thing is to be consistent. If you use the short form for operating your plant, stick with it. Sludge age indicates the growth rate and hence the metabolic state of the biological process. Microorganisms in old sludge with a longer SRT are in a starvation phase as there are too many of them competing for the same food. In high-rate systems, the SRT is relatively short, signifying the logarithmic growth phase and thus producing young sludge.

Sludge age and F/M ratio are inversely related. As the F/M is reduced, sludge age increases, resulting in higher BOD removal due to improved settlement and longer reaction time. However, due to higher aeration time, relatively more oxygen is required to remove the same amount of BOD.

Example Problem 24.3

An aerator is 7.0 m deep with a holding capacity of 5000 m^3. The MLSS concentration is 2500 g/m^3 and the return activated sludge solids concentration (RSSS) is 5000 mg/L. If 500 m^3 of the activated sludge is wasted daily, find the sludge age.

Given: $V_A = 5000\ m^3$
$MLSS = 2500\ g/m^3$
$WSSS = RSSS = 5000\ g/m^3$
$Q_{WS} = 500\ m^3/d$

SOLUTION:

$$SRT = \frac{MLSS}{WSSS} \times \frac{V_A}{Q_{WS}} = \frac{5000\ m^3.d}{500\ m^3} \times \frac{2500\ mg/L}{5000\ mg/L} = 5.00 = \underline{5.0\ d}$$

24.4.5 Substrate Utilization Rate

Another parameter commonly used to design aeration units is called specific substrate utilization rate, U. It is very much parallel to the F/M ratio except BOD removed rather than BOD applied is used, and the conversion factor of BOD into microbial mass factor is used to translate BOD into new growth.

$$U = \frac{M_{BODR}}{m_{MLSS}} = \frac{Q \times (BOD_{PE} - BOD_{FE})}{MLSS \times V_A}$$

Under steady-state conditions, when MLSS is maintained at a given level, the mass of solids wasted must be equal to new growth minus constant endogenous respiration rate, k_e. Based on this, it can be shown that U and SRT are related.

$$\frac{1}{SRT} = \alpha \times U - K_e = \frac{\alpha \times Q \times (BOD_{PE} - BOD_{FE})}{MLSS \times V_A} - K_e$$

α = maximum yield coefficient, g/g
U = specific substrate utilization rate, /d

The values of α and k_e are usually constant for a given municipal wastewater, with typical values of 1.0 and 0.06/d, respectively. The value of SRT adopted for the design controls the quality of the secondary effluent and the settlement of the mixer liquor in the secondary clarifier. As discussed before, for a selected SRT, the amount of activated sludge to be wasted can be worked out as they are related parameters.

$$\boxed{\text{Sludge wastage rate, } Q_{WS} = \frac{V_A \times MLSS}{WSSS \times SRT}}$$

Example Problem 24.4

Design a conventional activated sludge plant (step aeration) to treat municipal wastewater with a population of 44 000 contributing an average flow of 180 L/c·d containing a BOD of 230 mg/L. Assume 35% removal of BOD in the primary clarification and a desired effluent quality of 25 mg/L.

Given: Q = 180 L/c·d
BOD_{raw} = 230 mg/L
E_I = 35%
BOD_{FE} = 25 mg/L
Population = 44,000 people

SOLUTION:

$$BOD_{PE} = BOD_{raw} \times (1 - E_I) = \frac{230\ mg}{L} \times (1 - 0.35) = 149.5 = \underline{150\ mg/L}$$

$$E_{II} = \frac{BOD_{PE} - BOD_{FE}}{BOD_{PE}} = \frac{149.5 - 25}{149.5} = 0.833 = \underline{83\%}$$

$$Design\ Q = \frac{180\ L}{p.d} \times 44000\ p \times \frac{m^3}{1000L} = 7920 = \underline{7900\ m^3/d}$$

For this level of removal, use an F/M of 0.30/d and an MLSS of 2200 mg/L. For these selected values, the volume under aeration can be figured out:

$$V_A = \frac{Q \times BOD_{PE}}{MLSS \times (F/M)} = \frac{7920\ m^3}{d} \times \frac{d}{0.30} \times \frac{150g/m^3}{2200g/m^3} = 1800 = \underline{1800\ \ m^3}$$

$$AP = \frac{V_A}{Q} = 1800\ m^3 \times \frac{d}{7920\ m^3} \times \frac{24\ h}{d} = 5.45 = \underline{5.5\ h}\ (typical,\ so\ okay)$$

$$Check,\ SRT = \left[\frac{\alpha \times Q \times (BOD_{PE} - BOD_{FE})}{MLSS \times V_A} - K_e\right]^{-1}$$

$$= \left[1 \times \frac{7920\ m^3}{d} \times \frac{1}{1800\ m^3} \times \frac{(150 - 25)}{2200} - \frac{0.06}{d}\right]^{-1} = \left(\frac{0.19}{d}\right)^{-1} = 5.26 = \underline{5.3\ d}$$

Assuming that an aeration tank is 3.5 m deep with 0.5 m of free board, using a width of 5.0 m, the length of the aeration tank is calculated as follows:

$$L = \frac{V_A}{B \times d} = \frac{1800\ m^3}{4.0\ m \times 5.0\ m} = \underline{90\ m}$$

It may be recommended to have two tanks each of length 45 m. Thus, the total width of the unit is:

$$B = \# \times tank\ width + baffle\ wall = 2 \times 5.0\ m + 0.30\ m = \underline{10.3\ m}$$

= 2B = 10.3 m (each tank 5.0 m wide) d = 4.0 m

Example Problem 24.5

An activated sludge plant operating data is given as V_A = 2100 m^3, WSSS = 5100 mg/L and MLSS = 2000 mg/L. Based on this data, determine the waste sludge rate to achieve a target SRT of 5.0 d.

Given: V_A = 2100 m^3
WSSS = 5100 mg/L
MLSS = 2000 mg/L

SOLUTION:

$$Q_{WS} = \frac{V_A}{SRT} \times \frac{MLSS}{WSSS} = \frac{2100\ m^3}{5.0\ d} \times \frac{2000 mg / L}{5100 mg / L} = 146 = \underline{150\ m^3 / d}$$

Example Problem 24.6

The mixed liquor in a 1600 m^3 aeration tank has an MLSS of 2050 mg/L. The waste sludge solid concentration is 4900 mg/L. If the target MLSS is 2000 mg/L, determine the additional volume of waste sludge to be wasted.

Given: V_A = 1600 m^3
WSSS = 4900 mg/L
ΔMLSS = 2050 - 2000 = 50 mg/L

SOLUTION:

$$\Delta V_{WS} = V_A \times \frac{\Delta MLSS}{WSSS} = 1600\ m^3 \times \frac{50 mg / L}{4900 mg / L} = 16.3 = \underline{16\ m^3}$$

Example Problem 24.7

Given a return sludge and MLSS concentration of 9800 and 4800 g/m^3, respectively, and an aeration capacity of 1250 m^3, calculate the volume of sludge to be wasted to reduce the MLSS to 3700 g/m^3. Determine the pumping rate in L/s if a) wasted continuously over a 24-hour period and b) wasted in a batch over a four-hour period.

Given: Current MLSS = 4800
WSSS = 9800
Desired MLSS = 3700
$V_A = 1250\ m^3$

SOLUTION:

$$\Delta m_{MLSS} = V_A \Delta MLSS = 1250\ m^3 \times (4800 - 3700) \frac{kg}{m^3} = 1375.0 = 1380\ kg$$

$$V_{WS} = \frac{\Delta m_{MLSS}}{WSSS} = 1375\ kg \times \frac{m^3}{9800\ g} \times \frac{1000\ g}{kg} = 140.3 = \underline{140\ m^3}$$

$$Q_{WS+}(over\,24\ h) = \frac{140.3 m^3}{24\ h} \times \frac{h}{3600\ s} \times \frac{1000\ L}{m^3} = 1.624 = 1.6\ L/s$$

$$Q_{WS+}(Over\,4\ h) = \frac{140.3 m^3}{4.0\ h} \times \frac{h}{3600\ s} \times \frac{1000\ L}{m^3} = 9.74 = \underline{9.7\ L/s}$$

24.5 FINAL CLARIFICATION

Loading and sludge-settling characteristics will affect clarification, concentration and the removal function of the clarifier.

24.5.1 Hydraulic Loading

As discussed previously, hydraulic parameters used to measure the performance of a clarifier include hydraulic detention time (HDT or t_d), surface loading/overflow rate (v_o) and weir loading (WL). A clarifier is hydraulically overloaded when the incoming flow exceeds the hydraulic capacity of the clarifier. Low weir loading rates are recommended for secondary clarifiers. To achieve this, many clarifiers are designed with an inboard effluent channel to provide extra weir length.

24.5.2 Solids Loading

Solids loading rate (SLR) is the maximum rate of solids that can be applied to a clarifier. This is an important consideration in the operation of a secondary clarifier. The allowable solids loading rate is governed by the volume of the clarifier, the settling characteristics of the sludge and existing hydraulics. Excessive solids loading results in solids accumulated in the upper layer, which are carried out over the weirs by the overflow velocity. This would cause billowing clouds of particles at the weirs. With slow-settling sludge, the recirculation rate is generally high. However, the increase in return rate must not allow the solids loading to exceed the allowable value. A typical solids loading rate falls in the range of 100–150 kg/m^2·d

$$\boxed{SLR = \frac{M_{SS}}{A_S} = \frac{(Q + Q_{RS})\,MLSS}{A_S}}$$

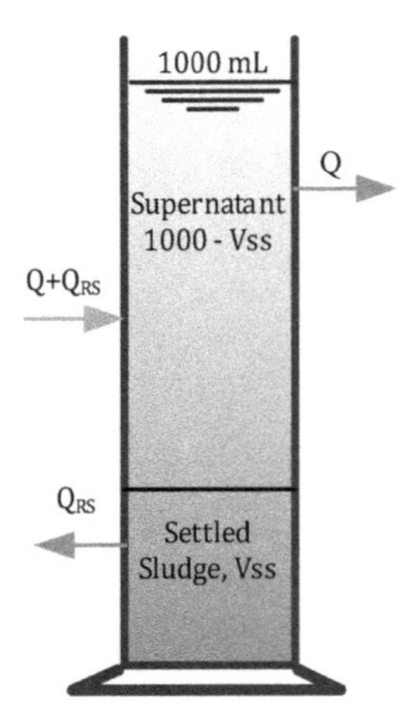

FIGURE 24.4 Settlometer Test

24.5.3 Sludge Settlement

Under actual operating conditions, sludge settling characteristics are known by running a settlement test. The sludge volume index (SVI) is computed based on the settling test and the recirculation rate is selected accordingly.

Settlement Test

One of the most common tests for monitoring the operation of an aeration system is the settlement test, as shown in **Figure 24.4**. This procedure involves determining the MLSS concentration by running a solids test. Sludge settlement is measured by observing the volume of settled sludge in a graduated cylinder filled with 1 litre of mixed liquor. The volume of settled solids after 30 minutes of settling is used to calculate the index. The sludge volume index (SVI) is the volume in millilitres occupied by one gram of settled suspended solids. Sludge density index (SDI) is the inverse of SVI expressed as percent solids.

$$SVI = \frac{V_{SSL}}{MLSS} \text{ as } mL/g \qquad SDI(\%) = \frac{1}{SVI} \times 100\%$$

The necessary conversion of units is made to express the results in mL/g. In general, a figure in the range of 50 to 150 indicates a good settling sludge. For MLSS up to 2500 mg/L, an SVI of 50 mL/g indicates a heavy, dense sludge. However, this sludge may fail to provide good clarification. This type of sludge is considered to be an old sludge that exhibits endogenous metabolic properties. By contrast, sludge with an MLSS of up to 2500 producing an SVI of 200 mL/g may result from young sludge. For the same mass of solids, this sludge will occupy four times more volume than sludge with an SVI of 50. Sludge solids concentration or denseness is related to SVI. A low SVI indicates a dense sludge and values greater than 150 indicate bulking sludge. Some authors use the SDI to indicate sludge settlement. The SDI represents the maximum concentration of return sludge that could be achieved by maintaining proper sludge pumping rates.

24.6 MATHEMATICAL RELATIONSHIPS

Applying the principle of ratio and proportion, the return rate can be estimated from the SVI as found in the settling test. By applying the principle of mass balance around a final clarifier, the solids concentration of return sludge can be predicted for a given return rate.

24.6.1 Return Rate and SVI

Referring to **Figure 24.4**, if the settling of mixed liquor in a graduated cylinder and clarifier are considered identical, then the ratio that return sludge flow (Q_{RS}) is to settled sludge volume (V_{SS}) what the final effluent flow (Q) is to the volume of supernatant (1000 - V_{SS}). This assumes that the sludge recirculation is at the required rate. A lesser flow value allows the solids to accumulate in the clarifier and results in eventual loss in the final effluent.

$$R_{hyp} = \frac{Q_{RS}}{Q} = \frac{V_{Sl}}{1000 - V_{Sl}}$$

If the sludge is returned at a rate indicated by the hypothetical relationship, the solids concentration in the return sludge will equal the SDI, which will indicate the minimum return rate. If the return rate exceeds this value, it will result in diluted sludge.

24.6.2 Return Ratio and Sludge Thickness

The primary function of the secondary clarifier is the clarification and thickening of settled solids. It concentrates the incoming stream of mixed liquor solids and leaves the clear supernatant to flow over the weirs. Under steady conditions, the mass of solids entering the clarifier must equal the solids drawn in the underflow. For a given return sludge rate, Q_{RS} or return ratio, $R = Q_{RS}/Q$, the expected consistency of the return sludge can be predicted by performing the mass balance around the final clarifier as shown below:

$$\frac{(Q + Q_{RS})}{Q} = \frac{RSSS}{MLSS} \quad or \quad \frac{1 + R}{R} = \frac{RSSS}{MLSS} \quad or \quad R = \frac{MLSS}{RSSS - MLSS}$$

For example, if the sludge is recycled at a rate of 50% (R = 0.5) of the wastewater flow rate, the return sludge will be three times as concentrated as mixed liquor. In other words, when sludge is pumped at a rate of half the wastewater flow, return sludge solids concentration can be expected to be about three times that of mixed liquor solids. If RSSS is observed to be significantly less, it will indicate the solids are not completely drawn from the clarifier. Keep in mind that the minimum return ratio must be equal to or greater than the hypothetical return ratio.

When the return ratio is close to the hypothetical return rate, the return sludge solids content should be the same as SDI. Usually, the return rate is kept at a rate greater than the hypothetical rate so that the solids do not become septic; hence solids concentration is proportionally lower than SDI.

Example Problem 24.8

The volume of settled solids in a 30-minute settling test is 300 mL/L. A solids test on the same sample of mixed liquor yielded the concentration of total solids to be 3000 mg/L. Calculate the SVI, SDI and minimum return rate, Q_{RS}.

Given: V_{ssl} = 300 mL/L
MLSS = 3000 mg/L = 3.0 g/L

SOLUTION:

$$SVI = \frac{V_{SSl}}{MLSS} = \frac{300\ mL}{L} \times \frac{L}{3000\ mg} \times \frac{1000\ mg}{g} = 100.0 = \underline{100\ mL/g}$$

$$SDI = \frac{1}{SVI} = \frac{g}{100\ mL} \times \frac{1.0\ g}{mL} \times 100\% = 1.00 = 1.0\%$$

$$R_{hyp} = \frac{V_{sl}}{1000 - V_{Sl}} = \frac{300\ ml}{L} \times \frac{L}{(1000 - 300)\, mL} \times 100\% = 42.8 = \underline{43\%}$$

Example Problem 24.9

A solids test is performed on a 50 mL sample of mixed liquor using a dish weighing 0.300g. After drying, the sample weighed 0.450 g. If the return rate is maintained at 40%, calculate the expected solids concentration of the return sludge.

Given: V = 50 mL
A = 0.450 g
B = 0.450 g

SOLUTION:

$$MLSS = \frac{m_{SS}}{V} = \frac{(0.450 - 0.300)\, g}{50.0\ mL} \times \frac{1000\ L}{m^3} \times \frac{1000\ mL}{L} = 3000.0 = 3000 g/m^3\ (mg/L)$$

$$RSSS = MLSS\left(\frac{1+R}{R}\right) = \frac{3000\ mg}{L} \times \frac{(1+0.40)}{0.40} = 10500.0 = 11000\ mg/L = \underline{1.1\%}$$

24.7 VARIATIONS OF ASP

There are many types of aeration systems. Some involve subtle differences, such as rates and points of air or wastewater applications, detention times, reactor shapes and types and methods of introducing air or oxygen. Others involve more drastic differences, such as sorption and settling prior to biological oxidation.

24.7.1 Conventional Aeration

This process is similar to the earliest activated sludge systems and is suitable for treating medium to large flows. The aeration basin is a long rectangular tank with air diffusers along the bottom for oxygenation and mixing. Long aeration basins are generally designed as plug flow reactors. In plug flow reactors, the wastewater and return sludge are combined at the head end, and the mixed liquor moves along the length of the tank to provide an aeration period of six to eight hours. As a given plug of flow has no contact with the incoming wastewater, the BOD loading in terms of F/M ratio is greatest at the head end, and bio growth is in the starvation phase at the exit end.

24.7.2 Step Aeration

Air is provided uniformly in step aeration. BOD loading is evened out by introducing wastewater at intervals or steps along the first portion of the tank, as illustrated in **Figure 24.5**. Diffused aeration

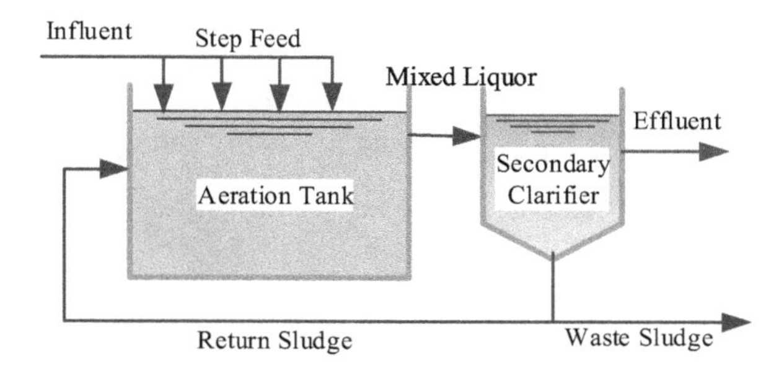

FIGURE 24.5 Step Aeration

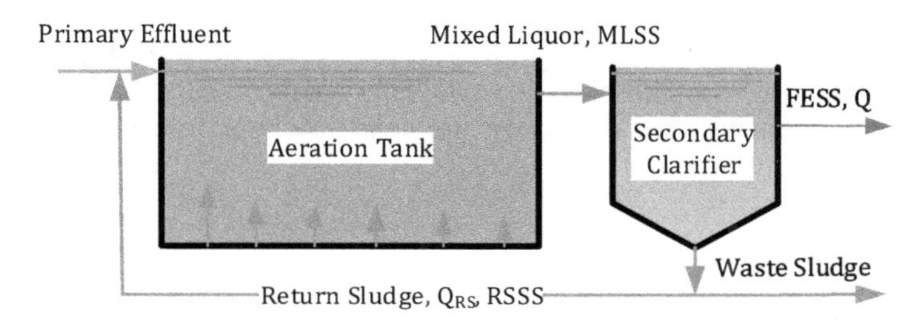

FIGURE 24.6 Tapered Aeration ASP Process

is commonly employed in such systems. The liquid depth is greater than 2.5 m to provide adequate mixing and oxygen transfer.

In spiral-flow aeration, a large number of diffusers are attached to the air header along one side of the tank to provide a rolling action. Because of wide hourly variations in loading from small communities, the conventional plug flow system can experience problems of biological instability. In such cases, aeration basins are usually designed as completely mixed flow reactors. Two or more completely mixed flow reactors attached in series will provide a combination of mixed and plug flow patterns.

24.7.3 Tapered Aeration

In tapered aeration, as shown in **Figure 24.6**, the air supply is tapered along the length of the tank to provide the greatest aeration at the head end where oxygen demand is at a maximum.

24.7.4 Contact Stabilization

In this process, aeration is divided into two portions: a contact zone and a stabilization zone. Wastewater flows into the contact or aeration zone, whereas return sludge flows into the stabilization or re-aeration zone, as shown in **Figure 24.7**. The effluent from the aeration zone flows into the clarifier. Note that no wastewater flow is introduced into the re-aeration zone. The contact zone has an aeration period of two to three hours, while re-aeration is four to six hours or more. Normal operating sludge circulation is 100%. Aeration tanks may be rectangular to simulate plug flow or completely mixed.

In the contact zone, microorganisms quickly adsorb food. Everything is then settled in the final clarifier before being recycled to a re-aeration zone for stabilization of the food. Since the aeration time is much longer, the food adsorbed in the contact zone is stabilized to a greater extent. In the

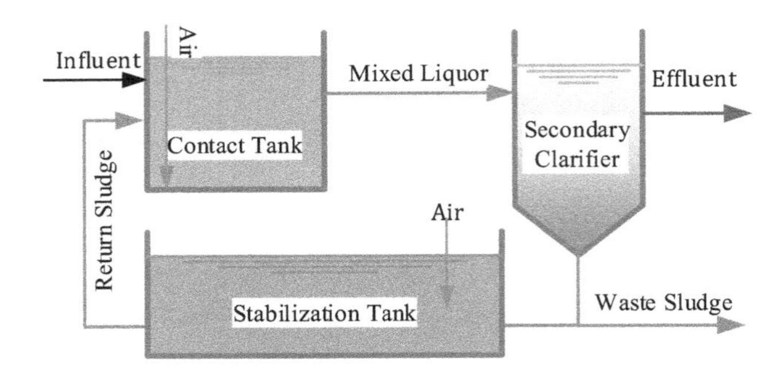

FIGURE 24.7 Contact Stabilization

stabilization zone, microorganisms are primarily in the endogenous growth phase. Due to a low growth rate, the volume of sludge produced is relatively small. This process is more resistant to hourly flow variations. However, this may cause problems if a large fraction of the BOD is in soluble form.

24.7.5 High-Rate Aeration

High-rate aeration systems operate in the logarithmic growth phase, thus producing young sludge. Operating at a high F/M ratio, these systems reduce the cost of construction by providing reduced aeration capacity. The aeration period is three to four hours, whereas the MLSS concentration is maintained at a level as high as 4000 mg/L. Due to rapid growth rate, oxygen requirements are quite high and sludge settlement is poor. These problems are overcome by modifying the process equipment. Using a combination of compressed air and mechanical aeration enhances oxygen transfer. To absorb the shock loading, the aeration tanks are completely mixed. The reduced settling is offset by the hydraulic thickening action of a rapid sludge return clarifier. BOD loading exceeding 1.3 $kg/m^3 \cdot d$ can depress DO and the carryover of pin floc in the final effluent during the peak flow period.

24.7.6 High-Purity Oxygen System

High-purity oxygen systems are employed to treat high strength wastes and to produce high-quality effluents. The aeration tank is divided into stages by means of baffles and covered in an enclosure. A slight pressure is maintained in the space between the cover and the top of the liquid level. Successive aeration chambers are connected so that liquid flows through the submerged ports and head gases pass freely from stage to stage with only a slight drop in pressure. A disadvantage of this system is its high operating cost. In addition to increased BOD removal efficiency, oxygen systems have a reduced volume of sludge and effective odour control. A comparison is shown in **Table 24.2**.

24.7.7 Extended Aeration

Extended aeration is characterized by having no primary treatment and a long aeration period of up to 36 hours. It is used for small towns or trailer parks. The capacity of the process is small due to the tank volumes needed to supply the required retention time and the large oxygen demand. Wasting provisions are not usually provided for small plants. The MLSS can increase for several months, after which the air is turned off and the floc is allowed to settle. The sludge is then pumped to a digester or hauled away. The MLSS ranges from 1000 to 10000 mg/L. The SRT is high because the same sludge is used over and over. The F/M ratio is quite low as the microorganisms are starving and

TABLE 24.2
Comparison of Plant Types

Parameters	Conventional	Contact	Extended
Aeration period, h	6–8	3–8	18–24
BOD loading (g/m³·d)	500–800	Same	80–250
F/M (1/d)	0.2–0.4	0.2–0.4	#0.1
SRT, d	2–5	2–5	> 15
Nitrification	> 5 days	> 5 days	> 15 days
Return rate	> 80%	25% > 100%	> 80%
BOD removal	1 mg/L	> 1 mg/L	> 1 mg/L
Nitrification	> 2 mg/L	> 2 mg/L	> 2 mg/L
Primary sedimentation	Yes	No	No
Grit removal	Yes	Yes	Yes
Sludge wasting	Yes	Yes	Yes
Sludge re-aeration	No	Yes	No
Sludge recirculation	Yes	Yes	Yes
Mechanical aeration	Yes	Yes	Yes
Diffused aeration	Yes	Yes	Yes

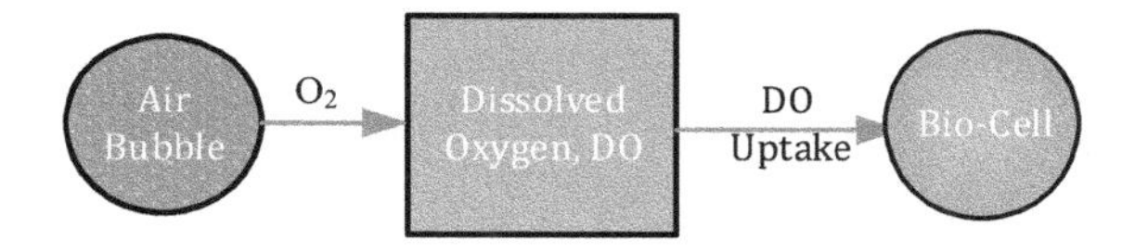

FIGURE 24.8 Oxygen Transfer and Utilization

therefore very competitive for food. This results in a highly treated effluent with 85% BOD removal and low sludge production. This process often has high concentrations of ash floc due to the high sludge age. These types of systems are available as pre-packaged plants.

A variation of the extended aeration process is the oxidation ditch. This has an oval ditch for the aeration tank in which the wastewater is pumped and circulated by mechanical aerators or pumps at a velocity of 0.2 to 0.4 m/s. The ditch is usually 1.2 to 1.8 metres deep. The ditch configuration is in the form of a racetrack with a surface beater for aerating. This type of plant requires minimal supervision as the process is mechanical.

24.8 OXYGEN TRANSFER

As shown in **Figure 24.8**, oxygen transfer into mixed liquor is a two-phase process. In the first phase, oxygen from the air is transferred or dissolved into the mixed liquor. In the second phase, dissolved oxygen is taken up or utilized by the microorganisms in the biochemical oxidation of the waste organic matter. The rate of oxygen uptake depends on several factors, including the temperature of the wastewater and the food to microorganism ratio. Increases in temperature and organic loading would increase the uptake rate, causing DO in the aeration tank to drop. **Table 24.3** shows a typical uptake rate for selected types of activated sludge processes.

Whatever aeration system is used, there is always some way of adjusting its output. This is usually done by increasing the airflow in diffused aeration systems and raising the tank level in mechanical aeration systems. The DO level in the aeration rate will increase, decrease or remain stable depending on how the oxygen transfer rate (OTR) compares with the oxygen uptake or utilization rate (OUR).

TABLE 24.3
Oxygen Uptake Rates

Activated Sludge Process	OUR, $g/m^3.h$
Extended aeration	< 10
Conventional	30
High rate	100

When the DO level in the aeration tank remains stable, this indicates the OTR equals the OUR. In the early morning hours, OUR runs low, thus DO concentration starts increasing until it reaches an equilibrium value. During the peak organic loading period, an increase in OUR causes a drop in DO level. Indirect variation in DO concentration in the aeration tank also indicates the rate of biochemical reaction. When toxins are present in wastewater, the DO level will remain high even when BOD loading is relatively high. The biochemical reaction rate drops due to the presence of toxins, resulting in an increase in DO and a drop in OUR.

24.8.1 Mass Transfer Equation

The rate of oxygen transfer from air bubbles into solution is determined by the transfer coefficient *K* and the oxygen deficit maintained in the aeration basin. It can be expressed by the following mass transfer equation.

$$\boxed{Oxygen\ transfer\ rate, OTR = KD = K\left(\beta \times DO_{sat} - DO\right)}$$

K = mass transfer coefficient, 1/h
β = oxygen saturation coefficient of wastewater, 0.8–0.9
DO_{sat} = saturation DO for clean water, mg/L
DO = actual dissolved oxygen concentration, mg/L
D = DO deficit, mg/L
OTR = oxygen transfer rate, mg/L.h

The transfer coefficient factor depends on the wastewater characteristics and, more importantly, on the physical features of the aeration system, including the type of diffuser (fine or coarse), liquid depth (shallow or deep), degree of mixing and basin configuration. The transfer coefficient factor will be higher if a fine diffuser is installed in deep tanks. A high OTR can be achieved by maintaining low DO levels in the mixed liquor. This means that for the same aeration equipment, higher oxygen transfer efficiency can be achieved.

24.8.2 Oxygen Transfer Efficiency

Oxygen transfer efficiency is the ratio of the mass of oxygen transferred to oxygen supplied. The mass of oxygen supplied can be worked out from the airflow rate and the oxygen content of the air. At standard temperature and pressure, this value is 0.279 kg O_2/m^3 of air. The aeration transfer efficiency is typically in the range of 5% to 20%. Diffused aeration systems are usually designed based on air requirements per kg of BOD, as shown in **Table 24.4**.

For mechanical aeration systems, the equipment should be capable of transferring at least 1.0 kg of oxygen per kg of BOD applied to the aeration tank.

TABLE 24.4
Air Requirements

Activated Sludge Process	Air Requirement
Extended	125 m^3/kg of BOD
Conventional	95 m^3/kg of BOD
High rate	4 m^3/kg of BOD

Example Problem 24.10

An aeration system has a transfer coefficient K of 3.0/h for a wastewater at 20°C with a β of 0.85. What is the rate of oxygen transfer when the DO in the aeration tank is a) 2.0 mg/L and b) 3.0 mg/L?

Given: K = 3.0/h
DO = 2.0 mg/L
DO_s = 9.2 mg/L at 20°C

SOLUTION:

$$OTR = K\left(\beta \times DO_{sat} - DO\right) = \frac{3.0}{h}\left(9.2 \times 0.85 - 2.0\right) mg/L = 17.46 = 18\ mg/L.h$$

$$OTR = K\left(\beta \times DO_{sat} - DO\right) = \frac{3.0}{h}\left(9.2 \times 0.85 - 3.0\right) mg/L = 15.8 = \underline{16\ mg/L.h}$$

24.9 OPERATING PROBLEMS

When activated sludge is operating well, it usually produces effluent containing less than 15 mg/L of solids and BOD each. However, as this is a biological process, it does not take much for something to go wrong. The key is to detect the problem early and take necessary action before it is too late. Here are some of the operating problems commonly encountered in the operation of an activated sludge plant.

24.9.1 Aeration Tank Appearance

Under normal operating conditions, the colour of mixed liquor is medium brown and has an earthly odour. A dark blackish colour indicates septic conditions, which may be due to inadequate operation or improper discharge of recycle streams. Another possible reason might be high strength industrial waste. All these situations may require an increase in air supply.

Turbulence

Much can be said about air distribution by observing turbulence patterns in the aeration tank. One tank may be receiving more air than another, indicating adjustment is necessary. Partially plugged diffusers may create high turbulence at some spots and dead spots in others. A high localized diffuser is usually the result of a broken or missing diffuser.

Foaming

Another observation to provide a clue as to the operation of the activated sludge process is foaming in the aeration tank. A small amount of white to light brown coloured foam is an indication of good operating conditions. During the initial start-up of the plant, or when mixed liquor is relatively dilute, it is normal to expect thick billows of white foam. It is logical to reduce sludge wasting to allow the build-up of mixed liquor solids. Dense, dark brown foam usually indicates old sludge. As a first step to correcting this problem, sludge wasting can be increased.

24.9.2 Secondary Clarifier Appearance

The first sign of impending clarification problems is an increase in the turbidity and solids content of the effluent. If attention is not paid, this can lead to serious problems. Bulking is by far the most common problem faced when operating an activated sludge plant. The first sign of sludge bulking is a rise in the sludge blanket. The trend in the sludge volume index will be upwards. This is where the real value of the settling test lies.

A slide of the mixed liquor should be checked for filamentous growth under a microscope. An abundance of hair-like structures indicates filamentous growth. Non-filamentous bulking is very rare; in this case, the sludge contains a large amount of water trapped in the floc.

If the bulking problem goes out of whack, some sort of chemical control may be applied. This includes the application of coagulant and flocculent aids to enhance settling. Some operators try killing filamentous growth with controlled chlorination. However, this may lead to poor effluent quality, and the solution is only temporary. It is best to reduce the load to the plant if possible.

Rising sludge at the top of the final clarifier is usually confused with bulking. The sludge settles well at the bottom of the clarifier. However, after settling it becomes lighter and clumps of it rise to the surface. The sludge is usually dark grey and rising gas bubbles are associated with it. This problem is caused by denitrification and septicity. It usually happens with old and well oxidized sludge. Again, the sludge wasting or loading to the aeration unit may need to be adjusted.

Pin floc refers to the very small floc, usually less than 1 mm in diameter. Some pin floc will always be there. An excess of pin floc is usually caused by over-oxidized sludge or unfavourable hydraulic conditions in the aeration tank.

Straggler floc refers to very fluffy, almost transparent and buoyant solids, typically 3–5 mm in diameter. Straggler floc is usually accompanied by clear effluent. In most cases, this is due to new growth when the SRT is on the low side.

Deflocculation occurs when sludge breaks up into very small particles that settle poorly, resulting in turbid effluent. The turbidity is caused by small particles of broken floc. Deflocculation usually occurs due to the presence of inhibiting substances, such as toxins or acid wastes.

Grease balls of varying sizes up to 10 cm in diameter can sometimes be found floating in aeration tanks or secondary clarifiers. They are formed by the joining together of grease particles in wastewater due to gentle rolling action in the aeration tank or clarifier. Grease balls have to be removed physically.

Solids washout refers to solids flowing out with the effluent even when bulking is not a problem. When the sludge blanket is well below the top surface, but close to the weirs it is high, it causes the sludge solids to flow with the effluent.

Toxic substances presence can be found if sudden changes are observed in process, for example, colour, DO, types of dominating organisms and plant removals. Prevention is the best solution. Any mishap like this should be followed by a thorough investigation.

Discussion Questions

1) What best describes biomass in an activated sludge process?
2) Compare two activated sludge control parameters, F/M ratio and SRT.
3) Explain, with the help of a sketch, the components of an activated sludge process plant.

4) Discuss the advantages and disadvantages of the various activated sludge processes.
5) Based on the mass balance, derive the relationship for finding the expected concentration of return sludge solids, RSSS, in terms of mixed liquor solids concentration, MLSS, and return rate ratio, R?
6) Aeration is an important function in an activated sludge process. Describe the various methods of aeration.
7) Briefly describe the main problems encountered in the operation of an activated sludge process.
8) Which parameter is unique to a secondary clarifier, and how is it used to control the clarifier's operation?
9) Assuming a settlement test on mixed liquor exactly simulates the settling in a secondary clarifier, how can the rate of sludge return be decided?
10) Explain why the extended rate process does not need primary clarification.
11) Compare pin floc and straggler floc problems in secondary settling.
12) In the operation of the activated sludge process, DO levels in the aeration tank are monitored. What typical pattern of DO should be expected on a normal day, and how should the aeration rate be adjusted?
13) During the peak BOD loading hours of the day, DO in the aeration tank should be suppressed. If this is not the case, what inferences can be drawn, and how can they be corrected?
14) Discuss how the modifications of step aeration and tapered aeration improve the efficiency of the activated sludge process.
15) List the factors affecting oxygen transfer efficiency. What role does a plant operator have in achieving higher efficiency?

Practice Problems

1) Design an activated sludge process for a town with a daily average flow of 10 ML/d. Assume primary BOD removal is 40% of the incoming BOD of 250 mg/L. The secondary effluent BOD should not exceed 15 mg/L. Based on the pilot studies, α is 0.5 and K_e is 0.05/d. Assuming an MLSS of 2200 mg/L, RSSS of 8000 mg/L and SRT of 10 d, find the following:
   a) Aeration capacity. (2.1 ML)
   b) Sludge wasting rate. (58 m^3/d)
   c) Sludge return ratio. (38%)
2) What is the required aeration capacity to treat a primary effluent of 6.5 ML/d with a BOD of 120 mg/L? The process should be operated at an F/M ratio of 0.20/d while maintaining an MLSS of 2000 mg/L. What is the required length of the aeration tank assuming a width and depth each of 5.0 m? (2.0 ML, 0 m)
3) Design a conventional activated sludge to serve a population of 45,000 at 200 L/c·d containing a BOD of 200 mg/L. Assume 35% primary removal and a desired final effluent BOD of 20 mg/L.
4) For a conventional activated sludge plant, the operational data is as follows: Q = 25 ML/d, BOD_{raw} = 320 mg/L, primary BOD removal = 35%, secondary effluent BOD_{II} = 25 mg/L, aeration capacity, V_A = 9500 m^3, MLSS = 2200 mg/L, WSSS = 0.90%, wastage rate, Q_{WS} =200 m^3/d. Calculate the F/M, SRT, and overall BOD removal. (0.25/d, 12 d, 92%
5) A conventional activated sludge plant receives a daily wastewater flow of 36 ML/d with a BOD of 260 mg/L. The aeration capacity is 11 ML and the return sludge solid

concentration is 9800 mg/L. The return sludge is wasted at 230 m^3/d to maintain a mixed liquor solid concentration of 2500 mg/L. Assuming a primary BOD removal of 35%, work out the aeration period, F/M ratio and SRT, neglecting the solids leaving in the effluent. (7.3 h, 0.22/d, 12 d)

6) Design an activated sludge process to produce secondary effluent with a BOD not exceeding 20 mg/L. Assume the BOD of the primary effluent is 150 mg/L and the daily wastewater flow is 15 ML/d. Assume yield coefficient = 0.65, k_e = 0.05/d, SRT = 10 d, R = 50% and MLSS = 2000 mg/L. (V_A = 4.2 ML, RSSS = 6000 mg/L, Q_W = 140 m^3/d, F/M = 0.27/d)
7) Determine the aeration capacity to treat a wastewater flow of 30 ML/d with a BOD of 250 mg/L and a desired effluent BOD of 10 mg/L. Assume yield coefficient σ = 0.40, k_e = 0.03/d, SRT = 30 d and MLSS = 3.0 kg/m^3. (15 000 m^3)
8) A conventional aeration tank is required to treat a flow of 4.0 ML/d of primary effluent. Raw wastewater BOD is 200 mg/L and 35% removal is expected in the primary treatment. The MLSS concentration is to be maintained at 2000 mg/L and an F/M ratio of 0.22/d is specified. Compute the capacity of the aeration tank. If the side water depth is 4.0 m and the length is three times the width, how long should the tank be? (1180 m^3, 30 m)
9) A suspended solids test on a sample of mixed liquor from an aeration tank indicates an MLSS of 1850 mg/L. In a 30-minute settling test, the sludge volume is measured to be 150 mL/L.
   a) Compute the SVI. (81 mL/g)
   b) Assume the settling test exactly simulates settling in the secondary clarifier. What minimum return ratio is recommended? (18%)
   c) If R is 25%, what is the expected concentration of solids in the return sludge? (0.93%)
10) The mixed liquor in a 1.5 ML capacity aeration tank operates at an MLSS of 2200 mg/L. The return sludge recirculation ratio is maintained at 40%. If the target MLSS is 2000 mg/L, what additional volume of sludge needs to be wasted? (39 m^3)
11) An extended aeration plant has an influent BOD of 200 g/m^3 and an aeration period of 36 hours. What size tank would be required to serve a trailer park with 15 hook-ups, averaging 2.0 people per trailer? Find the operating F/M ratio if the system is to maintain an MLSS of 1500 mg/L. Due to less convenience available at the trailer, assume a per capita wastewater production of 200 L/person.d. (9.0 m^3, 0.089/d)

25 Stabilization Ponds

Oxidation ponds are also known as lagoons and, more appropriately, stabilization ponds. They are the most common secondary treatment process in small rural communities. They are popular in rural areas due to the availability of large areas of land at low cost, and their simplicity of operation.

Some authors distinguish lagoons from ponds in that oxygen is provided by artificial aeration in a lagoon. The lagoons or stabilization ponds are shallow, impervious or watertight basins formed by excavating the topsoil and building earthen dikes. These basins are then lined with clay to prevent leakage. The adjoining groundwater is monitored for any possible contamination.

Non-aerated stabilization ponds can be aerobic, anaerobic or facultative. Aerobic ponds are shallow so as to maintain aerobic conditions. This is common for treating secondary effluents, thus providing tertiary treatment.

In facultative ponds, oxygen requirements are met by the transfer of oxygen at the air-water interface and by photosynthesis within algae and wind aeration. The water temperature is also a factor due to the solubility of oxygen. These types of lagoons are 0.9 to 1.5 m deep to accommodate oxygen requirements. In Northern climates, depths of 1.5 m are usual to prevent freezing of the entire depth. Anaerobic ponds are relatively deep and are usually employed to treat strong industrial wastes. A scum can form at the top and keeps odours under control.

25.1 FACULTATIVE PONDS

As the name implies, facultative ponds use both aerobic and anaerobic microorganisms. As shown in **Figure 25.1**, wastewater enters the pond and heavy solids settle out at the inlet where anaerobic bacteria break down the complex organics. These organics become organic acids upon which the aerobic bacteria in the surface layer feed and finally convert to gases and nutrients.

Algae consume nutrients and some by-products of the anaerobic reaction to release oxygen into the water. This reaction is called photosynthesis. As the name indicates, sunlight is required for this bio-reaction to take place. Thus, algae can grow only during daylight hours. Therefore, the dissolved oxygen (DO) level is highest during the middle of the day when photosynthesis is at its peak. In normal operation, DO levels can reach as high as 20 mg/L. Since carbon dioxide is consumed by the algae during photosynthesis, the pH of the water rises as photosynthesis progresses. During normal operating conditions, both DO and pH will show diurnal variation.

In Northern climates, photosynthesis may not happen at all due to the blocking of sunlight by the frozen layer. In such cases, wastewater is stored till spring when algae are re-established. Facultative lagoons are the most used systems in rural areas for the treatment of municipal waste.

Advantages

- Low initial and operating costs.
- Can withstand hydraulic and BOD shock loads.
- Can be easily redesigned and reconstructed for any modification.
- No highly skilled supervision is required.

Disadvantages

- More land area is needed due to low permissible loading.
- Odour problems, especially after the spring thaw.

DOI: 10.1201/9781003231264-25

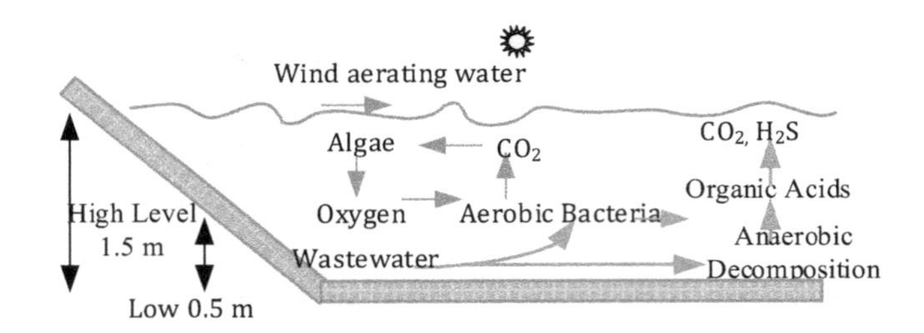

FIGURE 25.1 Facultative Pond Biology

- Suspended solids are usually higher due to algae leaving with effluent.
- Difficult to maintain uniform quality of effluent.

Where the facultative lagoons are organically overloaded, some artificial aeration is used to maintain aerobic conditions in the top layer. However, aeration is limited to allow some settling and keep the bottom layer anaerobic. This idea was originally used to upgrade overloaded facultative lagoons where expansion was not possible. Aerated facultative lagoons can handle as much as 10 times more loading. Detention time can be as small as two to five days compared to 10 to 30 days in a normal facultative lagoon.

25.2 LOADING PARAMETERS

Design parameters for lagoons depend on whether they are aerobic or anaerobic. Anaerobic lagoons are dependent on detention time for their treatment; therefore, volumetric loadings, rather than surface loading, are used. The organic loading in anaerobic lagoons is measured in relation to the volume available. Volumetric loadings are expressed as mass rate of BOD per unit volume, $g/m^3 \cdot d$. Aerated lagoons use organic loading based on water surface area. Typically, the BOD loading rate in lagoons is 2–5 $g/m^2 \cdot d$. Hydraulic loadings on lagoons are usually expressed as surface loading or overflow rate. In Indian conditions, as day temperature is high, organic loading as high as 3–5 times may be afforded.

- Surface loading rate is typically in the range of 10–15 mm/d.
- Higher BOD loading can be afforded in warmer climates.
- Hydraulic loading can vary over a wide range depending on the load applied, the depth of the wastewater, evaporation, and seepage losses.

Example Problem 25.1

A facultative pond for a small town consists of a 6-ha primary cell and two smaller cells of 3-ha each. The average daily wastewater flow is 1.0 ML/d containing a BOD of 200 mg/L. Calculate the BOD loading based on the area of the primary cell.

Given: Q =1.0 ML/d

BOD = 200 mg/L = 200 kg/L

Total area = 12 ha

Primary = 60 000 m^2

SOLUTION:

$$BODLR = \frac{Q \times BOD}{A_S} = \frac{1.0\ ML}{d} \times \frac{200\ kg}{ML} \times \frac{1000\ g}{kg} \times \frac{1}{600000\ m^2} = 3.33 = \underline{33\ g/m^2.d}$$

Example Problem 25.2

A facultative pond has an average length of 220 m with an average width of 140 m. given that the flow rate to the pond is 1100 m^3/d, and is operated at a depth of 1.8 m, what is the detention time in days?

Given: Q = 1100 m^3/d
L = 220 m
W = 140 m
d = 1.8 m

SOLUTION:

$$t_d = \frac{V}{Q} = \frac{L \times W \times d}{Q} = 220\ m \times 140\ m \times 1.8\ m \times \frac{d}{1100\ m^3} = 50.0 = \underline{50\ d}$$

25.2.1 BOD Removal

As wastewater flows through an oxidation pond, BOD is removed by biological oxidation. As the flow pattern is basically plug flow, BOD removed can be related to detention time by knowing the BOD rate constant K. Since BOD is a first-order reaction, BOD reaction rate slows as less and less BOD remains. BOD remaining at any time t is a function of the initial BOD.

$$BOD_t = BOD_i \times e^{-kt} \quad or \quad t = \frac{1}{k} \times \ln\left(\frac{BOD_t}{BOD_i}\right)$$

BOD_t = BOD in the effluent leaving after time t
BOD_i = BOD in the influent entering the pond
t = detention time
k = BOD rate constant

For a given percent removal, BOD_t/BOD_i is known. For example, for 90% removal, the ratio of influent and effluent BOD would be 10 and for 95% removal BOD_t/BOD_i would be 20. Another thing to keep in mind is that k is based on natural log and is usually given for 20°C. For other operating temperatures, use the following empirical relationship: K: $k_T = k_{20} \times (1.047)^{T-20}$

Some authors treat a stabilization pond as a completely mixed flow reactor. However, in facultative ponds, only the liquid portion can be treated this way since solids settled at the bottom are not re-suspended. Since the mass of solids settled cannot be quantified, mass balance can only be done for soluble food or a fraction of soluble BOD, S. Assuming conversion as a first-order reaction, mass balance can be written as:

$$Q\ S_0(in) = QS(out) + VkS(consumed) \quad or \quad \frac{S}{S_0} = \frac{1}{1 + k(V/Q)} = \frac{1}{1 + kt}$$

S/S_0 = fraction of soluble BOD remaining
K = reaction rate coefficient
t = hydraulic detention time

Usually, there is more than one pond connected in series to achieve the desired effluent quality. The effluent of one pond becomes influent to the next in series. Hence, the expression for soluble BOD remaining in the nth pond becomes:

$$\frac{S}{S_0} = \frac{1}{1+(kt/n)^n}$$

Because of the shape of the reactor and the inlet/outlet design, wind action flow conditions in ponds are somewhere in-between plug flow (dispersion = 0) and completely mixed flow (dispersion = ∞). Graphical relationships between the term t and substrate removals have been developed and can be used in sizing ponds.

Example Problem 25.3

Find out the detention time required to remove 90% of BOD in a stabilization pond with a minimum operating temperature of 12°C. Assume a BOD rate constant of 0.22/d at 20°C. Determine the pond area needed to treat a flow from 9500 p at 180 L/c.d. Assume the pond is operated to maintain a flow depth of 1.5 m.

Given: K_{20} = 0.22/d
BOD_i/BOD_e = 10
T = 12°C
Q = 180 L/p·d
Pop. = 9,500
d = 1.5 m

SOLUTION:

$$Q = \frac{180\ L}{p.d} \times 9500\ p \times \frac{m^3}{1000L} = 1710 = \underline{1700\ m^3/d}$$

$$K_T = K_{20}(1.047)^{T-20} = 0.22/d \times (1.047)^{12-20} = 0.152 = 0.15/d$$

$$t = \frac{1}{K}\ln\left(\frac{BOD_i}{BOD_e}\right) = \frac{d}{0.152} \times \ln 10 = 15.1 = \underline{15\ d}$$

$$A = \frac{V}{d} = \frac{Q \times t}{d} = \frac{1710\ m^3}{d} \times \frac{15.1d}{1.5\ m} \times \frac{ha}{10000\ m^2} = 1.726 = \underline{1.7\ ha}$$

Example Problem 25.4

Design a rectangular stabilization pond for treating wastewater from a subdivision of 6,500 people, contributing at 130 L/c.d, containing a BOD of 260 mg/L.

a) Assume the width of the pond is one-third its length and a BOD loading of 20 g/m².d can be afforded.
b) Find the detention time if the pond is operated by maintaining a water depth of 1.4 m.
c) Select the size of the inlet pipe to flow at an average velocity of 1.0 m/s. Assume most of the flow is contributed over a period of 12 hours.

Given: Q = 130 L/c·d
BOD = 260 mg/L
L = 3 × W
BOD = 20 g/m².d
6500 p
d = 1.4 m
v = 1.0 m/s

SOLUTION:

$$Q = \frac{130\ L}{p.d} \times 6500\ p \times \frac{m^3}{1000L} = 845 = \underline{850\ m^3\ /\ d}$$

$$BODL = \frac{845\ m^3}{d} \times \frac{260\ g}{m^3} \times \frac{kg}{1000\ g} = 219.7 = \underline{220\ kg\ /\ d}$$

$$A = \frac{BODL}{BODLR} = \frac{219.7\ kg}{d} \times \frac{15.1d}{1.5\ m} \times \frac{ha}{10000\ m^2} = 1.726 = \underline{1.7\ ha}$$

$$W = \sqrt{\frac{A}{3}} = \sqrt{\frac{1726\ m^2}{3}} = 23.96 = \underline{24\ m}$$

Select a pond of size 72 m × 24 m × 2.4 m (1 m f.b.).

$$t = \frac{Q}{V} = 72\ m \times 24\ m \times 1.4\ m \times \frac{d}{845\ m^3} = 2.86 = \underline{3.0\ d}$$

$$A = \frac{Q}{v} = \frac{845\ m^3}{d} \times \frac{s}{1.0\ m} \times \frac{d}{12h} \times \frac{h}{3600s} = 0.0195\ m^2$$

$$D = \sqrt{\frac{4A}{\pi}} = \sqrt{\frac{4 \times 0.0195\ m^2}{\pi}} = 0.157 = \underline{160\ mm}$$

The diameter of the outlet pipe may be taken as 50% more, that is, 160 × 1.5 = 240 mm.

$$\frac{S}{S_0} = e^{-kt} = e^{-\frac{0.23}{d} \times 2.86d} = 0.52$$

$$\frac{S}{S_0} = \frac{1}{1+kt} = \frac{1}{1+\frac{0.23}{d} \times 2.86d} = 0.60$$

25.2.2 Winter Storage

BOD removal in ponds is very much dependent on climatic conditions. During winter, bacterial activity and algae growth are both severely slowed by cold temperatures. In areas with snow, this is evident from the strong odours in the spring thaw.

In facultative ponds, operating water depths range from 0.5 to 1.5 m. The minimum depth is needed to prevent the growth of aquatic weeds. In Northern climates, the water level in the pond is lowered to the minimum level before the winter sets in. Discharge in the winter is minimized or completely stopped and the incoming wastewater is stored until spring. The pond area should be large enough to store the wastewater over the winter months.

Example Problem 25.5

A stabilization pond of 12-ha receives an average daily flow of 1200 m^3/d. Since the lagoon is operating in colder climates, discharge is stopped in the early winter and the water level is dropped to 0.60 m. Estimate the number of days of winter storage available between 0.60 m and 1.5 m water levels, assuming an evaporation and seepage loss of 2.5 mm/d.

Given: d_{sto} = 1.5 m - 0.60 m = 0.9 m
A_S = 12 ha = 120 000 m^2
Q = 1200 m^3/d d_{loss} = 2.5 mm/d

SOLUTION:

$$Q_{loss} = \frac{2.5\ mm}{d} \times \frac{m}{1000\ mm} \times 120000\ m^2 = 300.0 = 300\ m^3/d$$

$$Q_{net} = 1200 - 300.0 = 900.0 = 900\ m^3/d$$

$$t_{sto} = \frac{V_{sto}}{Q_{net}} = \frac{A_S \times d_{sto}}{Q_{net}} = 0.9\ m \times 120000\ m^2 \times \frac{d}{900\ m^3} = 120.0 = \underline{120\ d}$$

Example Problem 25.6

A sewage lagoon with discharge control has a total surface area of 22-ha. Average daily flow is 3200 m^3/d. Calculate the minimum water depth for a retention time of 90 days, assuming the difference between evaporation and seepage and precipitation is 1.0 mm/d.

Given: A_S = 22 ha = 220 000 m^2
Q = 3200 m^3/d
t_{stor} = 90 d
d_{sto} = ?

SOLUTION:

$$Q_{loss} = \frac{1.0\ mm}{d} \times \frac{m}{1000\ mm} \times 220000\ m^2 = 220.0 = \underline{220\ m^3/d}$$

$$t_{sto} = \frac{V_{sto}}{Q_{net}} = \frac{A_S \times d_{sto}}{(Q_{in} - Q_{loss})} \quad or \quad d_{sto} = \frac{t_{sto}(Q_{in} - Q_{loss})}{A_S}$$

$$d_{sto} = 90\, d \times \frac{(3200 - 220)\, m^3}{d} \times \frac{1}{220000\, m^2} = 1.20 = \underline{1.2\, m}$$

25.3 LAGOON DEPTH

Depending on the depth of water in a lagoon, the bioreaction may be aerobic, anaerobic or facultative. Lagoons of less than one metre in depth may be completely aerobic if there are no solids on the bottom because of the depth of sunlight penetration. This type of lagoon would rely on algae to convert the waste materials and add sufficient oxygen to maintain aerobic conditions. Lagoons with depths greater than 1.2 m, as in facultative lagoons, would allow for greater conservation of heat from incoming wastes. This enhances biological activity as the ratio between lagoon volume and lagoon area is more favourable.

In facultative lagoons, depths over 1.2 m provide physical storage for dissolved oxygen accumulated during the day. Since photosynthesis only occurs in sunlight, stored dissolved oxygen is important for maintaining aerobic conditions, particularly during the colder months when nights are long.

Shallower lagoon depths are generally used in warmer climates where freezing is less of an issue. A deeper pool is recommended for locations where winters are severe. Large volumes allow some room for freezing without totally compromising the total volume available in the lagoon and keep the process active during the winter months. Not only does the frozen layer provide insulation, it also allows for the development of a multi-layered reaction within the lagoon. The deeper layers of the lagoon will act in an anaerobic manner.

25.4 ALGAE

Though the growth of algae is important, if discharged with plant effluent it has negative effects, including increased turbidity, suspended solids and biochemical oxygen demand. For this reason, algae need to be monitored carefully to ensure they perform properly without affecting the overall lagoon performance.

Algal blooms refer to the rapid mass growth of algae. This usually happens seven to twelve days after wastes have been introduced into the lagoon. After another week, the bacterial decomposition of bottom solids will usually become established, limiting the food produced for the algae.

This is generally revealed by the sight of bubbles coming to the surface near the lagoon inlet, where most of the sludge deposits occur. These blooms are all associated with the equinox and solstice when the sunlight patterns change the reactions within the lagoon.

25.4.1 Algae Growth Factors

As surface water warms, algae tend to remain in the surface layer where they breed and grow in excess sunlight. Algae consume carbon from water in the form of bicarbonate, causing a rapid increase in pH (often pH 10–11), which may cause the precipitation of phosphate and iron from the nutrients in the solution. As discussed earlier, pH and DO show diurnal variation in a well-operating facultative lagoon.

25.4.2 Temperature

Any biological process is seriously affected by temperature. A bio-reaction typically accelerates with an increase in temperature. As the seasons change, the air temperature and the amount of solar radiation added into the lagoon will cause the water temperature to fluctuate, which will affect the operation of the lagoon.

Every reaction within the lagoon has an optimum temperature where the reaction is most efficient. Since the biological processes are affected by other factors such as hydraulic loading, it is important that each lagoon be individually treated and monitored. The potential for a lagoon to require continued operation in winter is the main reason Northern lagoons tend to be in the 3.0–4.0 m depth. This added depth will keep the whole lagoon from freezing over. If a lagoon is totally frozen, no reactions will be taking place.

25.5 BERMS

The selection of the steepness of the berm slope depends on several variables. A steep berm minimizes waterline weed growth. To prevent erosion, berm materials must be rocky or protected by riprap. A gentle slope will erode the least and is easier to operate equipment open when performing routine maintenance. The usual slope on a berm for a lagoon is 1:4. The other important aspect of a berm slope is its correlation with evaporation within the lagoon. A steeper slope will result in less evaporation from the lagoon than a shallow slope. If high winds are expected in the area where the lagoon is to be constructed, the lagoon should be arranged so that the winds will blow across the short width of the lagoon rather than the length to reduce berm erosion caused by waves. The width of the lagoon will give the water less time to build up speed and will increase the wave height.

25.6 DAILY MONITORING

One of the most effective types of monitoring in the operation of a lagoon is daily observation. There are no equipment requirements, and with some operational experience the operator can determine the condition of the lagoon using the following observations.

25.6.1 Visual Monitoring

Just as a physician can make a good diagnosis based on key observations like pulse rate, pH can lead to good indications about the health of the lagoon. Several signs indicate the level of pH in a lagoon. For example, a deep green sparkling colour generally indicates a high pH and satisfactory dissolved oxygen content. A dull green colour or lack of colour generally indicates a declining pH and lowered dissolved oxygen content. A grey colour indicates the lagoon is being overloaded or not working properly.

25.6.2 Water Level

Lagoons are usually equipped with a post that has markings on it to check water depth. Depth should be recorded at the same time each month to monitor changes. This provides information for the normal operation of the lagoon, for changes and trends, and acts as a record in case of a problem.

Wastewater lagoons sometimes need added water to maintain their minimum depth. The method for adding water must be controllable because flows during the year will affect the depth in the lagoon. Periods of high or low flows will require a varying amount of supplemental water to maintain the correct operating depth of the lagoon. The best lagoon depth is only determined through experience with the lagoon. However, it is a good idea to leave a 0.5 m free board above the maximum height of the water to the top of the berm in case of heavy storms.

25.6.3 Water Colour

As shown in **Table 25.1**, algae colour is directly related to pH and dissolved oxygen, and is a good indication of the health of the pond. A change to a less desirable colour has a cause and may require correction. A record of the colour and depth of the water should be made daily.

TABLE 25.1
Colour of Algae (Visual Monitoring)

Color	Conditions	Symptoms or Cause
Dark sparkling green	Good	pH and dissolved oxygen (DO) ideal.
Dull green to yellow	Not so good	DO and pH are dropping. Blue-green algae are becoming predominant.
Grey to black	Very bad	Lagoon is septic. Anaerobic conditions prevail.
Tan to brown	Okay if brown algae	Erosion or inflow of surface water.

25.6.4 Sampling

Sampling is the key activity in monitoring the operation of a lagoon or any other facility. Samples are required to monitor pH, temperature and dissolved oxygen. Since these parameters are subject to quick change, samples should be read as soon as possible. It is preferable to use portable meters so that values can be read at the sampling location wherever practical. Good records should be maintained for future reference and evaluating performance.

Some precautions should be taken into consideration when collecting a sample. When sampling for DO, it is important to avoid getting any atmospheric oxygen into the sample. If possible, use an electric meter and probe, being careful not to allow the membrane on the end of the probe to be exposed to the atmosphere during the actual DO measurement of the water sample.

Sample Location

Samples should always be collected from the same point or location. Typical locations for raw wastewater (influent) samples are the wet well of the influent pump station or at the inlet control structure. Samples of lagoon effluent should be collected from the outlet control structure or a well-mixed point in the outfall channel. It is a good idea to take multiple samples.

The influent, effluent, lagoon cells and sand filtration components of the lagoon should be tested regularly. This will provide the operator with a definite knowledge of what is happening in every part of the lagoon. Without this type of testing, the operator may not catch a problem until after it has disrupted the entire lagoon process. To maintain representativeness, samples should be collected from a point 2.5 m out from the water edge and 0.3 m below the water surface. Samples collected after high storms or winds may not be representative because solids will be stirred up after this activity.

25.7 OPERATIONAL PROBLEMS

Lagoon operation usually does not require much skill. Here are some common problems that might be encountered during the operation and maintenance of a lagoon.

25.7.1 Scum Control

Accumulation of scum is common in the spring when the water warms and vigorous biological activity resumes. If scum is not broken up it will dry and become crusted, providing a home for blue algae, creating odours. Scum also blocks sunlight, reducing the production of oxygen by algae.

There are many ways to break up scum. In a natural process, wind will help break up floating scum blocks and once broken into smaller pieces the scum will usually settle to the bottom of the lagoon. When the scum is near the edges of a lagoon, the operator can rake it out. Sometimes the scum must be broken up, which the operator can do with jets of water from pumps or tank trucks, or outboard motors on boats in large lagoons.

25.7.2 Odour Control

Most public complaints relate to odours, so plant operators need to do their best to keep them under control. Most odours are caused by overloading or poor housekeeping practices and can be remedied by taking corrective measures. If a lagoon is overloaded, loading must be stopped and influent diverted to other lagoons, if available, until the odour problem stops. The lagoon should then be gradually loaded again. Another reason for odours might be an unexpected plant shutdown. In such cases, it is recommended that an emergency odour control plan be available.

Odours occur during the spring warm-up in colder climates because biological activity has been reduced during cold weather. When water warms, microorganisms become active and use up all the available dissolved oxygen, and odours are produced under anaerobic conditions.

Floating aerators and heavy chlorination might help treat odours, but these treatments are usually very expensive. Re-circulation from an aerobic lagoon to the inlet of an anaerobic lagoon will reduce or eliminate odours.

Chemicals that mask odours are also used and should be ordered before the spring thaw when the odour problem is expected to be more severe. Some facilities opt for using sodium nitrate as a source of oxygen for microorganisms rather than sulphate compounds. Once the sodium nitrate is mixed into the lagoon it acts very quickly because many common organisms may use the oxygen in nitrate compounds instead of dissolved oxygen.

25.8 LAGOON MAINTENANCE

Lagoons can be maintained in two ways: artificial or natural. The artificial method tries to regulate every single aspect of the process of the lagoon, surrounding berm and environment. This process commonly removes unplanned plant growth and animal life. The disadvantages to this form of maintenance are that it can cause extra work for the operator, and one forced change may require a further change to balance the lagoon again.

The primary advantage is aesthetics. With this kind of maintenance, the lagoon will look like a controlled operation rather than a natural wetland. The natural method will allow for natural solutions to some of the problems faced when operating a lagoon. The natural method tries to use naturally occurring vegetation and animals to control the operations of a lagoon. This method also allows for unexpected plant or animal growth to remain as long as it does not affect the treatment process or pose a health and safety hazard. The major disadvantage to this form of maintenance is that the lagoon can look unkept and may not leave a good impression on visitors.

25.8.1 Lagoon Weeds

There are two methods used for dealing with weed growth that can occur in and around a lagoon. One method is by conducting daily inspections and immediately removing young plants. Some types of weeds harbour mosquitoes, hinder lagoon circulation and lead to scum accumulation, so must be removed.

A more natural solution is to plant vegetation that does not upset the lagoon's operation as much. Some options include bullrushes and duckweed. Duckweed can be a very useful growth in the lagoon because it can trap mosquito larvae and prevent them from developing, and it can be harvested and dried to act as a soil conditioner. Apart from duckweed's ability to control mosquitoes, it can also uptake phosphorus in the wastewater, thus reducing phosphorus loading in the effluent.

25.8.2 Berm Erosion

Berm slope erosion is caused by wave action or surface run-off from precipitation and is probably the most serious maintenance problem. If allowed to continue, it can result in a narrowing of the berm crown, making accessibility with maintenance equipment difficult.

If the berm is composed of erodible material, one long-range solution is the use of bank protection, such as stone riprap or broken concrete gravel. A semi-porous plastic sheet can be used with riprap that allows the two-way movement of air and water but prevents the movement of soils. This sheet also discourages weed growth and digging by crayfish.

The other long-range solution is to plant grasses and plants that will help anchor the berm slopes against erosion. Portions of the lagoon berm or dike not exposed to wave action should be planted with a low growing spreading grass to prevent erosion by surface run-off. Grazing animals should not be allowed to control vegetation because they may damage the berm near the waterline and complicate erosion problems. Plants with long roots can also damage the berm and cause berm failures and costly repairs. Berm tops should be crowned so that rainwater will drain over the side in a sheet flow. Otherwise, the water may flow a considerable distance along the berm crown and gather enough flow to cause erosion when it finally spills over the side and down the slope. If the berm is used as a roadway, make sure it is paved or well gravelled.

25.8.3 Mosquitoes

Mosquitoes will breed in sheltered areas of standing water where there is vegetation or scum to which egg rafts of the female mosquitoes can become attached. These egg rafts are fragile and will not withstand the action of disturbed water surfaces such as caused by wind action or normal currents. One solution to mosquitoes is to stock the lagoon with mosquito fish that will eat the mosquito larvae. Another is to use duckweed in the pond because mosquito larvae will not survive when covered in duckweed. Another solution is to encourage mosquito-eating birds to live in the area. One of the more effective ways of doing this is to plant low bushes around the lagoon. This gives the birds a landing place when they are hunting mosquitoes. Placing birdhouses in the area can also encourage purple martins and swallows to live in the area. Dragonflies will also eat mosquitoes, and if they are introduced into the lagoon, they can help reduce the number of mosquitoes surrounding a lagoon.

25.8.4 Daphnia

Minute shrimp-like animals called daphnia may infest the lagoon from time to time during the warmer months of the year. The daphnia will reproduce in great numbers, usually appearing in the lagoon three to seven days after an algae bloom. These predators live on algae and, at times, will appear in such numbers as to almost clear the lagoon of algae. During the more severe infestations there will be a sharp drop in the dissolved oxygen of the lagoon accompanied by a lowered pH because of the reduced algae. This is a temporary condition because the predators will overload the algae supply, causing a mass die-off of daphnia that will be followed by the rapid growth of algae. When the algae concentration in a lagoon is low under these conditions, lagoons operated on a batch basis may find this a good time for the release of water due to a low suspended solids value.

25.8.5 Screenings

During storm periods, especially during spring thaw in colder climates, screens should be inspected at least once or twice a day. The screenings should be disposed of in a sanitary manner to avoid odours and fly breeding. There are many ways to dispose of screenings and organic matter deposited in the grit channels. One is by burial, where the operator can dig a trench and dispose of the screening, or the screening can be trucked to a sanitary landfill.

Discussion Questions

1. Explain the role of algae in a facultative pond.
2. Compare an oxidation pond with an oxidation ditch.

3. Explain the term symbiosis as applied to facultative stabilization ponds. Which microorganisms are responsible for the symbiotic relationship?
4. Make a comparison between an aerated lagoon and a stabilization pond.
5. Explain why stabilization ponds are more common in rural settings.
6. Explain the mechanism by which BOD is removed in a facultative pond.
7. How are scum and odour controlled in a lagoon system?
8. Describe the advantages and disadvantages of natural and artificial maintenance of a lagoon.
9. Related to lagoon maintenance, explain mosquito control, weed control, berm erosion and screenings.
10. In a facultative lagoon, which two parameters show diurnal variation and why?
11. List the parameters that must be monitored by sampling to assist in operating a facultative system?
12. How can the colour of algae growth in a facultative pond be used to monitor the health of its operation?
13. Visual monitoring is important in the operation of a lagoon system. In the operation of a facultative pond, what interpretations can be made based on the colour of algae?
14. Explain in brief the principles and applications of aerobic, anaerobic and aerated lagoons.
15. In colder climates, no effluent is discharged over the winter months. How does this affect the operation of a facultative lagoon?

Practice Problems

1. Select the size of a rectangular stabilization pond for treating wastewater from a new development with 5000 people, contributing at 140 L/c.d with a BOD of 280 mg/L. Due to warmer climatic conditions, a BOD loading of 25 g/m^2.d can be afforded. (130 m × 65 m)
2. A 1-ha pond lagoon is operated by maintaining a water depth of 1.3 m. What detention time is required if the pond is treating flow contributed by 6000 people at a rate of 120 L/c.d? (18 d)
3. Using the data from Practice Problem 2, select the size of inlet pipe to flow at an average velocity of 1.0 m/s. Assume most of the flow is contributed over a period of 10 hours. (160 mm)
4. What detention period must be provided to achieve 90% BOD removal in a stabilization pond? Assume a BOD reaction rate constant of 0.20/d. (12 days)
5. Design an oxidation pond to serve a community of 10 000 people assuming a per capita wastewater contribution of 190 L/c.d containing a BOD of 280 g/m^3. As this community is based in a tropical region, a BOD loading of 30 g/m^2.d can be afforded.
   a. Assume two primary and one secondary pond, all with a length twice as much as their width. (# = 3, each 110 m × 55 m)
   b. Assume a per capita sludge production of 75 L/c.a on an annual basis. If the bottom 0.40 m of depth is for the storage of sludge, how long would it last before the sludge needs to be removed? (10 years)
6. Based on allowable loading of 25 g/m^2.d, what pond area is needed to treat a wastewater flow of 1.5 ML/d with a BOD content of 260 mg/L? Determine the minimum operating depth for providing a retention time of 15 days with a net loss of water of 2.0 mm/d. (1.6 ha, 1.4 m)

7. Design a facultative pond to serve a community of 8000 people with a per capita wastewater discharge of 160 L/c.d with a BOD of 250 mg/L. Water regulations require that the effluent BOD does not exceed 30 mg/L. Assuming a BOD rate constant of 0.16/d at the operating temperature and allowable BOD loading of 20 g/m^2.d, determine:
   a. The pond area required. (1.6 ha)
   b. The detention period achieved. (13 days)
   c. The pond capacity/depth. (17 ML, 1.1 m)
8. For the data of Practice Problem 7, it is planned to have a parallel-series system consisting of a total of six ponds with four primary and two secondary ponds. Each row contains two primary and one secondary pond. Assuming the length of each pond is 2.5 times the width, select the size of each pond. (82 m × 33 m)
9. Design a rectangular stabilization pond for treating wastewater from a subdivision of 5500 people, contributing at 110 L/c.d with a BOD of 210 mg/L. The pond will have two primary cells in parallel and a common secondary cell. Assume the width of each cell is one-third of the length and, due to warmer conditions, a BOD loading of 25 g/m^2.d can be afforded. Select the size of each cell. (#3, 2 primary, 1 secondary, each 72 m × 24 m)
10. Due to an increase in odour problems in a facultative lagoon, sodium nitrate is to be applied at 55 kg/ha·d in the wake of a motorboat. How many kg of sodium nitrate will be needed to treat a lagoon surface measuring 130 m × 200 m? (140 kg/d)
11. Using the data from Practice Problem 9:
    a. Find the detention time if the pond is operated by maintaining a water depth of 1.4 m. (12 days)
    b. Select the size of the inlet pipe to flow at an average velocity of 1.0 m/s. Assume most of the flow is contributed over a period of 12 hours. (#2, 100 mm)

26 Attached Growth Systems

26.1 TRICKLING FILTERS

Trickling filters, one of the oldest forms of biological treatment, can achieve a good quality effluent in three to four hours. Unlike the activated sludge process, trickling filters fall into the category of fixed growth systems. Trickling filters are essentially biological contact beds and are also called biological filters. A typical trickling filter plant flow diagram is shown in **Figure 26.1**.

The words filtration and filter are misnomers, as there is no straining or filtering action involved. Biological filtration consists of a tank filled with a fixed media on which the biological growth lives and the primary effluent is sprayed. There is also a thin anaerobic section at the bottom, even though trickling filters are an aerobic treatment.

Trickling filters alter the characteristics of sewage but do not remove solids. They convert non-settleable colloidal and dissolved solids to readily settleable solids. These largely organic solids are converted to living microscopic organisms or they become attached to the biological media. This build-up of solids is continuously unloading in small amounts, resulting in the need for a secondary clarifier. The falling of the slime layer attached to the filter media is called sloughing. The solids sloughing (breaking) off the media is called humus.

Wastewater is applied on the top of the filter by a water distribution system such as sprinklers. As the water trickles through the filter, organisms attached to the media pick up BOD from the passing water. The hydraulic loading needs to be very tightly controlled: too much flow can cause ponding and anaerobic conditions; too little may not be able to move the distributor arm of the sprinkler system. An important aspect of recirculation in trickling filter systems is maintaining the minimum hydraulic loading during the minimum flow period.

Although trickling filters are known for their ease of operation and sturdiness, the problem of plugging is a major disadvantage. Plugging occurs due to excessive organic growth, which plugs air passages. This results from organic overloading, which creates anaerobic conditions as well as odour problems. Other disadvantages include high cost of construction, large area requirements for set-up, the need to be covered in colder climates, they can cause odour and fly nuisance, and they need pumping. Advantages include ease of operation, they are quite forgiving, no aeration equipment is needed, their operational costs are less, and they produce high-quality effluent.

26.1.1 Combined Systems

Although the activated sludge process demands skilled operation, it offers flexibility for controlling the quality of effluent. Combined systems are becoming popular for taking advantage of the strengths offered by fixed growth and suspended growth systems. For example, combining a trickling filter with the activated sludge process has helped eliminate shock loads to the highly sensitive activated sludge process and provides high-quality effluent. Using a trickling filter alone cannot be expected to yield high removals.

26.2 MAIN COMPONENTS OF A TRICKLING FILTER

The main components of a biological filter are filter media, filter underdrains and wastewater distribution system. A brief discussion follows.

DOI: 10.1201/9781003231264-26

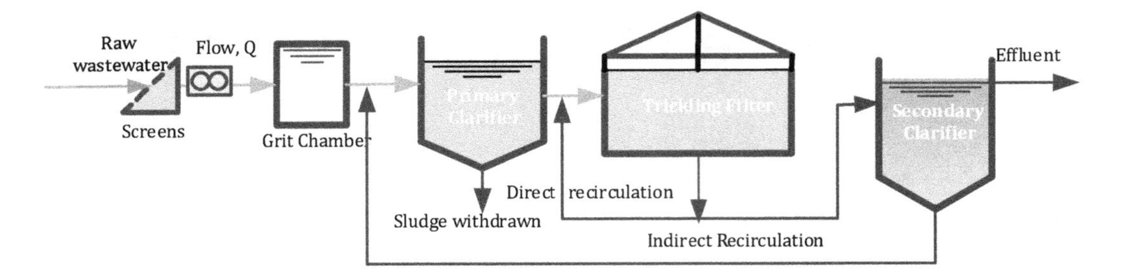

FIGURE 26.1 Trickling Filter Plant

26.2.1 Filter Media

The filter media supports biological growth. This slime growth, sometimes called a zoogleal film, contains microorganisms. The media may be rock, coal, brick or moulded plastics. The media should provide sufficient porosity (voids) for air to ventilate the filter. The media depth ranges from 1 m to 2.5 m for rock media filters and 5 m to 10 m for synthetic media, also called biological towers. The material used for the filter media should be clean, inert and range in size from 25–75 mm. Large size media is placed at the bottom above the underdrains.

26.2.2 Underdrains

An underdrain system carries away effluent and supplies an air source. The holes in the underdrain are determined by the amount of air required. The underdrain system is generally made of prefabricated blocks of concrete or vitrified clay. The flow from the underdrains leads to the main effluent channel graded to flow partially full to allow ventilation and the passage of air from the bottom to the top of the filter. This main effluent channel may be provided adjoining the central column of the distributor or along the periphery of the filter. The slope of the effluent channel should be sufficient to ensure a flow velocity of 0.90 m/s. Since the wastewater in the top part of the media is relatively warmer, it creates a natural circulation of air from bottom to top. Unless there is a serious plugging problem, this natural circulation is sufficient to maintain aerobic conditions.

26.2.3 Wastewater Distribution

A distributor arm fitted with orifices provides uniform hydraulic loading to the filter. The number of orifices in each section of the arm varies in proportion to the areal coverage. The arm is usually driven by the hydraulic action of the wastewater flowing out. Return rate is essential to arm rotation. The fixed-nozzle distribution system is not as common as the rotary type. The rotary type distributor requires less maintenance and provides uniform application. In fixed-nozzle distributors, the shape of the filter is usually square or rectangular, but in rotary type distributors, the filter enclosure is always circular. Loading on filters can vary quite a bit depending on the type of filter and the number of stages.

26.3 LOADING ON FILTERS

There are three types of filters: standard or low-rate, high-rate, and super high-rate or roughing filters. The main features, including loading, recirculation, depth, and type of filter media, are shown in **Table 26.1**. Organic loading on a filter is described as mass rate of BOD per unit volume of the

TABLE 26.1
Design Features for Trickling Filters

Design Features	Low Rate	High Rate	Roughing
Hydraulic loading, $m^3/m^2 \cdot d$	1–4	4–40	20–40
Organic loading, $g/m^3 \cdot d$	50–400	500–2000	8000–6000
Depth, m	1.8–3.0	0.9–2.5	4.5–12
Recirculation ratio	Usually absent	0.5–3.0	1–4
Filter media	Rock, gravel Slag	Rock, synthetic	Plastic media
Size of media, mm	25–75	25–60	25–60
BOD removal, %	80–85	65–80	Low
Quality of effluent	Nitrified	Partially nitrified	Not nitrified
Quality of sludge	Black, highly oxidized	Brown, not fully oxidized	Brown, not fully oxidized
Flexibility	Less flexible	More flexible	
Cost of operation	More	Less	
Land requirement	More	Less	
Supervision/operation	Less skilled	More skilled	

filter, much like in the activated sludge process. Hydraulic loading is defined as flow rate per unit surface area of the filter top surface.

$$BOD\ Loading,\ L = \frac{M_{BOD}}{V_F} = \frac{Q \times BOD_{PE}}{V_F}$$

26.3.1 Standard Filter

A standard or low-rate filter has an organic loading of 50 to 400 $g/m^3 \cdot d$ and a hydraulic loading of 10 to 40 $m^3/m^2 \cdot d$. This type of filter produces quite nitrified effluent with a BOD removal of 80% to 85% without any recirculation. Low-rate trickling rate filters are more suitable for the treatment of low-to-medium strength sewage.

26.3.2 High Rate

The basic difference between a high rate and a conventional filter is increased loading. High-rate trickling filters, single-stage and two-stage, are recommended for medium-to-high strength wastewaters. High-rate filters achieve BOD removals of only 65% to 80%, and there is little nitrification. Organic loading is 500 to 2000 $g/m^3 \cdot d$ with a hydraulic loading of 4–40 $m^3/m^2 \cdot d$. Due to increased flows, the biofilm is relatively thinner and is more efficient at supplying oxygen and nutrients to aerobic bacteria.

26.3.3 Roughing Filters

Super high-rate filters are sometimes used to treat high strength wastes or the pre-treatment of domestic wastes. They are characterized by low removal efficiencies and high loading rates. For this reason, they are also called roughing filters. Super high-rate filters use synthetic media, usually interlocking corrugated sheets of plastic, producing a non-clogging filter. Due to more efficient oxygen transfer, the depth of such filters can be as much as 10 m; hence the name biological towers.

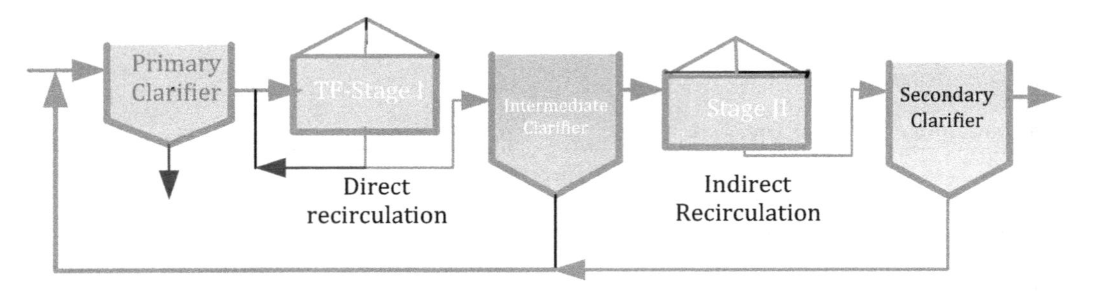

FIGURE 26.2 Recirculation and Staging in TF Filter Plants

26.4 RECIRCULATION

Though recirculation is practised in high-rate filters and not low-rate filters, recirculation of part of the settled (indirect recirculation) or filter effluent (direct recirculation) is one of the important features, as shown in **Figure 26.2**.

Return flows will be called Q_R. Direct recirculation represents the flow being returned directly to the tank they just exited. In the two-stage process, these recycling flows are further subdivided into Stage I and Stage II.

Indirect recirculation includes the recirculation of underflow from the final clarifier to the head unit during low flow periods. This help to maintain the minimum flows required to keep the distributor arm moving and prevent the filter surface from becoming dry. The advantages of recirculation are:

- It increases BOD removal efficiency.
- It maintains higher hydraulic loading that helps reduce plugging and aids in uniform organic load over the filter surface.
- It dampens the variations in strengths and flows of wastewater applied to filters
- It reduces the thickness of the biological film and flies breeding by forced sloughing.
- It keeps the distributor arm running during low flow periods just after midnight hours.
- It loads the lower parts of the filter more effectively.
- It freshens the sewage and thus helps to keep odours in check.

Recirculation Ratio

The recirculation ratio is defined as the ratio of recirculated flow, Q_R, to the influent flow, Q. It usually ranges from 0.5 to 3.0, and > 3 is used with high-strength wastewaters and super high-rate filters. The recirculation ratio is used to determine the required capacity of recirculation pumps and the hydraulic load placed on the filter.

26.4.1 Staging

High-rate or single-stage trickling filters are commonly used in industrial pre-treatment or in situations where there is very strong waste. Two-stage trickling filters are used when a BOD effluent of 30 mg/L is required for a strong sewage. The two-stage system, as shown in **Figure 26.2**, consists of two identical trickling filters in series with an optional intermediate settling tank in-between. The system is designed with several recirculation points from a single-stage to a two-stage trickling filter.

26.5 BOD REMOVAL EFFICIENCY

The BOD removal efficiency of biological filtration primarily depends on factors such as the depth of the bed, kind of media, temperature, recirculation, and organic loading. Empirical equations have been developed to predict BOD removal efficiency based on organic loading and recirculation ratios.

One of the most popular formulations evolved from filter plants at military installations in the USA. The equation given below applies to single-stage stone-media filters followed by a final clarifier and is for treating settled domestic wastewater with a temperature of 20°C. For single-stage or the first stage of a two-stage filter system, BOD removal can be estimated using the following empirical relationship.

$$E_I = \frac{100\%}{\left(1+0.014\sqrt{L/F}\right)} \qquad F = \frac{(1+R)}{(1+0.1R)^2} = \frac{(1+R)}{1+0.01R^2+0.2R} \sim \frac{1+R}{1+0.2R}$$

In this equation, L is the BOD loading on the filter expressed as g/m.³d, and F is based on the recirculation ratio. E_I is the percent BOD removal in the first stage. The term $0.01R^2$ in the denominator is relatively small and can be neglected. Therefore, expressions for percent BOD removal in the first and second stage are as follows:

$$E_I = \frac{100\%}{\left(1+0.014\sqrt{\frac{L(1+0.2R)}{(1+R)}}\right)} \quad \text{and} \quad E_{II} = \frac{100\%}{\left(1+\frac{0.014}{(1-E_I)}\sqrt{\frac{L}{F}}\right)}$$

The two-stage system consists of two identical trickling filters in series with an optional intermediate settling tank in-between. BOD removal is strongly dependent on wastewater temperature. Filters in cold climates operate at lower efficiencies for the same loadings. The following equation may be used to adjust removal efficiencies for temperatures above or below 20°C.

$$E_T = E_{20}(1.035)^{T-20}$$

Suffix T and 20 refer to temperature in degrees Celsius.

26.6 OPERATING PROBLEMS

The major operating problems with trickling filters are ponding or clogging of the filter media and fly and odour nuisance.

26.6.1 Ponding

The ponding of water on the filter surface due to plugging of the filter media can happen due to excessive BOD loading, inadequate hydraulic loading, and high fungus growth in the filter media. As a result of plugging, ventilation drops sharply and the filter becomes partially anaerobic. The problem can be remedied by raking or forking the filter surface. Washing the filters by applying a high-pressure water stream helps. Chlorinating the influent to the filter for periods of two to five hours during low flow periods has been successfully tried.

26.6.2 Fly Nuisance

Trickling filters are prone to fly infestations, especially those operated at low rates. Though these flies do not bite, they can bother the operating personnel. Heavy infestations are associated with thick films and high temperatures. They can be destroyed by flooding the filter and chlorinating the influent. Jetting the filter walls with high-pressure hoses and removing excessive growth should be done. High-rate filters are less prone to such infestations because of high hydraulic loading and continuous sloughing of the growth or humus.

26.6.3 Odour Nuisance

Odours are usually a result of anaerobic conditions due to the partial plugging of filter media. Odour problems are more serious when septic wastewaters are fed to low-rate filters. Odours can be eliminated by recirculating the filter effluent to create aerobic conditions. Aerating or chlorinating plant influent and maintaining a well-ventilated filter keep odours under control.

26.7 SECONDARY CLARIFICATION

Secondary sludge in trickling filter systems is quite different from that of the activated sludge process. Humus is commonly used to describe well-oxidized solids in trickling filter systems. Secondary sludge from tricking filter plants is dark brown, relatively inoffensive and flocculent in consistency. Bulking is usually not a problem, as is the case in activated sludge plants. Settling in the secondary clarifier of a trickling filter system is predominantly of the discrete type.

Example Problem 26.1

Design a single-stage trickling filter based on a hydraulic loading of 20 m/d and a filter media depth of 2.0 m. The wastewater influent is 10 ML/d and the settled wastewater BOD is 150 mg/L. The desired BOD of the filter effluent is not to exceed 30 mg/L. Make suitable assumptions.

Given: d = 2.0 m
Q = 10 ML/d
BOD_i = 150 mg/L
BOD_e = 30 mg/L
R = 2 assumed

SOLUTION:

$$A_S = \frac{(1+R)\times Q}{v_o} = \frac{3\times 10000\ m^3}{d}\times\frac{d}{20\ m} = 1500.00 = 1500\ m^2$$

$$L = \frac{M_{BOD}}{V_F} = \frac{10000\ m^3}{d}\times\frac{150\ g}{m^3}\times\frac{1}{1500\ m^2\times 2.0\ m} = 500.0 = 500\ g/m^3.d$$

$$E_{20} = \frac{100\%}{\left(1+0.014\sqrt{\frac{L(1+0.2R)}{1+R}}\right)} = \frac{100\%}{\left(1+0.014\sqrt{\frac{500(1+0.4)}{1+2}}\right)} = 82.16 = \underline{82\%}$$

$$BOD_e = BOD_i\left(1 - \frac{E}{100\%}\right) = \frac{150mg}{L}\left(1 - \frac{82\%}{100\%}\right) = 27\,mg/L < 30\,mg/L$$

$$Assuming\ two\ filters,\ D = \sqrt{\frac{4A}{\pi}} = \sqrt{\frac{4}{\pi} \times \frac{1500\ m^2}{2}} = 30.9 = \underline{31\ m}\ each$$

Example Problem 26.2

A trickling filter is sized so that BOD loading is 500 g/m³·d and the wastewater is recycled to provide a recirculation ratio of 1.0. Calculate the maximum allowable concentration of BOD in the primary effluent (filter influent) at a temperature of 16°C to produce an effluent BOD of 30 mg/L.

Given: BODL = 500 g/m³·d
R = 1.0
T = 16°C
BOD_e = 30 mg/L
BOD_i = ?

SOLUTION:

Ratio factor

$$F = \frac{(1+R)}{(1+0.1R)^2} = \frac{1+1.0}{(1+0.10)^2} = 1.650 = 1.65$$

$$E_{20} = \frac{100\%}{\left(1+0.014\sqrt{L/F}\right)} = \frac{100\%}{\left(1+0.014\sqrt{500/1.65}\right)} = 80.4 = \underline{80\%}$$

$$E_{16} = E_{20}(1.035)^{16-20} = 80\%(1.035)^{(16-20)} = 69.7 = \underline{70\%}$$

$$BOD_i = \frac{BOD_e}{(1-E)} = \frac{30\ mg/L}{(1-0.70)} = 100.0 = \underline{100\,mg/L}$$

Assuming a primary removal of 35%, the maximum allowable BOD in the raw wastewater is 150 mg/L. Since raw wastewater BOD is usually higher than 150 mg/L, during the winter months, filter effluent may not meet the target of 30 mg/L BOD in the plant effluent. It is suggested to add another stage or polish the filter effluent by adding a tertiary treatment of some kind, such as a polishing lagoon or using the water for land irrigation.

Example Problem 26.3

Size a high-rate trickling filter to treat a wastewater flow of 3.8 ML/d with a settled BOD of 150 mg/L. Assume a recirculation ratio of 1.6 and the plant effluent BOD is not to exceed 25 mg/L.

Given: Q = 3800 m³/d
$BOD_{settled}$ = 150 g/m³
BOD_{filter} = 25 g/m³
R = 1.6

SOLUTION:

$$E = \frac{(150 - 25)\ g}{m^3} \times 100\% = 83.33 = 83\%$$

$$L = \frac{Q \times BOD_{PE}}{V_F} = \frac{3800\ m^3}{d} \times \frac{150\ g}{m^3} \times \frac{1}{V_F} = \frac{5.7 \times 10^5}{V_F}\ g/d$$

$$F = \frac{1 + 1.6}{(1 + 0.16)^2} = 2.35 = 2.4$$

$$\sqrt{\frac{L}{F}} = \frac{1}{0.014} \times \left(\frac{1 - E}{E}\right) = \frac{1}{0.014} \times \left(\frac{1 - 0.83}{0.83}\right) = 14.6$$

$$L = F \times 14.6^2 = 2.35 \times 14.6^2 = 500\ g/m^3.d$$

$$V_F = \frac{5.7 \times 10^5\ g}{d} \times \frac{m^3.d}{500\ g} = 1140\ m^3$$

$$D = \sqrt{\frac{4}{\pi} \times \frac{V_F}{d}} = \sqrt{\frac{4}{\pi} \times \frac{1140\ m^3}{1.5\ m}} = 31.06 = \underline{31\ m}$$

Example Problem 26.4

A high-rate trickling filter is designed for an organic loading of 900 g/m³·d when treating a daily wastewater flow of 4.0 ML/d. What is the recirculation ratio required to achieve a minimum BOD removal of 75%?

Given: L = 900 g/m³·d
Q = 4.0 ML/d
E = 75%
R = ?

SOLUTION:

$$\frac{1}{E} = 1 + 0.014\sqrt{\frac{L(1 + 0.2R)}{(1 + R)}},rearranging$$

$$\frac{(1 + 0.2R)}{(1 + R)} = \frac{1}{L} \times \left[\frac{1}{0.014} \times \left(\frac{1 - E}{E}\right)\right]^2 = \frac{1}{900} \times \left[\frac{1}{0.014} \times \left(\frac{1 - 0.75}{0.75}\right)\right]^2 = 0.629 = 0.63$$

$$1+0.2R = 0.63(1+R) \ or \ 0.43R = 0.37 \ or \ R = \frac{0.37}{0.43} = 0.86 = 86\%$$

$$Q_R = R \times Q = 0.86 \times \frac{4.0 \ ML}{d} = 3.44 = \underline{3.4 \ ML/d}$$

Example Problem 26.5

Design the rotary and underdrainage systems for a 55 m diameter trickling filter unit to treat a daily average flow of 5.5 ML/d and a peaking factor of 2.5.

SOLUTION:

As a rule of thumb, the size of the central column is chosen so that the flow velocity does not exceed 2.0 m/s at peak flow and does not fall below 1.0 m/s at average flow conditions. Assuming a peaking factor of 2.5 times the average flow,

$$Q_p = 2.5 \times Q = 2.5 \times \frac{5.5 \ ML}{d} \times \frac{1000 \ m^3}{ML} \times \frac{d}{24h} \times \frac{h}{3600s} = 0.159 = 0.16 \ m^3/s$$

$$A_{min} = \frac{Q}{v_{max}} = \frac{0.159 \ m^3}{s} \times \frac{s}{2.0 \ m} = 0.0795 \ m^2$$

$$D = \sqrt{\frac{4A}{\pi}} = \sqrt{\frac{4 \times 0.0795 \ m^2}{\pi}} = 0.318 \ m = Select \ 300 \ mm \ (12 \ in \ NPS)$$

To satisfy the condition, we would need to select larger than calculated velocity. As peak flow is 2.5 times the average flow, selecting a larger size will cause the flow velocity to be significantly lower than 1.0 m/s.

$$v = \frac{Q}{A} = \frac{1}{2.5} \times \frac{0.159 \ m^3}{s} \times \frac{4}{\pi \times (0.30 \ m)^2} = 0.899 = 0.90 \ m/s$$

Distributor Arms

Assuming four arms, the discharge rate through each arm would be one-fourth. The length of the arm is half the diameter of the filter. Making adjustments for the column,

$$L = \frac{53.0 \ m - 0.30 \ m}{2} = 26.35 = 26.4 \ m$$

Let us provide three sections, the first two at 8.0 m each and the third at 10.4 m. The flow in these sections has to be adjusted in proportion to the filter area covered by these arm lengths.

$$A_1 = \frac{8.15^2 - 0.15^2}{26.55^2 - 0.15^2} = 0.094 = 9.4\%$$

$$A_2 = \frac{16.15^2 - 8.15^2}{26.55^2 - 0.15^2} = 0.275 = 27.6.\%$$

$$A_3 = \frac{26.55^2 - 16.15^2}{26.55^2 - 0.15^2} = 0.630 = 63\%$$

Full discharge through each arm would be one-fourth of 0.159 m^3/s. Full flow will flow through the first section and will keep reducing through each of the following sections. Guidelines suggest selecting a size to keep flow velocity of 1.2 m/s. Based on this, the diameter of three sections of pipe are:

$$D_1 = \sqrt{\frac{4Q}{\pi v}} = \sqrt{\frac{4}{\pi} \times \frac{1}{4} \times \frac{0.159\ m^3}{s} \times \frac{s}{1.2\ m}} = 0.205\ m = 200\ mm$$

$$D_2 = \sqrt{\frac{4Q}{\pi v}} = \sqrt{\frac{4}{\pi} \times 0.91 \times \frac{1}{4} \times \frac{0.159\ m^3}{s} \times \frac{s}{1.2\ m}} = 0.195 = 200\ mm$$

$$D_3 = \sqrt{\frac{4Q}{\pi v}} = \sqrt{\frac{4}{\pi} \times 0.63 \times \frac{1}{4} \times \frac{0.159\ m^3}{s} \times \frac{s}{1.2\ m}} = 0.163 = 175\ mm$$

Orifice Nozzle

Assuming each orifice is 12 mm in diameter and operates under a head of 1.5 m, the average discharge of each orifice can be found.

$$Q_{ori} = C_d A_o \sqrt{2gh} = 0.60 \times \frac{\pi}{4} \times (0.012m)^2 \sqrt{2 \times \frac{9.81\ m}{s^2} \times 1.5\ m} = 3.68 \times 10^{-4}\ m^3/s$$

$$\frac{\#}{arm} = \frac{Qp/arm}{Q_{ori}} = \frac{1}{4} \times \frac{0.159}{0.000368} = 107.9 = 108\ orifices/arm$$

The number of nozzles in each section of the distributor arm is directly proportional to the area coverage.

$$N_1 = A_1 \times N = 0.094 \times 108 = 10.1 = 10$$

$$N_2 = A_2 \times N = 0.275 \times 108 = 29.7 = 30$$

$$N_3 = A_3 \times N = 0.630 \times 108 = 68.0 = 68$$

Underdrainage System

A rectangular section effluent channel with a capacity to handle a peak wastewater flow of 0.159 m^3/s is fed by semi-circular laterals. The radial laterals are laid on a slope of 1 in 40 (2.5%) and will be in the form of underdrain block lengths containing semi-elliptical openings. For designing a rectangular channel section, the flow velocity is assumed to be 1.0 m/s at peak flow. Thus, the flow area required is 0.159 m^2. By choosing a channel width of 0.25 m, the depth of flow is calculated.

$$Depth,\ d = \frac{A}{B} = \frac{0.159\ m^2}{0.35\ m} = 0.454 = \underline{0.45\ m}$$

$$R_h = \frac{A}{P} = \frac{0.159\ m^2}{(0.35\ m + 2 \times 0.45m)} = 0.127\ m = 0.13\ m$$

Assuming n = 0.018, the required slope of the channel can be found using Manning's equation.

$$Channel,\ S = \left[\frac{v \times n}{R_h^{2/3}}\right]^2 = \left[\frac{1 \times 0.018}{0.127^{2/3}}\right]^2 = 5.075 \times 10^{-3} = \underline{0.50\%}$$

The laterals are laid on a slope of 1 in 40 (2.5%) and are designed to flow half full such that the filter is properly ventilated to keep the process aerobic. For partial flow conditions:

$$\frac{A}{A_F} = 0.25,\ \frac{v}{v_F} = 0.77,\ \frac{Q}{Q_F} = 0.194$$

If the flow velocity is chosen to be 0.75 m/s, the full-flow velocity is 0.75/0.77 = 0.97 m/s. The diameter of the section can be worked out using Manning's flow equation.

$$Lateral,\ D = \left[\frac{v_F \times n}{0.4\sqrt{S}}\right]^{1.5} = \left[\frac{0.97 \times 0.015}{0.4 \times \sqrt{0.025}}\right]^{1.5} = 0.110\ m = \underline{select\,150\ mm\ NPS}$$

$$Full\ flow,\ Q_F = v_F \times A_F = \frac{0.93m}{s} \times \frac{\pi}{8} \times (0.15\ m)^2 = 8.217 \times 10^{-3} = \underline{8.22\ L/s}$$

$$Actual\ flow,\ Q = Q_F \times 0.194 = 8.217 \times 10^{-3} \times 0.194 = 1.5941 \times 10^{-3}\ \underline{= 1.59\ L/s}$$

$$\# = \frac{Q}{Q/lateral} = \frac{159\ L/s}{1.59\ L/s/lateral} = 100.0 = \underline{100\ laterals}$$

26.8 ROTATING BIOLOGICAL CONTACTOR

Rotating biological contactors (RBCs) create a biological slime similar to that of the trickling filter, which is grown on plastic discs mounted on a long horizontal rotating shaft. A schematic of an RBC plant is shown in **Figure 26.3**. This is the most recent type of fixed culture system and came about in the earlier seventies. RBC systems can be adopted for small and medium towns for the treatment of domestic and industrial wastewaters. They differ from trickling filters in that the media is rotated into the settled wastewater and then into the atmosphere for oxygen. RBCs are placed between the primary and secondary clarifiers. There are no solids removed by the discs, just the breaking down of dissolved and suspended solids by bacteria. Stages can be added or removed with the use of baffles. The main disadvantage of RBCs is that they must be kept inside a building of some sort to prevent freezing, algae growth, UV radiation and rinsing due to rain.

The advantages of RBCs include ease of operation, high BOD removal, no air requirements, the ability to lend themselves to modular fabrication to suit a required effluent quality, low maintenance, the possibility of nitrification and denitrification, and the ability to work under shock loading.

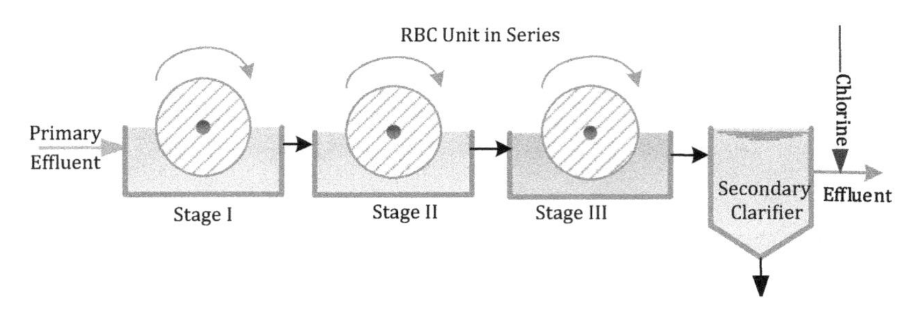

FIGURE 26.3 Rotating Biological Contactors

Some of the disadvantages of an RBC system include that structural damage may be expensive, the media must be sheltered, pumps may be necessary to move sludge from the bottom of the secondary clarifier to the primary clarifier, and it is necessary to avoid grease or oil coating on biofilm.

Biological discs are available in diameters of up to 3.7 m and may be assembled to form a drum 7.5 m in length. The spacing between sheets in the media used for BOD removal is 19 mm, and the spacing used for nitrification is 12 mm. A typical 7.5 m long drum with 3.7 m diameter discs will have a total surface area of 10000–15000 m^2. The submergence is about 40% and the typical operating speed is 1–10 rpm. The rotation of the disc unit ensures that the media are alternately in air and wastewater, resulting in the formation of biofilm. The peripheral speed must be limited to 0.30 m/s to avoid stripping the biomass.

As shown in **Figure 26.3**, a typical treatment system consists of a primary clarifier, an RBC unit and a secondary clarifier. Sludge from the secondary clarifier is usually pumped to the primary clarifier for storage. A cover is needed to protect the biofilm from heavy rain, frost and snow, and for safety. Recirculation in RBCs is generally not practised, and underflow from the final clarifier is allowed to settle in the primary clarifier. Biomass is similar to that of trickling filters.

26.8.1 Staging

RBC units are arranged in series or parallel formation. RBCs in series result in better quality effluent as long as influent organic loading is not too high. Large plants overcome this problem by placing their RBC shafts perpendicular to the flow. Small plants prefer RBCs in series as they can add baffles to create extra stages.

A series of four stages are normally installed for BOD removal. Additional stages may be required for introducing nitrification. Each stage acts as a completed reactor and the different stages combined act as a plug flow system. BOD loading decreases exponentially as wastewater moves from stage one to stage four.

26.8.2 Operation

Operation inspections are the most important process control tools. The first stage of an RBC should be uniformly brown and distributed in a thin layer with a dissolved oxygen level of 2.0 mg/L in order to be healthy. If the biomass is heavy and shaggy with white or grey patches then there is, or has been, an organic overload. It is common to get some sloughing of biomass as it gets washed off and carried to the final clarifiers for settling and ultimate removal. The discs are spaced to allow for sloughing while at the same time preventing plugging. This also allows air in as the wastewater trickles out. The sludge from an RBC can be filamentous, which settles slowly. The weir overflow rate in the secondary clarifier should be less than 100 $m^3/m \cdot d$.

26.9 PROCESS CONTROL PARAMETERS

Control parameters are hydraulic detention time, rotation velocity and arrangement of disc stages. The shafts are air-driven or mechanically driven.

26.9.1 Loading

BOD loadings are based on mass of BOD per unit surface area expressed in g/m²·d. Hydraulic loading rate is expressed as flow divided by the disc's surface. The commonly used units to express hydraulic loading are m³/m²·d.

$$\boxed{Disk\ surface\ A_S = \frac{\pi D^2}{4} \times \frac{2\ faces}{disc} \times \frac{L}{\Delta L}}$$

L = length of shaft (7.5 m standard)
ΔL = Disk spacing
D = Diameter of disk

Example Problem 26.6

A 5 m long RBC shaft is packed with 250 discs 3.6 m in diameter spaced 20 mm apart. Work out the total disc surface per shaft length. For a rotating speed of 1.5 rpm, determine the peripheral speed in m/s.

Given: D = 3.6 m
N = 1.5 rpm
250 discs/shaft

SOLUTION:

$$A_S = \frac{\pi (3.6\ m)^2}{4} \times \frac{2\ faces}{disc} \times \frac{250\ disc}{shaft} = 5086 = \underline{5100\ m^2}$$

$$v_p = \frac{1.5\ rev}{min} \times \frac{\pi (3.6\ m)}{rev} \times \frac{min}{60\ s} = 0.282 = \underline{0.28\ m/s}$$

26.9.2 Soluble BOD

Usually, the RBC unit is designed to remove soluble BOD. Total BOD in wastewater has two components: organic solids (particulate BOD) plus soluble and colloidal biodegradable matter. Soluble BOD can be thought of as the BOD of a filtered sample. Experience has shown that for a given wastewater, the BOD contributed by SS is proportional to the solid content. The constant of proportionality is typically in the range of 0.5 to 0.7 for municipal wastewater.

The K factor needs to be determined for a given operation. This can be done by observing total BOD, soluble BOD, and suspended solids concentration data over a certain period of time. The K value is the ratio of average particulate to BOD to average suspended solids concentration.

$$\boxed{Particulate\ BOD_{ptl} = BOD_{tot} - BOD_{sol} = k \times SS}$$

26.9.3 Organic Loading

Organic loading on RBC units is based on soluble BOD or total BOD. It is expressed as mass load of BOD in the primary effluent per unit disc surface.

$$BODLR = \frac{M_{BOD}}{A_S} = \frac{Q \times BOD_{PE}}{A_S}$$

A_S = total disc surface
Q = average flow
BOD = BOD of settled wastewater

Key Points:

- Typical loadings are 7.5 g/m²·d of soluble BOD or 15 g/m²·d of total BOD.
- In larger plants, RBC shafts are placed perpendicular to the direction of flow, thus each shaft acts as one stage of BOD removal.
- The various stages of RBC simulate plug flow, with maximum loading on the first stage. Thus, the loading on the first stage is an important consideration.
- If operated properly, the biological growth in the first stage should be fairly uniform and light brown. Subsequent stages should look similar except with an additional gold or reddish tone.
- Overloading conditions are evidenced by grey or white biomass.
- A loading of 60 g/m²·d of total BOD on the first stage should not be exceeded.
- For operating temperatures below 13°C, a temperature correction for additional disc surface should be applied at 15% for each 3°C below 13°C.
- Dissolved oxygen (DO) of first-stage effluent should not be allowed to fall below 0.5 mg/L and the DO of final stage effluent should preferably be more than 2.0 mg/L.

Example Problem 26.7

An RBC unit consists of 16 shafts, with each shaft having a disc surface of 5600 m². The installation has 16 shafts arranged with four rows of shafts of four stages each. On average, a primary effluent flow of 8500 m³/d containing 150 mg/L of BOD and 120 mg/L of SS is treated. Assuming k = 0.50, calculate the soluble BOD loading on the RBC process in g/m² d.

Given: # = 16 at 5600 m²/shaft
Q = 8500 m³/d
BOD = 150 mg/L
SS = 120 mg/L
k = 0.50

SOLUTION:

$$A_S = \frac{5600\ m^2}{shaft} \times 16\ shafts = 89600\ m^2$$

$$BOD_{ptl} = k \times SS = 0.50 \times 120\ mg/L = 60\ mg/L$$

$$BOD_{sol} = BOD_{tot} - BOD_{ptl} = 150\ mg/L - 60\ mg/L = 90\ mg/L$$

$$BODLR = \frac{Q \times BOD}{A_S} = \frac{8500\ m^3}{d} \times \frac{90\ g}{m^3} \times \frac{1}{89600\ m^2} = 8.53 = \underline{8.5\ g/m^2 \cdot d}$$

Example Problem 26.8

How many RBC shafts (1 hm^2/shaft) are required to treat a flow of 10 ML/d with a BOD content of 200 mg/L? Assume the primary removal of BOD is 35%. The designed BOD loading rate is 15 $g/m^2 \cdot d$.

Given: Disc surface = 1 hm^2/shaft
#of shafts = ?
Q = 10 ML/d
BOD_{raw} = 200 mg/L
BODL = 15 $g/m^2 \cdot d$
E_I = 35% = 0.35

SOLUTION:

$$M_{BOD} = Q \times BOD = \frac{10\ ML}{d} \times \frac{200\ mg}{L} \times (1 - 0.35) = 1300\ kg/d$$

$$\# = \frac{1300\ kg}{d} \times \frac{1000\ g}{kg} \times \frac{m^2.d}{15\ g} \times \frac{shaft}{10000\ m^2} = 8.67 = \underline{9.0\ shafts}$$

- A 3 × 3 system is suggested. If nitrifying, add another stage.
- To satisfy each of these conditions, three more shafts or 12 shafts in total are needed.

26.10 OPERATION OF RBC SYSTEM

Operation inspections are the most important process control tools. The operator needs to keep an eye on RBC movement, slime colour and appearance (**Table 26.2**). The first stage of an RBC should be uniformly brown and distributed in a thin layer with a DO level of 2.0 mg/L to be healthy. Since the first stage receives the maximum load, the DO level will remain suppressed. However, the DO level in the first stage should not be allowed to fall below 0.50 mg/L. If the biomass is heavy and shaggy with white or grey patches, then there is, or has been, an organic overload. In a nutshell, slime or bio-growth indicates the process condition.

It is common to get some sloughing of biomass as it gets washed off and carried to the final clarifiers for settling and ultimate removal. The discs are spaced to allow for sloughing while at the

TABLE 26.2
Operating Conditions and Slime Colour

Slime Colour	Process Condition
Grey, shaggy	Normal process
Reddish-brown, golden	Nitrification occurring
White chalky	High sulphur content
No slime growth	Severe temperature or pH changes

same time preventing plugging. This also allows air in as wastewater trickles out. As part of routine testing and sampling, the operator should observe DO content at various stages, pH, and suspended solids content. These results aid in assessing performance and adjusting the process.

Discussion Questions

1. Compare a trickling filter with a rotating biological contactor.
2. Discuss the working principles of a trickling filter.
3. Describe the various components of a trickling filter with the help of a sketch.
4. List the advantages and disadvantages of biological filtration.
5. With the help of a flow diagram, explain direct and indirect circulation in tricking filters.
6. What is the main purpose of each type of recirculation, direct and indirect?
7. Differentiate between a standard rate filter and a high-rate filter.
8. The advent of plastic media has led to the creation of super-rate filters. Comment.
9. How does a trickling filter differ from a rapid sand filter?
10. Sketch a flow diagram of a single-stage and two-stage trickling filter system. What modification to the NRC formula would be needed for BOD removal in the second stage?
11. Describe the different ways of introducing nitrification in an RBC biological system.
12. List the merits and demerits of an RBC system.
13. How might the colour of the slime on an RBC indicate the health of the RBC process?
14. Explain why trickling filters and RBCs are put under the category of fixed growth systems? Activated sludge falls into the category of suspended growth systems for biological treatment. Which of these three is the oldest and which is the most recent biological treatment?
15. Trickling filters are also called biological filters. What do the terms trickling, biological and filter signify?

Practice Problems

1. A town produces a sewage flow of 5.5 ML/d containing a BOD of 270 mg/L. After primary clarification, trickling filters are planned for secondary treatment. Assuming a primary BOD removal of 35%, work out the volume and depth of the filter unit based on a BOD loading rate of 150 $g/m^3{\cdot}d$ and a hydraulic loading of 4.0 $m^3/m^2{\cdot}d$. For the selected filter, calculate the BOD removal efficiency based on the NRC formula. (D = 42 m, # =2, d = 2.3 m, R = 1, 88%)
2. Select the diameter of a trickling filter to serve a population of 500 contributing wastewater at 130 L/c·d with a BOD content of 300 mg/L. Assume 40% of BOD is removed in primary clarification and the design BOD loading is 500 $g/m^3{\cdot}d$. Work out the hydraulic loading and check if it falls within range.
(4.0 m, 2.2 m deep, 8.0 $m^3/m^2{\cdot}d$ with 50% recirculation during low flow)
3. How many RBC shafts (1 hm^2/shaft) are required to treat a flow of 12 ML/d with a BOD content of 250 mg/L? Assume the primary removal of BOD is 40%. The design BOD loading rate is 15 $g/m^2{\cdot}d$. (12#, 3 × 4)
4. What diameter trickling filter is required to treat a flow of 20 ML/d with a BOD content of 240 mg/L. Assume a primary BOD removal of 35%, a recirculation ratio of 1.4 and a desired secondary effluent BOD of 30 mg/L. (55 m)

5. A standard-rate trickling filter has a diameter of 26 m and an average media depth of 2.1 m. The daily wastewater flow is 5200 m^3/d with an average BOD of 180 mg/L. During periods of low influent flow, 2.5 ML/d of underflow from the final clarifier is returned to the wet well.
   a. Calculate hydraulic loading, return ratio and BOD loading, assuming 35% BOD removal by the primary treatment. (15 m/d, 0.48, 550 $g/m^2 \cdot d$)
   b. Applying the NRC formula, determine the expected BOD removal assuming an operating temperature of 20°C. (78%)
   c. If the operating temperature drops to 16°C during the winter months, what BOD removal can be expected? (68%)
6. Over a period of three months, the average values of total BOD, soluble BOD and SS in the primary effluent are 190 mg/L, 95 mg/L and 150 mg/L, respectively. What fraction of suspended solids contribute to BOD? (63%)
7. Select the diameter of a high-rate filter to treat a wastewater flow of 5.0 ML/d with a recirculation rate of 150%. Assume the BOD in the primary effluent is 160 mg/L and desired BOD removal is 80%. (31 m, 2.0 m deep)
8. Using the data of Practice Problem 12.7, size a standard rate filter with no circulation. (40 m, 2.0 m deep)
9. Design a single-stage high-rate filter for treating the wastewater for a community of 38 000 people. The per capita wastewater production is 170 $L/c \cdot d$ with a BOD of 200 mg/L. The allowable BOD loading and hydraulic loading are 800 $g/m^3 \cdot d$ and 8.0 m/d, respectively. Assume a BOD removal of 30% by the primary treatment and a recirculation rate of 100%. Find:
   a. The volume of filter media based on organic loading. (1130 m^3)
   b. The surface area based on hydraulic loading. (810 m^2)
   c. The diameter and depth. (32 m, 1.4 m)
   d. What BOD is expected in the plant effluent. (76%)
10. A trickling filter is operated with an organic loading of 500 $g/m^2 \cdot d$ while maintaining a recirculation rate of 120%. What is the expected BOD removal if the operating temperature is 22°C? (87%)
11. A trickling filter is designed based on a BOD loading of 550 $g/m^2 \cdot d$. To meet effluent quality standards, required BOD removal is 80%. Find at what recirculation ratio the plant must be operated. (1.1)
12. Size a high-rate trickling filter to treat a wastewater flow of 3.5 ML/d with a settled BOD of 180 mg/L.
   a. What is the capacity of the trickling filter assuming a recirculation ratio of 1.4 and an expected BOD removal of 80%? (1100 m^3)
   b. What is the diameter of the filter assuming a depth of 1.5 m? (30 m)
13. An RBC unit consists of 12 shafts with three trains, each with four stages. Each shaft has a disc surface of 5500 m^2. On average, the primary effluent flow is 6500 m^3 /d containing 150 mg/L of BOD and 110 mg/L of SS. Assuming the particulate BOD is 50%, determine the soluble BOD loading. (9.4 $g/m^2 \cdot d$)
14. Determine the maximum allowable BOD loading on a high-rate trickling filter operated with 100% recirculation to remove 80% of the BOD. (530 $g/m^3 \cdot d$)
15. A trickling filter is designed for a BOD loading of 600 $g/m^3 \cdot d$. What is the minimum recirculation required to remove 80% of the BOD? (140%)

27 Anaerobic Systems

In anaerobic systems, the main bio-reaction takes place under anaerobic conditions, so there is no need to provide molecular oxygen by aeration. One of the main end products is methane, which can be used as an energy source in some cases. However, due to the production of foul gases, there is a problem of strong smells, hence such systems need to be well-ventilated.

The most common anaerobic systems are septic tanks, Imhoff tanks and anaerobic filters and reactors. One of the main advantages of such systems is energy savings, low capital costs and ease of operation.

27.1 SEPTIC TANKS

Septic tanks are onsite systems and are more common in rural areas, isolated buildings and institutions, hotels, schools, hospitals and small residential areas. To prevent contamination, septic systems must be set away from the source of drinking water and should not be located in swamp areas or areas prone to flooding. In addition, the soil should be porous to absorb effluent from septic tanks. A typical section of a septic tank is shown in **Figure 27.1**.

A septic tank is a watertight underground tank usually made of concrete. The capacity of the tank is such as to provide a long retention period of one to three days. Solids settle to the bottom where anaerobic sludge digestion takes place. Usually, a septic tank has two chambers separated by a baffle wall. The first chamber allows for the removal of grit and the second chamber is primarily for the settling of organic solids and anaerobic digestion. A sludge tank must be provided with a vent to eliminate foul gases produced as a result of anaerobic digestion. Digested sludge is pumped out periodically and is not allowed to exceed three years.

27.1.1 Design Considerations

Capacity

The capacity of a septic tank should be such to provide adequate detention time for solids to settle out and enough room for the storage of digested sludge until its withdrawal period. The recommended capacity for sewage flow is 90–150 L/c.d. Sludge and scum will accumulate in the tank at varying rates depending on the characteristics of the raw sewage. A typical value of sludge accumulation is 30 L/c.a. The minimum capacity of a septic tank is usually more than 2.0 m^3.

Free Board

A free board of 0.3 to 0.5 m is provided above the top of the sewage line to allow room for scum to accumulate. The scum layer prevents the spread of foul gases.

Inlet and Outlet

The design of the inlet and outlet should be such as to prevent short-circuiting. To achieve this, baffles, usually tees, should extend to the top level of the scum but below the ceiling of the tank. To avoid short-circuiting, the inlet should extend to a depth of about 30 cm below the sewage line. The outlet is submerged and is located 5–8 cm below the inlet.

DOI: 10.1201/9781003231264-27

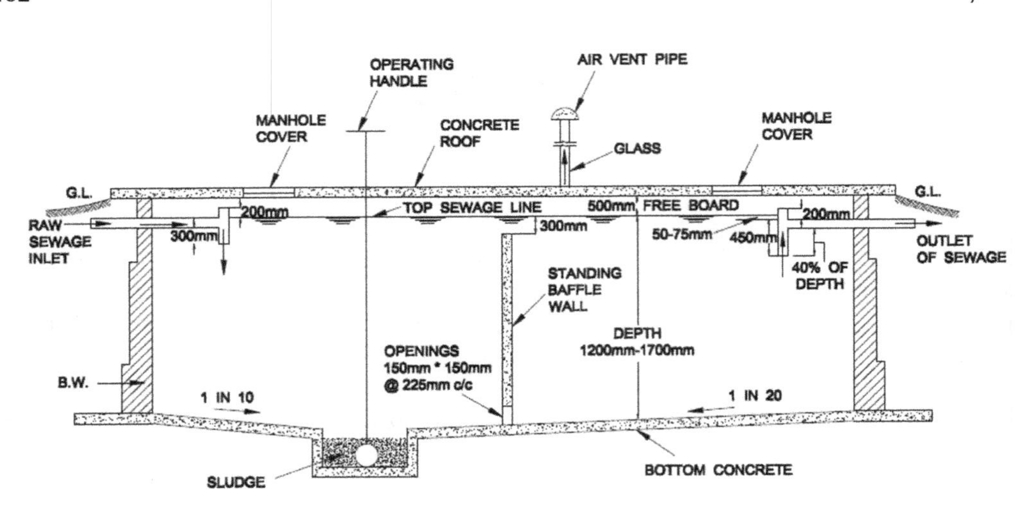

FIGURE 27.1 Section of a Septic Tank

Detention Time

As mentioned aboveearlier, the detention time for septic tanks varies from as small as 12 hours to as high as three days. However, 24 hours or one day is the most used when designing.

Shape of the Tank

Septic tanks are usually rectangular with a length 3–4 times their width. A minimum depth of 1.0 m is provided below the water level. The bottom of the tank is sloped for sludge to accumulate in sump form, where it is pumped out periodically.

Example Problem 27.1

Design a rectangular septic tank with a length to width ratio of 3:1 to serve a population of 100 discharging wastewater at 100 L/c.d. Assume a detention period of 30 hours and sludge withdrawal every second year at 50 L/c/2y.

Given: Q = 100 p at 100 L/c·d

t_d = 30 h

L, W = ?

SOLUTION:

$$V_I\left(water\right)=\frac{100\ L}{p.d}\times 100\ p\times 30\ h\times\frac{d}{24\ h}\times\frac{m^3}{1000\ L}=12.5\ m^3$$

$$V_{II}\left(sludge\right)=\frac{50\ L}{p}\times 100\ p\times\frac{m^3}{1000\ L}=5.0\ m^3\quad V_T=V_I+V_{II}=12.5\ m^3+5.0\ m^3=17.5\ m^3$$

Assuming the water depth in the septic tank is 1.5 m, and its length is three times its width, the width of the tank is:

$$W = \sqrt{\frac{V}{3d}} = \sqrt{\frac{17.5\ m^3}{3 \times 1.5\ m}} = 1.97 = 2.0\ m \quad and \quad L = 2.0\ m \times 3 = 6.0\ m$$

Assuming a free board of 0.4 m, the dimensions of the tank are 6.0 m × 2.0 m × 1.9 m.

Disposal of Tank Effluent

Effluent from the tank will still carry a large amount of organic load that needs to be taken care of. Since septic tank effluent carries a BOD in the range of 100–200 g/L, disposal practices are such as to provide some sort of biological treatment. Common methods of disposal are discussed in the following subsections.

27.1.2 Soil Adsorption System

When effluent is spread on pervious land, microorganisms in the soil reduce BOD by aerobic decomposition and add nutrients to the soil. This method is applicable if the soil is pervious and the groundwater is well below the surface.

The perviousness of soil is indicated by its percolation rate. The percolation rate is defined as the time in minutes required for the seepage of 1 cm depth of water. Therefore, the percolation rate for sandy soils will be quite low but high for clay soils.

Usually, the soils suitable for an absorption field or pit should have a percolation rate of less than 30 minutes and must not exceed 60 minutes. Soil absorption can be achieved in two ways: by a soak pit or a dispersion trench.

By knowing the percolation rate of the soil, the maximum allowable application rate of septic tank effluent in $L/m^2 \cdot d$ can be worked out using the following empirical relationship: $Q = \frac{204}{\sqrt{t}}$. For example, if the percolation rate is 10 minutes, the allowable application rate is $Q = \frac{204}{\sqrt{10}} = 64.5 = 65\ L/m^2.d$.

27.1.3 Absorption Field

An absorption field is constructed by laying trenches in the field. Trenches are laid with drainage tiles 75–100 mm in diameter to allow effluent to infiltrate into the soil. Dispersion trenches are 0.5–1.0 m deep and 0.3–1.0 m wide and excavated with a slight gradient. Trenches are provided with 150–250 mm of washed gravel or crushed stones. The length of the dispersion trench should be < 30 m and trenches should not be placed closer than 2.0 m to one another.

27.2 SOAK PIT

As seen in **Figure 27.2**, a soak pit is usually a circular covered pit through which the effluent is allowed to percolate and be absorbed into the surrounding soil. The pit may be empty or filled with stone aggregate. When the pit is empty, it is lined with brick or concrete such that joints are open to allow absorption into the surrounding soil. In addition, a backing of 75 mm thick coarse aggregate supports the pit surface below the inlet, and above the inlet should be plastered.

Whatever the subsoil dispersion system, it must not be closer than 20 m from any source of drinking water, such as a well, to mitigate the possibility of bacterial pollution of the water supply. In limestone or crevice rock formations, a soil absorption system is not recommended as there may be channels in the formation that carry contamination over a long distance. In such cases, and generally

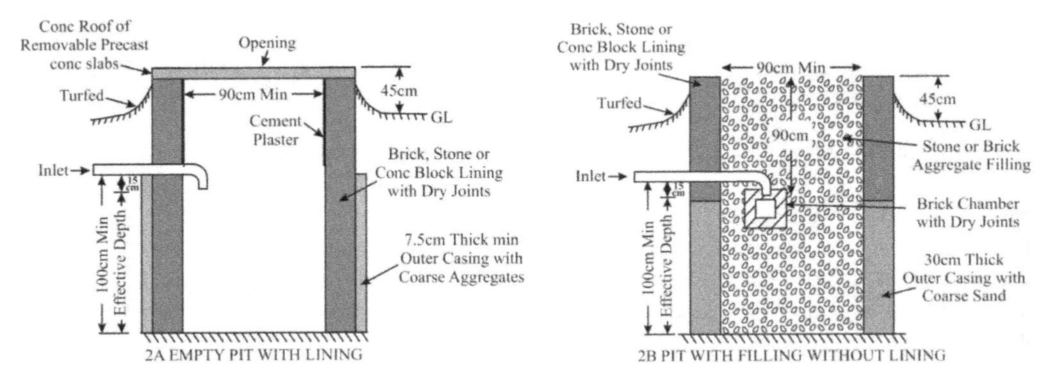

FIGURE 27.2 Schematic of a Soak Pit

Adapted from Indian standards code

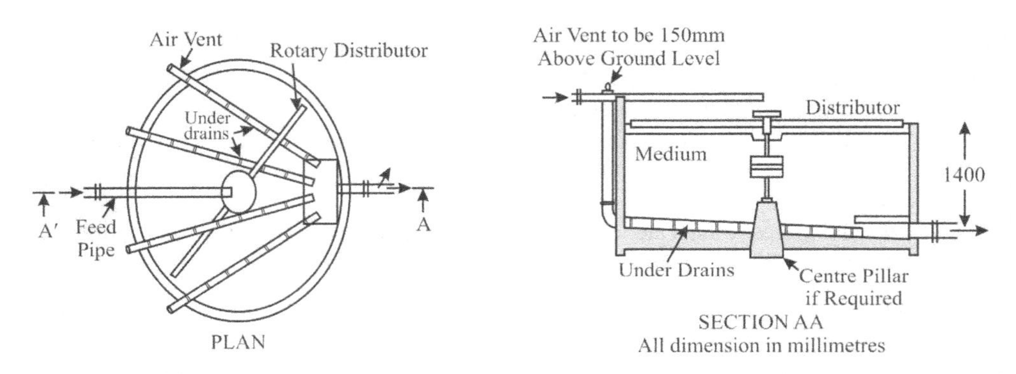

Note : Flexible joints may be required on inlet or outlet connection, where rigid pipes are used.

FIGURE 27.3 Biological Filters

where suitable conditions do not exist for the adoption of soil absorption systems, the effluent, where feasible, should be treated in a biological filter or upflow anaerobic filters.

27.3 BIOLOGICAL FILTERS

Biological filters are suitable for the treatment of septic tank effluent where the soil percolation rate exceeds 60 minutes, such as in waterlogged areas or where limited land area is available. In a biological filter (**Figure 27.3**), the effluent from the septic tank is brought into contact with a suitable medium, the surfaces of which become coated with an organic film. The film assimilates and biochemical oxidation takes place. The biological filter requires ample ventilation and an efficient system of undertrains leading to an outlet.

The volume of the filter medium must be sufficient, especially in small systems where more variations are expected. A volume of 1 m^3/head should be provided when serving less than 10 persons, and of 0.8 m^3/head for more than 10 persons. For populations larger than 50 people, the recommended volume is 0.6 m^3/head. The final effluent is either discharged into a city surface drain or evenly spread over a grass field. If the receiving water is suspected to be contaminated, disinfection should be provided before it is discharged.

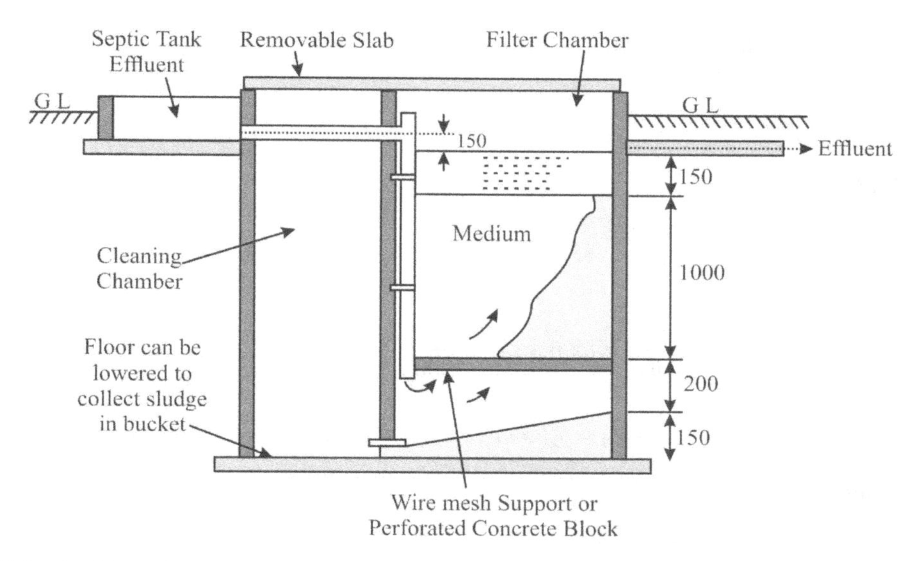

FIGURE 27.4 Single Chamber Upflow Filter

27.4 UPFLOW FILTERS

This method of septic effluent treatment is anaerobic and is used for areas with limited land or high-water table conditions, or clayey soils with high percolation rates. It is also called a reverse filter since the septic tank effluent is introduced at the bottom. The filter stone media remains submerged so that anaerobic conditions can prevail. The capacity of the unit is typically 40–50 L/c·d and high a BOD removal can be expected when operated properly.

A single-chamber upflow filter is shown in **Figure 27.4**. A double-chambered rectangular upflow filter is common for treating the effluent from septic tank units.

The filter material usually consists of 20 mm stones resting on a concrete slab with a false bottom. The septic tank effluent enters at the bottom through a 150 mm pipe and is fitted with a tee at the bottom. The other end of the tee towards the cleaning chamber is plugged during operation. When cleaning, the plug is removed to empty the filter into the cleaning chamber. During filter operation, treated water exits over a v-notch weir to maintain a water depth of 150 mm above the top of the filter media.

In a double-chamber filter, both chambers are filled with filter media. Septic tank effluent enters at the top of the media in the first chamber and flows upwards through the media in the second chamber. This allows more contact time and hence better BOD removals can be expected. A galvanized iron pipe at the bottom fitted with a valve leading to an adjacent chamber can be used to clean the filter when required.

Example Problem 27.2

Design a rectangular septic tank to serve a population of 85 discharging wastewater at 90 L/c.d. Assume the sludge is withdrawn every year and the tank has a length-to-width ratio of 2:1.

a. Size the soak pit based on a percolation rate of 1200 L/m^3·d.
b. If the tank effluent is to be discharged into an absorption field with a percolation rate of 5.0 min, what size should the trenches be?

Given: Q = 85 p at 90 L/c·d
Percolation = 1200 L/m^3·d
L:W:: 4:1

SOLUTION:

$$V_I = \frac{90\ L}{p.d} \times 85p \times 24\ h \times \frac{d}{24\ h} \times \frac{m^3}{1000\ L} = 7.65\ m^3$$

$$V_{II} = \frac{30\ L}{p.a} \times 85\ p \times 1\ a \times \frac{m^3}{1000\ L} = 2.55\ m^3$$

$$V_T = V_I + V_{II} = 7.65\ m^3 + 2.55\ m^3 = 10.1\ m^3$$

Assuming a water depth of 1.5 m:

$$W = \sqrt{\frac{V}{2d}} = \sqrt{\frac{10.1\ m^3}{2 \times 1.5\ m}} = 1.83 = 1.8\ m \ \ and\ \ L = 2 \times 1.83 = 3.66 = 3.7\ \ m$$

Assuming a free board of 0.4 m, the dimensions of the tank are 3.7 m × 1.8 m × 1.9 m.

a. The soak pit can be sized by assuming a percolation rate of 1200 L/m^3·d and a depth of 2.0 m.

$$D = \sqrt{\frac{1.27V}{d}} = \sqrt{\left(\frac{1.27}{2.0\ m} \times \frac{90\ L}{p.d} \times 85p \times \frac{m^3.d}{1200\ L}\right)} = 2.01 = 2.0\ m$$

b. The dispersion trench is based on the maximum percolation allowed, which is found from the following empirical equation:

$$q = \frac{204}{\sqrt{t}} = \frac{204}{\sqrt{5}} = 91.23 = 91\ L/m^2.d \qquad A_{trench} = \frac{7650\ L}{d} \times \frac{m^2.d}{91\ L} = 85\ m^2$$

Assuming three trenches, each 1.0 m wide, the length of trench (#3) required is:

$$L_{trench} = \frac{85\ m^2}{1.0\ m} \times \frac{1}{3} = 28.8 = 30\ m\ say$$

Hence, three trenches of 30 m × 1.0 m each will absorb the septic effluent.

27.5 IMHOFF TANKS

An Imhoff tank is an improvement over a septic tank in which sludge and incoming wastewater are kept in separate units or chambers. The upper chamber is called the sedimentation chamber and the lower chamber is called the digestion chamber, as shown in **Figure 27.5**. Imhoff tanks are much deeper than septic tanks.

The settlement of solids occurs in the upper chamber of an Imhoff tank and some digestion takes place in the lower chamber. A lip at the bottom of the upper chamber, as shown in

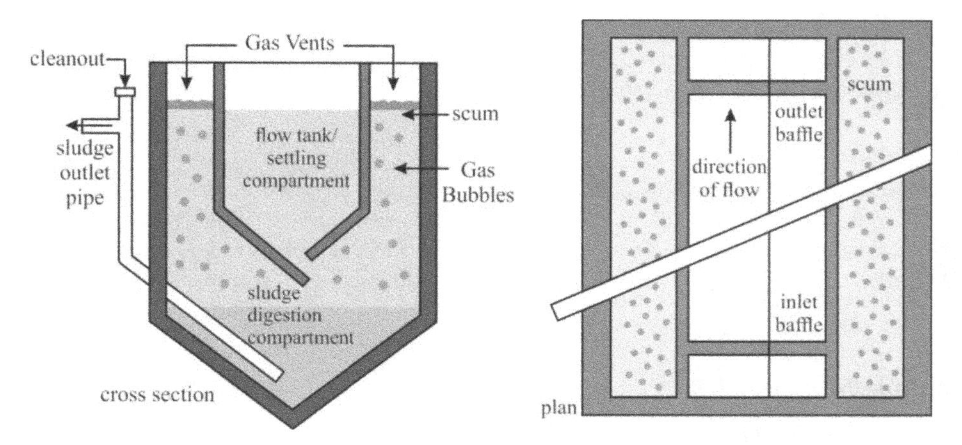

FIGURE 27.5 Imhoff Tanks

TABLE 27.1
Design Criteria for an Imhoff Tank

Design Parameter	Value
Detention time	2–4 hours
Length and length-to-width ratio	L < 30 m, 3:1–5:1
Tank depth	Total: 10 m, Sedimentation: 3–3.5m
Capacity of digestion chamber	60–100 L/capita

Figure 27.5, prevents solids from re-entering and the gases produced in the digestion chamber from mixing. A gas vent, also called a scum chamber, is provided above the digestion chamber and alongside the upper chamber to allow the escape of gases produced during digestion. To prevent scum or sludge from entering the sedimentation chamber, the sludge and scum layer must be maintained at least 45 cm above and below the slots, respectively. This free or clear zone is called the neutral zone.

The digestion chamber is divided into three or four interconnected compartments. The bottom of each compartment is made in the shape of an inverted cone or hopper with a side slope of 45° to allow the sludge to concentrate in the bottom.

The digested sludge is removed periodically through de-sludging pipes provided in each compartment to allow for the uniform distribution of sludge. Flow in the sedimentation unit is reversed intermittently. The availability of other systems coupled with the necessity for deep excavations has led to a decrease in the popularity of Imhoff tanks.

27.5.1 Design of Imhoff Systems

Sedimentation Chamber

Design parameters are shown in **Table 27.1**. The sedimentation tank is rectangular and its length does not exceed 30 m. It should be able to provide a minimum detention time of two hours. The flow-through velocity should not exceed 5 mm/s and the surface loading should be preferably < 30 $m^3/m^2 \cdot d$. The depth of this chamber should be kept shallow to allow solids to slide down to enter the slot. A free board of 45 cm is provided.

Digestion Chamber

The minimum capacity of a digestion chamber is kept as 60 L/capita. However, in warmer conditions this may be reduced as digestion gets accelerated.

Scum Chamber

The surface area provided for this chamber should be about 25–30% of the horizontal projection of the top of the digestion chamber. The area should be sufficient to allow the easy exit of gases produced due to digestion without causing any problems such as foaming. The minimum width of a vent should be 0.60 m. The design procedure is further illustrated in Example Problem 27.3.

Example Problem 27.3

Design an Imhoff tank to serve a population of 25 000 discharging wastewater at 140 L/c.d. Make the appropriate assumptions wherever necessary.

Given: Q = 25 000 people at 140 L/c·d

SOLUTION:

$$Q = \frac{140\ L}{p.d} \times 25000\ p \times \frac{m^3}{1000\ L} = 3500\ m3/d\ Assuming\,a\,detention\ time\ of\ 2\ hours$$

$$V = \frac{3500\ m^3}{d} \times 2\ h \times \frac{d}{24\ h} = 291.6 = 290\ m^3$$

Two tanks are more desirable. Assuming a length-to-width ratio of 4:1 and effective depth of 2.2 m:

$$Width, W = \sqrt{\frac{V}{2 \times 4d}} = \sqrt{\frac{291.6\ m^3}{8 \times 2.2\ m}} = 4.07 = 4.0\ m, \quad L = 4 \times 4.0\ m = 16\ m$$

$$v_O = \frac{1}{2} \times \frac{3500\ m^3}{d} \times \frac{1}{4.0\ m \times 16\ m} = 27.3 = 27\ m\,/\,d < 30\ so\ okay$$

The bottom is on a slope of 1:1.25; thus, the height of the sloping bottom is 1.25 × 4.0/ 2 = 2.5 m.

$$y = 2.2\ m - 0.50 \times 2.5\ m = 0.95\ m$$

Adding a free board of 0.5 m, the total depth of the sedimentation chamber is found:

$$Sedimentation,\ d = 0.50\ m + 0.95\ m + 2.5\ m = 3.95\ m = 4.0\ m$$

Provide a neutral zone of 0.50 below the depth of 4.0 m. The tank is 16 m long but below the 4.0 m depth it is subdivided into a number of compartments, say four, each 4.0 m in length. A gas vent has to be provided on both sides of the sedimentation chamber. Its width should be about 50% of the total width. Hence, the total width of the tank:

$$Top\ width,\ TW = 1.50 \times W_{sed} = 1.5 \times 4.0\ m = 6.0\ m$$

Assuming the chamber walls are 15 cm thick, the width of the gas vent on each side:

$$W_{vent} = \frac{1}{2} \times (6.0\ m - 4.0\ m - 2 \times 0.15\ m) = 0.85\ m$$

The capacity of the digestion chamber at 40 L/p·d:

$$V_{Digestion} = \frac{1}{2tanks} \times \frac{tank}{4\ units} \times \frac{40\ L}{p.d} \times 25000p \times \frac{m^3}{1000\ L} = 125\ m^3$$

Assuming the depth of each hopper to be 1.5 m, and the side slopes 1:1:

$$Bottom,\ A_{II} = (4.0\ m - 2 \times 1.5\ m) \times (6.0\ m - 2 \times 1.5\ m) = 1.0\ m \times 3.0\ m = 3.0\ m^2$$

$$Top = 4.0\ m \times 6.0\ m = 24\ m^2$$

$$V_{II} = \frac{h}{3}\left(A_I + A_{II} + \sqrt{A_I A_{II}}\right) = \frac{2.0\ m}{3}\left(24\ m^2 + 3\ m^2 + \sqrt{24\ m^2 \times 3\ m^2}\right) = 23.65 = 24\ m^3$$

$$V_I = V_{unit} - V_{II} = (125 - 24)\ m^3 = 101\ m^3$$

$$d_I = \frac{V_I}{A_I} = \frac{101\ m^3}{6.0\ m \times 4.0\ m} = 4.20 = 4.2\ m$$

$$d_{digestion} = 4.2\ m + 1.5\ m + 0.50\ m = 6.2\ m$$

$$d_{tank} = 6.2\ m + 4.0\ m = 10.2\ m$$

27.6 UPFLOW ANAEROBIC SLUDGE BLANKET

Upflow anaerobic sludge blanket (UASB) units require no special media since the sludge granules themselves act as the media and stay in suspension. A typical UASB reactor is shown in **Figure 27.6**.

The anaerobic unit does not need to be filled with stones or any other media. The upflowing sewage forms millions of small granules or particles of sludge that are held in suspension and provide a large surface area on which organic matter can attach and undergo biodegradation. A high solid retention time (SRT) of 30–50 days or more occurs within the unit. No mixers or aerators are required. The gas produced can be collected and used if desired. Anaerobic systems function satisfactorily when temperatures inside the reactor are above 18–20°C. Excess sludge is removed from time to time through a separate pipe and sent to a simple sand bed for drying.

27.6.1 Design Approach

Generally, UASBs are considered where the temperature in the reactors will be above 20°C. At equilibrium conditions, the sludge withdrawn must be equal to the sludge produced daily. The sludge produced daily depends on the characteristics of the raw wastewater since it is the sum of the following:

1. The new VSS produced due to BOD removal, the yield coefficient being assumed as 0.1 g VSS/ g BOD removed.

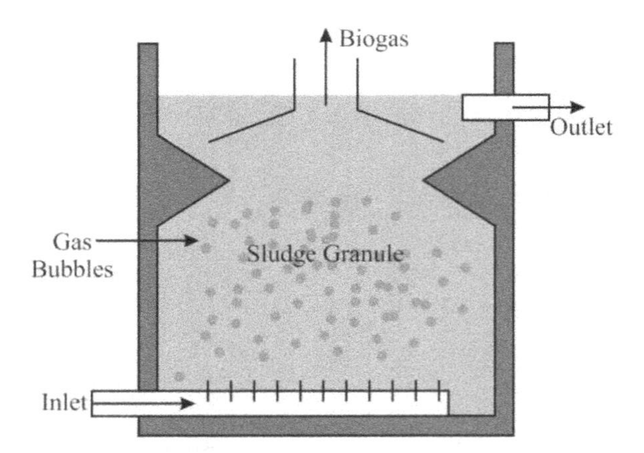

FIGURE 27.6 A Typical UASB Reactor

2. The non-degradable residue of the VSS coming in the inflow assuming 40% of the VSS is degraded and the residue is 60%.
3. The ash received in the inflow, or non-volatile fraction, TSS-VSS mg/l.

At steady-state conditions, the reactor volume must be chosen so that the desired SRT value can be achieved. This is done by solving for hydraulic retention time (HRT) from the SRT equation assuming the depth of the reactor, the effective depth of the sludge blanket, and the average concentration of solids in the sludge blanket (7.0%).

The full depth of the reactor for treating a low BOD municipal sewage is often 4.5 to 5.0 m, of which the sludge blanket itself may be 2.0 to 2.5 m in depth. For high BOD wastes, the depth of both the sludge blanket and the reactor may have to be increased so that the organic loading on the solids may be kept within the prescribed range. Once the size of the reactor is fixed, the upflow velocity can be determined from the expression: height of the reactor/HRT.

The average HRT can be calculated using the average flow rate, while the peak flow rate gives the minimum HRT at which minimum exposure to treatment occurs. To always retain any flocculent sludge in the reactor, the upflow velocity should not exceed 0.5 m/h at average flow or 1.2 m/h at peak flow. At higher velocities, carryover of solids might occur and effluent quality may deteriorate. The feed inlet system is designed next to determine the required length and width of the UASB reactor.

The settling compartment is formed of sloping hoods for gas collection. The depth of the compartment is 2.0 to 2.5 m and the surface overflow rate is kept at 20 to 28 $m^3/m^2.d$ (1 to 1.2 m/h) at peak flow. The flow velocity through the aperture connecting the reaction zone with the settling compartment is limited to no more than 5 m/h at peak flow. It is important to ensure the proper working of the gas-liquid-solid separator (GLSS), the gas collection hood, the incoming flow distribution for spatial uniformity and the outflowing effluent.

27.6.2 Physical Parameters

A single module can handle wastewater flows of 10 to 15 ML/d. For large flows, a number of modules could be provided. Some physical details of a typical UASB reactor module are shown in **Table 27.2**.

Process Design Parameters

A few process design parameters for UASBs are listed in **Table 27.3**. The values furnished in **Table 27.3** are specific to municipal sewages with a BOD of about 200–300 mg/l and temperatures above 20°C.

TABLE 27.2
Physical Details of UASB Reactor

Reactor configuration	Rectangular or circular. Rectangular is preferred.
Depth	4.5 to 5.0 m for sewage.
Width or diameter	Limit lengths of inlet laterals to around 10–12 m to facilitate uniform flow distribution and sludge withdrawal.
Length	As necessary.
Inlet feed	Gravity feed from the top (preferred for municipal sewage) or pumped feed from the bottom through manifold and laterals (preferred for soluble industrial wastewaters).
Sludge blanket depth	2 to 2.5 m for sewage. More depth is needed for stronger wastes.
Deflector/GLSS	This is a deflector beam that, together with the gas hood (slope 60), forms a gas-liquid-solid separator (GLSS) that allows the gas to go to the gas collection channel at the top while the liquid rises into the settler compartment and the sludge solids fall back into the sludge compartment. The flow velocity through the aperture connecting the reaction zone with the settling compartment is generally limited to about 5m/h at peak flow.
Settler compartment	2.0–2.5 m in depth. Surface overflow rate equals 20–28 $m^3/m^2/d$ at peak flow.

TABLE 27.3
Design Parameters of UASB Reactor

HRT	8–10 hours at average flow (minimum 4 hours at peak flow).
SRT	30–50 days or more.
Sludge blanket concentration (average)	15–30 kg VSS/m^3. About 7.0% TSS.
Organic loading on sludge blanket	0.3–1.0 kg COD/kg VSS day (even up to 10 kg COD/ kg VSS/day for agro-industrial wastes).
Volumetric organic loading	1–3 kg COD/m^3 day for domestic sewage (10–15 kg $COD/m^3/d$ for agro-industrial wastes).
BOD/COD removal efficiency	Sewage 75–85% for BOD. 74–78% for COD.
Inlet points	Minimum 1 point per 3.7–4.0 m^2 floor area.
Flow regime	Either constant rate for pumped inflows or typically fluctuating flows for gravity systems.
Upflow velocity	About 0.5 m/h at average flow, or 1.2 m/h at peak flow, whichever is lower.
Sludge production	0.15–0.25 kg TS per m^3 sewage treated.
Sludge drying time	Seven days (in India).
Gas production	Theoretical 0.38 m^3/kg COD removed. Actual 0.1–0.3 m^3 per kg COD removed.
Gas utilization	Method of use is optional. 1 m^3 biogas with 75% methane content is equivalent to 1.4 kWh electricity.
Nutrients (nitrogen and phosphorus) removal	5 to 10% only.

Example Problem 27.4

Design a UASB reactor for an average daily flow of 5.0 ML/d containing a COD of 450 mg/L. Make the following assumptions:

ii. Design hydraulic residence time = 8 hours.
ii. Design COD loading 1–3 kg /m³·d.
iii. Rise rate in the reactor = 0.5 m/h.
iv. Overflow rate in the settling chamber = < 30 m³/m²·d.
v. Flow area covered by each inlet =1–3 m².

Given: Q = 5.0 ML/d = 5000 m³/d
COD = 450 mg/L
HRT = 8.0 h
v_{up} = 0.50 m/h
A = 2.0 m²/inlet

SOLUTION:

$$V = Q \times t = \frac{5000\,m^3}{d} \times 8.0\,h \times \frac{d}{24\,h} = 1666.6 = 1700\,m^3$$

$$Loading = \frac{5.0\,ML}{d} \times \frac{450\,kg}{ML} \times \frac{1}{1700\ m^3} = 1.32 = 1.3\,so\,okay$$

$$Reactor, H = v_{up} \times t = \frac{0.5\,m}{h} \times 8.0\,h = 4.0\,m \qquad A = \frac{V}{H} = \frac{1700\,m^3}{4.0\,m} = 425\,m^2$$

Assume the total height of the reactor is 5.5 m, and a square section:

$$Reactor, L = \sqrt{\frac{425\,m^2}{2}} = 14.57 = 15\,m$$

Hence, each reactor is 15 m × 15 m × 5.5 m. For an allowable overflow rate of 40 m/d, for average flow:

$$Surface\,settling, A_s = \frac{Q}{v_o} = \frac{1}{2} \times \frac{5000\,m^3}{d} \times \frac{d}{40\,m} = 62.5 = 63\,m^2$$

Discussion Questions

1. Compare septic tanks with Imhoff tanks as systems of wastewater treatment.
2. Briefly describe the various methods of disposing and treating septic tank effluent.
3. Discuss the advantages and disadvantages of Imhoff tanks.
4. Describe the type of wastes and conditions for which high-rate anaerobic reactors would be recommended.

5. What is the main concept on which UASB works? Give a brief account of the design approach.
6. Compare UASB with anaerobic sludge reactors.
7. What is the typical BOD of septic tank effluent? Is it good enough to discharge without any further treatment? If not, which treatment should be recommended to make it acceptable from an environmental protection point of view?
8. How can the suitability of a given method for treating septic tank effluents be determined?
9. Though Imhoff tanks are superior to septic tanks, their popularity is on the decline. What might some possible causes be?
10. One of the biggest problems of anaerobic systems for treating domestic wastewater is odours. What things can worsen the problem?
11. Scum is allowed to build up in the operation of septic and Imhoff tanks. Explain.
12. Write down the steps for performing a percolation rate test.
13. In the design of septic units, two chambers are preferred over a single chamber. Explain.
14. What considerations should be made when designing the inlet and outlet of a septic tank?

Practice Problems

1. A septic tank with a soak pit for absorbing tank effluent is required to serve a resort community of 500 people with a daily wastewater production of 150 L/c.d.
   a. Design the tank allowing a minimum detention time of 12 hours in the septic tank and a length-to-width ratio of 1:3 with an effective depth of 2.0 m. Assume annual sludge accumulation of 15 L/c.a. (9 m × 3 m × 2.5 m)
   b. Two circular soak pits are planned to treat effluent from the septic tank. Determine the size of the soak pit assuming that the absorption value of soil is 1500 L/m^3.d. (4.0 m in diameter, 2.0 m deep)
2. Design a septic tank based on the following data: population served = 400, daily wastewater production = 100 L/c.d.
   a. What flow capacity is required to provide a detention period of 24 hours and the storage of sludge at 30 L/c.a? (52 m^3)
   b. If soil absorption is used as effluent disposal, find the length of each of the ten 1.0 m wide trenches in soil with a percolation rate of 8 minutes. (10#, 55 m × 1 m)
3. Design a septic tank to serve the population from a small community of 100 people with the following assumptions:
   a. The detention time is 30 hours and the water supply is 130 L/p·d.
   b. The sludge is removed every year and produced at 30 L/p·d.
   c. The length-to-width ratio is 1:4 and the depth is 1.5 m. (#2, 5.2 m × 1.3 m × 2.0 m)
4. Design a UASB reactor for an average daily flow of 3.5 ML/d containing a COD of 600 mg/L. Make the following assumptions:
   a. The design hydraulic residence time is 10 hours.
   b. The design COD loading is 1–3 kg /m^3·d.
   c. The rise rate in the reactor is 0.5 m/h.
   d. The overflow rate in the settling chamber is < 30 m^3/m^2·d.
   e. The flow area covered by each inlet is 1–3 m^2. (#2 each 13 m × 13 m × 6.0 m)

5. Design an Imhoff tank to serve a population of 20,000 discharging wastewater at 180 L/c.d. Make appropriate assumptions wherever necessary. (Sedimentation: #2, 40 m × 4.0 m × 3.6 m, Digestion: #8, 5.0 m × 5.8 m × 5.1 m)
6. Calculate the storage capacity required for each of 10 compartments of an Imhoff digestion unit to serve a population of 25,000 discharging wastewater at 160 L/c.d with an SS content of 220 mg/L. Assume an SS removal of 50% and raw sludge SS content is 3.0%. After digestion, the volume of sludge reduces to one-third and the sludge is stored for a period of 100 days. (100 m^3/compartment)

28 Biosolids

During the processing of sludges, solids removed as slurry is called sludge or, more appropriately, biosolids. In a wastewater treatment plant, the solids slurry collected at the bottom of the primary clarifier is called primary or raw sludge. Similarly, the sludge pumped from the bottom of the secondary clarifier is called secondary sludge. In some plants, secondary sludge is pumped into the primary clarifier. In such cases, sludge pumped out of the primary clarifier will be a combination of primary and secondary sludge solids.

28.1 PRIMARY SLUDGE

Primary sludge is raw sludge. Since it is produced as a result of the settling of solids under gravity, primary sludge is septic or anaerobic, dark and offensive. Solids are concentrated without undergoing any biological or chemical treatment. The solids content of this sludge will depend on the characteristics of the raw wastewater, the level of solids removal by primary treatment, and any chemical addition upstream, such as alum addition for phosphorus removal. Due to chemical coagulation, the removal of solids will be enhanced and result in increased raw sludge production.

Since primary sludge is anaerobic and lacks molecular oxygen, sludge left too long in the clarifier may get gasified and become buoyant.

28.2 SECONDARY SLUDGE

Secondary sludge is not offensive as it is produced by an aerobic process. In activated sludge plants, secondary sludge is golden brown, light and has an earthly odour. The solids in secondary sludge are partially oxidized compared with the solids in raw sludge. To stabilize this sludge, it needs to undergo further biodegradation, known as digestion. Digested sludge with minimal metal content can be used as a soil conditioner in agricultural lands. The general characteristics of secondary sludge from activated sludge plants follow.

28.3 SLUDGE THICKENING

Sludge thickening is important for reducing the volume of sludge by removing water. For example, a sludge with 3% solids (97% water), when thickened to 6% solids, is reduced in volume by a factor of two. This makes further processing or disposal easier and efficient. Depending upon the characteristics of the sludge, two types of thickening are commonly used.

28.3.1 Gravity Thickener

A gravity thickener is designed to further concentrate sludge before being sent to additional sludge handling and treatment processes, such as digestion, conditioning, and dewatering. Thickening the sludge results in a reduced load on these subsequent processes.

A gravity thickener works on the same principle as a clarifier. The calculation of hydraulic loading is important in determining whether the process is underloaded or overloaded. Hydraulic loading for a gravity thickener is expressed as flow per unit surface area, and the solids loading rate on the thickener is calculated in the same fashion as kg of solids entering daily per square metre

DOI: 10.1201/9781003231264-28

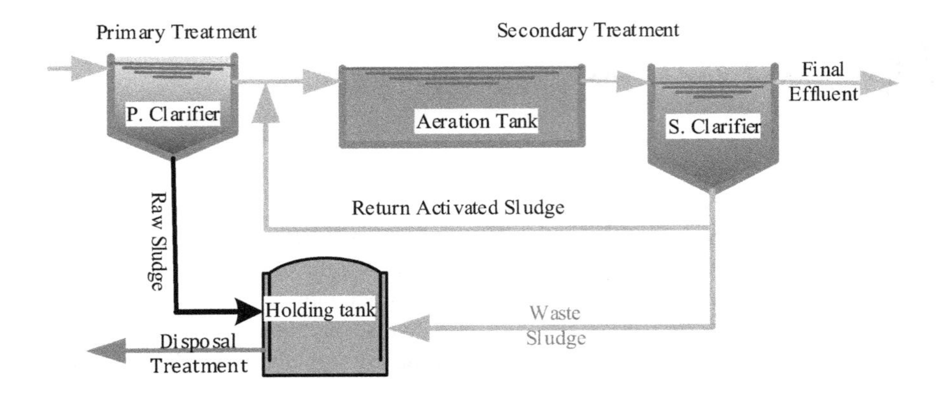

FIGURE 28.1 Primary and Secondary Sludge Flow Streams

area. The sludge detention time refers to the length of time the solids remain in the gravity thickener, which depends on the volume of the sludge blanket and the pumping rate of the sludge from the bottom of the thickener.

$$\text{Solids loading, } SLR = \frac{M_{SS}}{A_S} = \frac{Q_{Sl} \times SS}{A_S} \quad \text{Solids detention, } SDT = \frac{V_{Sl}}{Q_{Sl}} = \frac{A_S \times d_{Sl}}{Q_{Sl}}$$

Concentration factor is another way of determining the effectiveness of the gravity thickener. This parameter indicates the factor by which sludge has been thickened. A concentration factor of three will mean that the thickened sludge is three times as concentrated as the influent sludge.

Example Problem 28.1

A primary sludge flow of 410 m³/d with a solids content of 3.5% is pumped into a 12.5 m diameter gravity thickener. Calculate the hydraulic and solid loading rate.

Given: Q = 410 m^3/d
SS = 3.5% = 35 kg/m^3
D = 12.5 m

SOLUTION:

$$HLR = \frac{Q_{Sl}}{A_S} = \frac{Q_{Sl}}{0.785D^2} = \frac{410\ m^3}{d} \times \frac{1.27}{(12.5\ m)^2} = 3.34 = \underline{3.3\ m^3/m^2.d}$$

$$SLR = \frac{Q_{Sl} \times SS}{A_S} = \frac{410\ m^3}{d} \times \frac{35\ kg}{m^3} \times \frac{1.27}{(12.5\ m)^2} = 116 = \underline{120\ kg/m^2.d}$$

Example Problem 28.2

The sludge entering a gravity thickener contains 3.1% solids. The effluent from the thickener contains 1200 mg/L solids. What is the solids removal efficiency of the thickener? Calculate the concentration factor if the thickened sludge is 7.4% solids.

Given: $SS_i = 3.1\%$
$SS_e = 1200\ mg/L = 0.12\%$
$SS_{thick} = 7.4\%$

SOLUTION:

$$PR = \frac{(SS_i - SS_i)}{SS_i} = \frac{(3.1-0.12)\%}{3.1\%} \times 100\% = 96.12 = 96\%$$

$$Concentration\ factor,\ CF = \frac{SS_{thick}}{SS_{fed}} = \frac{7.4\%}{3.1\%} = 2.38 = 2.4\times$$

Example Problem 28.3

Given the data below, determine the change in sludge blanket solids.

Parameter	Feed Sludge	Effluent	Thickened Sludge
Q, m^3/d	630	240	290
SS, %	3.5	0.01	8.0
kg/m^3	35	0.1	80

SOLUTION:

$$M_{feed} = Q \times SS = \frac{630\ m^3}{d} \times \frac{35\ kg}{m^3} = 22050\ kg/d$$

$$M_{eff} = Q \times SS = \frac{240\ m^3}{d} \times \frac{0.10\ kg}{m^3} = 24\ kg/d$$

$$M_{thick} = Q \times SS = \frac{290\ m^3}{d} \times \frac{80\ kg}{m^3} = 23200\ kg/d$$

$$\Delta M = M_{in} - M_{out} = 22050 - 24 - 23200 = -1174\ kg/d$$

The minus sign means the sludge blanket will drop.

28.4 MASS VOLUME RELATIONSHIP

The solids concentration of sludges varies depending on the point of production, the removal of solids and the specific gravity (SG) of sludge. As mentioned earlier, for sludge with a solids concentration of less than 10%, it is safe to assume that the SG of the sludge is very close to that of water. That is to say, each litre of wet sludge would contain 10 g of dry solids. Solids in sludge consist of volatile and non-volatile or fixed solids. Non-volatile solids, being inorganic, are relatively heavier. The specific gravity of the total solids in sludge can be found as follows:

$$\frac{1}{SG_{TSS}} = \frac{VSS_f}{SG_{VSS}} + \frac{FSS_f}{SG_{FSS}} \qquad \frac{1}{SG_{SL}} = \frac{SS_f}{SG_{SS}} + \frac{WC_f}{1} \quad or \quad SG_{SL} = \frac{SG_{SS}}{\left(SS_f + SG_{SS} \times WC_f\right)}$$

Sub f = decimal fraction
TSS = total suspended solids
VSS = volatile suspended solids
FSS = fixed suspended solids
SL = sludge slurry
SS = dry solids
WC = water content

Example Problem 28.4

Determine the volume of wet sludge before and after digestion produced for every 100 kg of dry solids as feed sludge having the characteristics shown in the following table:

Parameter	Feed Sludge, 1	Digested Sludge, 2
SS, %	5.5	11
VF, %	65	40% destroyed
SG_{FS}	2.5	2.5
SG_{VS}	1.0	1.0

SOLUTION:

$$\frac{1}{SG_{TSS}} = \frac{VSS_f}{SG_{VSS}} + \frac{FSS_f}{SG_{FSS}} = \frac{0.65}{1} + \frac{0.35}{2.5} = 0.79 \quad or \quad SG_{TSS} = 1.265$$

$$SG_1 = \frac{SG_{SS}}{\left(SS_f + SG_{SS} \times WC_f\right)} = \frac{1.265}{\left(0.055 + 1.265 \times 0.945\right)} = 1.0116 = 1.012$$

$$V_1 = \frac{M_{SS}}{SS} = 100\ kg \times \frac{m^3}{55\ kg \times 1.012} = 1.79 = 1.8\ m^3$$

$$M_{SS} = fixed + volatile = \frac{100\ kg}{d} \times \left(0.35 + 0.65 \times 0.60\right) = 74\ kg$$

$$VF = \frac{M_{VSS}}{M_{TSS}} = \frac{39\ kg}{74\ kg} = 0.527 = 53\%$$

$$\frac{1}{SG_{TSS}} = \frac{VSS_f}{SG_{VSS}} + \frac{FSS_f}{SG_{FSS}} = \frac{0.53}{1} + \frac{0.47}{2.5} = 0.718 \qquad SG_{TSS} = 1.39$$

$$SG_2 = \frac{SG_{SS}}{\left(SS_f + SG_{SS} \times WC_f\right)} = \frac{1.39}{\left(0.11 + 1.39 \times 0.89\right)} = 1.0318 = 1.032$$

$$V_2 = \frac{M_{SS}}{SS} = 74\ kg \times \frac{m^3}{110 \times 1.032} = 0.651 = \underline{0.65}\ m^3$$

28.5 SLUDGE STABILIZATION

The process of stabilization makes sludge innocuous. This further reduces the BOD and volatile solids content of the sludge. In addition to sludge digestion, which will be discussed in detail, other common methods of sludge stabilization include wet air oxidation, chemical oxidation, lime treatment and composting.

28.6 SLUDGE DIGESTION

Sludge produced during wastewater treatment is not fully stabilized and must be before final disposal. For non-stabilized sludge, the most common disposal method is landfill, whereas stabilized sludge can be used for conditioning agricultural soils or other similar uses. Sludge can be stabilized using an anaerobic or aerobic process.

28.6.1 Anaerobic Sludge Digestion

Anaerobic digestion is more common and will be discussed in more detail. Anaerobic sludge digestion takes place in two distinct stages. Single-stage digestion, where both stages take place in the same unit, is shown in **Figure 28.2**.

In the first stage of digestion, acid-forming bacteria break down large organic compounds into organic acids. As this is an anaerobic process, the end products include methane, carbon dioxide, some hydrogen sulphide and organic acids. If, for some reason, the second stage is not working or going at a slow pace, accumulated acids will lower the pH and pickle the raw waste to stop any further decomposition. This condition is commonly known as souring of the digester.

In the second stage of digestion, gasification occurs and methane-forming bacteria convert organic acids to carbon dioxide and methane. Whereas acid-forming bacteria are sturdier, methane formers are very sensitive to pH, temperature and loading changes. Any rapid change will inhibit the

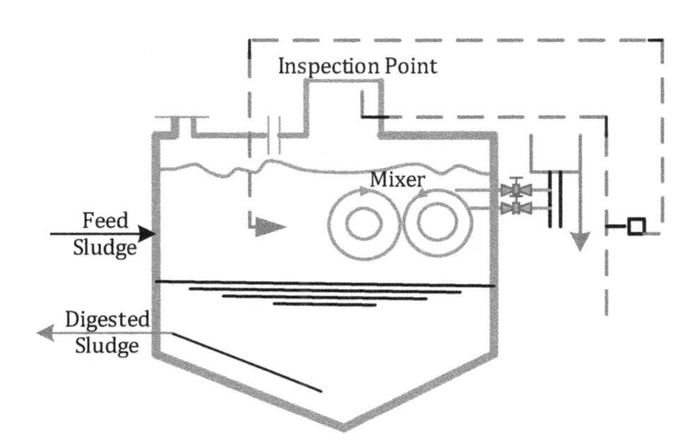

FIGURE 28.2 Single-Stage Anaerobic Digester

methane formers and result in the souring or pickling of the digestion process. Noting the sensitivity of the process, a certain number of tests must be done as part of daily operation. This includes tests for volatile acid concentration, temperature, pH, alkalinity, gas composition, the volatile fraction of solids, scum blanket depth, and suspended solids in the supernatant. In addition, records of the quantity of sludge and gases plus the mixer schedule should be maintained.

28.6.2 Aerobic Sludge Digestion

Aerobic sludge digestion is commonly used for the digestion of activated sludge. Since this is an aerobic process, the contents of the sludge are kept aerated. Aerobic digestion can be thought of as an aeration process for the sludge to stabilize by further reducing its BOD and volatile solids. A comparison of anaerobic and aerobic sludge digestion is shown in **Table 28.1**.

28.6.3 Anaerobic Digester Capacity

The determination of digester tank volume is a critical step in the design of an anaerobic system. The digester volume must be sufficient to prevent the process from failing under all accepted conditions. The digester capacity must also be large enough to ensure that raw sludge is adequately stabilized and there is room for storage if needed. The relationship between volatile solids reduction and detention time is shown in **Figure 28.3**.

$$V = V_1 + V_2 = \frac{(Q_1 + Q_2)}{2} \times T_1 + Q_2 \times T_2 = \frac{(Q_1 + 2Q_2)}{3} \times T_1 + Q_2 \times T_2$$

As shown in **Figure 28.3**, during digestion, the reduction in volume is assumed to be linear. Some authors are of the view that this reduction occurs non-linearly in a parabolic fashion. In that case, the second expression is used.

TABLE 28.1
Aerobic versus Anaerobic Digestion

Characteristic	Aerobic	Anaerobic
Source of sludge	Waste activated sludge	Primary and secondary
End products	CO_2 and cellular protoplasm	CH_4 + Unused Organics + Small portion of cellular protoplasm
Growth	Limited by availability of carbon	Limited by lack of H ion acceptor
Completion	Close to complete	Complete degradation not possible

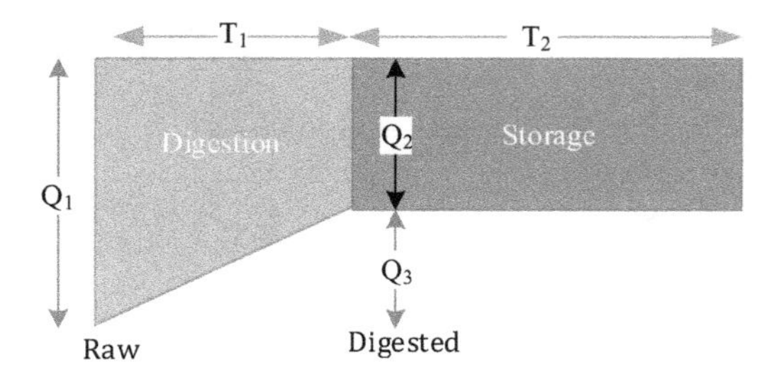

FIGURE 28.3 Digestion and Storage Capacity

Process Failure

Process failure is defined as the accumulation of volatile acids that results in a decrease in pH, also known as pickling. When the volatile acids–alkalinity ratio becomes greater than 0.5, acid-forming bacteria take over and the second stage virtually stops, and no methane is produced. Once the digester turns sour, it usually takes several days to return to normal operation after the corrective actions are taken. It is important to note that digestion is a very sensitive process, and skilled operators are required to control the process.

Two-Stage Digestion

In a high-rate digester, two stages are usually used, thus separating the mixing and storage. This produces better conversion and allows for better control. However, two tanks are needed to do the job. The capacity of each tank can be worked out as follows:

$$V_I = Q_I \times T_1 \qquad V_{II} = \frac{(Q_I + 2Q_{II})}{3} \times T_{II} + Q_2 \times T_2$$

Time, T_I, is the solids retention time in stage I, typically 10–15 days. Time, T_{II}, is the digestion period in stage II, and T_2 is the storage period required during winter or the monsoon period. The volume of sludge is reduced since some of the solids are converted to gases and sludge can thicken in the digester. Two-stage digestion offers the following advantages:

- Digestion happens more efficiently and quickly.
- The quality of supernatant is much better.
- The second stage is more for polishing and storage, so heating and mixing is mostly needed in the first-stage tank.
- Since there is no mixing in the second stage, scum is free to build up to the desired depth.
- The total cost of a two-stage digester may be less than two single-stage digesters.

28.6.4 Volatile Solids Reduction in Digestion

During digestion, volatile solids are oxidized and converted to carbon dioxide and methane, as discussed before. Thus, the total amount of volatile solids is reduced. However, the concentration of solids in the sludge increases due to thickening. Due to these factors, the volume of digested sludge is significantly less than the feed sludge. The reduction in volatile fraction can be worked out by observing the volatile fraction of the solids in the feed sludge and the digested sludge, as shown in the expression and **Figure 28.4**.

$$\textit{Volatile solids reduction}, VR = \frac{(VF_1 - VF_2)}{VF_1 - VF_1 \times VF_2}$$

VF = volatile fraction
VF_1 = volatile fraction of feed sludge
VF_2 = volatile fraction of digested sludge

Gas Composition

During digestion, volatile solids are converted to gases. Sludge gases are normally composed of 60% to 70% methane and 25% to 35% carbon dioxide by volume. Smaller quantities of other gases

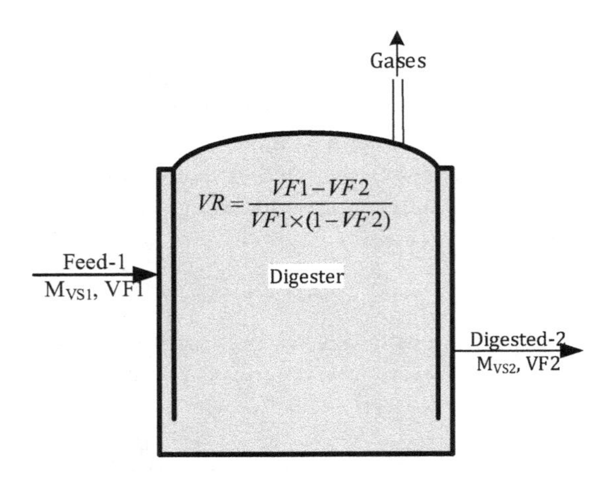

FIGURE 28.4 Volatile Solids Reduction During Digestion

such as hydrogen sulphide, hydrogen, nitrogen and oxygen are produced. The combustible constituent is primarily methane. The concentration of hydrogen sulphide primarily depends on the sulphate content of the feed sludge. Hydrogen sulphide is corrosive and causes problems during the burning of the gas. In terms of volatile acids destroyed during digestion, the average gas production is 0.9 m^3/kg of volatile acids destroyed at a normal operating pressure of 1.5 to 2.0 kPa. Sludge gases are collected under positive pressure to prevent them from mixing with air and causing an explosion. The gas may be collected directly from under a floating cover or from a fixed cover by maintaining a constant water level.

Example Problem 28.5

What is the percent reduction in volatile matter in a primary digester if the volatile content of the raw sludge is 69% and the volatile content of the digested sludge is 51%?

Given: $VF_1 = 69\% = 0.69$
$VF_2 = 51\% = 0.51$

SOLUTION:

$$VR = \frac{(VF_1 - VF_2)}{VF_1 - VF_1 \times VF_2} = \frac{0.69 - 0.51}{0.69 - 0.69 \times 0.51} = 0.532 = \underline{53\%}$$

28.6.5 Digester Solid Mass Balance

The principle of mass balance dictates that what enters a process or combination of processes in series must equal what goes out. If you can account for 90% of the solid material leaving your plant in the form of sludge, liquid (effluent) or gas (digester gas), then you have control of your plant. This accounting process provides a good check on your metering devices, sampling procedures and analytical techniques. The mass entering the digester is sludge that is comprised of solids and water. During the digestion process, part of the volatile solids is converted to gases such as methane and carbon dioxide, and the rest leave the digester as supernatant and digested sludge.

Example Problem 28.6

Work out the digester capacity required for a low-rate anaerobic digester required to serve a population of 50,000. Assume the per capita production of sludge is 90 g/c·d with a solids content of 4.0% and volatile fraction of 70%. Assume a digestion period of 25 days and a storage period of 20 days, a volatile reduction of 50% during digestion, and the solids content of the digested sludge is 8.0%.

Given: $SS_1 = 4.0\%$
$SS_2 = 8.0\%$
$T_1 = 25$ d.
$T_2 = 20$ d
$VF_1 = 70\%$
$VF_2 = 50\% \times 70\% = 35\%$

SOLUTION:

$$Q_I(Feed) = \frac{M_{SS}}{SS} = \frac{90\ g}{p.d} \times 50000\ p \times \frac{kg}{1000\ g} \times \frac{m^3}{40\ kg} = 112.5 = 110\ m^3/\text{d}$$

$$M_{SS}(Digested) = \frac{90\ g}{p.d} \times 50000 p \times \frac{kg}{1000\ g} \times (0.30 + 0.70 \times 0.50) = 2925\ kg/d$$

$$Q_{II}(Digested) = \frac{M_{SS}}{SS} = \frac{2925\ kg}{d} \times \frac{m^3}{80\ kg} = 36.56 = 37\ m^3/d$$

$$V(Digester) = \frac{(112.5 + 2 \times 36.56)\ m^3}{3.d} \times 25\ d + \frac{36.56\ m^3}{d} \times 20\ d = 2278 = \underline{2300\ m^3}$$

Example Problem 28.7

Work out the digester capacity required for a high-rate anaerobic digester required to serve a population of 50,000 people. Assume the per capita production of sludge is 90 g/c·d with a solids content of 4.0% with a volatile fraction of 70%. The sludge is kept for a period of 10 days at stage I before it is transferred to stage II. In the second stage, assume a digestion period of 15 days and a storage period of 20 days, a volatile reduction of 50% during digestion, and the solids content of the digested sludge is 8.0%.

Given: $SS_1 = 4.0$
$SS_2 = 8.0\%$
$T_I = 10$ d
$T_{II} = 15$ d
$T_2 = 20$ d
$VF_1 = 70\%$
$VF_2 = 50\% \times 70\% = 35\%$

SOLUTION:

$$Q_I = \frac{M_{SS(I)}}{SS} = \frac{90g}{p.d} \times 50000 p \times \frac{kg}{1000\ g} \times \frac{m^3}{40\ kg} = 112.5 = 110\ m^3/d$$

$$M_{SS(II)} = \frac{4500\ kg}{d} \times (0.30 + 0.70 \times 0.50) = 2925\ kg/d$$

$$Q_{II} = \frac{M_{SS}}{SS} = \frac{2925\ kg}{d} \times \frac{m^3}{80\ kg} = 36.56 = 37\ m^3/d$$

$$V_I\,(Stage\ I) = Q_I \times T_I = \frac{112.5 m^3}{d} \times 10\ d = 1125 = \underline{1130\ m^3}$$

$$V_{II}\,(Stage\ II) = \frac{(112.5 + 2 \times 36.56)\ m^3}{3d} \times 15\ d + \frac{36.56\ m^3}{d} \times 20\ d = 1659.3 = \underline{1700\ m^3}$$

28.7 DEWATERING OF SLUDGE

Digested sludge, also known as blended sludge, usually contains more than 90% water. It would be expensive to carry this watery sludge for disposal. Hence, sludge is dewatered before disposal. Depending on the method of dewatering, the sludge water content can be brought to as low as 20–30%, thus reducing the volume many times. In India, the most common practice is to spread the sludge on drying beds and let it dry in the open. When the space required for dry beds is not available, the sludge is conditioned, followed by mechanical dewatering. Mechanical dewatering methods include vacuum filters, filter presses or centrifugation.

28.7.1 Sludge Drying Beds

This method can be used in places where adequate land is available and dried sludge can be used for soil conditioning. Sludge drying beds are affected by weather, sludge characteristics, a system design that includes the depth of the sludge layer, and the frequency with which the beds are scraped after drying.

Area of Beds

The area needed for the dewatering and drying is dependent on the volume of the sludge. The cycle time between two successive dryings depends on the characteristics of the sludge, including its water content, drainage ability and rate of evaporation, and the acceptable water content in the dried sludge.

Drying Beds

A sludge drying bed usually consists of a bottom layer of gravel of uniform size laid over a bed of clean sand. Graded gravel is placed on the underdrain in layers of up to 30 cm with a minimum of 15 cm above the top of the underdrain. The top layer should be at least 3.0 cm thick and consists of gravel 3–6 mm in size. Gravel should be covered with a 20–30 cm thick layer of clean sand of effective size 0.50–0.75 mm and a uniformity coefficient of less than 4.0. The finished surface should be level. Open joints under drainpipes 100–150 mm in diameter are laid with a spacing of 6.0 m or less. The pipes are laid on a grade of 1%. The beds are about 15 m × 30 m and are surrounded by about 1.0 m high brick walls above the sand surface.

Example Problem 28.8

Design sludge drying beds for digested sludge as discussed in the previous Example Problem 28.7. The population served is 50,000 and the average production of digested sludge is 37

m^3/d. Assume 25 cm of digested sludge is spread every couple of weeks. Though drying in summer is done over 10 days, a drying cycle of 14 days is assumed to compensate for wet weather.

Given: $Q_{dsl} = 37\ m^3/d$
$t = 14\ d$
$d = 25\ cm/cycle$
Area of each bed = 30 m × 15 m

SOLUTION:

$$A = \frac{37\ m^3}{d} \times \frac{cycle}{0.25\ m} \times \frac{14\ d}{cycle} = 2072\ m^2 \quad \# = 2072\ m^2 \times \frac{bed}{30\ m \times 15\ m} = 4.6 = 5\underline{beds}\ say$$

Making 100% allowance for space for storage, repairs and the resting of beds, 10 beds are required, each 30 m × 15 m.

28.7.2 Mechanical Methods of Dewatering Sludge

Mechanical methods of dewatering include filter press, vacuum filtration and centrifuges.

Vacuum Filtration

Rotary vacuum filtration consists of a cylindrical drum covered with a filter media that rotates partially submerged in a vat of sludge. The physical mechanisms that take place during vacuum filtration may be divided into three phases.

The first phase, which refers to the cake pick-up or form phase, occurs when a segment of the drum rotates into the sludge. A vacuum is applied to that segment, and filtrate is drawn through the media and discharged. Concurrently, sludge solids are deposited on the media to form a partially dewatered cake. As the sludge cake increases in thickness, its resistance to the passage of filtrate increases.

The second phase, cake-drying, occurs as the drum segment leaves the sludge and before the cake is removed. As the drum leaves the sludge, the cake is still under vacuum, and additional moisture is drawn out.

The third phase, cake discharge, occurs after an acceptable cake dryness has been achieved without vacuum. All operations are continuous such that all three phases occur simultaneously on different portions of the drum.

Precoat vacuum filtration is like conventional filtration, except a precoat is applied prior to filtration. The precoat is normally diatomite, the siliceous skeletal remains of single-cell aquatic plant life called diatoms. These diatoms form a permeable coating on the filter allowing filtrate to pass through easily while trapping sludge solids. Use of a precoat produces filtrate of very high quality.

Belt Filter Press

The continuous belt filter press was originally developed and, in subsequent years, modified and improved in West Germany. Installation of the latest and best models in the United States has only recently experienced popularity. These systems were developed to overcome the sludge pick-up problem occasionally experienced with rotary vacuum filtration. A combination of sludge conditioning, gravity dewatering and pressure dewatering is utilized to increase the solids content of the sludge.

With all units, the infeed sludge is mixed with a polymer (or other chemicals) and placed onto a moving porous belt or screen. Dewatering occurs as the sludge moves through a series of rollers that squeeze the sludge to the belt or between two belts, much like an old washing machine wringer. The sludge cake formed is then discharged from the belt by a scraper mechanism. Three processing zones occur along the length of the unit:

1. A drainage zone, which is analogous to the action of a drying bed.
2. A pressure zone, which involves the application of pressure.
3. A shear zone, in which shear is applied to the partially dewatered cake.

The shearing action is accomplished by positioning the support rollers of the filter belt and the pressure rollers of the pressure belt so that the belts and the sludge between them form an s-shape curve. This condition creates a parallel displacement of the belts relative to each other due to the difference in radius. The use of chemicals to condition the sludge prior to dewatering is necessary regardless of the type of belt filter press selected. Proper sludge conditioning results in the flocculation of small particles into larger particles of sufficient size and strength to bridge the openings in the filter media and thus be retained on the belt.

Pressure Filtration

Of the several types of pressure filters available, the most widely used consists of a series of vertical plates held rigidly in a frame that are pressed together between a fixed and moving end. Mounted on the face of each plate is a filter cloth to support and contain the cake produced.

Pressure filters do not produce a cake by pressing and squeezing. Instead, sludge is fed into the press "batch mode" through feed holes in trays along the length of the press. Pressures of up to 1600 kPa are applied to the sludge causing water to pass through the cloth while the solids are retained, forming a cake on the surface of the filter cloth. The sludge feed is stopped when the chambers between the trays are filled. Drainage ports are provided at the bottom of each chamber where the filtrate is collected, taken to the end of the press and discharged.

The dewatering phase is complete when the flow of filtrate through the filter cloth nears zero. At this point, the sludge feed pump is stopped, and any back pressure in the piping is released. Each plate is then turned over the gap between the plates and the moving end to allow for cake removal. Filter cake usually drops below onto a conveyor for further removal.

The main advantage to using a filter press for sludge dewatering is the reduced sludge disposal costs associated with producing a drier cake. However, a detailed cost analysis should be performed to determine if these savings are sufficient to offset its high capital cost.

Centrifugation

A centrifuge is essentially a sedimentation device in which the solids–liquid separation is enhanced using centrifugal force. This is accomplished by rotating the liquid at high speeds to subject the sludge to increased gravitational forces. There are three types of centrifuges available for sludge dewatering.

Continuous Solid Bowl Centrifuge

This centrifuge consists of two principal elements: a rotating bowl, which is the settling vessel; and a conveyor, which discharges the settled solids. The rotating bowl is supported between two sets of bearings and includes a conical section at one end. This section, which is not submerged, forms the dewatering beach or drainage deck. Sludge enters the rotating bowl through a stationary feed pipe extending into the hollow shaft of the rotating screw conveyor and is distributed through ports into a pool within the bowl. As the bowl rotates, centrifugal force causes the slurry to form an annular pool, the depth of which is determined by the effluent weirs.

Basket Centrifuge

The basket centrifuge, or imperforate bowl-knife discharge unit, is a batch dewatering unit introduced primarily for use as a partial dewatering device for small operations.

Disc Centrifuge

The disc centrifuge is a continuous flow variation of the previously described basket centrifuge. The incoming sludge is distributed between a multitude of narrow channels formed by stacked conical discs.

28.7.3 Sludge Conditioning

Whatever the methods of dewatering, sludge needs conditioning to facilitate the extraction of moisture held by sludge solids. Coagulants and coagulant aids such as alum, lime and polymers are commonly used. Because of its high alkalinity, digested sludge needs a very high dose of coagulant if not elutriated.

Polymers are less affected by pH and work better with finely dispersed solids. It is difficult to specify a formula suitable for a polyelectrolyte. Even though there are quite a few such polyelectrolytes in the market, it is best to carry out actual laboratory-scale testing before launching into procurement.

In general, polyelectrolytes are available in powder and viscous liquid form. Both are the same as far as usage is concerned because the powder will immediately become viscous when added to water. The choice of chemical for conditioning depends on many factors, including pH, non-volatile fraction and temperature. Optimum doses are determined by doing tests in a laboratory.

Elutriation of sludge is done to reduce chemical dosage and is carried out by washing the sludge followed by decantation.

28.8 DISPOSAL OF SLUDGE

In Indian conditions, digested sludge is usually disposed of on land as manure for soil or as a soil conditioner. In developed countries, unstabilized sludges are sent to sanitary landfills, incinerated, or barged into the sea.

The most common method of disposal is to utilize the digested sludge as a fertilizer. Ash from incinerated sludge is used as a landfill. In some cases, wet sludge, raw or digested, as well as supernatant from a digester, can be constructed as lagoons as a temporary measure, but such practice may create problems such as odour nuisance, groundwater pollution and other public health hazards. Wet or digested sludge can be used as sanitary landfill or for mechanized composting with city refuse. Burial is generally resorted to for small quantities of putrid sludge. In Ontario, Canada, the disposal of sludge has to be done as per the hazardous waste (handling and management) rules of the Ministry of the Environment, Conservation and Parks, as shown in **Table 28.2**, and faecal coliform limits are not violated.

TABLE 28.2
Ceiling Concentration of Heavy Metals

Chemical	Conc., mg/kg	Chemical	Conc., mg/kg
As	75	Cr	500
Cd	85	Se	100
Cu	4300	Zn	7500
Hg	57	Mb	75
Ni	420	Pb	840

28.8.1 Sludge as Soil Filler

Raw sludge is unstabilized and considered hazardous, and its use as a soil filler directly on land for raising crops as a means of disposal is not desirable. The application of sewage sludge to soils should take into consideration the following guiding principles:

1. Sludge from open-air drying beds should not be used on soils where it is likely to come into direct contact with vegetables and fruits.
2. Sludge from drying beds should be ploughed into the soil before raising crops. Top dressing soil with sludge should be prohibited.
3. Dried sludge may be used for lawns and for growing deep-rooted cash crops and fodder grass where direct contact with the edible part is minimized.
4. Heat-dried sludge is the safest from a public health point of view. Though deficient in humus, it is convenient in handling and distribution. It should be used along with farmyard manure.
5. Liquid sludge, either raw or digested, is unsafe to use. It is unsatisfactory as fertilizer or soil conditioner. If used, it must be thoroughly incorporated into the soil and the land should be given rest so that the biological transformation of organic material can take place. It should be used in such a way as to avoid all possible direct human contact.

In general, digested sludge is indelicate but has definite value as a source of slowly available nitrogen and some phosphate. It is comparable to farmyard manure except for its deficiency in potash. It also contains essential elements for plant life and minor nutrients in the form of trace metals. Sludge humus also increases the water-holding capacity of soil and reduces soil erosion, making it an excellent soil conditioner. This is especially true in arid regions by providing much-needed humus content, which results in greater fertility. Dewatered cake is typically stored before additional treatment (e.g. heat-drying) or being hauled off-site for use or disposal. Most flammable liquids would have been removed during dewatering, and methane-generating microorganisms do not thrive in dry aerobic environments.

The amount of storage needed depends on the end-use of the sludge cake. Often, biosolids will only be held for a few days or weeks before being treated further or hauled off-site. In this case, they are typically stored in large roll-off containers, dump trailers, concrete bunkers with push walls, or bins with augurs. However, if the biosolids are to be applied on land or surface disposed, long-term storage may be required. In these cases, they are often stockpiled on concrete slabs or other impervious pads. When designing long-term storage facilities, buffering, odour control and accessibility should be considered. Whether or not the storage facility should be open or covered should also be determined.

Dried solids are typically stored on-site or at a land-application site before their disposal or beneficial use. They may be stored in stockpiles or silos. Because dried solids contain a significant amount of combustible organic material that can be released as dust, temperature control is important. If silos are used, they should be designed to promote cooling and maximize heat dissipation. Therefore, tall, narrow silos are better than wide ones. Narrow silos also make fires easier to control. However, if the silo is too narrow, relief venting will be problematic. If multiple silos are used, there should be procedures to ensure they are emptied cyclically to avoid exceeding safe residence times. The stored product's thermal stability should also be considered in case of prolonged plant shutdown or if silo blockage occurs.

28.8.2 Incineration

The purpose of incineration is to burn out the organic material at a high temperature, the residual ash being generally used as landfill. During this process, all the gases released from the sludge are burnt off and the organisms are all destroyed. Dewatered or digested sludge is subjected to temperatures of between 650°C to 750°C. Cyclone or multiple hearth and flash furnaces are used with proper heating arrangements with temperature control and drying mechanisms. Dust, fly ash and soot are collected for use as landfill.

Incineration has the advantage of being free from odours and provides a great reduction in the volume and weight of materials to be disposed of finally, but the process requires high capital and recurring costs, the installation of machinery and skilled operation. Controlled drying and partial incineration have also been employed to dewater sludge before being put on drying beds.

28.8.3 Sanitary Landfill

The decomposition of organic solids in a landfill may cause odour if sufficient cover is not available. Surface water contamination and the leaching of sludge components to the groundwater must be considered. Decomposition may result in soil settlement, resulting in surface water ponds above the fill. The typical depths of soil cover over the fill area are 20 cm after each daily deposit and 60 cm over an area that has been filled completely.

The surface topography should be finished to allow rainfall to drain away and not infiltrate into the solid landfill. Landfill leachate requires long-term monitoring and should satisfy the relevant water pollution control standards for land applications. Vegetation must be established quickly on completed areas to provide erosion control. It is general practice not to crop the landfill area for several years after completion.

28.8.4 Disposal in Water or Sea

This is not a common method of disposal because it is contingent on the availability of a large body of water adequate to permit dilution at some seacoast sites. The sludge, either raw or digested, must be barged to sea far enough to provide the required dilution and dispersion. The method requires careful consideration of all factors, including flora and fauna, for proper design and the siting of outfall to prevent any coastal pollution or interference with navigation.

28.8.5 Sludge Composting

Sludge composting is a method in which microorganisms decompose degradable organic matter in sludge under aerobic conditions and create stable material that is easy to handle, store and use for farmland. Sludge compost is a humus-like material without detectable levels of pathogens that can be applied as a soil conditioner and fertilizer to gardens, food, and feed crops and farmland. Sludge compost provides large quantities of organic matter and nutrients such as nitrogen and potassium to the soil. It improves the soil texture and elevates soil cation exchange capacity (an indication of the soil's ability to hold nutrients), all characteristics of a good organic fertilizer. Sludge compost is safe to use and generally has a high degree of acceptability by the public.

Sludge composting involves mixing dewatered sludge with a bulking agent to provide carbon and increase porosity. The resulting mixture is piled or placed in a vessel where microbial activity causes the temperature of the mixture to rise during the first-phase active composting period. The specific temperatures that must be achieved and maintained for successful composting vary based on the method and use of the end product. After the first phase, the active composting period of the material is cured. In the second phase, it becomes compost and is distributed. Sludge composting methods are divided into aerated static pile, windrow and in-vessel.

Aerated static pile

Dewatered sludge cake is mechanically mixed with a bulking agent and stacked into long piles over a bed of pipes through which air is transferred to the composting material. After the first-phase composting, i.e. active composting, as the pile starts to cool down, the material is moved into the second-phase composting, i.e. a curing pile. The bulking agent is often reused in this composting method and may be screened before or after curing so that it can be reused.

Windrow

Dewatered sludge cake is mixed with a bulking agent and piled in long rows because there is no piping to supply air to the piles; they are mechanically turned to increase the amount of oxygen. This periodic mixing is essential for moving the outer surfaces of the material inward so that they are subjected to the higher temperatures deeper in the pile. A number of turning devices are available. As with aerated static pile composting, the material is moved into the second-phase composting, i.e. curing piles, after the first-phase composting, which is active composting. Several rows may be placed into a larger pile for curing.

In-Vessel

There are two types of in-vessel composting reactor: vertical and horizontal. A mixture of dewatered sludge cake and bulking agent is fed into a silo, tunnel, channel or vessel. Augers, conveyors, rams or other devices are used to aerate, mix and move the product through the vessel to the discharge point. Air is generally blown into the mixture. After first-phase composting, i.e. active composting, the finished product is usually stored in a pile for second-phase composting, i.e. curing prior to distribution.

All three composting methods require the use of bulking agents. Wood chips, sawdust and shredded tires are commonly used, but many other materials are suitable.

28.8.6 Applicability

The physical characteristics of most sewage sludge allow for its successful composting. However, many characteristics, including moisture content, volatile solids content, carbon content, nitrogen content and bulk density, will impact the design decisions for the composting method. Both digested and raw solids can be composted, but some degree of digestion (or similar stabilization) is desirable to reduce the potential for the generation of foul odours from the composting operation. This is particularly important for aerated static pile and windrow operations.

The carbon and nitrogen content of sewage sludge must be balanced against that of the bulking agent to achieve a suitable carbon-to-nitrogen ratio of between 25 and 35 parts carbon to one part nitrogen.

Site characteristics make composting more suitable for some sewage treatment plants than others. An adequate buffer zone from neighbouring residents is desirable to reduce the potential for nuisance complaints. In urban and suburban settings, in-vessel technology may be more suitable than other composting technologies because the in-vessel method allows for the containment and treatment of air to remove odours before release. The requirement for a relatively small amount of land also increases the applicability of in-vessel composting in these settings.

Design Considerations

Important matters to consider when designing a sludge composting system are as follows:

- Sewage sludge characteristics such as moisture content, volatile solids content, carbon content, nitrogen content and bulk density.
- Sludge composting should maintain uniform aerobic conditions during composting. Proper air supply procedures are required, such as the turning of piles or adequate aeration, and the selection of a proper bulking agent.

- Compaction of the composting mass should be avoided to maintain sufficient pore space for aeration.
- A bulking agent appropriate for sludge characteristics such as size, cost/availability, recoverability, carbon availability, pre-processing requirements, porosity and moisture content should be selected.
- The metal content of the sewage sludge should be considered to ensure a market for the final product.
- An odour control system should be considered, especially for an in-vessel composting system.
- Composting detention time and temperature should be determined by considering the quality of product compost. An aerated static pile or an in-vessel system should be kept at 55°C for at least three days, and for a windrow, 55°C for at least 15 days with five turns.

Mixed Composting of Sewage Sludge and Municipal Solid Waste

Dewatered sludge can be mixed with grounded organic municipal solid waste and be used as a good soil conditioner (compost). However, this process needs proper policy guidelines, stringent regulations, standards, and above all, community awareness.

Discussion Questions

1) Show that for sludge with a solids concentration of less than 10%, it is safe to assume that the density of the sludge is close to that of water.
2) What is the purpose of thickening the sludge? Describe the various methods of thickening.
3) Write a short note on sludge composting.
4) Briefly describe the various methods of sludge disposal.
5) Most plant personnel believe that sludge processing is a difficult and expensive process. Explain.
6) What is the dewatering of sludge? What processes can be used to dewater sludge?
7) In a country like India, what is the most common method of sludge dewatering? Discuss its merits and demerits.
8) Compare the following:
   a) Primary sludge versus secondary sludge.
   b) Activated sludge plant sludge versus trickling filter plant sludge.
   c) Aerobic versus anaerobic sludge digestion.
9) What is the purpose of conditioning sludge? What role does elutriation play in sludge conditioning?
10) What disposal methods can be used to dispose of undigested sludge?

Practice Problems

1) What is the percent reduction of volatile matter in a primary anaerobic digester if the volatile content of the raw sludge is 71% and the volatile content of the digested sludge is 49%? (61%)
2) Size a low-rate anaerobic digester required to serve a population of 35 000. Assume the per capita production of sludge is 100 g/c·d (dry solids) with a total solid content of 3.5% of which 65% are volatile. Assume a digestion period of 25 days and a storage period of 30 days. Half of the volatile solids are destroyed during digestion and the digested sludge solids content is 7.0%. (2410 m^3)

3) Calculate the digester capacity required for a high-rate anaerobic digester to serve a population of 75 000. Assume the per capita production of sludge is 95 g/c·d with a solid content of 3.5% of which 70% are volatile. The sludge is retained in stage I for a period of 10 days and then transferred to stage II. During stage II, assume a digestion period of 15 days and a storage period of 20 days. During digestion, half the volatile solids are destroyed and the sludge is thickened to 7.5% solids. (2040 m^3, 2900 m^3)
4) Prove that a sludge concentration of 1.0% is the same as a mass concentration of 10 kg/m^3 or 10 g/L.
5) Design an anaerobic digester to treat a blend of primary and secondary sludge. The daily production of feed sludge is 250 m^3/d. The solids content of the feed sludge and digested sludge are 5.0% and 9.5%, respectively. Based on the prevailing temperatures, a digestion period of 25 days is considered and a minimum storage period of 40 days is required. Assume the density of a wet sludge of less than 10% solid concentrations is the same as that of water. Make other assumptions as appropriate. (70% volatile, 60% reduction, 5200 m^3)
6) By how many times is the volume of sludge reduced when the moisture content is reduced from 95% to 90%? (2 times)
7) A wastewater pollution control plant produces on average 1350 kg of dry solids as sludge with a solids concentration of 5.0%. The solids are 70% volatile with an SG of 1.05 and 30% fixed with an SG of 2.5. Find the SG of dry solids and that of the sludge produced, and hence the volume of sludge produced at this plant before digestion. (27 m^3/d)
8) Calculate the solids concentration in the sludge fed to a digester that consists of a mixture of a 3.5% primary sludge flowing at 31 m^3/d and 5.0% thickened secondary sludge flowing at 21 m^3/d. (4.1%)
9) Design sludge drying beds for digested sludge. The population served is 65 000 and the average production of digested sludge is 0.60 L/c.d. Assume 35 cm of digested sludge is spread every couple of weeks. Though drying in summer is done over 12 days, a drying cycle of 14 days is assumed to compensate for wet weather. (7 beds, each 30 m × 15 m)
10) At a given facility, during digestion, 53% of volatile matter is destroyed. Determine the expected volatile content in the digested sludge when sludge with a volatile content of 69% is fed. (51%)
11) In a sewage treatment plant, the average primary SS removal is 55%. Estimate the primary sludge production for treating a wastewater flow of 11 ML/d with an SS content of 250 mg/L. Assume the moisture content of the primary sludge is 96%. (38 m^3/d)
12) Blended sludge containing 750 kg of dry solids is fed to a digester for stabilization. The characteristics of the feed sludge and the digested sludge are shown below. Determine the volume of wet sludge before and after digestion. (15 m^3, 4.6 m^3)

Parameter	Feed Sludge, 1	Digested Sludge, 2
SS, %	5.0	10
VF, %	60	60% destroyed
SG_{FS}	2.5	2.5
SG_{VS}	1.0	1.0

13) Using the information in Practice Problem 12, work out the volume of sludge cake produced after dewatering to a water content of 65%. (1200 L)
14) Sludge containing 600 kg of dry solids is fed to a digester for stabilization. The characteristics of the feed sludge and the digested sludge are shown below. Determine the volume of wet sludge before and after digestion. (9.9 m^3, 2.8 m^3)

Parameter	**Feed Sludge, 1**	**Digested Sludge, 2**
SS, %	6.0	12
VF, %	65	65% destroyed
SG_{FS}	2.5	2.5
SG_{VS}	1.0	1.0

29 Advanced Wastewater Treatment

Conventional wastewater treatment consists of preliminary and primary treatment followed by secondary treatment, which is usually a biological process. Early treatment objectives were concerned with the removing suspended and floatable material, treating biodegradable organics and eliminating pathogenic organics. In **Table 29.1**, the residual constituents in treated wastewater are shown.

As a result of wastewater plant effluent limits becoming more stringent, a major effort was undertaken to achieve a more effective and widespread treatment of wastewater. As a result, the required degree of treatment has significantly increased, and additional treatment objectives and goals have been added. The recently added treatment objectives are the removal of nutrients, suspended solids and BOD, and toxic substances. This phase of wastewater treatment is known as advanced wastewater treatment. Advanced treatment can be part of primary or secondary treatment or a third stage called tertiary treatment.

Thus, advanced wastewater treatment refers to the processes and methods that remove more contaminants from wastewater than are usually taken out by conventional treatments and techniques. Several of the pollutants or contaminants not taken out by secondary biological methods can adversely affect aquatic life in the receiving waters, accelerate eutrophication in lakes, and hinder the reuse of surface water for domestic needs. The renovation and reuse of wastewater are becoming more important as water needs cannot be met by the relatively fixed natural supply available.

29.1 SUSPENDED SOLIDS REMOVAL

Suspended solids removal is one of the most important and common applications of advanced wastewater treatment. The effluent from the secondary treatment plant may contain a suspended solids concentration ranging from 20 to 40 mg/L, depending upon the type of treatment method. The suspended solids in treated wastewater are in part colloidal and in part discrete, ranging from 10 to 100 microns. Because of their size, these solids are not easily settleable. In the following paragraphs, some of the commonly used advanced treatments are briefly discussed.

29.1.1 Micro-Screening

Micro-screening utilizes a special woven metallic or plastic filter fabric mounted on the periphery of a revolving drum provided with continuous backwashing. The drum rotates about a horizontal axis at a variable low speed (up to 4 rpm). The size of openings of the filter fabric may range from 23 to 60 μm. Wastewater enters through the open upstream end of the drum and flows radially outward through the microfabric, leaving behind suspended solids. The solids retained on the fabric are washed through a trough, which recycles the solids to the sedimentation tank. Water for backflushing is drawn from the filtered water effluent.

29.1.2 Ultrafiltration

Ultrafiltration is similar in operation to reverse osmosis and requires the membrane to be far coarser and the pressure lower. Ultrafiltration membranes are made of a thin film cast from organic

DOI: 10.1201/9781003231264-29

TABLE 29.1
Residual Constituents in Treated Wastewater

Constituents	Effect	Critical Conc. Mg/L
Suspended solids	May cause sludge deposits or interfere with receiving water quality.	Variable
Organics	May deplete oxygen resources.	Variable
Volatile organic compounds	Toxic to humans; carcinogenic; form photochemical oxidants.	Varies by individual constituent
Nutrients	Increases chlorine demand; can be converted to nitrates and deplete DO.	Any amount
Ammonia		
Nitrate	Stimulates algal and aquatic growth; can cause methemoglobinemia in infants.	0.3 (for quiescent lakes)
Phosphorus		
	Stimulates algal and aquatic growth.	0.015 (for quiescent lakes)
Other inorganics	Interferes with coagulation and softening; increases hardness and dissolved solids.	250
Ca & Mg		
Chloride	Imparts salty taste; interferes with agricultural and industrial processes; cathartic action.	75–200
Surfactants	Causes foaming and may interfere with coagulation.	1.0–3.0

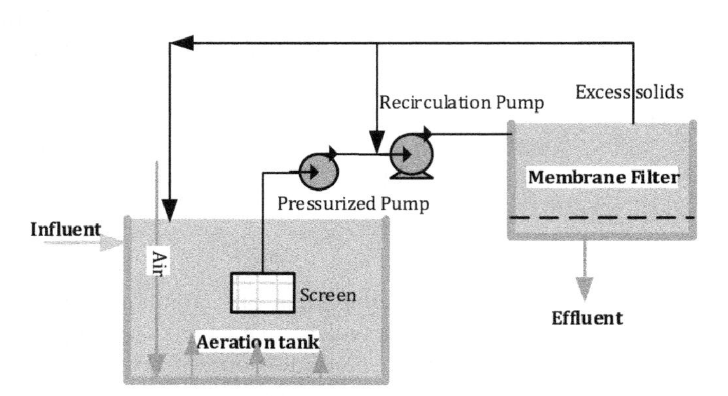

FIGURE 29.1 Combination of Ultrafiltration and Biodegradation

polymer solutions that range in thickness from 0.1 mm to 0.3 mm. The film is anisotropic, i.e. it has an extremely thin separation layer (of thickness 0.1 to 10 um) on a relatively porous structure, consisting of pores of closely controlled size ranging from 1 to 10 nm. Ultrafiltration membranes may be packed as either a plate or a tube device.

Ultrafiltration is used in conjunction with biological oxidation processes such as the activated sludge process. When the effluent from the activated sludge process is passed through an ultrafiltration system, the membranes filter out the biological cells while allowing the passage of treated effluent. The retained solids are returned to the activated sludge reactor. This leaves a high concentration of biological solids in the activated sludge reactor, resulting in the quick degradation of organics, prevention of fouling of the membrane surface and a reduction in the size of the activated sludge reactor.

29.1.3 Granular Media Filtration

Although granular media filtration is one of the principal unit operations used in potable water treatment, the filtration of effluents from wastewater treatment processes is a relatively recent

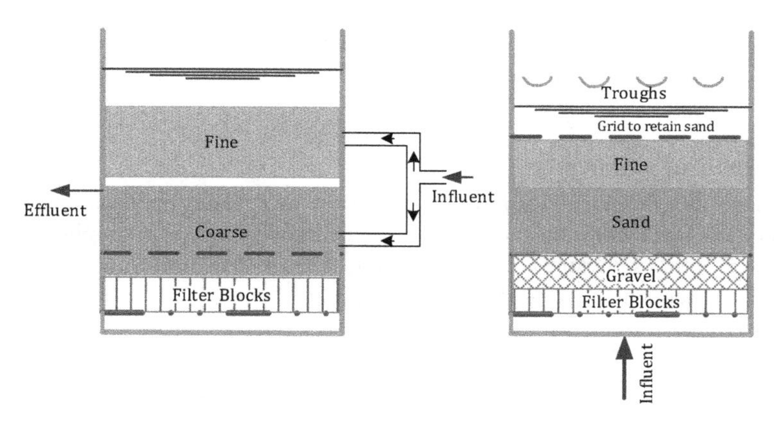

FIGURE 29.2 Gravity Filter for Wastewater Filtration

practice. It is used for the supplemental removal of suspended solids, including particulate BOD from wastewater effluents of biological and chemical treatment processes, and the removal of chemically precipitated phosphorus.

The design of gravity filtration systems must consider the higher suspended solids content and fluctuating rate of wastewater flow not common in water treatment. Multi-media filters are recommended for in-depth filtration and greater solids holding capacity, resulting in higher filter runs. Efficient backwashing requires an auxiliary scour. The filters may be either gravity or pressure units, depending on the size of the treatment plant. Chlorination prior to filtration prevents growth within the filter.

29.2 CONTROL OF NUTRIENTS

Discharges containing nitrogen and phosphorus may accelerate the eutrophication of lakes and reservoirs and stimulate the growth of algae and rooted aquatic plants in shallow streams. Other effects of excessive nitrogen are depletion of dissolved oxygen, toxicity toward aquatic life, etc.

29.2.1 Phosphorus Removal

The principal sources of phosphorus in wastewater are domestic sewage, agricultural return water and land run-off. Phosphorus, like nitrogen, contributes to the eutrophication of surface water. It is often cited as the culprit responsible for the simulation of aquatic plants, since the concentration of phosphorus necessary to support algal blooms is only 0.005 to 0.05 mg/L as P. Phosphorus in wastewater may be present in three forms: orthophosphate, polyphosphate and organic phosphorus. Phosphorus may be removed biologically and chemically. Sometimes, chemicals may be added to biological reactors. In other cases, phosphorus may be chemically precipitated, usually in primary clarification.

Biological Phosphorus Removal

The presence of phosphate in raw waste in a greater proportion than that required for bacterial growth (BOD:N:P = 100:5:1) may result in an effluent phosphate concentration sufficient to cause eutrophication in the receiving stream. Taking phosphates out of solution by photosynthesis has led to the concept of removing nutrients from wastewater by growing algae in stabilization ponds and then separating the cells from suspension by physical or chemical means. The activated-algae and bacterial-assimilation processes have the potential to remove additional phosphorus, but if stoichiometric ratios are maintained, it would be necessary to add both carbon and nitrogen to the latter and the resultant sludge would be perhaps five times the amount produced in ordinary plants.

Chemical Phosphorus Removal

As stated earlier, phosphorus in wastewater may exist as organic phosphate, polyphosphate or orthophosphate, the latter consisting of four different ionic forms. For simplicity, phosphorus is considered to be present as phosphate ion (PO_4^{3+}). Phosphate can be removed by chemical precipitation. The principal chemicals used for this purpose are lime, alum and ferric chloride.

$$5Ca^{2+} + 3HPO_4^{2-} + 4OH^- = Ca_5\left(PO_4\right)_3 \downarrow + 3\ H_2O$$

$$Al_2\left(SO_4\right)_3.14.3H_2O + 2PO_4^{3-} = 2AlPO_4 \downarrow + 3SO_4^{2-} + 14.3H_2O$$

$$FeCl_3 + PO_4^{3-} = FePO_4 \downarrow + 3Cl^-$$

For the efficient removal of phosphorus, an excess of alum is required. Unused alum reacts with alkalinity to increase the solids in the sludge.

$$Al_2\left(SO_4\right)_3.14.3H_2O + 6HCO_3^- = 2Al\left(OH\right)_3 \downarrow + 3SO_4^{2-} + 6CO_2 + 14.3H_2O$$

Since polyphosphates and organic phosphorus are less easily removed than orthophosphorus, adding aluminium or iron salts after secondary treatment (where organic phosphorus and polyphosphorus are transformed into orthophosphorus) gives the best removal.

Example Problem 29.1

Calculate the theoretical sludge production for the hypothetical biological-chemical treatment of phosphorus removal by adding liquid alum (600 g/mole) at a rate of 90 mg/L into the aeration tank. Assume an influent sludge solids (SS) concentration of 240 mg/L and 50% is

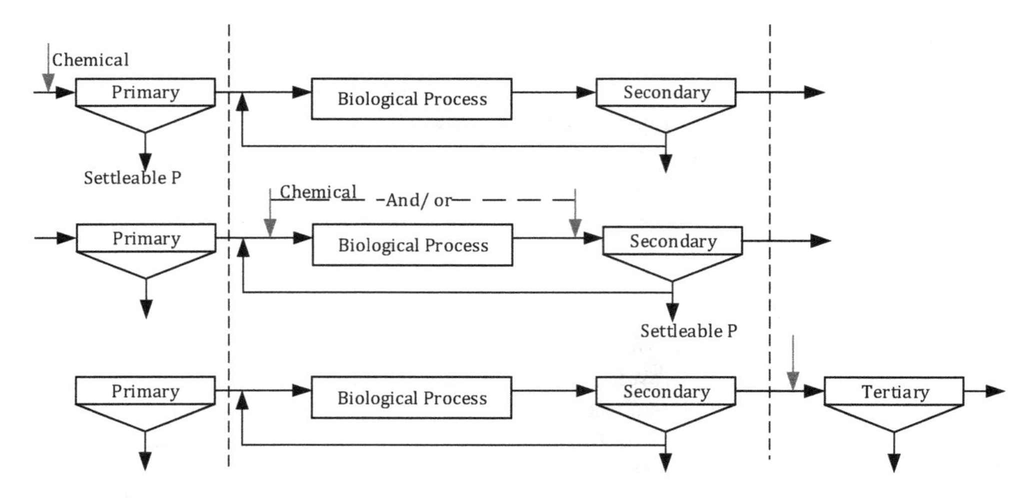

FIGURE 29.3 Alternative Points of Chemical Addition for Phosphorus

removed during the primary settling. Assume for the activated sludge process an SS removal of 60 mg/L and additional removal of SS of 15 mg/L due to the addition of alum. This would produce an effluent with an SS of 15 mg/L or less. Assume an incoming P of 7.5 mg/L, a primary removal of 1.0 mg/L and a secondary removal during biological uptake of 1.6 mg/L to produce secondary effluent with P not exceeding 1.0 mg/L.

Given: $P_i = 7.5$ mg/L
$P_e = 1.0$ mg/L
$SS_i = 240$ g/m^3
$SS_e = 15$ g/m^3
Alum = 90 g/m^3

SOLUTION:

The total SS removed includes the SS removed in primary treatment, in secondary treatment as $Al(PO_4)$, biological removal as $Al(OH)_3$, and additional removal due to precipitation.

$$SS(primary) = \frac{240\ g}{m^3} \times 50\% = 120\ g/m^3$$

$$P(reacting\ with\ alum) = 7.5 - 1.0 - 1.0 - 1.6 = 3.9\ g/m^3$$

$$SS(as\ AlPO_4) = \frac{AlPO_4\ mole}{P\ mole} \times P = \frac{122\,g/mole}{31\,g/mole} \times \frac{3.9\ g}{m^3} = 15.34 = 15\ g/m^3$$

$$Alum(unreacted) = 90 - \frac{Alum\ mole}{2 \times P\ mole} \times P = \left(90 - \frac{600}{2 \times 31} \times 3.9\right) g/m^3 = 52.2 = 52\ g/m^3$$

The alum that reacted with P will react with alkalinity.

$$SS(as\ Al(OH)_3) = \frac{2 \times Al(OH)_3\ mole}{Alum\ mole} \times Alum = \frac{2 \times 78}{600} \times 52\ g/m^3 = 13.5 = 14\ g/m^3$$

The total sludge solids production equals the sum of the primary removal, secondary removal as aluminium phosphate and aluminium hydroxide, and additional removal due to alum.

$$SS = SS_{pri} + SS_{sec} + SS_{pho} + SS_{OH} + SS_{add} = 120 + 60 + 15 + 14 + 15 = 234 = 230\ g/m^3$$

Biological-Chemical Phosphate Removal

This method utilizes an activated sludge system and an anaerobic cell in which phosphorus taken up in the aeration tank is released to the liquid phase producing a concentrated phosphates stream and a phosphate deficient sludge that is returned to the aeration tank where the uptake of phosphorus is repeated. Phosphorus is precipitated by dosing the concentrated phosphate stream with lime.

29.3 NITROGEN REMOVAL

Nitrogen in wastewater is found as organic nitrogen, ammonia nitrogen, nitrate nitrogen and very small minor amounts of nitrite. All these forms of nitrogen in wastewater effluents are potentially harmful. When present as organic nitrogen or ammonia, nitrogen exerts oxygen demand in accordance with the following equation:

$$NH_4 + 2O_2 \rightarrow NO_3^- + 2H^+ + H_2O$$

Nitrogen enters the aquatic environment from both natural and man-made sources. Natural sources include precipitation and biological fixation. Man-made sources, which contribute nitrogen in domestic wastewater, include faeces, urine and food processing discharges. The largest single source of nitrogen is urea, which, together with ammonia, comprises approximately 85% of the nitrogen from human excreta. The following are popular methods for the removal of nitrogen from wastewater.

29.3.1 Biological Nitrification-Denitrification

Nitrogen can be removed from wastewater by the progressive biological oxidation of nitrogen compounds to nitrites and nitrates, followed by conversion into nitrogen gas. Thus, nitrogen removal is nitrification followed by denitrification. In nitrification, ammonia is oxidized to nitrites and then to nitrates by aerobic nitrifying autotrophic bacteria. In denitrification, nitrates are reduced to nitrogen gas by either autotrophic or heterotrophic anaerobic bacteria. The nitrogen of interest in advanced wastewater treatments is organic, inorganic and gaseous nitrogen. As a first step, bacterial decomposition releases ammonia by deamination of nitrogenous organic compounds:

$$\boxed{Organic\ N\ \overrightarrow{Deamination}\ \ NH_4^+ + O_2\ \overrightarrow{Nitrification}\ \ NO_3^-\ \overrightarrow{Denitrification}\ \ N_2 \uparrow}$$

Nitrosomonas in the first stage and Nitrobacter in the second stage use energy derived from reactions for cell growth and maintenance. The biological processes used for nitrification are identified as aerobic suspended growth and aerobic attached growth. Inorganic nitrogen, along with carbon dioxide, is taken up by vegetation during photosynthesis.

Suspended-Growth Process

In the suspended-growth process, nitrification is achieved in an activated sludge process along with the usual carbonaceous oxidation, with some modifications to the design. Nitrifying bacteria are very sensitive to pH (varying from 7.8 to 8.9 with an optimum value of 8.4) and have a very slow rate of growth compared to heterotrophic bacteria. Nitrification rate drops rapidly in temperatures below 10°C. Thus, nitrification is difficult to achieve in colder climates. It has been shown that to complete nitrification in a conventional process, a higher level of mixed liquor volatile suspended solids (MLVSS) and a longer solids retention time (SRT) should be maintained. In addition, the dissolved oxygen (DO) level should be maintained above 2.0 mg/L.

Biological Denitrification

Biological denitrification is achieved under anaerobic conditions by heterotrophic microorganisms that utilize nitrate as a hydrogen acceptor. The nitrate present in the waste process essentially requires a fully nitrified influent, an anaerobic condition and a supply of proper substrate. A balanced amount of substrate is supplied to reduce the nitrate in the process of stabilization of the supplied substrate under anaerobic conditions. In the denitrification process, the heterotrophic bacteria found in activated sludge reduce NO_2^- and NO_3^- to nitrogen gas. Organisms metabolize methanol according to the following reaction:

$$6NO_3^- + 5CH_3OH \rightarrow 3N_2 \uparrow + 5CO_2 + 7H_2O + 6OH^-$$

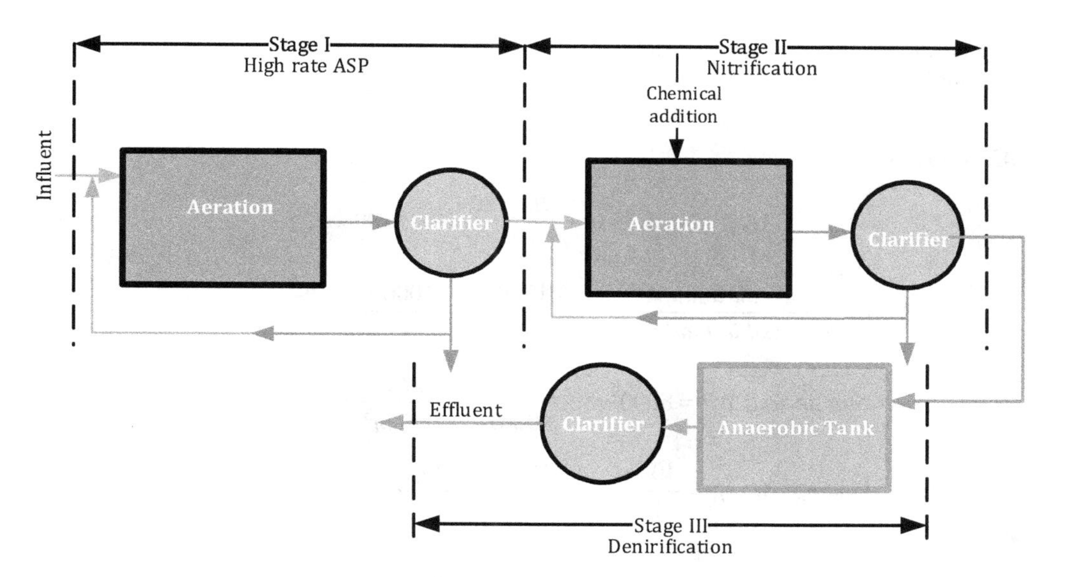

FIGURE 29.4 Three-Stage Nitrification-Denitrification Process

As per this equation, 5 moles of methanol are required to completely reduce 6 moles of nitrate to molecular nitrogen. The overall reaction requires methanol for the reduction of nitrite, nitrate and DO in nitrified wastewater.

29.3.2 Three-Stage Nitrification-Denitrification

In this three-stage process, nitrification and denitrification are carried out separately (**Figure 29.4**).

The recommended MLVSS is 1500–2000 mg/L, and the SRT is longer than seven days. The loading on the nitrification basin is expressed as $g/m^3 \cdot d$. Typical loading is 15–200 $g/m^3 \cdot d$ for operating temperatures greater than 15°C. Correction should be applied for the peaking factor of 1.2–1.6 for the average nitrogen load. A DO of 3.0 mg/L is desirable and must not fall below 1.0 mg/L.

Stage II (Separate-Stage Denitrification)

Carbon oxidation nitrification-denitrification occurs in a separate reactor and sludge is collected in a separate reactor, thus the name separate-stage system.

The design of a suspended-growth denitrification system is similar to that of an activated sludge system. Complete mix and plug flow reactors can be used. The nitrogen released during this process can attach to biological solids, thus a nitrogen release step is required using aeration. Factors affecting the denitrification process include nitrate concentration, carbon, temperature and pH.

Example Problem 29.2

With an allowable loading of ammonia nitrogen of 150 $g/m^3 \cdot d$, calculate the required aeration volume for biological nitrification (stage I) following conventional secondary treatment.

Given: Q = 40 ML/d
NH_3-N = 20 mg/L
BOD = 40 mg/L
MLVSS 1600 mg/L

Permissible loading of NH_3-N = 150 g/m^3·d
Peaking factor = 1.5

SOLUTION:

$$M_{NH_3-N} = 1.5 \times \frac{40\ ML}{d} \times \frac{20\ kg}{ML} = 1200\ kg/d$$

$$V_A = \frac{Peak\ loading}{Permissible\ loading\ rate} = \frac{1200\ kg}{d} \times \frac{1000\ g}{kg} \times \frac{m^3}{150\ g} = 8000\ m^3$$

$$Aeration\ period,\ AP = 8000\ m^3 \times \frac{d}{40\ ML} \times \frac{ML}{1000\ m^3} \times \frac{24\ h}{d} = 4.8 = 5.0\ h$$

$$BOD\ loading,\ M_{BOD} = \frac{40\ kg}{ML} \times \frac{40\ ML}{d} \times \frac{1000\ g}{kg} \times \frac{1}{8000\ m^3} = 200\ g/m^3 \cdot d$$

29.4 TREATMENT METHODS FOR REMOVAL OF TOXINS

Refractory organics are compounds resistant to microbial degradation in a conventional biological treatment process.

29.4.1 Carbon Adsorption

Carbon adsorption is used to remove nitrogen, sulphides and heavy metals. Both granular and powdered carbon is used in this process. After treatment, the effluent BOD ranges from 2 to 7 mg/L and the COD ranges from 10 to 20 mg/L. Types of carbon contactors include upflow column, downflow column, fixed bed and expanded bed.

The economic application of activated carbon depends on an efficient means of regenerating the carbon after its adsorptive capacity has been reached. Granular carbon can be regenerated by heating it in a multiple hearth furnace at 800°C, wherein the adsorbed organics are volatilized and released in gaseous form from the carbon surface. With proper control, granular carbon can be restored to near virgin adsorptive capacity with only 5% to 10% weight loss.

Carbon adsorption combines the use of powdered activated carbon with the activated sludge process. It facilitates system stability during shock loads, the reduction of refractory priority pollutants, colour and ammonia removal, and improved sludge settleability.

29.4.2 Chemical Oxidation

Chemical oxidation is used to remove ammonia, reduce the concentration of residual organics, and reduce the bacterial and viral content of wastewater. Typical chemical dosages for chlorine and ozone for the oxidation of organics in wastewater are used.

29.5 IMPROVED TREATMENT TECHNOLOGIES

To meet more stringent requirements for wastewater effluent quality, engineers have come up with innovations to meet the challenge of upgrading old sewage treatments while also dealing with often limited space for expansion. These technologies include the sequencing batch reactor, membrane bioreactor process, ballasted flow reactor, biological aerated filter, and integrated fixed-film activated sludge process.

29.5.1 Sequencing Batch Reactor (SBR)

As the name implies, an SBR employs a non-steady process in which the reactor is filled with raw wastewater for a discrete time and operated in a batch mode. Two or more SBRs are used so that when one is filling, the other one is operating. There is no primary treatment, and the aeration tank performs bio-oxidation, settling and clarification. After a given period of aeration, aeration is turned off, the solids settle, and the clarified water is decanted off. No sludge recirculation is needed as the sludge remains in the aeration tank. Since it is a batch process, operating parameters can be tightly controlled to meet the desired quality. With the advent of better control systems, SBR systems can remove as much as 95% of BOD and take up less area than conventional activated sludge plants.

29.5.2 Membrane Bioreactor Process (MBP)

In an MBP, aeration, secondary clarification and filtration occur in a single reactor. Filtration is achieved by microfiltration membranes modules submerged in the bioreactor. Vacuum pumps pull the effluent through the membranes and leave solids behind in the tank.

29.5.3 Ballasted Floc Reactor (BFR)

The process in a BFR is similar to ballasted flocculation in potable water treatment. A coagulant is added to the influent to initiate coagulation and flocculation. Sand and polymers are then added to make heavier floc that settles readily. The addition of polymers makes the organics stick on the surface of sand particles that can easily settle out. Settled sludge is pumped to a hydroclone where the sand is separated and recycled back to the BFR.

29.5.4 Biological Aerated Filters (BAF)

In a BAF reactor, the basin media is submerged and serves as a contact surface for biological activity and a filter to separate solids. The primary effluent flows up through the basin, and fine-air bubble diffusion is used for aeration. Routine backwashing is used to remove captured solids from the media. Two reactors can be used if both nitrification and BOD removal are intended. Organic loading in a BAF can be six times that of a conventional activated sludge process, and its footprint is much smaller. BAFs are more common in Europe.

29.5.5 Integrated Fixed-Film Activated Sludge (IFAS)

This technology is particularly used in plants being upgraded to introduce nitrification, especially where there is no space for expansion. In an IFAS process, small plastic sponges or rings, called carriers, are suspended in the aeration tank. The carriers increase the total biomass undergoing nitrification and allow the system to operate at higher BOD loading.

29.6 WATER RECYCLE AND REUSE

As the demand for fresh water increases due to rises in population and changes in lifestyle, the need for water conservation and the reuse of water cannot be overstressed. Global warming is further adding to the challenge of water needs for survival. Terms like reclamation, recycling and reuse are commonly used when referring to the development and expansion of natural water sources. Though these terms are used interchangeably by some engineers, distinction should be made between reclamation and reuse. Water reclamation is defined as the treatment of water to meet predefined water quality criteria; water reuse refers to the use of treated water for beneficial uses, such as urban irrigation needs including in parks and supplementing agricultural irrigation. Environmentally speaking, it makes sense to recycle the water.

29.6.1 Water Conservation

In a municipal water supply, water conservation is one way to increase supply. Water conservation applies to all water users. Broadly, water conservation measures can be categorized as soft measures, such as public education, and hard measures, such as enforcing by-laws and regulations. Public education might include the use of low-flow toilets and appliances, and encouraging native vegetation for landscaping.

When soft measures do not work, restrictions on, e.g.,. car washing and lawn watering, and increasing fees are implemented. Some of these measures do result in increased water conservation. However, water reuse and reclamation are more appropriate for wastewater producers. Municipal wastewater reuse is especially common in arid regions where there is limited rainfall and evaporation loss is high. Environmental factors such as global warming, water scarcity and cost factors are expected to increase the reuse of wastewater.

29.6.2 Reuse of Processed Wastewater

One of the major uses of processed wastewater in the US is in agricultural irrigation and industry as cooling water and process waters. Since the public are not generally exposed to these reuse waters, high-quality biological treatment with or without disinfection is often considered satisfactory.

Other uses of processed wastewater are in urban landscaping, irrigation and the recharging of aquifers. As mentioned earlier, such uses are gaining popularity in developing countries too. For these applications, however, tertiary treatment with disinfection is necessary.

29.7 WATER QUALITY AND REUSE

Protection of public health is a primary concern in establishing water quality standards for water reuse. Some important environmental considerations include the protection of groundwater, soils and crops. In the selection of quality standards for water reuse, cost must be considered. For example, if the reuse is for growing fodder crops, conventional treatment may be considered satisfactory. On the other hand, if the reuse is for recharging aquifers or augmenting groundwater, the treatment must meet much stricter requirements.

The most common applications of wastewater recycling involve non-potable purposes. After some level of treatment, polished wastewater effluents can be used for watering golf courses and recreational fields, lawn and landscape irrigation, crop irrigation and other agricultural purposes, and to augment, enhance and sustain natural water bodies, aquatic environments and ecosystems. It can also be used in industry for cooling and boiler water make-up, and other purposes.

29.7.1 Urban Landscape

For urban irrigation use, treated wastewater must meet some quality standards. For urban reuse, the suggested water quality is achieved by biological treatment, filtration, and disinfection with turbidity levels not exceeding 2 NTU, and no detectable faecal coliforms. These criteria are usually applicable to unrestricted public areas, such as parks and other recreational areas.

29.7.2 Reclaimed Wastewater

Reclaimed wastewater refers to wastewater treated to its original quality. There are two kinds of planned wastewater recycling protocols for drinking water use: indirect potable reuse (IPR) and direct potable reuse (DPR).

IPR is used for intentional and unintentional groundwater recharge by seepage of treated wastewater spread or impounded on the ground surface. The recommended treatment is specific for each site and based on soils, percolation rate, the thickness of the unsaturated soil profile, natural groundwater quality and dilution.

In general, reclaimed water should meet drinking water standards and contain no measurable levels of pathogens after it has percolated through the vadose (unsaturated) zone. Direct groundwater recharge is the injection of reclaimed water into potable aquifers. The recommended processing is secondary treatment, filtration, disinfection and advanced wastewater treatment such as chemical precipitation, carbon adsorption and reverse osmosis. The most inclusive water quality criteria are to meet drinking water standards before injection and retention underground for an extended period of time prior to being pumped from wells. There are many facilities around the world that make indirect use of reclaimed wastewater. For example, in the US, the Fred Harvey Water Reclamation plant in El Paso, Texas, and the Groundwater Replenishment System plant in Orange County, California. Similar facilities exist in Toowoomba, Australia, and Singapore.

DPR can be defined as the intentional introduction of reclaimed wastewater directly into the source water intake of a potable water plant. There are no criteria or standards yet for DPR implementation, although research is geared to take DPR to the next level. Another issue is the public acceptance of DPR. As the need for sustainable water supply is growing, DPR seems to be the most economical and viable solution.

Discussion Questions

1. In the chemical precipitation of phosphorus, coagulant can be added as part of primary, secondary or tertiary treatment. Discuss the advantages and disadvantages of each.
2. List the common characteristics of secondary effluent from a conventional activated sludge plant. What contaminants are most likely to cause pollution if the plant effluent is discharged with limited dilutional flow? Which contaminants adversely affect a lake or reservoir receiving plant effluent?
3. What are the potential benefits of flow equalization?
4. What pollutants are removed by granular carbon columns?
5. Verify by the appropriate calculations that the mass ratio of alum to phosphorus is 9.7 and of alum to alkalinity as $CaCO_3$ is 2.
6. What are the necessary conditions for performing biological nitrification with BOD removal in an aeration tank of an activated sludge process?
7. Why is methanol or another carbon source needed in biological denitrification?
8. What is the difference between water reuse and water reclamation?
9. Why is water conservation gaining popularity in modern times?
10. What are the various beneficial uses of processed wastewater?
11. What are the different types of membrane filtration, and which one is more suitable for processing secondary effluents? What are the advantages and disadvantages of membrane filtration?
12. What are IPR and DPR? Is DPR the way for the future? Discuss.
13. List five alternative technologies for providing tertiary treatment when space is limited and more removals are required.
14. Search the internet and find the scheme of processes used to reclaim wastewater at the Groundwater Replenishment System plant in Orange County, California, USA.

Practice Problems

1. Calculate the theoretical sludge production for the hypothetical biological-chemical treatment of phosphorus removal by dosing alum at a rate of 80 mg/L into the aeration tank. Assume an influent SS concentration of 200 mg/L and 45% are removed during the primary settling. Assume for the activated sludge process an SS removal of 50 mg/L and the additional removal of SS of 15 mg/L by alum addition. This would produce an effluent with SS of 15 mg/L or less. The raw wastewater phosphorus concentration is 7.0 mg/L. Assume that the primary removal is 1.0 mg/L and the secondary biological removal is 1.5 mg/L to produce a secondary effluent with P not exceeding 1.0 mg/L. (180 mg/L)
2. Based on an allowable loading of ammonia nitrogen of 150 $g/m^3 \cdot d$, calculate the required aeration volume for biological nitrification (stage I) following conventional secondary treatment. The wastewater characteristics are as follows: Q = 40 ML/d, NH_3-N = 20 mg/L, MLVSS 1600 mg/L, peak load = 1.5 × (8000 m^3)
3. In the chemical precipitation of phosphorus, an excess dosage of liquid alum at 40 mg/L is applied. This excess alum reacts with alkalinity. Determine how much alkalinity is consumed and inorganic solids are produced as aluminium hydroxide? (20 mg/L, 10 mg/L)
4. Calculate the area of gravity filters to remove SS from a wastewater effluent with an average daily flow of 1.5 ML/d and a peak 4-h discharge of 3.3 ML/d. The nominal filtration rate should exceed neither 180 $m^3/m^2 \cdot d$ with all filters operating nor 350 $m^3/m^2 \cdot d$ during peak flow hours with one out of four filters out of service (13 m^2)
5. A secondary wastewater treatment plant is being upgraded to a tertiary plant to meet a more stringent requirement of < 10 mg/L of BOD in the plant effluent. Average BOD removals at this plant are 35% in primary and 85% in secondary units. What should be the minimum BOD removal of the tertiary unit knowing the plant influent contains 240 g/L of BOD? Determine the BOD content of the primary effluent and secondary effluent. (57%, 160 mg/L, 23 mg/L)

30 Industrial Wastewater Treatment

Water is generally used by all types of industry and discharged in various water resources. Untreated or partially treated industrial wastewater can pollute the receiving waters to the extent of making them unfit for domestic, recreational and commercial purposes. Rapid growth in developing countries like India has created multifarious methods for the disposal of industrial wastes. Sites of untreated wastewater discharges on land that produce strong odours and contaminate groundwater are common. In some cases, damage to water supply sources can result in chronic diseases and make them unfit for drinking purposes. In **Table 30.1**, the most common effects of industrial pollutants are shown.

30.1 INDUSTRIAL WASTEWATER DISCHARGES

From the view of treatment, industrial dischargers can be broadly categorized as direct or indirect. Direct dischargers include industrial effluents that treat their wastewater and directly discharge it into the receiving waters. Industries such as steelmaking, electroplating, pulp and paper and tanneries fall into this category. The onus is on the industry to meet the standards stipulated by the regulating body. Indirect dischargers include those industries such that their wastewaters contain conventional pollutants and are directly discharged into municipal sewers to be treated with domestic sewage, for example, dairy, brewery and food processing. The typical characteristics of some industrial wastes are shown in **Table 30.2**.

Here are some key points related to industrial discharges:

- Industrial discharges are usually stronger in terms of BOD than domestic wastes.
- The total solids are usually greater but vary in character from colloidal to dissolved organics.
- Discharges from chemical and material industries are usually deficient in nutrients.
- Industries such as metal plating require their wastewater to be pre-treated.
- Industries must consider segregation of wastes, flow equalization and reduction of waste strength.
- Process changes, equipment modifications, by-product recovery and plant reuse of wastewater can result in cost savings for the industry.

30.1.1 Pre-Treatment of Industrial Waste

As mentioned aboveearlier, the pre-treatment of industrial waste can result in savings and make the waste more suitable for further treatment. The main purpose of pre-treatment is:

1. The reduction of waste strength and volume.
2. Equalization and neutralization.

Strong and weak waste may be separately collected, treated and disposed of, as mixing them creates large volume and difficulty. Thus, reduction of waste, strength and volume is necessary. Wastewater may be conserved by reusing or recycling. Many substances such as chromium, potash and silver

DOI: 10.1201/9781003231264-30

TABLE 30.1
Industrial Waste Characteristics

Undesirable Characteristics	Effect
Soluble organics	Depletion of oxygen in river water
Suspended solids	Disturbs aquatic life, forms benthic deposits
Phenol and trace organics	Bad taste and odour
Colour and turbidity	Unaesthetic, affects photosynthesis
Nitrogen and phosphorus	Undesirable algal growth, increases eutrophication
Oil and floating matter	Slows re-aeration

TABLE 30.2
Characteristics of Selected Industrial Wastewaters

Industry	BOD (mg/L)	COD (mg/L)	pH	TSS (mg/L)	N (mg/L)	Cl (mg/L)	Grease (mg/L)/
Milk processing	1000	1900	8	1000	50	-	-
Meat packing	1400	2100	7	3300	150	-	500
Synthetic textiles	1500	3300	5	8000	30	-	-
Chlorophenolic manufacture	4300	5400	7	53 000	-	27 000	-

can be recovered easily for economic gains. Fluctuations or variations in input quality can also be controlled.

30.2 INDUSTRIAL WASTEWATER TREATMENT

The characteristics of wastewater vary from industry to industry; therefore, each industry or group of industries should conduct a thorough investigation to choose the best treatment method for its effluent. A brief description of the common industries is presented in the following paragraphs.

30.2.1 Dairy Plant

Plants producing liquid milk and products with a short shelf life, such as yoghurts, creams and soft cheeses, tend to be located on the outskirts of urban centres close to consumer markets. Plants manufacturing items with a longer shelf life, such as butter, milk powders, cheese and whey powders, tend to be situated in rural areas closer to the milk supply. Most large processing plants tend to specialize in a limited range of products. High-rate trickling filters and activated sludge plants efficiently used as aeration can reduce BOD. Oxidation plants are also designed.

30.2.2 Cane Sugar Industries and Distilleries

Lime is added to crushed sugarcane, and the juice of the sugarcane is filtered and crystallized to obtain sugar. In distilleries, mother liquor (molasses) is acidified with sulphuric acid, supplemented with nitrogen and phosphorus, fermented with yeast and finally distilled for obtaining ethanol. About 3000L/day of wastewater is generated during the production of one ton of sugar. Bagasse from the sugar plant is used as fuel and for the preparation of fibre balls.

30.2.3 Fertilizer Industry

Fertilizer units mainly produce nitrogenous or phosphate compounds. Fluoride, oil, grease and arsenic can also be produced. Approximately 500 L of liquid effluent is generated in the production of one ton of urea. Carbon is removed and used again. Oil is removed by oil separators. Air stripping helps recover the acid, and sludge is removed in lagoons or settling tanks.

30.2.4 Paper Industry

Cellulose containing wood material is cooked in caustic soda and sodium sulphide mixture under controlled temperature and pressure. Spent liquor (black liquor) is used to recover the caustic soda, and the pulp is washed and bleached with chlorine. The pulp is used in paper manufacturing by adding dies, alum and talc to fibres. Sedimentation and floatation are reliable treatments. Lagooning is an economical treatment for the removal of 50% BOD. Activated sludge treatment is also a satisfactory treatment.

Biological treatment (aerobic) is effective. Neutralization activated sludge units and oxidation ponds are also suggested units of treatment.

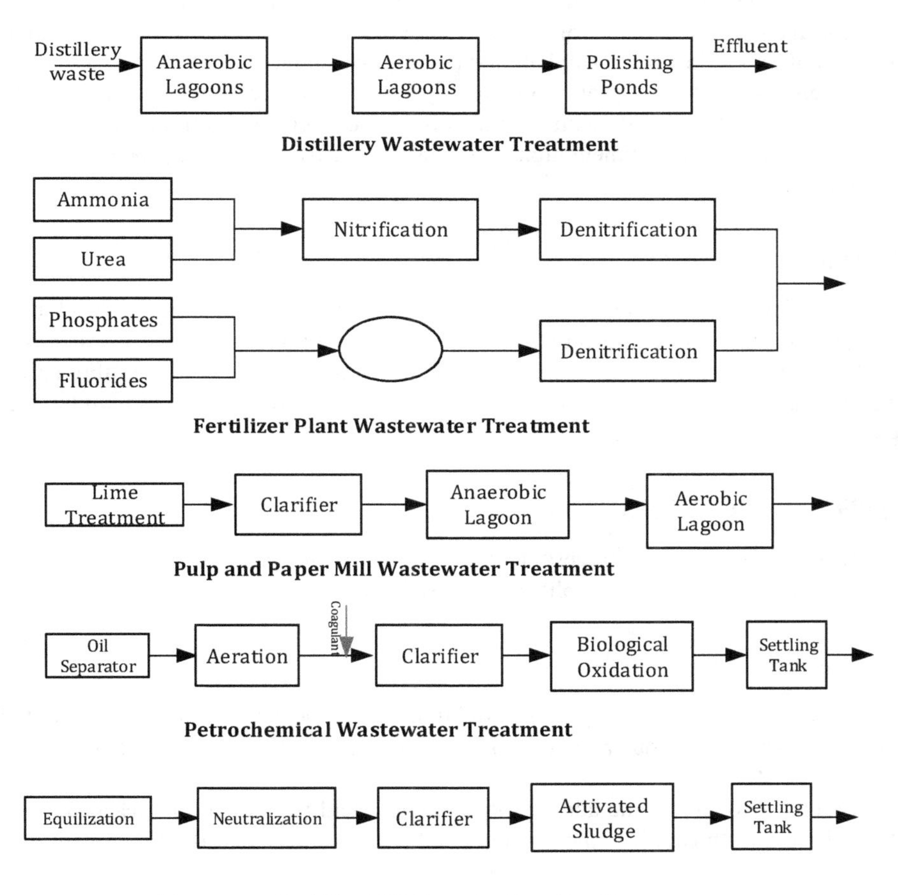

FIGURE 30.1 Flow Schemes for Industrial Wastewaters

30.2.5 Pharmaceutical

Basic drugs and various formulations are manufactured in pharmaceutical units. Antibiotic plants yield wastewater with a pH value of 4.5, whereas sulfa drug units create wastewater with a pH of 9. The BOD varies from 1000 to 10000 mg/l and the total solids exceed between 10000 and 50000 mg/L. Neutralization, extended aeration, chemical oxidation, filtration and oxidation ponds are employed for treating pharmaceutical waste. Hazardous waste is generally incinerated. When the total dissolved solids level is high, multiple effect evaporators are used.

30.2.6 Steel Plants

Steel plants use coke ovens, blast furnaces and other furnaces, and perform the operations of casting, rolling, pickling, heat treatment and finishing. Water is used for a different purpose here and the and waste generated can contain a number of chemicals, such as lead, zinc, chromate, phosphate, chloride, sulphide, iron andsilica. The froth floatation process removes fine coal particles. Chemical oxidation and biological treatments are used.

30.2.7 Tanneries

The processes involved in the operation of tanneries includes soaking raw hides in water, treating them with lime and sodium sulphide for the removal of hair and flesh, further treating them in a vegetable tannin solution of chromium sulphate for tanning, washing, neutralizing, and finally treating them with an emulsion of sulphonated oils and dyeing. The wastewater can contain a high level of suspended and dissolved solids with high alkalinity. Recovery of chromium is done. Treatment processes include screening, sedimentation tanks, lagoons, activated sludge process, trickling filters and oxidation ditches.

30.2.8 Textile

Textile mills produce cotton, wool and synthetic fibres. The desizing, scouring, bleaching, mercerizing, dyeing and printing operations of cotton mills generate liquid wastes to the tune of 3000m^3/day. The pH value of cotton textile waste is 8 to 10, and 3 to 5 from a synthetic fibre unit. Primary treatment consists of segregation, recovery, recycling, screening, neutralization with lime, equalization, coagulation and clarification.

30.3 SPECIAL PROCESSES

Industrial wastewaters need a process to remove chromium, phenol, mercury and nitrogen. Adsorption is used to remove colour, phenol, etc. Activated carbon is used the most.

30.3.1 Removal of Chromium

Hexavalent chromium is toxic (0.1mg/l threshold limit) and is present in tanning, electroplating, fertilizer production and other industries. Methods for the removal of chromium include:

1. Cr_6 is reduced to Cr_3 by the addition of sulphuric acid (to reduce the pH to between 2 and 3) and ferrous sulphate. It is then precipitated in a settling tank after being neutralized with sodium hydroxide or calcium hydroxide.
2. Cationic resins can be used to recover chromium in the form of sodium chromate or chromic acid. The regeneration of resin is done with sodium chloride and sulphuric acid.
3. Lime coagulation and adsorption with activated carbon.
4. Reverse osmosis.

30.3.2 Removal of Phenol

Phenol is released by chemical, petroleum, pharmaceutical, plastic, metallurgical, printing and textile industries. Methods for the removal of phenol include:

1. Oxidizing agents such as hydrogen peroxide, potassium permanganate and sulphur dioxide are used. Aeration and biological action are employed.
2. Steam stripping, adsorption, ion exchange and solvent extraction.

30.3.3 Removal of Mercury

Chloralkali plants, the mining industry, paper and pulp mills and pesticide units are the main sources of mercury. Methods for the removal of mercury include:

1. Precipitation by adding H_2S or Na_2S.
2. Adding ferric chloride.
3. Adsorption with the use of activated carbon, clays and silica gel.
4. Mercury salt in solution may be reduced by the addition of active metals such as iron.
5. Ion-exchange using natural or synthetic resins.

30.3.4 Removal of Nitrogen

Nitrogenous compounds cause eutrophication. Methods for the removal of nitrogen include:

1. Reducing the amount of ammonia by adjusting the pH and utilizing bacteriological and algal symbiosis. This is a low-cost treatment.
2. A simple and economical treatment is to adjust the pH to between 10 and 11 by adding lime.
3. Air stripping.
4. Nitrification under aerobic conditions at pH 8 to 9 and denitrification under anaerobic conditions.
5. Ion exchange or reverse osmosis.

30.3.5 Removal of Dissolved Solids

Methods for the removal of dissolved solids include:

1. Ion exchange. Certain compounds such as resin can exchange one of their ions from the surrounding medium due to ionization potential and electrolytic action.
2. Dialysis. The separation of solutes by means of their unequal diffusion through a semi-permeable membrane.
3. Reverse osmosis. The flow of solvent from a strong concentrated solution to a dilute salt concentration.
4. Algae use. In oxidation ponds, algal growth utilizes minerals such as Na, K, P, Ca and Mg for their food and activity.
5. Multiple-effect evaporation.

30.4 COMMON EFFLUENT TREATMENT PLANTS

As mentioned above, some industrial wastewaters can be treated by conventional methods, hence those industries can be allowed to discharge into municipal sewers systems. For example, wastewaters from dairy, brewery, milk processing and food processing plants are amenable to biological treatment. The effluent includes industrial wastewaters and domestic sewage generated from

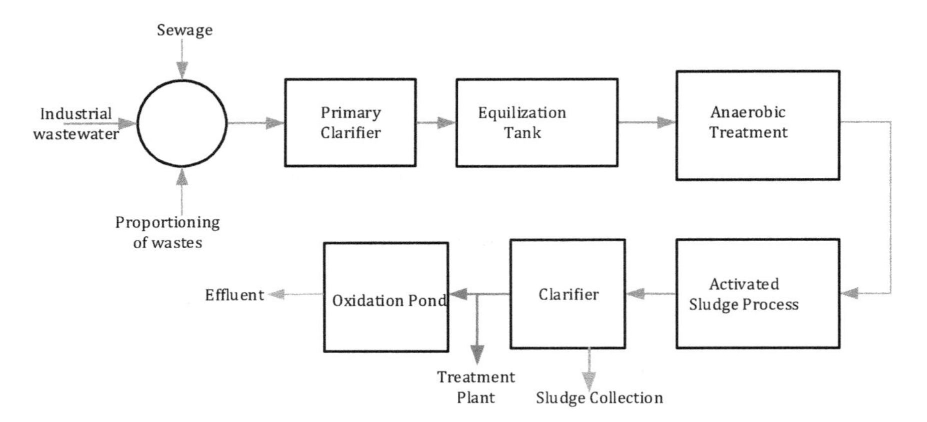

FIGURE 30.2 CETP Flow Diagram

the estate. This arrangement allows small and medium-scale industries to dispose of their effluents. Otherwise, it may not be economical for these industries to treat their wastewaters, or there may be space constraints. Such an arrangement is a win-win both for the community and industry. In addition, it helps to prevent pollution by the discharge of untreated or partially treated wastewaters into the receiving waters.

Some of these industries may require pre-treatment at the facility to reduce the strength of the wastewater, correct the pH or remove a specific pollutant before it is discharged into sewers. Combined wastewaters from domestic and industrial sources are usually expected to have a high strength, and the flow hydrograph may show a different pattern compared to a typical municipal plant. A typical process scheme for a combined effluent treatment system is shown in **Figure 30.2**.

As seen in the flow diagram, the equalization of wastewater is an integral part of a treatment scheme. If this is not done, the wastewater will show tremendous variation in strength and flows.

Equalization may help to freshen wastewater and afford uniform and balanced loading. This would allow the designer to size the tanks and devices more economically.

A common effluent treatment plant (CETP) is designed based on:

- The quality and flow rate of the wastewater.
- The effluent standard required by the CETP.
- The possibility of recycling and reusing the treated wastewater.
- The availability of land, manpower, energy and expertise in specific treatment methods.
- The willingness of the industries located in the industrial estate to contribute towards the capital and operating expenses of the CETP.

Discussion Questions

1. List the harmful effects of the disposal of industrial wastes without adequate treatment.
2. Discuss the merits and demerits of combining industrial and domestic wastes before treatment and disposal.
3. Name the pollutants from the paper and textile industries.
4. Describe the process available for treating liquid waste from a steel plant.
5. Describe the use of aerobic and anaerobic lagoons in the treatment of industrial waste.
6. Write a short note on the characteristics of tanneries waste.
7. Describe the process of the removal of chromium and mercury from industrial waste.

8. Why is there a need for special processes to treat industrial effluent, and what are these processes?
9. Write a short note on the pre-treatment of industrial waste.
10. What is the concept of a common effluent treatment plant (CETP)? On what basis it is designed? Draw the flow diagram of a CETP.
11. Describe the process of the removal of phenol and nitrogen from industrial waste.
12. Classify the pollutants generated from the various types of industries.
13. What are the methods of treatment used to remove solids from industrial waste?
14. Discuss the positive and negative effects of an industry of your choice on society and the environment. How should its effluent be treated before disposal? Give a suggestion for reducing the number of effluent treatment steps and its economic cost.

31 Sources of Air Pollution

Human beings need oxygen, water and food for survival. On average, a person needs at least 12 kg of air every day to live but only 1.5 kg of water and 0.70 kg of food. A person can live without food for weeks and without water for days, but only 5 minutes without air. That shows the importance of cleaner air. However, though the atmosphere is invisible and contains the air, it is as susceptible to pollution as water and land environments.

The gaseous envelope that surrounds the earth is called the atmosphere. Under ideal conditions, the atmosphere has a qualitative and quantitative balance that maintains the well-being of the environment. When the balance of the components of atmospheric air is disturbed, the air is said to be polluted. Some specific definitions of air pollution are given below:

- Air pollution is the presence of chemicals in the atmosphere in concentrations high enough to harm organisms, ecosystems and materials, or alter the climate.
- The American Medical Association define air pollution as the excessive concentration of foreign matter in the air that adversely affects the well-being of individuals or causes damage to the property. According to the Engineers Joint Council in the US, "Air pollution means the presence in the outdoor atmosphere of one or more contaminants, such as dust, fumes, gas, mist, odor, smoke, or vapor, in quantities, of characteristics, and of duration such as to be injurious to human, plant, or animal life or to property, or which unreasonably interfere with the comfortable enjoyment of life and property."

31.1 COMPOSITION AND STRUCTURE OF THE ATMOSPHERE

The atmosphere is a life-supporting gas mantle comprising several gases in varying quantities. The background concentrations of the various components of dry atmospheric air are given in **Table 31.1**. Clean, dry air contains 78.09% nitrogen by volume and 20.94% oxygen. The remaining 0.97% consists of a gaseous mixture of carbon dioxide, helium, argon, neon, krypton and nitrous oxide, as well as very small amounts of some other organic and inorganic gases whose content varies with time and place. In addition to these gases, air may contain water vapour and varying amounts of dust, pollen and other very small particulates from natural sources.

The average temperature and pressure of atmospheric air vary with altitude (**Figure 31.1**). Based on this, the atmosphere is further subdivided into five different zones called troposphere, stratosphere, mesosphere, ionosphere and thermosphere. The atmospheric layer below the tropopause is known as the troposphere. It is the region of turbulent and large-scale vertical motions of air masses where the prevailing westerlies (in the middle latitudes) blow in both the Northern and Southern hemispheres.

North–South flow also occurs by a number of processes in both the troposphere and the stratosphere, and thus leads to air exchange between hemispheres. Most of the heat in the troposphere is received from the ground rather than directly from the sun. The troposphere, which is roughly 12 km in depth, contains about 85% of the mass of the atmosphere. As most living things exist in this sphere, it is of greatest importance as regards air pollution and control. The density of air, which is about 1.23 kg/m^3 at sea level, decreases significantly with an increase in altitude. In fact, above the troposphere, there is not enough oxygen to support life.

DOI: 10.1201/9781003231264-31

TABLE 31.1
Chemical Composition of Dry Atmospheric Air

Constituent	Percent by Volume	Concentration in Parts Per Million
Nitrogen (N_2)	78.084	780,840
Oxygen (O_2)	20.946	209,460
Argon (Ar)	0.934	9,340
Carbon dioxide (CO_2)	0.036	360
Neon (Ne)	0.00182	18
Helium (He)	0.000524	5
Methane (CH_4)	0.00015	1
Krypton (Kr)	0.000114	1
Hydrogen (H_2)	0.00005	0.5

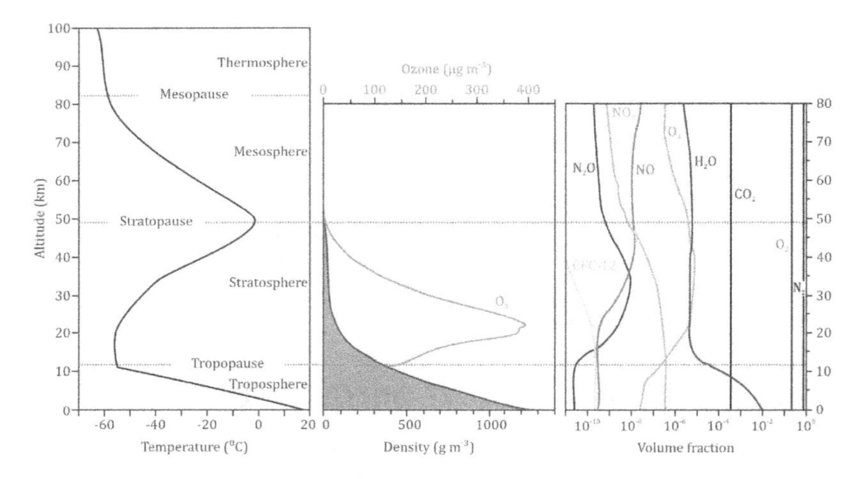

FIGURE 31.1 Variations in Atmospheric Parameters

The layer of air above the troposphere is called the stratosphere. It extends upwards from the troposphere to an altitude of about 50 km. In the stratosphere, little vertical mixing takes place. Although the stratosphere contains less matter than the troposphere, its composition is similar. Much of the atmosphere's ozone is concentrated in a portion of the stratosphere called the ozone layer. This layer keeps about 95% of the sun's harmful UV radiation from reaching the earth's surface. The layers of the atmosphere that exist above the stratosphere are essentially unaffected by air pollution.

31.2 UNITS OF MEASUREMENT

When expressing the relative quantity of gases in the air mixture, units such as part per million (ppm) or parts per hundred (pph) on a volume basis can be used. Parts per hundred is more commonly called percent (%). For example, oxygen in the atmosphere is 21% or 210000 ppm, but it is more convenient to express it as a percentage. However, the concentration of carbon dioxide, which is close to 0.04%, may be more conveniently expressed as 400 ppm on a volume basis. Gaseous pollutants are also expressed in a similar fashion.

$$1\,ppm = \frac{1\ volume\ of\ gaseous\ pollutant}{10^6\ volumes\ of\ air\left(pollutant + air\right)}$$

The concentration of other pollutants is usually expressed in terms of mass per unit volume of air. On a mass basis, the concentration of a pollutant is expressed as micrograms of pollutant per cubic metre of air, $\mu g/m^3$. For a gaseous pollutant, concentrations expressed in ppm on a volume basis can be converted to a mass basis by knowing the volume of gas per unit mole. As evident from gas laws, volume per unit mole is influenced by the temperature and pressure of the gas. According to Avogadro's law, one mole of any gas occupies the same volume as one mole of any other gas at the same temperature and pressure. At standard conditions of temperature (273 K or 0°C) and pressure (1 atmosphere or 760 mm Hg), this volume is 22.4 L/mol. However, most regulations for ambient air quality are referenced at 25°C and 1 atm, with a corresponding volume of 24.5 L/mol. The L/mol at various other ambient air conditions can be converted using the following formula.

Volume of I mole of gas at 25°C (298 K)

$$V_2 = V_1 \times \frac{T_2}{T_1} \times \frac{P_1}{P_2} = \frac{22.4\,L}{mol} \times \frac{298\,K}{273\,K} \times \frac{1\,atm}{1\,atm} = 24.451 = 24.5\,L/mol$$

Example Problem 31.1

The exhaust gas from an automobile contains 1.5% by volume of CO. What is the concentration in g/m^3 at 25°C and a pressure of one atmosphere?

SOLUTION:

$$CO = \frac{12\,g}{mol} + \frac{16\,g}{mol} = 28\,g/mol$$

$$C_{m/v} = \frac{1.5\%}{100\%} \times \frac{mol}{24.5L} \times \frac{28g}{mol} \times \frac{1000L}{m^3} = 17.1 = 17g/m^3$$

Example Problem 31.2

The primary air quality standard for NO_x expressed as NO_2 as an annual average is 100 $\mu g/m^3$. What is the equivalent concentration in parts per million at 25°C and 1 atm?

SOLUTION:

$$NO_2 = \frac{14\,g}{mol} + 2 \times \frac{16\,g}{mol} = 46\,g/mol$$

$$C_{V/V} = \frac{100\,\mu g}{m^3} \times \frac{24.5\,L}{mol} \times \frac{mol}{46\,g} \times \frac{1000\,m^3}{1ML} \times \frac{g}{10^6\,\mu g}$$

$$= 0.0532 = 0.053\,L/ML = 0.053\,ppm$$

31.3 CAUSES OF AIR POLLUTION

As said earlier, air pollution is any atmospheric condition in which certain substances are present in such concentrations that they produce undesirable effects on man and his environment. The main causes of air pollution are briefly described in the following sections.

31.3.1 Population Growth

As a result of high population density, there is a higher rate of fuel consumption.

31.3.2 Rapid Industrialisation

This is one of the major causes. Common sources of atmospheric pollution include smoke from factories, coke ovens, furnaces and steam engines; exhaust fumes from power plants; chemical fumes from oil refineries, zinc refineries, chemical industries, metallurgical plants, iron and steel plants, and incineration plants. Industrial pollution is the most complex due to the large number of chemicals emitted into the atmosphere and the modifications these emitted pollutants can undergo by reaction with one another in the presence or absence of sunlight. Some critics call air pollution "the price of industrialization".

31.3.3 Transportation Facilities

The intensive increase in transportation services such as motor vehicles, rail trains, aeroplanes, etc., throughout the world is another major contributing factor to air pollution. The exhaust fumes from automobiles pollute the atmosphere considerably in urban areas. Air pollution caused by automobiles is sometimes described as "a disease of wealth".

31.3.4 Radioactive Substances

The evolution of radioactive gases and suspended radioactive dust from atomic explosions and accidental discharges from nuclear reactors are very dangerous sources of air pollution.

31.3.5 Natural Causes

The natural sources that cause air pollution include organic compounds from vegetation, ground dust, salt spray from oceans, cosmic dust, and the evolution of hydrogen sulphide from natural sources. These natural sources are beyond the control of man.

31.4 SOURCES OF AIR POLLUTION

There are two main sources of air pollution: natural and man-made. Natural air pollution sources consist of windblown dust, salt particles from sea water and dust of meteoric origin. Man-made sources include combustion processes, petroleum operations and mineral processing.

31.4.1 Primary and Secondary Air Pollutants

Primary air pollutants are those that are emitted directly from identifiable sources. Important primary pollutants include particulate matter, sulphur oxides and sulphur compounds, nitrogen oxides, carbon monoxide, etc.

Secondary air pollutants are those that are formed in the atmosphere because of interaction between two or more primary pollutants or by reactions with normal atmospheric constituents, with or without photoactivation. These are chemical substances that are often more harmful than the original basic chemicals that produce them. Examples of secondary pollutants are sulphuric acid (H_2SO_4) or acid mist, ozone (O_3), formaldehyde, peroxyacetyl nitrate (PAN) and photochemical smog.

31.4.2 Stationary and Mobile Sources

An air pollutant may also be classified based on its source as stationary or mobile. Alternatively, emitting sources can be classified as point, area or line sources.

Automobiles moving on roads create mobile sources of emissions of undesirable gases and particulates into the atmosphere. Two- and three-wheelers emit highly toxic gases such as carbon monoxide in quantities twice that emitted by other vehicles. Industrial installations form stationary sources of air pollution. Key examples include asphalt plants, heating installations, cement manufacturing units, steel plants, fertilizer manufacturing plants, mineral acid manufacturing units, paper and pulp plants, power plants, municipal incinerators and sewage treatment plants.

Point sources are the large stationary sources mentioned above. Area sources are small stationary sources and mobile sources with indefinite routes. Line sources are mobile sources with definite routes and include highway vehicles, railroad locomotives and channel vessels.

The distinction between stationary and mobile sources is important because of the different pollution control technology applied to each type, as well as the different kinds of contaminants they emit.

31.5 AIR POLLUTANTS

Air pollutants can be natural or man-made. Pollutants produced as a result of human activities are called anthropogenic. However, air pollution may also result from natural sources such as volcanoes and forest fires and can be far more severe and longer-lasting than air pollution from human activities.

31.5.1 Natural Pollutants

Natural contaminants consist of natural fog, pollen grain, bacteria, and products of volcanic eruptions. Out of these, pollen grains are the most significant since they can be carried away by the wind to a great distance. They are discharged into the atmosphere from natural vegetation such as trees, plants, grasses and weeds. These airborne pollutants produce allergic responses in sensitive individuals, causing asthma, hay fever, bronchitis and dermatitis.

31.5.2 Aerosols

Aerosols refer to the dispersion of solids or liquid particles of microscopic size in gaseous media, such as smoke, fog or mist. An aerosol can also be defined as a colloidal system where the dispersion media is a gas and the dispersed phase is a solid or liquid. In general, the term particulate represents all atmospheric substances that are not gases. They can be suspended droplets or solid particles, or a mixture of the two. The term aerosol is used during the time it is suspended in the air. Thus, particulate matter is an air pollutant only when it is an aerosol. A particulate matter is a nuisance both as an aerosol causing reduced visibility and as settled or deposited matter resulting in the soiling of surfaces and corrosion. Particulates can be composed of inert or extremely reactive materials ranging in size from 0.01 μm to 100 μm. The various types of aerosols are dust, smoke, mists, fog and fumes.

Dust is a commonly used term applied to solid particles larger than colloidal particles that are capable of temporary suspension in air or other gases. These are formed by the natural disintegration of rock and soil or by mechanical processes such as grinding and spraying, with their size varying from 1 μm to 200 μm.

Smoke consists of finely divided particles resulting from incomplete combustion or other chemical processes. They consist predominantly of carbon particles with a size of less than 1 μm.

Mist is made up of liquid droplets that are formed by condensation in the atmosphere. Mist particles are 40–500 μm in size.

Fog is a loose term applied to liquid-dispersed aerosols that exist in the air as condensation. Fog refers to visible aerosols in which the dispersed phase size of particles ranges from 1 μm to 40 μm.

Fumes are solid particles generated from a gaseous state. Their particle size varies from 0.1 μm to 1 μm.

31.5.3 Criteria Pollutants

Air quality standards in the US characterize five primary pollutants and one secondary pollutant as criteria pollutants. The five primary criteria pollutants include: three gases, sulphur dioxide, nitrogen oxide and carbon monoxide; very small amounts of solid or liquid particulate matter; and lead particulates. Except for perhaps lead, the primary pollutants are emitted in industrialized nations in tons and tons per year. Ozone gas is the secondary criteria pollutant.

31.5.4 Sulphur Dioxide

One of the most damaging oxides emitted by pollution sources is sulphur dioxide (SO_2). SO_2 is a colourless gas and is moderately soluble in water, where it forms sulphuric acid. It is easily oxidized in clean air to make sulphur trioxide.

31.5.5 Oxides of Nitrogen

Oxides of nitrogen are the second-most abundant atmospheric contaminants in many cities next to sulphur dioxide. From an air-pollution standpoint, the most important nitrogen oxides in the atmosphere are nitric oxide (NO) and nitrogen dioxide (NO_2), and these are often referred to jointly in air pollution literature as NO_x. At high concentrations, these oxides are very toxic, and many people have been killed by accidental emissions. They are also important because of the role they play in the chemistry of photochemical smog. Other oxides of nitrogen are N_2O, N_2O_3, N_2O_4 and N_2O_5. The main sources of man-made NO_x are coal, oil, natural gas and motor-vehicle fuel consumption.

31.5.6 Carbon Monoxide

Carbon monoxide (CO) is an odourless and colourless gas that is highly poisonous and is classified as an asphyxiant. The major sources of carbon monoxide in the environment are man-made sources arising from the incomplete combustion of carbonaceous fuels and natural sources such as the atmospheric oxidation of methane and formaldehyde derived from biological activity and the decay of chlorophyll in plants and algae and other biological sources in oceans, and the photochemical oxidation of terpenes. It is estimated that about 420 million tons of carbon monoxide are discharged into the atmosphere each year through human activities, of which 99.5% is contributed by transportation alone. Global emissions from natural sources amount to an estimated 3400 million tons per year, which is approximately eight times more than the emissions from all human activities. About 88% of this total of 3400 million tons comes from the atmospheric oxidation of CH_4 and formaldehyde, and the remaining comes from other sources. Carbon monoxide, when inhaled, passes through the lungs and diffuses directly into the bloodstream where it combines with haemoglobin forming carboxyhaemoglobin, COHb. The affinity of CO for haemoglobin is 210 times greater than that of oxygen, and as a result, the amount of haemoglobin available for carrying oxygen for body tissue is considerably reduced. This may result in death by asphyxiation (lack of oxygen).

31.5.7 Hydrogen Sulphide

Hydrogen sulphide is a foul-smelling gas principally produced from petroleum refinement, coal coking, natural gas refinement, viscose rayon manufacture, and black liquor evaporation in the kraft pulping process. The sources of its natural emission are anaerobic biological decay processes on land, in marshes and the ocean. Volcanoes and natural water springs emit hydrogen sulphide to some extent. Other sulphur compounds of interest in air pollution are methyl mercaptan (CH_3SH), dimethyl sulphide (CH_3SCH_3), dimethyl disulphide (CH_3SSCH_3) and their higher molecular homologues. The atmospheric concentration of hydrogen sulphide and other reduced forms of sulphur can create objectionable odours and cause direct effects on materials, plants, animals and humans. Repeated exposure to even low concentrations can irritate the mucous membrane, eyes and respiratory tract.

31.5.8 Hydrogen Fluoride and Other Fluorides

Fluorides may be found in the air as gases or particulate matter, and industrial emissions will often contain fluoride compounds in both states. Industries that emit fluoride compounds include the mining and processing of phosphate rock and other fluoride-containing minerals, the fertilizer-manufacturing industry, the aluminium and steel industries, the glass-making industry and the chemical industry. Types of fluoride compounds include hydrogen fluoride (HF), silicon tetrafluoride, hydrofluosilicic acid, sodium fluoride, aluminium fluorite, etc. Out of these, hydrogen fluoride is an important air contaminant even in extremely low concentrations of 0.001–0.10 ppm by volume. It is extremely irritant and corrosive. It is more significant in terms of injury to vegetation and animals than injury to humans.

31.5.9 Chlorine and Hydrogen Chloride

Chlorine is found in polluted atmospheres in various forms such as elemental chlorine, hydrogen chloride and other organic and inorganic chlorides. Various sources of chlorine and its compounds are the manufacturing processes in which chlorine is used to produce other chemicals, water purification plants and wastewater treatment plants. Elemental chlorine produces respiratory irritation and damage to vegetation, while hydrogen chloride causes corrosion as well as damage to vegetation.

31.5.10 Hydrocarbons and Photochemical Oxidants

Hydrocarbons are compounds containing only hydrogen and carbon. They are generally divided into two categories: aliphatic hydrocarbons, which include alkanes (methane), alkenes (olefins) and alkynes; and the group of aromatic hydrocarbons such as benzene, etc. These are chiefly released into the atmosphere by automobile exhausts, but they are also released into the atmosphere in incinerator smoke, oil refinery fumes and the evaporation of gasoline at service stations. However, at the concentrations usually found in urban air, hydrocarbons cause no adverse effect on human health. Aromatic hydrocarbons are more reactive than aliphatic ones and irritate mucous membranes. The major oxidant produced in photochemical smog is ozone, which is poisonous and smelly. It exists in great abundance under natural conditions in the upper atmosphere. It is likely that combustion and sunlight are involved in the production of ozone. Many other oxidants produced in photochemical smog, particularly peroxyacyl nitrates, cause eye irritation. Oxidants such as peroxyacetyl nitrate and peroxybenzonyl nitrate irritate the nose and throat and cause chest constriction.

31.5.11 Aldehydes

Although not present in as high a concentration as hydrocarbons, aldehydes are important constituents of smog because they contribute to eye irritation and possibly other physiological effects, and participate in chemical reactions that produce unpleasant products. At higher concentrations than typically occur in the atmosphere, the mucous membranes of the nose and throat are also irritated. Aldehydes are produced by the combustion of gasoline, diesel oil, fuel oil and natural gas.

Automobiles are by far the most significant source of aldehydes in urban pollution. Not only do they emit aldehydes directly, they also emit hydrocarbons that are converted to aldehydes in smog by photochemical reactions. However, essentially all burning of organic fuels (such as in municipal incineration, home heating, power plants, etc.) also introduces aldehydes into the atmosphere. Aldehydes are also introduced in the atmosphere by nature in the form of forest and brush fires initiated by lightning and in the form of essential oils emitted by plants.

31.6 EFFECTS OF AIR POLLUTION

In general, the effects of the polluted atmosphere can be classified under the following headings: effects on certain materials, effects on plants, effects on animals, effects on human health, and effects on the physical features of the atmosphere.

31.6.1 Materials

Air pollutants affect certain materials by abrasion, deposition and removal, direct chemical attack, indirect chemical attack and corrosion.

31.6.2 Plants

Air pollution has long been known to have an adverse effect on plants. The air pollutants that affect plants are sulphur dioxide, hydrogen fluoride, hydrogen chloride, chlorine, ozone, oxides of nitrogen, ammonia, mercury, ethylene, hydrogen sulphide, hydrogen cyanide, PAN, herbicides and smog. The most obvious damage caused by air pollutants to plants and vegetation occurs in the leaf structure. The stomata of the leaf get clogged, thereby reducing the intake of CO_2, thus affecting photosynthesis. The effects range from a reduction in growth rate to the death of the plant.

31.6.3 Animals

The effect of pollutants on farm animals takes place in two steps: accumulation of air pollutants in the vegetation, plants and forage; and the subsequent poisoning of the animals when they eat contaminated vegetation. The key contaminants that affect livestock are fluorine, arsenic and lead. These pollutants originate either from industries situated nearby or from dusting and spraying. Out of these contaminants, fluorine contamination is the most prominent since cattle and sheep are more susceptible to it. Symptoms of advanced fluorosis include lack of appetite, general ill health due to malnutrition, lowered fertility, reduced milk production and growth retardation. Arsenic in dusts or sprays on plants can poison cattle, leading to salivation, thirst, vomiting, uneasiness, feeble and irregular pulse, and respiration issues. Lead contamination of the atmosphere takes place on account of various activities such as smelters, coke ovens and other coal-based industries. Prostration, staggering and an inability to rise are the prominent symptoms of lead poisoning in animals, as well as a complete loss of appetite, paralysis of the digestive tract and diarrhoea.

31.6.4 Human Health

The inhalation of undesirable gases from the atmosphere has marked adverse effects on human health. Their adverse effect can be divided into two classes: acute effects and chronic effects. Acute effects manifest themselves immediately upon short-term exposure to air pollutants of high concentrations, while chronic effects become evident only after continuous exposure to low levels of air pollution. The following is a list of health effects caused by air pollutants: ear, nose and throat irritation; irritation of the respiratory tract; odour nuisance due to hydrogen sulphide and ammonia mercaptans, even at low concentrations; chronic pulmonary diseases such as bronchitis and asthma are aggravated by high concentrations of SO_2, NO_2, particulate matter and photochemical smog; asthmatic attacks initiated by pollen; cancer caused by carcinogenic agents; respiratory diseases caused by dust particles (silicosis is caused by silica dust from cement factories and asbestosis is caused by asbestos plants); lead poisoning due to the entry of lead into the lungs; bone fluorosis and the mottling of teeth caused by hydrogen fluoride; death by asphyxiation due to carbon monoxide, which also increases stress on persons suffering from cardiovascular and pulmonary diseases; an increase in mortality rate and morbidity rate due to air pollution in general; and radioactive fallout may cause cancer, shortening of life span and genetic defects.

31.6.5 Carcinogens

These are cancer-causing agents. An increase in lung cancers in cities has been reported due to aromatic hydrocarbons emitted by automobile emissions. Common examples of such organic carcinogens are benzo(a)pyrene, benzo(c)phenanthrene and benzo(c)pyrene, of the polynuclear aromatic hydrocarbons group. Another class of organic carcinogens leading to bladder cancers are aromatic amines. Possible inorganic carcinogens include arsenic, beryllium, cadmium, chromium, cobalt, lead, mercury, nickel, selenium, silicon, silver, zinc and asbestos. Exposure to these can occur in industrial situations, in water, in insecticide residues on food, and as airborne gases or dusts.

31.6.6 Radioactive Gases

Radioactive materials from the explosion of nuclear devices are the main causes of radioactivity in the environment. Another major source of radioactive gases and particulates is nuclear power reactors and related fuel-handling facilities and fuel reprocessing plants. To date, the most serious effects of gaseous releases have been in the underground mining of uranium ore. The major generation of radioactive gaseous wastes occurs in nuclear reactors, though these gases are not necessarily released directly into the environment.

31.6.7 Smog

Smog is a contraction of two words: smoke and fog. Smog can be photochemical or coal-induced. The conditions for the formation of photochemical smog are air stagnation, abundant sunlight and high concentrations of hydrocarbon and nitrogen oxides in the atmosphere. It is restricted to highly motorized areas in metropolitan cities. Photochemical smog was first observed in Los Angeles in the mid-1940s, and since then the phenomenon has been detected in most major metropolitan cities around the world. Smog is caused by the interaction of some hydrocarbons and oxidants under the influence of sunlight giving rise to dangerous peroxyacetyl nitrate (PAN). Its main constituents are nitrogen oxide, PAN, hydrocarbons, carbon monoxide and ozone. The main ill effects of photochemical smog are:

1. It causes coughing and chest soreness.
2. It irritates the eyes.
3. It kills leaf tissue.
4. It degrades rubber and cellulose.
5. It reduces visibility.

Fog from burning coal covers urban areas at night or on cold days when the temperature is below 10°C and calm meteorological conditions prevail. This fog consists of smoke, sulphur compounds and fly ash. Prolonged exposure to such smog may result in a high mortality rate, especially among the elderly and those with a history of chronic bronchitis, asthma, bronchopneumonia or other lung or heart diseases.

31.6.8 Atmosphere

As defined earlier, air pollution is any atmospheric condition in which certain substances are present in such concentrations that they produce undesirable effects on the environment.

31.6.9 Visibility

Visibility is reduced due to the concentration and physical properties of particulate pollutants present in the atmosphere. The measurement of prevailing visibility is standard meteorological practice. Stormy winds raise dust particles resulting in a decrease in visibility. In unsaturated humidity conditions, hygroscopic particles pick up moisture, and as they increase in size, visibility is affected. Due to the angle of the sun, visibility observations in polluted areas show strong directional variations. Other meteorological factors such as inversion, height and wind speed, the presence of hygroscopic particles and relative humidity also affect visibility. Fog and photochemical smog reduce visibility considerably.

31.6.10 Urban Atmosphere

Urban air pollution is mainly caused by smoke, dust, fog and other aerosols, which all affect weather conditions. The polluted area becomes cloudier and foggier, resulting in the reduction of solar radiation by an extent of about 30%. The area may have 5% to 10% more precipitation since air pollutants can add to the condensation of the nuclei of the cloud system.

31.6.11 Atmospheric constituents

Due to air pollution, the balance between the various constituents of air is disturbed. Atmospheric carbon dioxide is the main source of organic carbon in the biosphere. There has been a steady increase in the percentage of CO_2 in the atmosphere due to combustion and other factors causing air pollution. CO_2 is interpreted as one of the factors responsible for a rise in ambient temperature. Due to continuous air pollution, lead aerosol concentration is now 30 times more than in pre-industrial days. Freezing nuclei are formed in large numbers when automobile exhaust gases are exposed to minute quantities of iodine vapours. These nuclei are the main basis of weather modification cloud seeding operations.

31.6.12 Global Effects

Unlike other forms of pollution, air pollution has no frontiers. Pollutants emitted in one place can get transported to another location. Therefore, the problems of air pollution exist on all scales, from local to global. The major global issues of air pollution today are acid rain, global warming and ozone layer depletion.

Acid rain normally occurs when acid-forming gases, usually generated by burning fossil fuels, dissolve in water vapour or moisture. The following types of reaction may occur:

$$SO_2 \Rightarrow H_2SO_4$$

$$NO_x \Rightarrow HNO_3$$

Global warming is perhaps the most dangerous situation the modern world is facing. It is a phenomenon where gases such as CO_2, CH_4 and N_2O form a layer in the troposphere that allows the sun rays to enter the earth's surface but prevent the hot gases from escaping, thus increasing the atmospheric temperature.

Ozone in the stratosphere protects the earth from the UV fraction of solar radiation. Certain chemicals such as chlorofluorocarbons (CFCs) can bind the ozone in an atmospheric reaction and deplete this protective layer.

31.7 AIR SAMPLING AND MEASUREMENT

To evaluate air quality and to design efficient pollution control equipment and devices, it is necessary to determine pollutant emission rates and the types of pollutants in the gas and surrounding air. Before such measurements can be made, appropriate samples must be collected. Pollutant concentration is usually expressed as mass in a given volume of air. Measurement of mass is primarily carried out in an analytical laboratory, while the determination of volume is done in the field at the time of sampling. The analytical instruments available today can detect very small amounts of pollutant, even within the range of a few molecules. There are also so many methods and devices that measure volume, velocity and flow rate. The choice of technique for sampling and measuring mass and volume depends on the properties of the sampled air and the specific pollutant to be analyzed.

There are three distinct kinds of air sampling: source, ambient and indoor sampling. A brief discussion follows.

31.7.1 Source Sampling

Source sampling is performed at the location of pollution discharge, such as exhaust from a chimney, a ventilation system or the tailpipe of an automobile. Source sampling is called stack sampling at power plants, waste incinerators or factories equipped with chimneys.

One of the basic purposes of source sampling is to evaluate the pollutant concentration in the discharge and to use the results to determine compliance.

31.7.2 Ambient Air Sampling

Ambient air sampling pertains to outdoor air quality. Samples are collected from the air after the pollutants from various sources are thoroughly dispersed. Ambient air provides broad area or background air quality data to assess health effects and compliance, and to predict the effects of proposed new sources of air pollution.

31.7.3 Indoor Sampling

Indoor air sampling includes industrial hygiene sampling and residential sampling. Industrial or occupational sampling is done in workplaces to protect the health of workers who may be exposed to pollutants. Residential sampling, on the other hand, provides data regarding indoor air quality to protect the health of residents.

31.7.4 Air Monitoring

A distinction should be made between sampling and monitoring. Air monitoring refers to the collection of pollutant data effectively on a continuous basis. Monitoring equipment such as stack-gas monitors, as well as ambient or indoor monitoring devices, are available. However, continuous monitoring devices are not sensitive enough to detect extremely low concentrations of pollutants.

Due to the dangers of even small amounts of toxins, sampling followed by laboratory analysis has gained importance. This is particularly true for indoor sampling and emission sampling.

31.7.5 Particulate Sampling Devices

The collection and measurement of suspended particulates in the air depend on principles very different from those used to analyze gases. The mass and size of the particulates dictate the sampling process. The three general methods for collecting and measuring the particulates in air pollution, which rely on physical properties, are the gravimetric, filtration and inertial techniques.

The gravity method is the simplest and can measure the amount of settleable particulates (dust and fly ash) in the air. A basic device called a dust fall bucket is an example of a first-generation method used to determine how much particulate matter settles down. In this technique, an open bucket containing water to trap and hold the particles is exposed at a suitable location, such as a building rooftop. After a collection time of usually 30 days, the water is evaporated and the particulates weighed. The results can be expressed as g/m^2.month or kg/ha.month. The total amount of dust that will settle out of the atmosphere in a typical urban area can be quite high.

Example Problem 31.3

The mass of a 15 cm diameter bucket is 120.00 g when empty. After 30 days of exposure, the bucket plus collected particulates have a combined mass of 120.35 g. Calculate the concentration of the dust fall.

SOLUTION:

$$M_p = \frac{(120.35 - 120.00)g}{month} = 0.35\,g\,/\,month$$

$$v_p = \frac{M_p}{A} = \frac{4M_p}{\pi D^2} = \frac{0.35\,g}{month} \times \frac{4}{\pi(0.15\,m)^2} = 19.806 = 19.8\,g\,/\,m^2.\,month$$

$$= \frac{19.806\,g}{m^2.month} \times \frac{10000\,m^2}{ha} \times \frac{1\,kg}{1000\,g} = 198.06 = 198\,kg\,/\,ha.month$$

Dust fall buckets have several limitations, including a long sampling period needed to get results. They have been largely replaced by more accurate sampling devices known as high-volume samplers (**Figure 31.2**). These devices, which reduce the sampling time to 24 hours, use filtration rather than gravity to capture particulates. A high-volume sampler draws a large volume of air through a glass fibre or membrane filter. The filter is weighed before and after sampling the airflow, which decreases as particulates accumulate on the filter surface. The results are usually expressed as µg/m^3 of total suspended particulate (TSP). TSP concentrations for urban centres are quite high compared to rural areas.

Example Problem 31.4

The airflow through a high-volume sampler was recorded at 1.65 m^3/min at the beginning of sample collection and 1.14 m^3/min after 24 hours of continuous sampling. The filter weighed

FIGURE 31.2 High-Volume Sampler

10.000 g and 10.215 g before and after sample collection, respectively. What is the TSP level of the sample?

SOLUTION:

$$\bar{V} = \frac{(1.65+1.14)}{2}\frac{m^3}{min} \times 24\ h \times \frac{60\ min}{h} = 2008.8\ m^3$$

$$TSP = \frac{m}{\bar{V}} = \frac{(10.215-10.000)\ g}{2008.8\ m^3} \times \frac{10^6 \mu g}{g} = 107.03 = 107 \mu g / m^3$$

High-volume samplers may be used to obtain samples of particulates for the analysis of metals, organics, sulphates and nitrate compounds. The measurement of particulate levels in ambient air can also be done with dichotomous samplers, which are modified high-volume devices. They hold two filters for the separation and collection of PM_{10} and $PM_{2.5}$. The subscript number indicates the largest size of particulate matter in μm that the filter will collect.

Inertial samplers operate on the principle that when airflow direction is changed abruptly, the inertia of the particulates will cause them to hit an impaction surface, on which they can be trapped. The impaction surface may be a glass fibre mat or a solid surface coated with oil and is weighed before and after sampling. The impaction surface may simply be a hard surface parallel to the gas flow, as in a cyclone collector, in which the impacted particulates fall to the bottom of the collector for removal and weighing.

Another type of inertial sampling device used to collect particulates such as pollen grains and bacteria is the cascade impactor. This sampler captures particulates on separate slides that are placed in the air stream. The small orifices through which the air flows progressively decrease in size, therefore increasing the flow velocity. The particulates on the slides are observed and measured with a microscope.

Smoke is an aerosol of tiny particulates of materials that have not been completely burned. A Ringelmann smoke chart may be used to determine if the plumes of smoke from individual chimneys are within the desired range, reading the visual appearance and optical density of the smoke. The density of a plume is compared to standard shades of grey on the chart between white (0) and black (5). In many countries, power plants are expected to have Ringelmann readings not exceeding 1.

31.7.6 Gas Sampling

Due to the size of the molecules of the gases, gravimetric and filtration techniques cannot be applied when sampling gases. Instead, the process of either absorption or adsorption is used. Absorption can be either a physical or chemical process. The chemical process uses a liquid that reacts with the gas to form an easily detectable product. A typical absorption sampler is a bubbler or impinger tube. The air is pumped through a small glass diffuser and bubbled up through the liquid. For example, if a known volume of air containing sulphur dioxide is bubbled through hydrogen peroxide, sulphuric acid is formed, which can be measured by titration.

A three-gas sampler can be used to sample three different gases at the same time.

A similar device, called a sequential sampler, can collect up to 12 samples in sequence for a fixed time interval. This type of sampling is more useful to determine the peak concentration on a daily basis.

Adsorption is a physical process in which the gas molecules are attracted to the surface of a solid. The adsorbent is usually a porous material such as activated carbon. Desorption is the reverse of adsorption. Industrial hygiene surveys for the detection of organic compounds in the workplace often use charcoal adsorption tubes and solvent desorption. Gas chromatography or similar techniques are used to analyze the sample. In colourimetric tubes, the tubes are precoated with chemicals that react with the gas and change colour proportional to the concentration of the gas.

Whole-air samples collect a grab sample of air by drawing a volume of air by the vacuum. If the gas under study is insoluble in water or another liquid, the volume of air can be withdrawn using a displacement bottle. As the liquid flows out, an equal volume of air is withdrawn into the bottle. One of the limitations is that the sample size is limited, thus this method is not applicable when the concentration of gas is low.

31.8 AIR QUALITY INDEX

An Air Quality Index (AQI) is used to inform the public about the quality of air and trends in air quality. Previously this was known as the Pollutant Standards Index (PSI). In addition to its use as a public information tool, governing bodies, pollution control boards and local officials use an AQI to help determine what precautionary steps need to be taken if pollution levels rise to an unhealthy range.

An AQI is determined for each of the criteria pollutants: particulates, carbon monoxide, nitrogen dioxide and ozone. This is done by converting the measurement of a pollutant concentration in the air to a number on a scale of 0 to 500. There are five intervals on the AQI scale, each of which is related to the potential health effects for each of the criteria pollutants.

As shown in **Table 31.2**, the most significant number on the AQI scale is 100 since this corresponds to the standards established under the Clean Air Act in the US. A 75 ppb reading for sulphur dioxide, for example, would translate to an AQI of 100. Similarly, a 12 $\mu g/m^3$ measured concentration of PM2.5 would be equivalent to an AQI level of 100. Usually, the AQI with the higher value is reported. For example, if the AQI reported is 110 for ozone, that means the concentration of ozone is in the unhealthy range and the other pollutants are below that value. Although in North America, most urban centres rarely experience an AQI above 100, in developing countries like India, it is very common to exceed this standard. In addition to the exhaust from automobiles, air temperature and stubble burning by farmers has made things go from bad to worse.

TABLE 31.2
Air Quality Index

AQI	Health Effect
0–50	Good
51–100	Moderate
101–200	Unhealthful
201–300	Very unhealthful
301+	Hazardous

In the cities of developing countries, the severity of air pollution is often quite noticeable to residents in the form of dense haze and smoke. However, the use of a number like an AQI can clearly emphasize the extent of the air pollution. AQI levels exceeding 100 may trigger preventive actions by state or local officials. These could include health advice for susceptible people to limit some outdoor activities. A level above 300 will probably trigger a warning.

Discussion Questions

1. Name and describe the major layers of the atmosphere.
2. Explain the variation of air temperature and pressure with height using a diagram.
3. What are the causes and effects of air pollution?
4. What are primary and secondary pollutants? Give examples.
5. Define dust, smoke, mist, fumes and vapours, and their characteristics.
6. What are the principal sources of SOx, NOx and CO emissions to the atmosphere?
7. What two chemical compounds are necessary ingredients for producing photochemical smog?
8. Describe the purpose and types of air sampling.
9. How does air monitoring differ from air sampling?
10. Describe first-generation and second-generation air-volume samplers for determining the concentration of particulate matter.
11. Describe the global effects of air pollution.
12. Describe the various methods of gas sampling.
13. What is the basis of AQI? What is the significance of an AQI level of 100?
14. What is the purpose of reporting AQI levels in urban centres?

Practice Problems

1. A gaseous chemical has a concentration of 41.6 $\mu mol/m^3$ in the air at 1 atm pressure and a temperature of 293 K. the universal gas constant R is 82.05 x 10^{-6} m^3.atm)/ mol. K. Assuming that ideal gas law is valid, what is the concentration of the gaseous chemical in ppm? (1 ppm)
2. The primary air quality standard for SO_2 as an annual average is 80 $\mu g/m^3$. What is its equivalent in ppm at 25°C and 1 atm? (31 ppb)
3. Air quality standards limit eight-hour carbon monoxide levels to 9 ppm. Express this concentration in mg/m^3 at 25°C and 1 atm pressure. (10 mg/m^3)

4. The volume of ideal gas at standard temperature and pressure is 22.4 L/mol. At what temperature will the volume be 25 L/mol? (31.7°C)
5. The exhaust of an automobile contains 0.10% of nitrogen dioxide at a temperature of 90°C. Express this concentration as g/m^3. (1.5 g/m^3)
6. National standards limit one-hour average CO levels to 40 μg/L at 25°C and 1 atm pressure. What is this concentration expressed as ppm and percent on a volume basis? (35 ppm, 0.0035%)
7. The standards for annual mean nitrogen dioxide levels are 100 $\mu g/m^3$ at 25°C and 1 atm pressure. How many ppm? (0.053 ppm)
8. The exhaust of an automobile contains 1.8% by volume of carbon monoxide at a temperature of 85°C. Express this concentration as g/m^3. (17 g/m^3)
9. A 150 mm diameter bucket weighs 125.00 g and 125.37 g before and after 30 days' collection of particulate matter. Compute the dust fall in kg/ha.month. (209 kg/ha.month)
10. The airflow through a high-volume sampler was recorded at 1.54 m^3/min at the beginning of sample collection and 1.14 m^3/min after 24 hours of continuous sampling. The filter weighed 9.980 g and 10.215 g before and after sample collection. What is the TSP level in the sample? (122 $\mu g/m^3$)

32 Meteorological Aspects of Air Pollution

Air pollutants emitted from various sources get transported, dispersed or concentrated in the atmosphere due to meteorological or topographical conditions. This process depends upon meteorological factors such as prevailing winds (direction and speed), temperature and pressure conditions, moisture (humidity), precipitation and topographical conditions. These air pollutants should be transported and diluted in the atmosphere before they undergo various physical and photochemical transformations and ultimately reach their receptors. Many historical air pollution episodes indicated that meteorological conditions restricted the dispersion of pollutants, resulting in dangerous levels near the source of emission. The following meteorological parameters influence the dispersion of air pollutants.

Primary Parameters

1. Wind direction and speed.
2. Temperature.
3. Atmospheric stability.
4. Mixing height.

Secondary Parameters

1. Precipitation.
2. Humidity.
3. Solar radiation.
4. Visibility.

The three dominant mechanisms identified as influencing the dispersion of a pollutant in the atmosphere are:

1. The average wind speed with which the air transports the pollutant in the downwind direction.
2. The turbulence or velocity fluctuations that disperse the pollutant in all directions.
3. Mass diffusion due to concentration gradients. Moreover, the general aerodynamic characteristics (size, shape and weight) that affect the settling of non-gaseous pollutants (particulates) to the ground or buoyed upwards influence the dispersion of pollutants.

32.1 ATMOSPHERIC STABILITY

32.1.1 Environmental Lapse Rate (ELR)

The effective dispersion of pollutants in the atmosphere depends primarily on the degree of stability of the atmosphere. Atmospheric stability is defined as its tendency to resist vertical motion or to suppress existing turbulence. This tendency directly influences the atmosphere's ability to disperse pollutants emitted into it. Thermodynamic considerations show that the stability of the atmosphere depends upon the rate of the change of ambient temperature with altitude. In a well-mixed dry air,

DOI: 10.1201/9781003231264-32

the temperature decreases by about 0.66°C per 100 m. The negative of this vertical temperature gradient is known as lapse rate and is expressed by:

$$\frac{dT}{dZ} = -\left(\frac{n-1}{n}\right) \times \frac{g}{R}$$

Where R is the gas constant for air.

Based on the meteorological data in the troposphere up to about 10 km, the environmental lapse rate (or normal lapse rate) is 6.6°C per km. Hence, for a polytropic model, we have:

$$\frac{-6.6^0 C}{km} = -\left(\frac{n-1}{n}\right) \times \frac{g}{R}$$

From the above equation, n comes out to be 1.23, which is applicable only for the troposphere. Above the troposphere is the stratosphere, the lower region (up to 25 km) of which has a constant temperature or, say, n = 1 (**Figure 32.1**). In the upper region of the stratosphere the temperature increases with altitude as a result of ozone formation.

32.1.2 Adiabatic Lapse Rate (ALR)

The lapse rate of a parcel of dry air as it moves upwards in a hydrostatically stable environment and expands slowly to the lower environmental pressure without the exchange of heat is known as adiabatic lapse rate. The dry adiabatic lapse rate (DALR) is given by:

$$\left(\frac{dT}{dZ}\right)_{ad} = \frac{-9.86^\circ C}{km} = -\frac{n-1}{n} \times \frac{g}{R}$$

The dry adiabatic lapse rate is 1^0 C/100 m. Dry adiabatic lapse rate is important in defining atmospheric stability. A comparison of the adiabatic rate to the environmental lapse rate indicates the stability of the atmosphere. Atmospheric stability is a measure of the ability of the atmosphere to disperse pollutants emitted into it.

32.1.3 Stability Conditions

Several possible environmental lapse rates are compared with the adiabatic lapse rate in **Figure 32.2**.

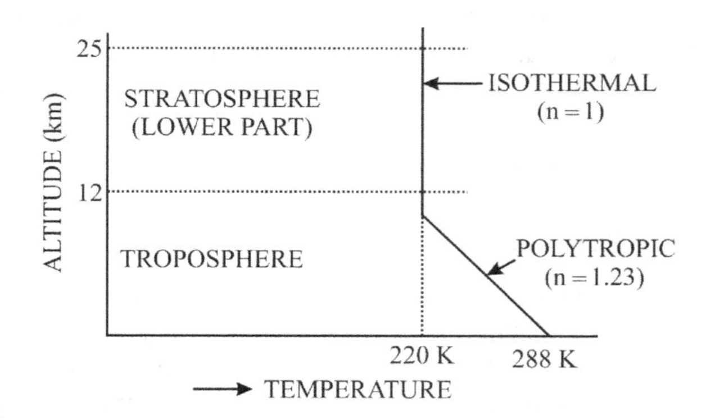

FIGURE 32.1 Temperature Altitude Profile

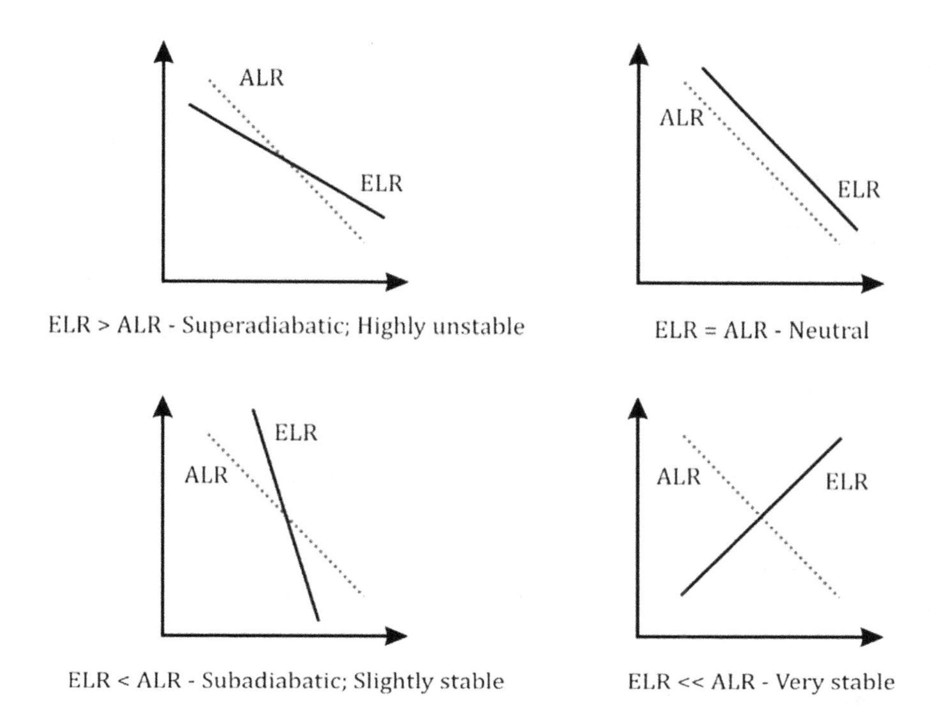

FIGURE 32.2 Atmospheric Stability Conditions

When the prevailing lapse rate, ambient lapse rate or environmental lapse rate is greater than the dry adiabatic lapse rate, the atmosphere is said to be superadiabatic or unstable. This is because any perturbation in the vertical direction tends to be enhanced. Therefore, it is a favourable atmospheric condition with respect to the dispersion of pollutants.

When the environmental lapse rate is approximately the same as the dry adiabatic lapse rate, the stability of the atmosphere is said to be neutral. Any parcel of air carried rapidly upwards or downwards will have the same temperature as the environment at the new height. Hence there is no tendency for any further vertical movement.

When the environmental lapse rate is less than the dry adiabatic lapse rate, a rising air parcel becomes cooler and denser than its surroundings and tends to fall back to its original position. Such an atmospheric condition is called stable, and the lapse rate, which is subadiabatic, is negative. In such conditions, a dense cold stratum of air at ground level gets covered by lighter, warmer air at a higher level. This phenomenon is called inversion.

During inversion, vertical air movement is stopped and the pollution is concentrated beneath the inversion layer, i.e., the denser air at ground level. Thus, atmospheric inversion influences the dispersion of pollutants by restricting vertical mixing; hence pollutants in the air do not disperse. Inversion is a frequent occurrence in the autumn and winter months. There are radiation, subsidence and double inversions.

Radiation Inversion

This type of inversion usually occurs at night and results from the normal diurnal cooling cycle. After sunset, the ground loses heat by radiation and cools the air in contact with it. Consequently, a temperature inversion is set up in the first few hundred metres above the surface between the cool low-level air and the warmer air above. If the air is moist and its temperature is below the dew point, fog will form. This type of inversion is more common in winter than summer because of

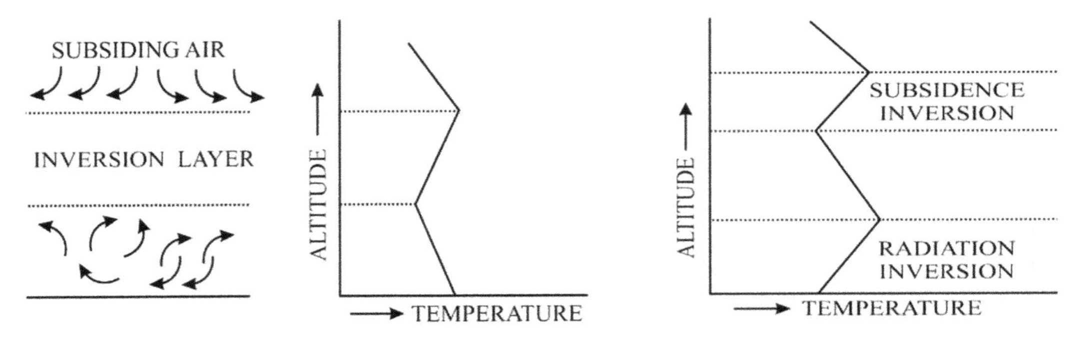

FIGURE 32.3 Subsidence and Combination

the longer nights. However, the next morning, the sunlight will destroy the inversion as the earth is warmed.

Subsidence Inversion

This is one of the most common types that occurs at modest altitudes and often remains for several days. It is usually associated with subtropical anticyclones (i.e., high-pressure areas surrounded by low-pressure areas) where the air is warmed by compression as it descends in a high-pressure system and achieves a temperature higher than that of the air underneath.

An inversion will result if the temperature increase is sufficient. The air circulating the area descends slowly at a rate of about 1000 m per day. This acts as a lid to prevent the upwards movement of contaminants. The inversion height may vary from the ground surface to 1600 m; when this drops to less than 200 m, extreme pollution occurs.

It is possible for both the radiation and subsidence inversions to occur simultaneously, as illustrated in **Figure 32.3**. The joint occurrence of the two types of inversions leads to a phenomenon called stack plume trapping, which will be discussed later.

32.1.4 Wind Velocity Profile

Wind speed, direction and turbulence play an important role in the dispersion of atmospheric pollutants. Wind, the movement of air near the earth's surface, is slowed by frictional effects proportional to the surface roughness. The variation of wind with altitude is shown in **Figure 32.4**.

Frequent knowledge of the wind speed at some height other than the standard height of the anemometer is necessary. The following power law has been found useful:

$$\frac{u}{u_1} = \left(\frac{z}{z_1}\right)^p$$

Where u = wind speed at an altitude z

u_1 = wind speed at an altitude z_1

p = positive exponent that depends on the stability conditions of the atmosphere, with a value ranging between 0 and 1.

Using Sutton's expression, $p = \frac{n}{2-n}$, where n is a parameter that defines the stability characteristics of the atmosphere, one can find approximate values of the exponent. A wind rose diagram

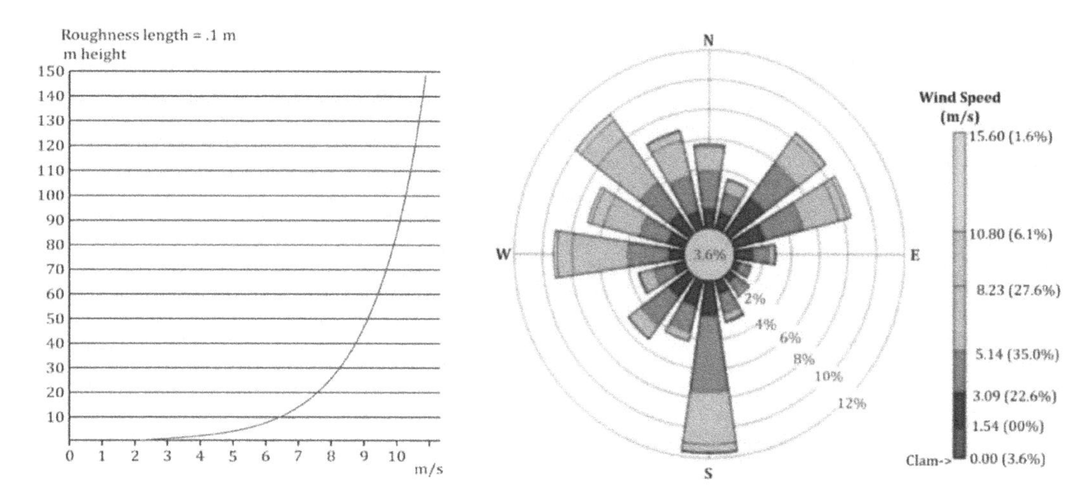

FIGURE 32.4 Vertical Wind Profile and Wind Rose Diagram

(**Figure 32.4**) yields a graphical representation of the direction, frequency and speed of the wind in a particular location.

32.1.5 Plume Behaviour

A plume is defined by the path taken by the continuous discharge of gaseous effluents emitted from a stack or chimney. The shape of the path and the concentration distribution of gaseous plumes depend upon the localized air stability. Typical plume patterns generally encountered in the lower atmosphere (< 300 m above ground) as illustrated in **Figure 32.5** are: looping, coning, fanning, lofting, fumigating and trapping plumes.

Looping Plume

This is a common type of plume behaviour that occurs under superadiabatic lapse rate (SALR) conditions with light to moderate wind speeds on a hot summer afternoon when large scale thermal eddies are present. The plume has wavy behaviour since it occurs in a highly unstable atmosphere. High turbulence helps the rapid dispersion of the plume, but a high concentration may occur close to the stack if the plume touches the ground.

Fanning Plume

This occurs under extreme inversion conditions in the presence of light wind. Most of the vertical dispersion is suppressed by extremely stable conditions, and the plume fans out in the horizontal direction. Strong concentrations at plume height are exhibited downwind of the stack.

Coning Plume

A coning plume occurs on a cloudy day or on nights with strong winds (velocity 32 km/h) when the lapse rate is neutral (adiabatic conditions). The plume shape is vertically symmetrical about the plume line. However, the plume reaches the ground at a greater distance than with looping.

Lofting Plume

This type of plume occurs when there is a strong SALR above a surface inversion.

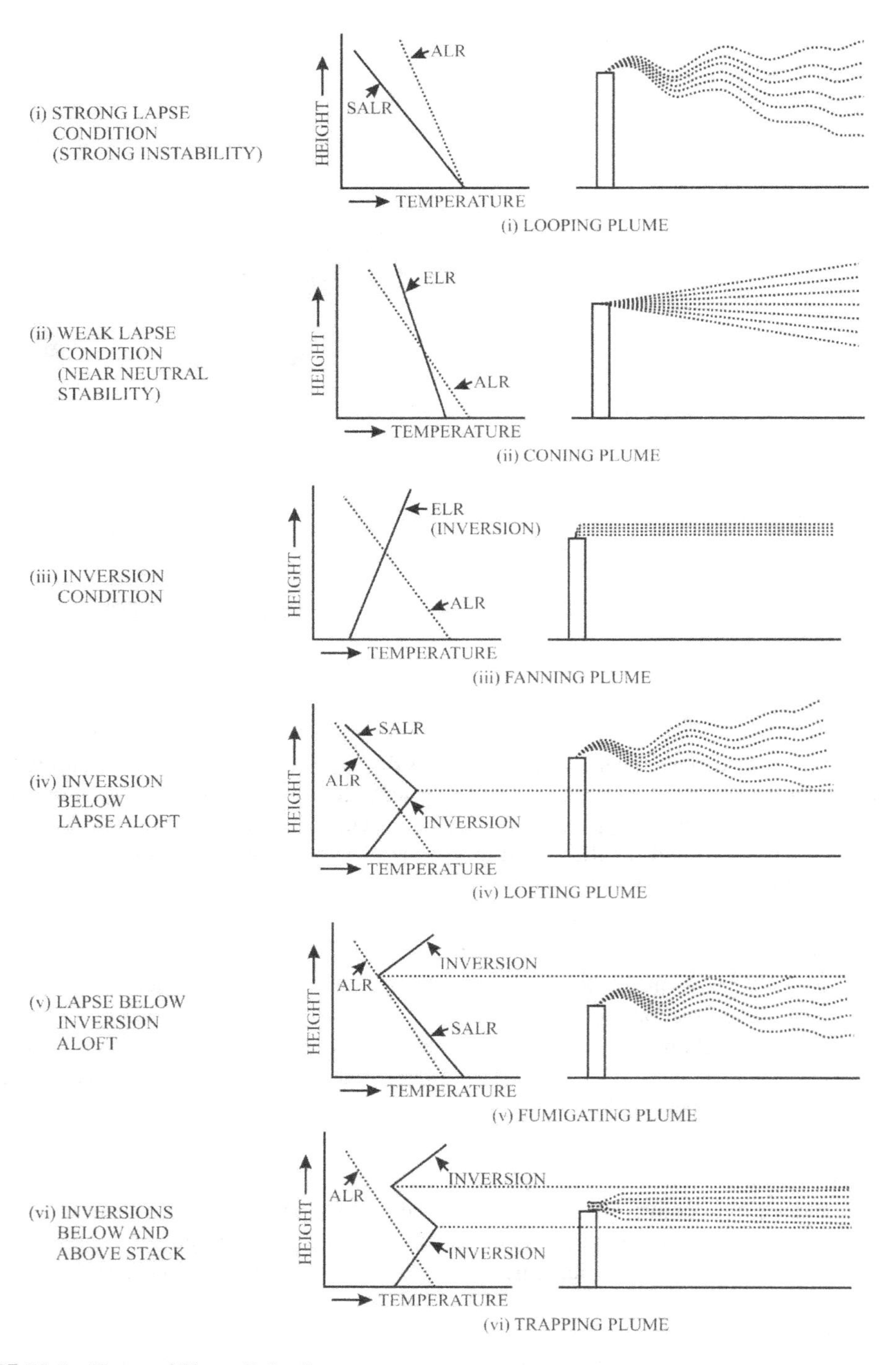

FIGURE 32.5 Types of Plume Behaviour

Fumigating Plume

The conditions for fumigation are just the inversion of a lofting plume. Fumigation takes place when an inversion layer occurs at a short distance above the top of the stack and superadiabatic conditions prevail below it.

Trapping Plume

This condition is achieved when the plume is caught between two inversion layers. As a result, the emitted plume can go neither up nor down and will be trapped between the two levels. The diffusion of the effluent will be severely restricted to the unstable layer between the two stable regions. A trapping plume is a bad condition for dispersion.

32.1.6 Maximum Mixing Depth

The dispersion of pollutants in the lower atmosphere is greatly aided by the convective and turbulent mixing that takes place. The maximum mixing depth (MMD) is the vertical extent to which this mixing takes place in a particular locality diurnally or seasonally. The greater the vertical extent, the larger the volume of the atmosphere available to dilute the pollutant concentration. MMD can be estimated by plotting maximum surface temperature and drawing a line parallel to the dry adiabatic lapse from the point of maximum surface temperature to the point at which the line intersects the environmental lapse rate.

Example Problem 32.1

The ground-level air temperature at a given location is 18°C and the normal maximum surface temperature for that month is known to be 30°C. At an elevation of 700 m, the temperature is measured as a) 15°C and b) 20°C. Calculate the MMD in these two conditions.

SOLUTION:

$$ELR(15°C) = \frac{(15-18)°\text{C}}{700\ m} \times \frac{1000\ m}{km} = -4.28 = -4.3°\text{C / km}\ \ or\ slope = 0.23\ km\,/\,°\text{C}$$

The dry adiabatic lapse rate is = -10°C/km. So, from **Figure 32.6**, the MMD is found where the lines intersect at an altitude of 2.1 km = 2100 m.

This can also be done by solving algebraic equations:

$$Z = 4.2 - 0.233Z\ \ and\ dry\ adiabatic\ Z = 3.0 - 0.1T\ \ solving\ simultaneously$$

$$1.33\ Z = 30 \times 0.233 - 4.2\ \ or\ \ Z = \frac{(6.99-4.2)}{1.33} = 2.097\ km = 2097\ m = 2100\ \text{m}$$

$$ELR = \frac{(20-18)°\text{C}}{700\ m} \times \frac{1000\ m}{km} = 2.857 = 2.86°\text{C / km}$$

$$Z = 0.35\ T - 6.3\ \ \ \ or\ \ \ \ 4.5\ Z = 30 \times 0.35 - 6.3\ \ or\ Z = \frac{4.2}{4.5} = 0.93\ km = 930\ m$$

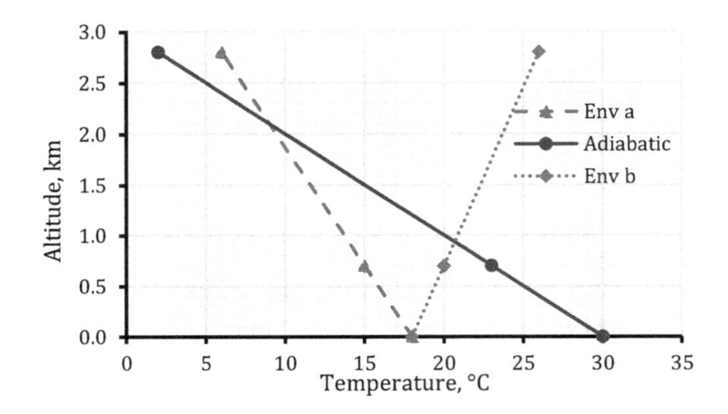

FIGURE 32.6 Mixing Depth (Ex. Prob. 32.1)

TABLE 32.1
Worksheet (Ex. Prob. 32.2)

Altitude	Temp.	LR
m	°C	°C/100 m
0	21	-
50	20	2
100	19	2
150	22	-6
200	21	2
250	20	2
300	19	2

Example Problem 32.2

Plot a graph of altitude versus prevailing temperature using the following data:

Altitude, m	0	50	100	150	200	250	300
Temp. °C	21	20	19	22	21	20	19

a. What kind of lapse rate exists between 0 and 100 m?
b. What kind of lapse rate exists between 200 and 300 m?
c. Does this data show a temperature inversion? If so, where?
d. Would fumigation occur if smoke at 20°C was discharged from a chimney at a height of 50 m? How about at a height of 250 m?

SOLUTION:

Environmental lapse rate (0–100 m)

$$ELR = \frac{(21-20)°C}{50\ m} \times \frac{100\ m}{100\ m} = 2°C/100\ \text{m}$$

The lapse rate for each range of altitude is calculated and shown in **Table 32.1**.

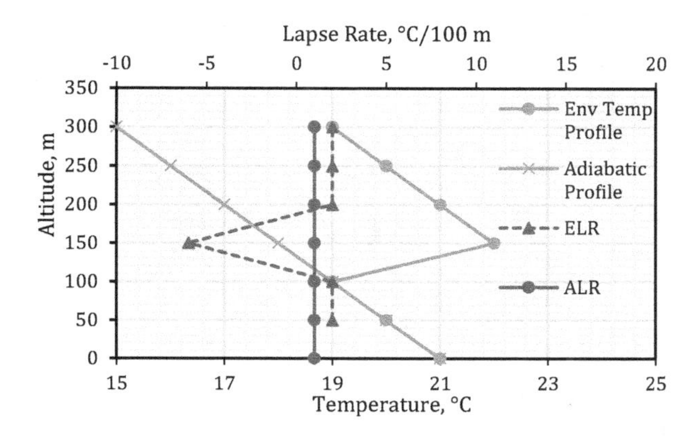

FIGURE 32.7 Altitude Versus Temperature

a. In the 0–100 m range, ELR is 2 > 1, hence superadiabatic.
b. In the 200–300 m range, ELR is 2>1, hence superadiabatic.
c. In the 100–150 m range, the environmental lapse rate is positive, indicating inversion.
d. Yes, at a height of 50 m, since it is superadiabatic below the inversion. No, at 250 m, since it is superadiabatic above the inversion.

32.2 DISPERSION MODELS

The ability to predict ambient concentrations of pollutants on the basis of dispersion from sources within a region is essential for meeting air quality standards. Dispersion is primarily caused by turbulent diffusion and bulk airflow. The turbulent diffusion models are based on Fick's law of molecular diffusion. Several models have been developed to estimate the concentration of pollutants in the plume at any distance x, y and z, in a horizontal downwind direction, a horizontal crosswind direction and a vertical direction, respectively. If K_x, K_y and K_z are eddy diffusion coefficients in the three directions, C is the concentration of the pollutant and v_x, v_y and v_z, are the respective velocity components. Then, from the continuity principle, one obtains the following diffusion equation in terms of time, t:

$$\frac{\partial C}{\partial t}+v_x\frac{\partial C}{\partial x}+v_y\frac{\partial C}{\partial y}+v_z\frac{\partial C}{\partial z}=\frac{\partial}{\partial x}\left(K_x\frac{\partial C}{\partial x}\right)+\frac{\partial}{\partial y}\left(K_y\frac{\partial C}{\partial y}\right)+\frac{\partial}{\partial z}\left(K_z\frac{\partial C}{\partial z}\right)$$

Using Fick's law above, and based on the knowledge of the shape of the concentration distribution (**Figure 32.8**) for a continuous single emission source, Gaussian developed the following simple statistical equation to compute the turbulent transport of gas or aerosols (particulates < 20 μm)
Pollutant concentration

$$C(x,y,z,H)=\frac{Q}{2\pi u\sigma_y\sigma_z}\left(\exp-\left(\frac{y^2}{2\sigma_y^2}\right)\right)\left\{\left(-\frac{(z-H)^2}{2\sigma_z^2}\right)+\exp\left(-\frac{(z+H)^2}{2\sigma_z^2}\right)\right\}$$

Where C = concentration of pollutant (g/m^3)
Q = pollutant emission rate (g/s)
u = mean wind speed (m/s) at height (H) of the stack

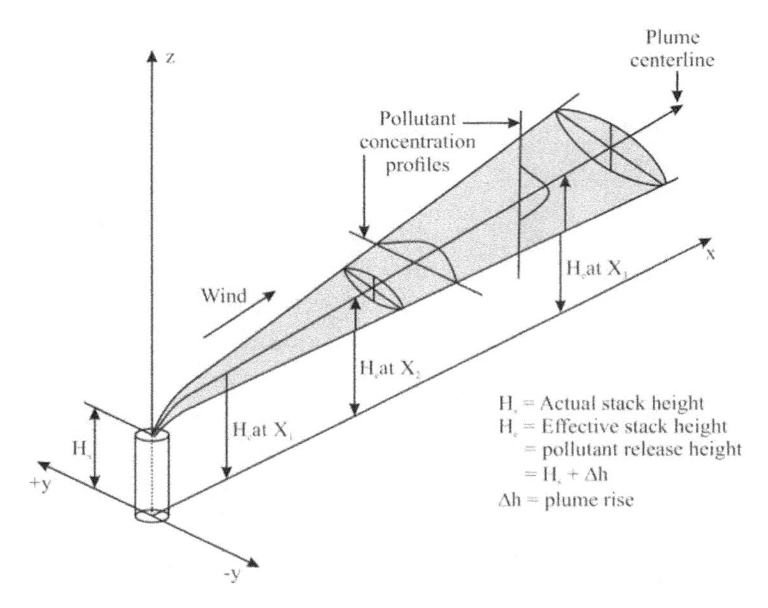

FIGURE 32.8 Gaussian Model

TABLE 32.2
Generalized Stability Categories

Surface Wind Speed (at 10m) (m.s-1)	Day° Incoming Solar Radiation			Night°	
	Strong	Moderate	Slight	Thinly Overcast or $\geq \frac{4}{8}$ Low Cloud	$\leq \frac{3}{8}$ Cloud
< 2	A	A–B	B		
2–3	A–B	B	C	E	F
3–5	B	B–C	C	D	E
5–6	B	C–D	D	D	D
> 6	C	D	D	D	D
A Extremely Unstable		C Slightly Unstable		E Slightly Stable	
B Moderately Unstable		D Neutral		F Moderately Unstable	

H = effective height of stack (m)
x, y = downwind and cross-wing horizontal distances (m)
z = level of computation of concentration (m)
σ_y = Plume's standard deviation in a crosswind direction (m)
σ_z = Plume's standard deviation in a vertical direction (m)

Modifications

If the concentration of the pollutant is desired along the centre line at ground level, the transport equation can be further modified as follows.

$$C(x, y, 0, H) = \frac{Q}{\pi u \sigma_y \sigma_z} \exp\left\{-\frac{1}{2}\left(\left(y / \sigma_y\right)^2 + \left(H / \sigma_z\right)^2\right)\right\}$$

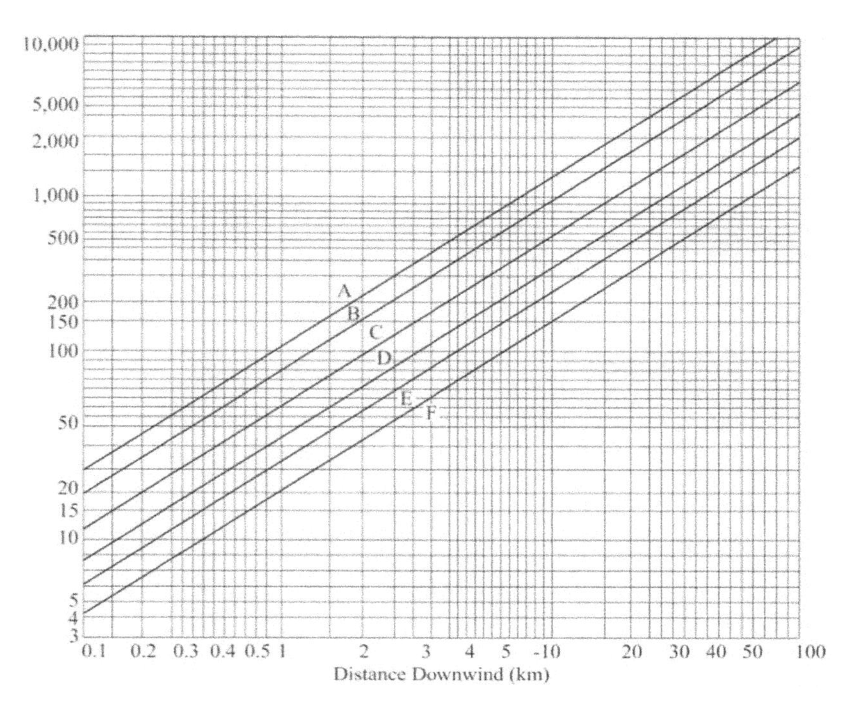

FIGURE 32.9 Diffusion Coefficient, σ_y

If the ground-level concentration is required along the centre line of the plume in the downwind horizontal direction (x-direction), y = 0.

$$C(x,0,0,H) = \frac{Q}{\pi u \sigma_y \sigma_z} \exp\left(-\frac{1}{2}\left(H / \sigma_z\right)^2\right)$$

If the source is at ground level (ground-level burning), the equation is further simplified by making the effective stack height equal to zero, and the equation for maximum ground-level concentration is simplified as follows.

$$C(x,0,0,0) = \frac{Q}{\pi u \sigma_y \sigma_z}$$

In determining the values of σ_y and σ_z, it can be seen that they are not only a function of downwind distances but also a function of atmospheric stability. The values of σ_y and σ_z for various downwind distances (x) with various stability categories are given in **Figures 32.9 and 32.10**, respectively. **Table 32.2** shows the generalized stability categories.

Example Problem 32.3

SO_2 is emitted at a rate of 160 g/s from a stack with an effective height of 60 m, wind speed = 6.0 m/s and atmospheric. The stability condition is D for an overcast day. Determine:

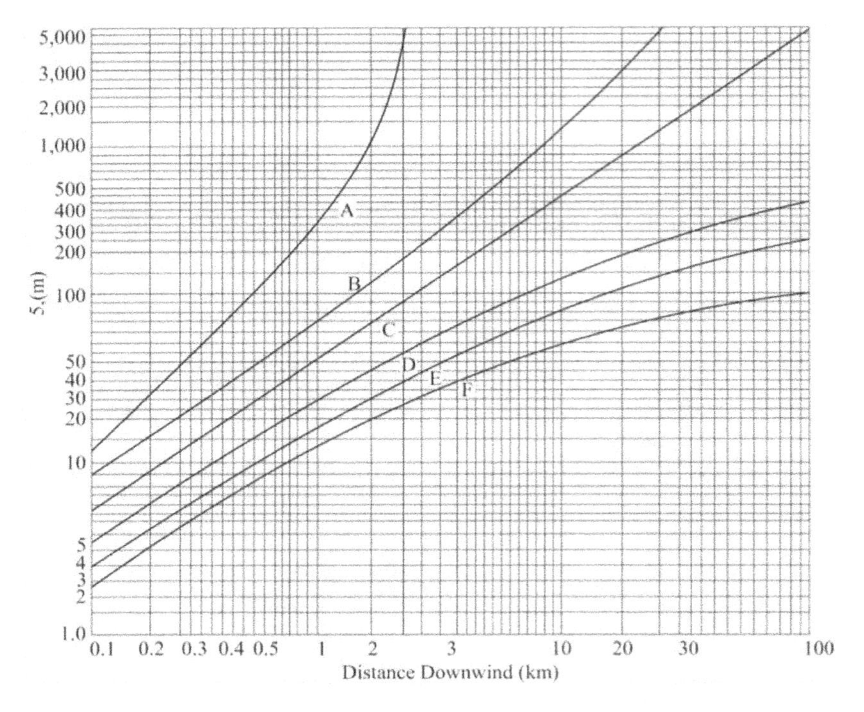

FIGURE 32.10 Diffusion Coefficient, σ_z

a. The ground-level concentration along the centre line at 500 m from the stack.
b. The concentration crosswind at 50 m from the centre line from the downwind distance of 500 m.

Given: Q = 160 g/s
H = 60 m
u = 6.0 m/s
Stability class = D
Read from the respective curve $\sigma_y = 18.5$ m $\sigma_z = 36$ m

SOLUTION:

$$C(x,0,0,H) = \frac{Q}{\pi\sigma_y\sigma_z u}\exp\left\{-\frac{1}{2}\left(\frac{H}{\sigma_z}\right)^2\right\}$$

$$C(500,0,0,H) = \frac{1}{\pi}\times\frac{160g}{s}\times\frac{1}{18.5m}\times\frac{1}{36m}\times\frac{s}{6m}\times e^{-\frac{1}{2}\left(\frac{60m}{36m}\right)^2}$$

$$= 3.178\times10^{-3} = 3.2\times10^{-3}\, g/m^3 = 3.2\, mg/\text{m}^3$$

50 m crosswind away from the centre line.

$$C(x,y,0,H)=\frac{Q}{\pi\sigma_y\sigma_z u}e^{-0.5\left\{\left(\frac{y}{\sigma_y}\right)^2+\left(\frac{H}{\sigma_z}\right)^2\right\}}$$

$$C(500,50,0,H)=\frac{1}{\pi}\times\frac{160\,g}{s}\times\frac{1}{36\,m\times18.5\,m}\times\frac{s}{6\,m}\times e^{-0.5\left\{\left(\frac{50}{36}\right)^2+\left(\frac{60}{18.5}\right)^2\right\}}$$

$$=2.525\times10^{-5}=2.5\times10^{-5}\,g/m^3=25.\mu g/m^3$$

32.2.1 The Maximum Ground-Level In-Line Concentration

In plume dispersion studies, it is important to determine the downwind distance for the maximum ground-level concentration and the maximum concentration at that point. From the characteristics of the σ_y and σ_z charts, it can be seen that the ratio σ_y/σ_z is nearly independent of the distance x. If this ratio is taken as constant and y is set equal to zero, the equation for C can be written solely as a function of σ_z, which is a function of x for a stability class. Hence, by the maximization technique of differential calculus, one can obtain the analytical information regarding the maximum concentration. The location of the maximum concentration is where $\sigma_z = H/\sqrt{2}$. Substituting this, maximum concentration on the ground can be found from the following equation.

$$C_{max}=\frac{0.12Q}{u\sigma_y\sigma_z}$$

32.2.2 Line Sources

In some situations, such as heavy traffic along a stretch of highway, the pollution problem may be modelled as a continuous emitting infinite line source. When the wind direction is normal to the line of emission, the ground-level concentration downwind is given by:

$$C(x,0)=\frac{2q}{\sqrt{2\pi}u\sigma_z}\left(exp-\left(\frac{H^2}{2\sigma_z^2}\right)\right)$$

Where q is the source strength per unit distance, g/s.m.

32.2.3 Plume Rise and Effective Height of Stack

The effective height, H, is equal to the sum of the actual height of the stack, h, and the rise of the plume, Δh, beyond the stack exit. The plume rise depends on many factors, including exit velocity, wind speed, diameter of stack, temperature of the plume, and lapse rate. Several empirical formulas are available to predict the plume rise.

General Equation

In general, plume rise can be expressed by $\Delta h=\frac{KQ_H^\alpha}{\mu u^\beta}$, where α, β and K are dimensional constants, Q_H is the heat emission rate from the stack and u is the average wind velocity. According to the Canadian Combustion Research Laboratory formula of the above form, α = 0.25, β = 1 and K = 66.4, where K is expressed in kcal/s and u in m/s.

TABLE 32.3
Regression Coefficients for Moses and Carson Equation

Atmospheric Stability	C_1	C_2
Superadiabatic	3.47	5.15
Neutral	0.35	2.64
Subadiabatic	−1.04	2.24

Moses and Carson Equation

$$\Delta h = \frac{C_1 v_s D}{u} + \frac{C_2 \sqrt{Q_H}}{u}$$

Δh = plume rise (m), v_s = stack exit velocity (m/s), u = wind speed (m/s), D = diameter of stack at exit (m), Q_H = heat emission rate (kJ/s), C_1, C_2= Plume rise regression coefficients that depend on atmospheric stability. Regression values for the three atmospheric conditions are shown in **Table 32.3**.

Holland's Equation

$$\Delta h = \frac{v_s D}{u}\left\{1.5 + 0.00268\, pD\frac{(T_s - T_a)}{T_s}\right\}$$

p = atmospheric pressure in millibars, T_s = stack gas temperature (K), T_a = air temperature (K)

This equation is applicable for neutral conditions. For unstable conditions, the calculated value of Δh should be increased by 10–20%, while for stable conditions, it should be decreased by 10–20%.

Davidson and Bryant Equation

This is another frequently used equation for computing plume rise. All symbols in this equation have the same meaning as discussed above.

$$\Delta h = D\left(\frac{v_s}{u}\right)^{1.4}\left\{1 + \frac{(T_s - T_a)}{T_s}\right\}$$

Briggs Formulae (Recommended by ISI)

1. For hot effluents with a heat release of the order 10^6 cal/s or more:

$$\Delta h = \frac{0.6 Q_H^{0.25}}{u}(12.4 + 0.09h)$$

Q_H = heat release in J/s
h = height of stack in m
u = wind speed in m/s

2. For effluents with a not very hot release, which can be counted as the momentum sources above:

$$\Delta h = \frac{3v_0 D}{u}$$

Where v_0 = afflux velocity (m/s)
u = wind speed (m/s)
D = stack exit diameter (m).

32.3 DESIGN OF STACK HEIGHT

The basic function of a stack is to provide natural draft for combustion. Stacks are designed to fulfil two additional requirements: that the pollutants be sufficiently dispersed and that the smoke should not re-enter the building under anticipated wind conditions. The ratio of the stack exit velocity to wind velocity should be greater than 1.5 to allow the effluent to break cleanly from the stack and to eliminate downwash. For design purposes, maximum local wind velocity conditions should be determined, and stack exit gas velocities selected proportionately. Higher exit velocities will give more plume rise and, therefore, can be used to reduce stack height, and hence the installation cost of that stack. However, this increases the pressure drop and therefore the required horsepower of blowers is increased, increasing the running costs. The position of the nearby buildings may cause mechanical turbulence and bring the plume to ground level, especially when the stack is downwind of the building and wind speeds are high. Hence, the stack height should be at least 2 to 2.5 times the height of the surrounding buildings. The diameter of the stack at the exit end can be determined on the basis of exhaust gas flow rate and the exit velocity required.

The Central Board for the Prevention and Control of Water Pollution, New Delhi, under its Emission Regulations, proposed two empirical equations to determine the minimum chimney height, h.

For a chimney emitting particulate matter

$$h_{min} = 74\left(Q_p\right)^{0.27}$$

h_{min} = minimum chimney height in m
Q_p = particulate matter in t/h

For a chimney emitting sulphur dioxide

$$h_{min} = 14\left(Q_s\right)^{0.33}$$

Q_s = SO_2 emission in kg/h

The maximum of the two heights as calculated by the two equations should be adopted. The application of these equations is illustrated in Example Problem 32.4.

The board further recommended that the calculated height should be subjected to following minimum values:

1. For chimneys for industries in general, except thermal power plants. h_{min} = 30 m
2. For thermal power plants > 200 MW and < 500 MW capacity, h_{min} = 220 m
3. For thermal power plants with capacity > 500 MW, h_{min} = 275 m

As per these standards, the minimum stack was to be 30 m, but the board later relaxed the provisions for chimneys of boilers and diesel generator sets to lower minimum values. For a boiler generating steam at the rate of:

1. < 2 t/h, h_{min} = 9.0 m
2. = 30 t/h, h_{min} = 30 m
3. >2 t/h, < 30 t/h, h_{min} = 9–30 m

For diesel generator sets, the minimum chimney height to be kept is only 1.5 m to 3.5 m more than the height of the building and can be worked out using this formula:

$$h = h_b + 0.2\,kVA$$

h_b = height of the building, m

Example Problem 32.4

An industry, on average, utilizes 0.5 ML of fuel oil per month. Based on the records, the quantities of various air pollutants emitted per ML of oil burn per annum are as follows:

Pollutant	SO_2	NOx	HC	CO	PM
Quantity, t/ML	55	7.5	0,4	0.5	3.1

Calculate the height of the chimney to allow for the safe dispersion of the pollutants.

SOLUTION:

Assuming an industry operates 300 days per year, the particulate matter generation rate is:

$$Q_p = \frac{3.1\,t}{ML} \times \frac{0.5\,ML}{mo} \times \frac{12\,mo}{a} \times \frac{a}{300 \times 24\,h} = 2.58 \times 10^{-3}\,t/h$$

$$h_{min} = 74\left(Q_p\right)^{0.27} = 74 \times \left(2.58 \times 10^{-3}\right)^{0.27} = 14.81 = 15\,m$$

Sulphur dioxide emission rate is;

$$Q_S = \frac{55\,t}{ML} \times \frac{1000\,kg}{t} \times \frac{0.5\,ML}{mo} \times \frac{12\,mo}{a} \times \frac{a}{300 \times 24h} = 45.8\,kg/h$$

$$h_{min} = 14\left(Q_s\right)^{0.33} = 14\left(45.8\right)^{0.33} = 49.47 = 49.5 = 50\,m$$

The maximum of the two heights is 50 m. Thus, the minimum height of the chimney should be 50 m.

Discussion Questions

1. What are the metrological parameters that influence the dispersion of air pollutants?
2. What is meant by lapse rate? How is it important in atmospheric studies?

3. Explain the relationship between adiabatic lapse rate, environmental lapse rate and atmospheric stability.
4. Describe the wind-speed profile with altitude.
5. A parcel of dry air rising has a temperature of 60°C at 15 m. Assuming a dry adiabatic lapse rate, determine the temperature at 100 m.
6. Name and explain the different atmospheric stability conditions.
7. Explain the phenomenon of inversions. Describe the different types of inversions.
8. Sketch and explain the different plume behaviour.
9. Using the generalized Gaussian dispersion model:

$$C(x,y,z,H) = \frac{Q}{2\pi u \sigma_y \sigma_z}\left[\exp-\left(\frac{y^2}{2\sigma_y^2}\right)\right]\left\{\exp\left[-\frac{(z-H)^2}{2\sigma_z^2}\right]+\exp\left[-\frac{(z+H)^2}{2\sigma_z^2}\right]\right\}$$

Prove that if the effective stack height is doubled, the maximum concentration downwind at ground level on the centre line is reduced by a factor of four.
10. In the troposphere, air temperature drops with an increase in altitude. However, in the upper part of the stratosphere, temperature increases with an increase in altitude. Explain.
11. Ozone is a secondary gaseous pollutant. However, there is a concern about the thinning of the ozone layer in the atmosphere. Explain.

Practice Problems

1. Plot a graph of altitude versus prevailing temperature using the following data:

Altitude, m	0	50	100	150	200
Temp. °C	19.0	19.5	20.0	20.5	19.0

   a. What kind of lapse rate exists between 0 m and 150 m? (Superadiabatic)
   b. What kind of lapse rate exists between 150 m and 200 m? (Subadiabatic)
   c. Do this data have a temperature inversion. If so, where? (0–150 m)
   d. Would fumigation occur if smoke at 20°C was discharged from a chimney at a height of 100 m? How about at a height of 200 m? (No, No)
2. At a given location, the ground-level air temperature is 19°C while the normal maximum surface temperature for that month is 28°C. At an elevation of 1.0 km, the temperature is 14°C. Calculate the maximum mixing depth. (1800 m)
3. Repeat Practice Problem 2, assuming the temperature at an elevation of 1.0 km is 20°C. (820 m)
4. An industry, on average, utilizes 16 kL of fuel oil per day. Based on its records, the quantities of various air pollutants emitted are as follows:

Pollutant	SO_2	NO_x	HC	CO	PM
Quantity, g/L	45	6.5	0.40	0.50	2.9

   What should the minimum height of the chimney be to allow for the safe dispersion of the pollutants? (44 m)
5. An industry, on average, utilizes 120 kL of fuel oil per week. Based on its records, the quantities of various air pollutants emitted are as follows:

Pollutant	SO_2	NOx	HC	CO	PM
Quantity, t/ML	25	4.5	0.50	0.30	5.5

What should the minimum height of the chimney be to allow for the safe dispersion of the pollutants? (36 m)

6. A thermal power plant burns 150 t/d of coal and discharges the effluent gases through an 80 m tall stack. The coal has a sulphur content of 5.0% and the wind velocity at the top of the stack is 30 km/h. Atmospheric conditions are moderately to slightly unstable. Determine the maximum concentration of SO_2 on the ground and the distance at which it occurs. (490 $\mu g/m^3$, 850 m)
7. For the data of Practice Problem 6, compute the concentration of SO_2 at a distance of 2.0 km downwind at the centre line of the plume and crosswind at a distance of 0.5 km on either side of the plume. (200 $\mu g/m^3$, 15 $\mu g/m^3$)
8. Due to a leak in an oil pipeline, H_2S is emitted at the rate of 2.0 g/min. This happened on a sunny summer day, with a wind speed of 3.5 m/s. What will the concentration of the gas be 1.0 km directly downwind from the stack? (130 ng/m^3)
9. A power plant has a stack with a 1.8 m diameter. Hot gases are emitted with an exit velocity of 12 m/s and a heat emission rate of 4.5 MJ/s. On average, the wind speed is 5.0 m/s and atmospheric stability is neutral. Applying the Moses and Carson empirical formula, estimate the plume rise. If the stack has a geometric height of 50 m, determine the effective stack height. (37 m, 87 m)
10. Find the effective height of the 155 m tall stack with a diameter of 100 cm. Assume the wind velocity is 3.0 m/s, the air temperature is 20°C, the barometric pressure is 1 bar, the exit velocity of effluent gases is 15 m/s and the gas temperature is 150°C. (167 m)

33 Air Pollution Control

The need to minimize or control air pollution has been evident for several hundred years. Today, it is known that smoke is more than a temporary annoyance; smoke and pollutants affect health and well-being, aesthetic sensibilities and climate on a global scale. This unit begins with an overview of air pollution control strategies.

33.1 ZONING

Air pollution can be effectively controlled by adopting a zoning system at the planning stage. Cumulative zoning in the past has resulted in less availability of land for industries. This system has since been modified to a permissible system, though this system has resulted in the crowding of industrial zones with other uses besides industry. The next improved system is the exclusive zoning system, which provides for the compatible uses of each zone, excluding other uses. In this system, a separate zone or area is set aside for industries (known as an industrial area or zone), reducing the ill effects of air pollution on urban dwellers. With proper zoning, the planning of the city can be done such that residential areas and heavy industries are not located too close to each other. This is achieved by providing a green belt between the industries and the township. Zoning for industrial areas can be done on the basis of functional requirements and performance characteristics. The functional requirements of industries include inter-industry linkages, railway sidings, groupings, land traffic generation utilities, etc. The performance characteristics of industries include traffic congestion, obnoxious and hazardous characteristics, and industrial nuisances such as dust, noise, smoke, odour, heat, fire, hazardous gases, etc. Separate areas must be earmarked according to their performances so that neat industries are placed away from obnoxious industries.

33.2 STACKS

Various aspects of the atmospheric dispersion of pollutants have already been discussed. The atmosphere, like a natural stream, possesses self-cleansing properties that continuously clean and remove pollutants from the atmosphere under natural conditions, provided the pollutants are discharged into the atmosphere judiciously so that effective dispersion takes place. If the pollutants are carried away some distance or taken to high altitudes, they are reduced in concentration by diffusion and dilution. Pollutants are taken to high altitudes by tall stacks. The height of the stack should be such that the maximum ground-level concentration, which varies inversely with the square of the stack height, is within permissible limits.

33.3 SOURCE PREVENTION METHODS

As the saying goes, prevention is better than cure. Source prevention is the best method for mitigating and controlling air pollution. Source prevention can be achieved through a change in raw materials, process changes, and equipment modification or replacement.

33.3.1 Raw Material

If one type of raw material currently in use results in an air pollution problem, and a substitute material, which may be of purer grade, does not, the substitution will be more desirable. A typical

DOI: 10.1201/9781003231264-33

example is the use of low-sulphur fuel in place of high-sulphur fuel. The raw material in current use may contain a certain ingredient that is not essential for the process but contributes to pollution; the non-essential ingredient should be removed through prior processing so that pollution can be minimized.

33.3.2 Process Modification

Atmospheric pollutant emissions can sometimes be reduced by adopting modified or new processes. A typical example is how the use of exhaust hoods and ducts over several types of industrial ovens has not only reduced pollutants but also resulted in the recovery of valuable solvents that could have become air pollutants. Similarly, volatile substances can be recovered by condensation, and non-condensable gases can be recycled for additional reactions.

33.3.3 Equipment Modification

Old equipment that contributes a great degree of air pollution can be modified or completely replaced. For example, basic oxygen furnaces, which are replacing open-hearth furnaces in the steel industry, pose lesser pollution problems. In many cases, newer types of equipment are less pollution prone. Newer types of equipment in the paper and pulp industry also cut down the quantity of pollutants emitted. Air pollutant emissions from industrial operations can also be reduced by proper equipment maintenance, housekeeping and cleanliness in facilities and premises. With the appropriate regulations and by-laws, industries can be required to install air pollution control equipment to meet standards.

33.4 EQUIPMENT FOR PARTICULATE POLLUTANTS

There are a variety of devices available for the control of particulate emissions. The most common include gravity settling chambers, inertial separators, cyclones, fabric filters, electrostatic precipitators and scrubbers or wet collectors.

33.4.1 Gravity Settling Chambers

A gravity settling chamber consists of a large rectangular expansion chamber in which dust is separated from the gas by reducing the velocity of the gas. As a result, the dust particles settle down in the chamber under gravity.

Gravity settling chambers are set horizontally, often on the ground, and can be constructed in brick and concrete. To reduce the size of the chamber, the gas velocity is kept between 0.5 m/s (to obtain good results) and 3 m/s to prevent the re-entrainment of settled particles. Hence, such chambers are capable of removing only large particles between 40 μm to 50 μm. Their efficiency is quite low for smaller particles, hence these devices are normally used as pre-cleaners prior to passing the gas stream through a high-efficiency collection device. Settling chambers are widely used to remove large solid particulate from natural draft furnaces, kilns, etc. They are also used in process industries, particularly the food and metallurgical industries, as a first step in dust control. The efficiency of a settling chamber can be improved if the height to be travelled by the particles is shortened.

33.4.2 Impact Separators

Such separation devices cause a sudden change in the direction of the gas stream and thereby separate the particles by inertia, impingement on a target or centrifugal force. Two types of equipment

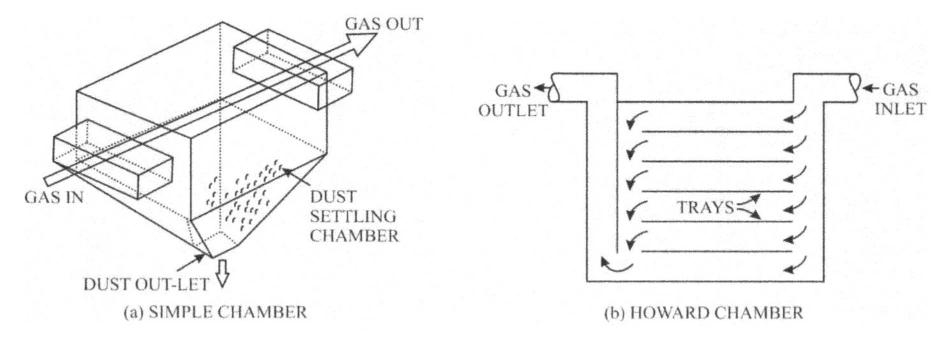

FIGURE 33.1 Gravity Settling Chamber

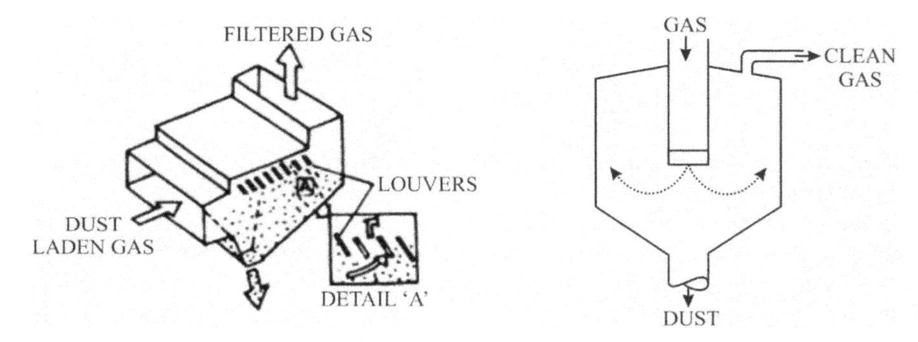

FIGURE 33.2 Louver Separator and Dust Trap Separator

utilizing this principle are inertial or impact separators and cyclonic separators. Inertial or impact separators employ the incremental changes in the direction of the carrier gas stream to exert greater inertial effects of the dispersoid, while cyclonic separators produce a continuous centrifugal force as a means of exerting the greater inertial effects of the dispersoid. Common impact separators include baffle separators, louver collectors and dust traps.

In a baffle separator, the gas stream is made to follow a tortuous flow path obtained by inserting staggered plates into the path of the gas stream. The device removes particles greater than 20 μm in size and is widely used for particulate removal in power plants and rotary kilns. A louver separator mainly consists of several blades set at an angle to the flow path of the gas stream. The blades are set to force a quick, sharp change in the direction of a large portion of the gas stream. At the point of this sharp change in direction, the dust particles are separated out and collected in the bed of the separator. This device is suitable for removing particles larger than 30 μm. In a dust trap, the dust-laden gas is introduced into a chamber through a central pipe and made to undergo a change in direction by 180°. The dust falls into the conical chamber.

33.4.3 Cyclone Separators

A cyclone separator depends upon centrifugal force for its action. It is a structure, without moving parts, where the velocity of an inlet gas stream is transformed into a confined vortex from which the centrifugal forces tend to drive the suspended particles to the wall of the cyclone body.

The centrifugal force on particles in a spinning stream is much greater than gravity, hence cyclones are effective in the removal of much smaller particles than gravitational settling chambers. They also require much less space to handle the same gas volume. A simple reverse-flow cyclone

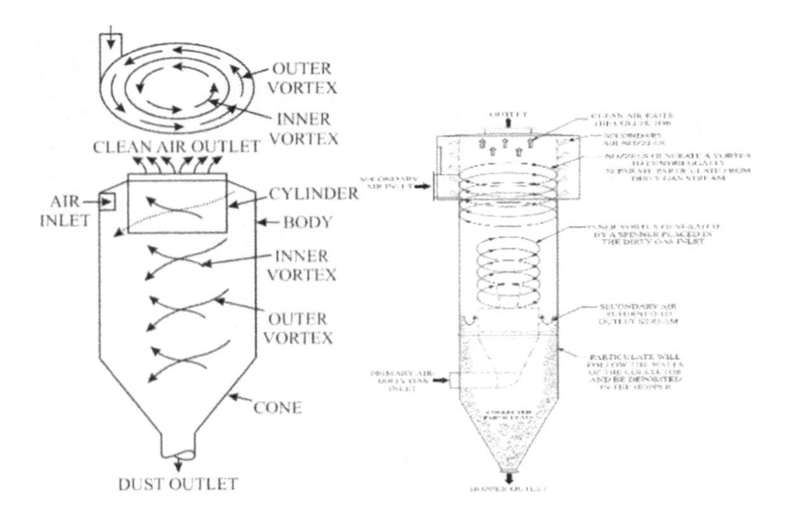

FIGURE 33.3 Reverse-Flow Cyclone

consists of a vertically placed cylinder with an inverted cone attached at its bottom and fitted with a tangential inlet near the top.

The outlet pipe for the purified gas is a central cylindrical pipe at the top, which is extended into the cylinder of the cyclone to prevent short-circuiting of the gas from the inlet to the outlet. The cyclone has an outlet at the bottom of the cone for discharging the separated particles. Upon entering the cyclone cylinder tangentially at its top, the particle-laden gas receives a rotating motion. The outer vertex so formed develops a centrifugal force that throws the particles radially towards the wall. The gas spirals downwards to the bottom of the cone, where the gas flow reverses to form an inner vortex that leaves through the outlet pipe situated at the top of the cyclone. Due to inertia, the dust particles tend to concentrate on the surface of the cyclone wall from where they are led to the receiver. During cyclonic separation, the carrier gas rotational velocity may exceed several times the average inlet gas velocity. Cyclones are cheaper in cost and best suited to dry dust particles of size 10–40 μm. The cyclones can handle a wide range of physical and chemical conditions of operations compared with other collection equipment.

33.4.4 Fabric Filters

A fabric filter consists of a tubular bag or an envelope suspended or mounted with open ends in such a way that when dirty gas passes through it, the particulate matter is arrested on the inside fabric surface of the bag.

The whole system, generally known as a baghouse, consists of a number of these bags and is constructed with several compartments so that one compartment may be isolated for cleaning as and when needed while the other compartments are operating. Cleaning is accomplished by shaking at fixed intervals, causing the finer particles collected on the bag fabric surface to fall into the hopper. The gas entering through the inlet pipe strikes a baffle plate which causes larger particles to fall into the hopper due to gravity. The gas then flows upwards into the cloth fabric tubes and then outwards through the fabric.

33.4.5 Electrostatic Precipitators

Electrostatic precipitation is a physical process by which particles suspended in a gas stream are charged electrically and under the influence of an electric field, separated from the gas stream. They are used to remove aerosol from large volumes of gas.

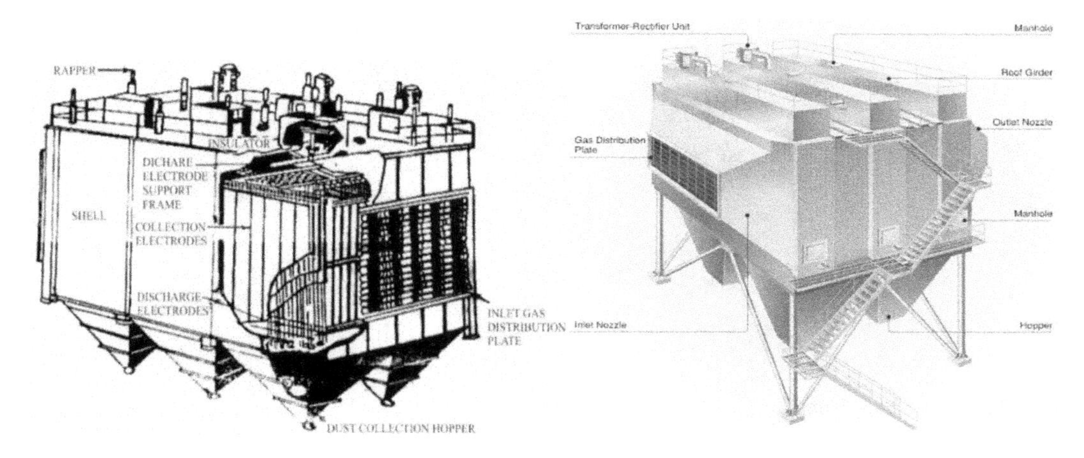

FIGURE 33.4 Baghouse Filter and Electrostatic Precipitator

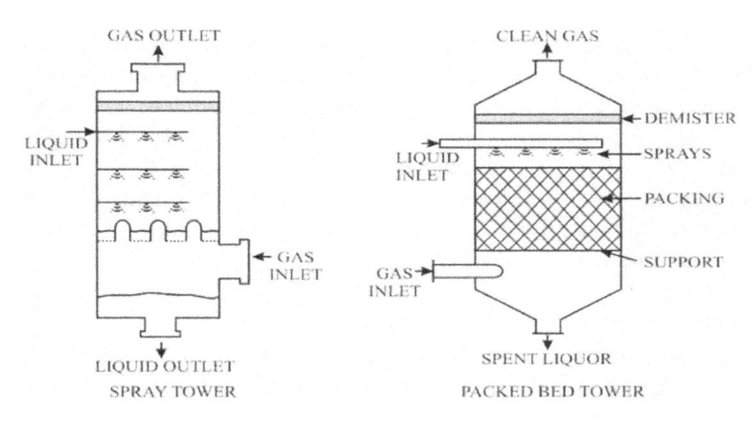

FIGURE 33.5 Spray Tower and Packed Bed Tower

33.4.6 Scrubbers or Wet Collectors

A scrubber is a device that utilizes a liquid to assist in the removal of particulates from the carrier gas stream. They are mixed phases of gas and liquid. The object of a scrubber is to transfer suspended particulate matter in the gas to the scrubbing liquid, which can be readily removed by the gas-cleaning device.

Spray Towers

A spray tower consists of a rectangular or round tower with a number of spray nozzles. The carrier gas flows upwards and particle collection results because of inertial impaction and interception on the droplets. Only moderate contact between the phases is affected, hence these are used to remove coarse dust (> 1–2 μm) where high efficiency is not required. They are therefore used as primary cleaners in treating blast furnace gas and for fly ash and cinder removal.

Packed Bed Towers

These use fibre glass (fine glass) packing or other packing. The polluted gas stream moves upwards and comes into contact with the scrubbing liquid stream, which moves downwards over the packing in a film. The particles are captured by inertial impaction while the gas stream takes a curved path through the pore space of the packing.

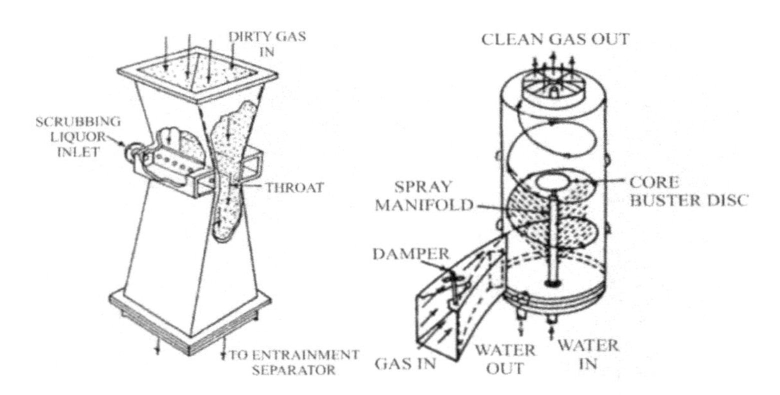

FIGURE 33.6 Venturi and Cyclone Scrubber

Venturi Scrubber

In a venturi scrubber, dust-laden gas is accelerated to a high velocity (60 to 100 m/s) while passing through the converging section and approaching the throat section. Water sprays are introduced just ahead of the venturi throat. The high-velocity gas impinges (impacts) upon the liquid stream in the throat, atomizing the liquid into a large number of fine droplets. The high-differential velocity between the gas and the atomized droplets causes liquor droplets and dust particles to collide, impact and agglomerate. After the particulates have been trapped within the liquor droplets, the resulting agglomerates are readily removed from the gas stream in the separator. Such scrubbers are used in removing mists and dust from the gases of pulp and paper industries, steel and non-ferrous metal industries and chemical industries.

Cyclone Scrubber

This is a modified form of dry cyclone in which the scrubbing liquor is introduced into the gas path by being sprayed through nozzles mounted on the walls of the scrubber or mounted on a pipe in the centre of the vessel. A spinning motion is initiated by introducing the flue gas tangentially or having the scrubbing liquid introduced (or sprayed) tangentially inward from the wall. The sprays assist in the collection of the dispersoid and tend to prevent re-entrainment. The scrubbing liquor with the collected particles runs down the walls to the scrubber bottom and out, while the cleaned gas leaves through the top.

Mechanical Scrubbers

These are high emerging scrubbers. Sprays are generated by mechanical means such as a rotary paddle, cage, drum or disc, or spray nozzles. The liquid is mechanically sheared to break into fine droplets for collection by inertial impaction between droplets and dust particles. Sprays are normally generated at right angles to the direction of the flue gas flow. In the mechanical-centrifugal collector (also called a dynamic wet scrubber), liquor is sprayed at the inlet to enhance collection efficiency. The collection mechanism is the impingement of dust particles on the rotating blades. In high-pressure spray scrubbers, the scrubbing liquid is sprayed at high pressure to increase the number and size of droplets and thereby increase the collection efficiency. Such scrubbers have a high initial cost and a high operating cost. They require a large quantity of water and considerable maintenance.

33.5 CONTROL OF GASEOUS POLLUTANTS

The emission of gaseous pollutants can be controlled by absorption, adsorption and combustion.

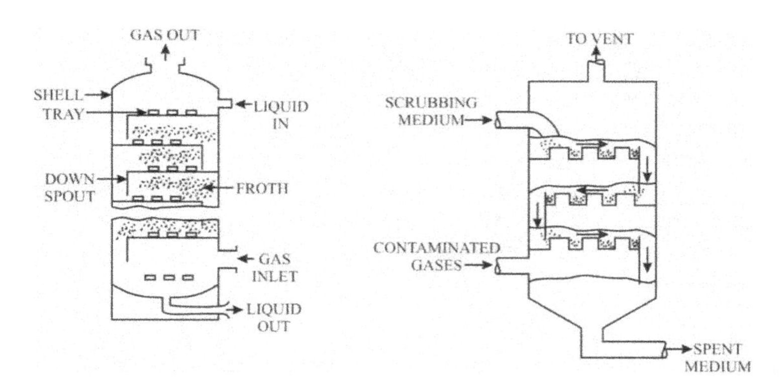

FIGURE 33.7 Plate Tower and Bubble Plate Tower

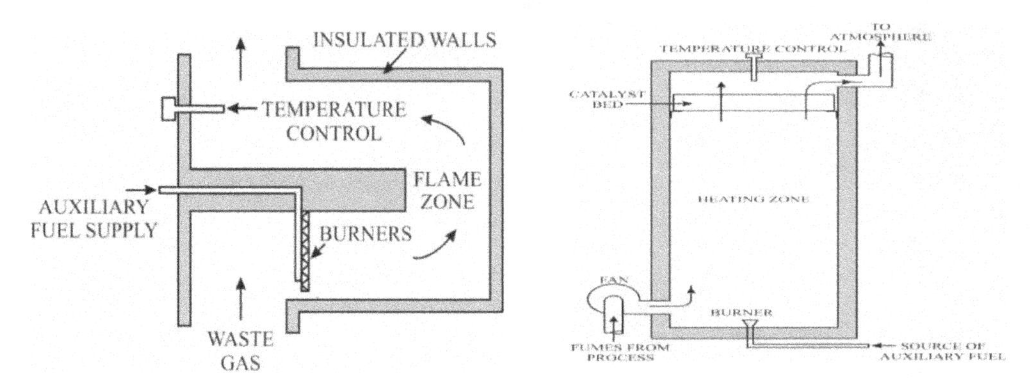

FIGURE 33.8 Direct Flame and Catalytic Incinerator

33.5.1 Absorption

Absorption involves the transfer of pollutants from a gas phase to a liquid phase across an interface in response to a concentration gradient, with the concentration decreasing in the direction of mass transfer. The absorption of a gaseous contaminant by a liquid occurs because the liquid is not saturated with the contaminant at the conditions existing in the absorber. The difference between the concentrations of the contaminant in the liquid and the actual concentration provides the driving force for the absorption. Hence, the more soluble a contaminant is in the liquid phase, the greater the overall efficiency.

33.5.2 Adsorption

An alternative to absorption by liquids is the adsorption of air pollutants on solids. Commonly used absorbers include activated carbon, molecular sieves, silica gel, activated alumina, etc.

33.5.3 Combustion or Incineration

This is used when the pollutants in the gas stream are oxidizable to an inert gas. Direct flame incineration is a control technique for combustible organic air pollutants. It is accomplished in the presence of a flame and sufficient oxygen by raising the temperature of the gases above their ignition temperature and then maintaining the temperature until the oxidation reactions are complete.

In catalytic incineration, a mixture of dilute organic gases and oxygen is exposed to a catalytic surface at a temperature high enough for the oxidation to occur and for a length of time sufficient for the oxidation to go to completion. Catalysts are usually solids that are neither reactants nor products of a reaction yet alter the rate of chemical reactions. The effect of the catalyst is to reduce the temperature required to oxidize the organic compounds, and hence the inlet gases need not be heated to ignition temperature. This requires less fuel than would be needed for direct flame incineration. The catalytic oxidation reaction occurs as a very rapid sequence of steps at the catalytic surface. Hence, the overall residence time required is much less than that in a direct flame incinerator. Due to this, the size of the unit is reduced.

Discussion Questions

1. Name and describe the four major layers of the atmosphere.
2. What is the variation of air temperature and pressure with height? Explain it with a diagram.
3. What are the causes and effects of air pollution?
4. What are primary and secondary pollutants? Give examples.
5. Define dust, smoke, mist, fumes and vapours and their characteristics.
6. What are the principal sources of SOx, NOx, and CO emissions into the atmosphere?
7. What two chemical compounds are necessary ingredients for producing photochemical smog?
8. Describe the meteorological parameters that influence air pollution dispersion.
9. What are the different types of atmospheric stability? Explain each with a diagram.
10. Explain temperature inversions and their types.
11. Describe the various types of plume behaviour occurring in the atmosphere with diagrams.
12. Write the Gaussian dispersion equation. What are its applications?
13. What is the reason for calculating the plume rise and effective height of a stack? Give any two formulas to calculate it.
14. What are the air control methods? Explain.
15. What equipment or devices are used for the control of particulate pollutants? Explain each with the help of a diagram.

34 Introduction to Solid Waste

Any non-liquid material that is discarded as useless (zero monetary value) and unwanted is considered to be solid waste. Municipal solid waste (MSW) is always generated in cities, towns and villages. At first glance, the disposal of solid waste may appear to be a simple and mundane problem. However, in the modern world, with increasing populations and increasing use of materials, the disposal of solid waste is becoming a more complex problem.

Solid waste, also known as dry refuse, includes house, trade and street refuse, and is practically in a dry state. The removal and disposal of refuse or solid waste is a very important aspect of environmental sanitation. Solid waste consists of garbage, ashes, rubbish and dust-like materials. According to EPA regulations, solid waste means any garbage or refuse, sludge from water and wastewater treatment plants or air pollution control facilities, and any other discarded material. In a more practical sense, solid waste is defined as waste produced by a community.

34.1 TYPES OF SOLID WASTE

Solid waste or dry refuse can be organic or combustible matter, and inorganic, mineral or non-combustible matter. Municipal solid waste can be classified into five main categories.

34.1.1 Recyclable Material

As the name implies, this includes things that can be recycled or reused. In Indian conditions, pickers usually segregate this type of waste from dumps and waste containers. Paper, cardboard and bottles are usually separated by householders and sold to the recyclers. Though it was not planned by the local government, a large amount of waste is recycled. This is one of the reasons why in developing countries such as India, the per capita waste ending up at landfill is 20–25% of that in countries like the US and Canada. However, single-use plastic is causing significant problems. In addition to littering the streets, it plugs storm sewer inlets and sewer manholes, making drainage worse during the rainfall season.

It should be noted that recycling is gaining momentum in developed countries. Homeowners segregate different types of wastes and put them in separate containers usually supplied by the municipality. Recyclable materials are placed in blue and grey bins and organic materials in green bins. General waste and recycled waste is collected by different trucks. Recycling has become so common that in many cities in the US and Canada, the frequency of collection of recycling waste is weekly, while garbage pickup has been reduced to biweekly.

34.1.2 Biodegradable Waste

Paper (which can also be recycled), food and kitchen waste, and green waste (flower, vegetables, fruits, leaves, etc.) fall into this category. A large portion of this waste can be composted, thus reducing the amount ending up at the landfill. At the landfill, this type of waste biodegrades under anaerobic conditions and produce methane, which can be used as an energy source.

Inert waste comprises dirt, rocks, construction and demolition waste. It is mostly inorganic and hence not biodegradable. Composite wastes include tetra packs and waste plastics such as toys and clothing.

DOI: 10.1201/9781003231264-34

34.1.3 Household Waste

Household waste consists of vegetable and animal waste matter, ashes, cinders, rubbish, and debris from cleaning and the demolition of structures arising from residential units.

34.1.4 Domestic Hazardous Waste

Being hazardous, this type of waste should not be part of municipal solid waste. However, due to ignorance, negligence and lack of regulation, especially in developing countries, it mostly ends up at the sanitary landfill. Domestic hazardous waste includes electronic articles, medication, light bulbs, fluorescent tubes, shoe polish, chemicals, paints, batteries, fertilizers and pesticide containers, spray cans, etc.

34.1.5 Street Waste

Street waste consists of empty packets and bottles, empty matches and cigarette boxes, fruit peels, tree leaves, street sweepings, etc.

34.1.6 Trade/Industrial wastes

This consists of solid wastes arising from factories, commercial and business centres, slaughterhouses, etc. Waste from industrial processes does not form part of municipal solid waste since many such wastes may be hazardous.

34.2 MUNICIPAL SOLID WASTE

The term municipal solid waste (MSW) is generally used to describe most of the non-hazardous solid waste from a city, town or village that requires routine or periodic collation and transport to a processing or disposal site. Sources of municipal waste include private homes, commercial establishments, institutions and industrial facilities. However, MSW does not include industrial process waste, construction waste, demolition waste or agricultural waste. The classification of different types of non-hazardous MSWs is illustrated in **Figure 34.1.**

34.2.1 Garbage

Garbage consists of all sorts of putrescible organic waste from kitchens, hotels and restaurants in the form of waste food articles, vegetables and fruit peelings. It is organic and decomposes quickly. It normally weighs from 450 to 900 kg/m^3. It should be handled carefully because flies, insects and rats can breed in it.

34.2.2 Rubbish

Rubbish consists of all non-putrescible wastes, excluding ashes. Common items that fall under this category are rags, paper pieces, paper packets, glass and plastic bottles, broken pieces of glass, broken crockery, stationery items and cardboard.

34.2.3 Trash

Trash includes bulky waste materials that generally require special handling and is generally not collected on a routine basis. In developing countries such as India, trash is minimal since articles

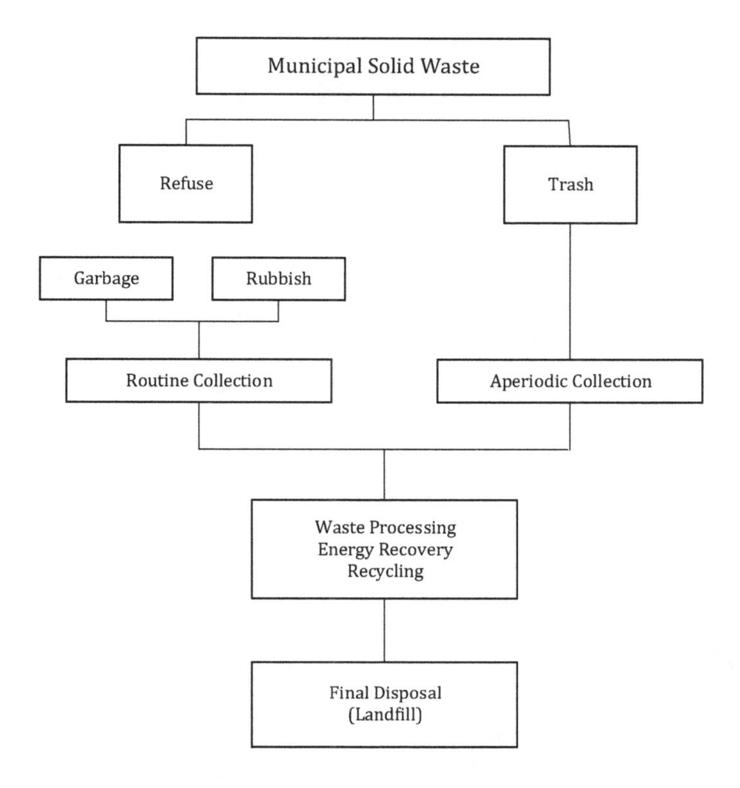

FIGURE 34.1 General Classification of MSW

thrown by one person are often reused by another person or repaired. However, the quantity of trash is significant in developed countries since repairs are expensive and new items are relatively cheaper. Another way of categorizing waste is on the chemical nature of the waste.

34.2.4 Organic Waste

This includes dry animal and vegetable refuse, cow dung, the excreta of birds, tree leaves, sticks, plastic bottles, paper waste and rags. This waste is subject to decay with time and evolves a highly offensive odour and gases that are detrimental to health. It is important to point out that most of the plastic waste biodegrades at a very slow rate.

34.2.5 Inorganic Waste

This consists of non-combustible materials such as grit, dust, mud, metal pieces, metal containers, broken glass, crockery, tiles and waste building materials. It is not subject to decay and is therefore not harmful to public health.

34.2.6 Hazardous Waste

This includes waste that poses a substantial danger to human, animal or plant life immediately or over a period of time. Such waste can be radioactive substances, toxic chemicals, biological wastes and flammable or explosive wastes.

34.3 SOURCES OF SOLID WASTES

Knowledge of the types and sources of solid wastes, along with their composition and the quantity of their generation, is essential to the engineering management of solid wastes.

34.3.1 Municipal Solid Waste

Municipal solid waste (MSW), also called trash, consists of everyday items such as product packaging, yard trimmings, furniture, clothing, bottles and cans, food, newspapers, appliances, electronics and batteries. Sources of MSW include residential waste (including waste from multi-family housing) and waste from commercial and institutional locations, such as businesses, schools and hospitals. **Table 34.1** shows the general sources and types of MSW.

34.3.2 Hazardous Wastes

Hazardous wastes are generated in many industrial activities. The type and characteristics of these wastes vary with respect to the specific industrial process. The handling and disposal of such wastes is also a critical issue in solid waste management. Hazardous waste is discussed in more detail in the next chapter.

34.4 QUANTITY AND COMPOSITION OF SOLID WASTES

The quantity and general composition of solid wastes generated is of critical importance in the systems of solid waste management. Unfortunately, reliable quantity and composition data is difficult to obtain because solid wastes vary from place to place and from season to season. The quantity of garbage depends upon the types of activities, food habits and standard of living. It also depends upon whether the town is residential, commercial or industrial. For example, per capita waste generation in rural areas may be significantly less than per capita water production in an urban set-up. **Table 34.2** compares the amounts and types of waste generated in a typical city of India versus a city in the US. One drastic difference is that solid waste in Indian conditions exists more as garbage, while in the US, it is highest in the form of rubbish. Typically, the MSW generation rate in an urban scenario ranges from as low as 0.5 kg to as high as 2.5 kg per capita per day. For example, on average, MSW in the US is 2 kg/c.d, which is twice as much as in Western Europe. In comparison, per capita waste generation in an Indian city is 0.4–0.5 kg/c.d.

TABLE 34.1
Sources and Types of Municipal Solid Waste

Source of MSW	Type of Solid Waste
Residential	Food waste, food containers and packets, cans, bottles, papers and newspapers, clothes, garden waste, e-waste, furniture waste.
Commercial centre (office, small shop)	Various types of papers and boxes, food waste, food containers and packets, cans, bottles.
Institutional (school, hospital)	Office waste, food waste, garden waste, furniture waste.
Industry	Office waste, cafeteria waste, processing waste.
City centre	Various types of garden waste, construction site waste, public waste.

Source: Franklin Association (1999)

TABLE 34.2
Composition and Properties of Refuse

Constituent	Average Composition % on Mass Basis	
	Typical Indian city	Typical US city
Garbage	55	25
Rubbish	15	55
Ashes	10	10
Dust silt sand	20	10
Unit weight, kg/m^3	350–600	100–300
Calorific value MJ/kg	6	15

Factors that affect the quantity of MSW generated include geographic location, season, collection frequency, use of equipment for onsite handling (e.g., kitchen waste grinders), population characteristics, the extent of salvaging and recycling and regulation.

34.4.1 Properties of Solid Wastes

The average composition of refuse by weight in India is about 25% cinder, 27% fine dust, 15% ashes, 4% empty tins and cans, 14% putrescible matter, 2% glass and crockery, 2% rags, 1% bone and 8% miscellaneous matter. Organic wastes make up, generally 55%, while inorganic wastes make up 45%.

34.4.2 Physical Composition

Apart from the identification of the individual components that make up MSW, analysis of particle size, moisture content and density of the solid waste are the physical compositional parameters that are normally analyzed.

34.4.3 Chemical Composition

Energy recovery options and the alternate processing of wastes largely depend on the chemical composition. The typical analyses are:

- Proximate analysis: moisture, volatile matter, ash and fixed carbon.
- Ultimate analysis: percent C, H, O, N, S and ash.
- Heating value: calorific value (kCal/kg).

34.5 WASTE GENERATION

The quantification of MSW is important as it is used to check compliance with waste regulations, choose equipment, and make collection and management decisions, and is critical to the proper design of a facility.

Two approaches are commonly used to quantify waste generation: materials balance methodology and load count analysis. The application of either approach requires that waste be categorized using the mass balance.

$$\textit{Generated waste} = \textit{disposed waste} + \textit{recycled waste} + \textit{diverted waste}$$

Disposal waste = solid waste collected and taken to disposal (landfill or incinerator).

Recycled waste = solid waste materials separated from recycling.

Diverted waste = solid waste materials not processed through municipal channels.

Example Problem 34.1

Estimate the daily solid waste generation in a city with a population of 0.5 million, assuming a waste generation rate of 0.25 kg/c.d. If a waste collection truck can compact waste to 400 kg/m^3, find the daily number of trips it will take to the landfill knowing that the volume capacity of the truck is 8.5 m^3/load.

SOLUTION:

$$M_{waste} = \frac{0.25\ kg}{person.d} \times 500000\ people \times \frac{t}{1000\ kg} = 125\ t/d$$

$$V_{waste} = \frac{M}{\rho} = \frac{125\ t}{d} \times \frac{m^3}{400\ kg} \times \frac{1000\ kg}{t} = 312.5\ m^3/d$$

$$\# = \frac{312.5\ m^3}{d} \times \frac{load}{8.5\ m^3} = 36.7 = 37\ \underline{loads/d}$$

Discussion Questions

1. Observe and record the solid waste components in your locality and compare them with the available data of major cities.
2. Give a brief definition of municipal solid waste (MSW). Define the terms refuse, garbage, rubbish and trash.
3. Compare the composition of MSW in an urban set-up in India with that of an urban set-up in the US.
4. What are the factors that affect the generation of solid wastes?
5. Identify the major sources of solid wastes and list the approximate characteristics.
6. List and contrast the approximate composition of solid wastes in rural, urban and industrial set-ups.
7. What are hazardous wastes?
8. What are the major parameters for assessing the physical and chemical composition of solid wastes?
9. Which method of solid waste treatment is better from an environmental perspective, landfilling or incineration?
10. What type of solid waste is generated in the town or city where you attend school?

35 Solid Waste Management

The overall objective of solid waste management is to minimize the nuisance or adverse environmental effects caused by the indiscriminate disposal of solid wastes. The decisions on choices of solid waste management require careful consideration of the following.

1. Materials flow in the society, which indicates how and where the solid wastes are generated.
2. Effective reduction in raw materials usage, which provides opportunities to reduce the generation of waste.
3. Reduction in solid waste quantities by reducing the amount of raw materials used to manufacture a product by increasing the useful life of the product.
4. Reusing solid waste materials and effectively recovering useful materials from solid wastes.
5. Possible energy recovery using suitable alternative energy-recovery technologies.
6. Coordination of day-to-day solid waste management activities.

35.1 COMMON METHODS OF SOLID WASTE DISPOSAL

The commonly practised technologies for solid waste management (SWM) can be grouped under three major categories: bioprocessing, thermal processing and sanitary landfill. The bioprocessing method includes aerobic and anaerobic composting. Thermal methods include incineration and pyrolysis. Sanitary landfills are generally used for the disposal of the final rejects coming out of the biological and thermal waste processing units. However, anaerobic biological oxidation happens in a sanitary landfill and methane gas is produced as a by-product. Other chemical reactions happen to varying degrees, and many contaminants from the waste end up in the leachate collected at the bottom of the sanitary landfill. A sanitary landfill is not just a waste dump but a properly designed and operated site for handling municipal solid waste.

35.2 GENERATION TO DISPOSAL

Six functional elements, from the point of generation to the final disposal of the solid waste, include waste generation; on-site handling, storage and processing; collection; transfer and transport; processing and recovery; and disposal. The interrelationship of these functional elements comprising a solid waste management system is illustrated in **Figure 35.1**.

35.2.1 Waste Generation

Waste generation involves activities in which materials are identified as no longer valuable and are either thrown away or gathered for disposal. The types and physical and chemical characteristics of solid wastes, as explained in the previous chapter, are of importance in the decision-making on solid wastes management systems.

35.2.2 Collection and Removal of Solid Wastes

The frequency of collection of solid waste depends upon the quantity of the waste and the season. In Indian conditions, waste is generally collected from individual houses in small containers or cans kept outside the premises of the house, from where it is removed daily by sweepers. Public dustbins are

DOI: 10.1201/9781003231264-35

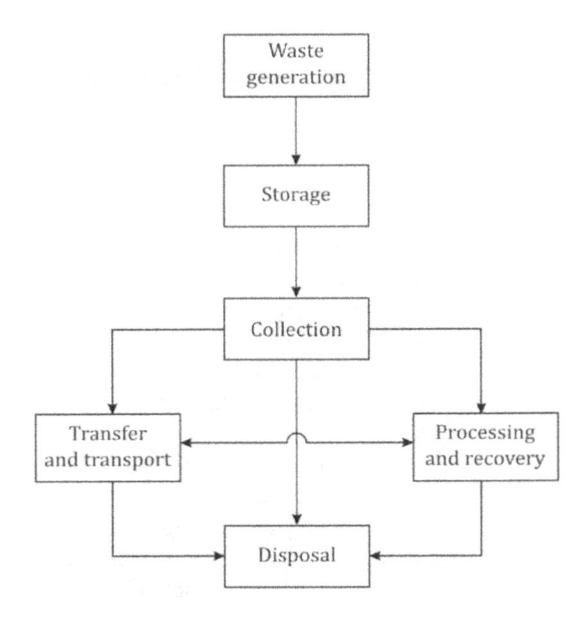

FIGURE 35.1 Interrelationship of Functional Elements

sometimes provided by the municipalities or local boards at convenient places on the sides of roads. Dry inert waste that has fallen on the public streets and roads, along with road sweepings, are usually collected once or twice a day by sweepers employed by the local authority. A portable galvanized iron receptacle with a closely fitting lid and a capacity of 20–100 L is generally used for this purpose. The containers or cans used for waste collection should be clean, without any dirt; otherwise, fresh refuse gets seeded with putrefactive organisms and starts giving off a foul smell. The frequency of refuse collection is kept such that the refuse may not start causing a nuisance by odour and fly breeding. The collection of refuse from business areas should be done during non-working hours.

In developed countries, proper containers are used, and waste is collected weekly or biweekly by municipal trucks specifically designed for solid waste collection.

35.2.3 Transport of Wastes

The waste collected in public dust bins located by the sides of roads is transported to the disposal site by means of auto rickshaws, trailers and trucks.

Auto Rickshaws

Auto rickshaws have three or four wheels and covered bodies. Since their capacity is limited to 1 to 4 tons, they are used only for narrow localities where other heavy vehicles cannot go.

Trailers

Trailers have a slightly larger capacity (2–3 tons). They are also used for localities where trucks cannot go. The loading of trailers is done manually; however, as they are of the tilting-tipping type, their unloading is done automatically with the help of hydraulically operated jacks.

Trucks

Trucks have a large capacity (5–10 tons). They are generally of the tilting-tipping type, so unloading is automatic. Special types of trucks capable of bodily lifting covered skip boxes (in place of ordinary dust bins) are now available and should be used to avoid the nuisance of flies. These trucks are of the kind used in western countries, though smaller in size.

Vehicles employed to transport refuse should be of such pattern and design that collected garbage does not fall once again onto the road during transport. The transport vehicle should be strong, durable and watertight. They should be made of steel with a smooth interior surface and round edges and corners so that they can be kept clean.

The wastes so transported may be temporarily stored in transfer stations before being finally loaded into transport vehicles that go to the disposal stations.

Example Problem 35.1

Assume that each pickup location produces 15 kg of waste per week and the packer vehicle can compact the waste to a density of 450 kg/m^3. Estimate the number of customers that have their waste collected prior to making a trip to the landfill. Assume the truck capacity is 12 m^3.

SOLUTION:

$$Customers\ served\ per\ trip = 12\ m^3 \times \frac{450\ kg}{m^3} \times \frac{Customer}{15\ kg} = 360\ customers$$

Example Problem 35.2

Assume that a garbage collection truck and its crew service customers at a rate of 4 customers/5.0 minutes and the actual time spent is 5.5 hours. Out of a total of 8.0 working hours, the rest of the time is used to make trips to the landfill site and for breaks. Estimate the number of customers that can be served per day.

SOLUTION:

$$= 5.5\ h \times \frac{4\ customers}{5\ min} \times \frac{60\ min}{h} = 264 = 260\ customers$$

35.3 DISPOSAL OF SOLID WASTE

Depending on the location, size of the community and availability of land, municipal solid wastes can be finally disposed of by various methods. A brief discussion of such methods is provided in the following paragraphs.

35.3.1 Controlled Tipping

Controlled tipping is useful where an adequate site for redevelopment is available. It consists of tipping the refuse in hollows to a depth of 1 to 2 m. While tipping, coarse material is tipped at the bottom while fine material is tipped on the top. These tips are covered with soil to provide a seal under which bacterial decomposition can take place. At the end of a roughly 12-month period, the decomposition is complete, and the tip settles down to a height of 30 cm only. Normally, an area of 0.3 to 0.5 m^2 per capita per year is required.

35.3.2 Landfilling

Landfilling is very common in urban centres. Garbage is dumped into low-lying areas or depressions available nearby. Dumping is done in layers of 1 to 2 m, and each layer is covered by 0.2 m deep of good earth. After two to three weeks, a second layer is laid. If the dry refuse is loosely packed it may give rise to health hazards. Hence, each layer should be compacted by the movement of dumping vehicles to aid in its settlement before starting to fill the second layer. Recently, due to environmental problems resulting from dumping wastes, new landfill sites are properly engineered and operated, as is the practice in developed countries.

35.3.3 Trenching

Trenching is generally adopted when low-lying areas are not available. Trenches 4 to 10 m long, 2 to 3 m wide and 1 to 2 m deep are excavated with a clear spacing of 2 m. They are then filled with refuse in 15 cm thick layers. On the top of each layer, a 5 cm thick layer of faecal sludge or animal dung is spread in semi-liquid form. On the top layer, protruding 30 cm above the ground surface, a 10 cm layer of good earth or other non-combustible material is spread to act as a seal so that flies do not get access and wind does not blow the refuse off. The dumped garbage is converted into a type of compost by fermentation carried out by anaerobic bacteria within a period of six months.

35.3.4 Dumping Into the Sea

Solid waste can also be disposed of by barging out a reasonable distance (15 to 20 km) into the sea. This is necessary to prevent the shores from refuse nuisance because sea waves can carry the refuse back to shore. The depth at the disposal point should be no less than 30 m. This method of solid waste disposal is limited to coastal towns and cities. Serious disadvantages of this method include:

1. Bulky and light matter in the refuse may float, spread out and return to shore during high tides.
2. During stormy weather and monsoons, it is not possible to send barges out into the sea.
3. In spite of best care, some portion of refuse may return to shore and spoil it.

35.3.5 Pulverization

With pulverization, dry refuse is pulverized into powder form without changing its chemical form. The powder can be used either as poor-quality manure or disposed of by landfilling.

35.3.6 Incineration

Incineration consists of burning refuse in an incinerator plant. It is commonly used to dispose of garbage from hospitals and industrial plants. Before incineration, non-combustible and inert material such as earth, broken glass, chinaware and metals are separated to reduce the load on the hearth. The by-product of this method is ash and clinker, which can be easily disposed of by landfilling. The heat generated by burning dry refuse can be utilized for raising steam power. The following points should be carefully observed during incineration:

1. The refuse charging should be thorough, rapid and as continuous as possible.
2. Each batch of refuse entering the furnace should be well-mixed.
3. Auxiliary burners are usually installed above the refuse to ignite it and establish a draft at the beginning of the cycle. This is even more necessary when the moisture content of the air is high.

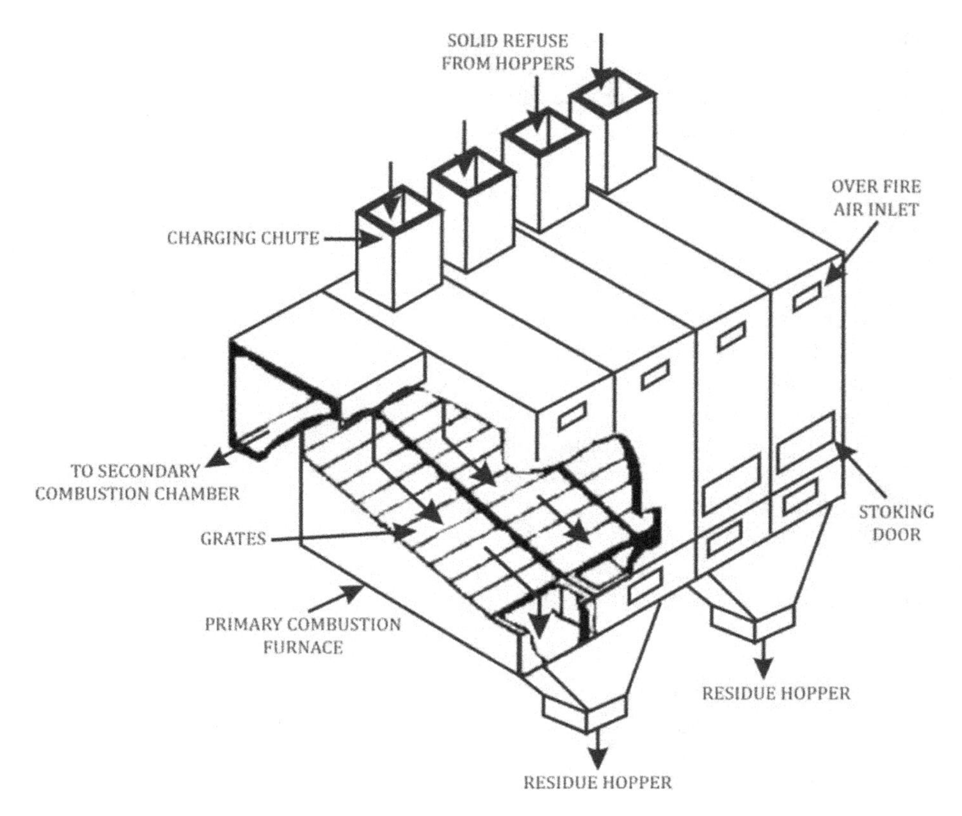

FIGURE 35.2 Multicell Incineration Furnace

4. The minimum temperature in the combustion chamber should be sufficient (> 670°C) for all organic matter to be incinerated and foul-smelling gases to be oxidized.
5. Afterburners are sometimes required, together with particulate removal devices such as settling chambers or scrubbers.

Furnaces can be vertical, circular, rotary, multicell and rectangular in their design. In a vertical circular furnace, the refuse is charged through a door in the ceiling and drops into a central cone grate surrounded by a circular grate. Primary combustion is supported by underfired air. In a rotary kiln furnace, the wastes are partially burnt in a rectangular furnace and then fed, via grates, to a rotating kiln. The rotary action exposes the unburnt material for combustion. Final combustion takes place in the chamber after the kiln discharge point. A multicell furnace has cells side by side. Each cell has grates for moving the refuse across them. Several cells have a common combustion chamber and residue hopper. In a rectangular furnace unit, two or more grates are arranged in tiers so that the refuse is agitated as it falls from one level to the next. Secondary combustion is usually employed. A pathological waste incinerator handles the organic wastes of human or animal origin and crematory furnaces. **Figure 35.2** shows the details of a multicell incineration furnace.

35.3.7 Composting

Composting is a method by which putrescible organic matter in solid waste is digested aerobically or anaerobically and converted into humus and stable mineral compounds. It is a hygienic method that converts refuse into manure that can be used to condition soil. Compost is widely used as manure and as a soil conditioner. Since a significant portion of waste is converted into gases during composting, the volume of waste is very much reduced. In addition, biological oxidation destroys many pathogens such that compost can be safely handled. In India, the faecal sludge of the conservancy system is normally also disposed of at the same timer as the refuse, producing valuable manure.

Composting by Trenching

In composting by trenching, trenches 4 to 10 in long, 2 to 3 m wide and 0.7 to 1 m deep are excavated with a clear spacing of 2 m. They are then filled with waste in layers of 15 cm. On the top of each layer, a 5 cm thick sandwiching layer of animal dung is spread in semi-liquid form. On the top layer, protruding 0.3 m above the original ground surface, a 10 cm layer of good earth is spread so that flies do not get access to the refuse, and so the refuse does not get blown off by the wind. Within two to three days, intensive biological action starts to biodegrade the organic matter present in the refuse. Considerable heat is generated in the process, and the temperature of the composting mass rises to about 75°C. This further prevents the breeding of flies. The refuse stabilizes in a 4–5 month period and changes into a brown, odourless, innocuous powdery form known as humus, which has manure value because of its nitrogen content. The stabilized mass is removed from the trenches, passed through a 12.5 mm sieve to exclude coarse inert materials such as stones, brickbats, broken stone, etc. The sieved material is then sold out as manure.

Open Windrow Composting

A windrow is a long pile of prepared organic waste. Before composting, a large proportion of mineral matter such as dust, stone and broken glass pieces are removed. The organic waste is then dumped on the ground in the form of 0.6 to 1 m high, 6 m long and 1 to 2 m wide piles at about 60% moisture content. The size of the windrow in developed countries is quite large. A typical windrow is 50 m long, 2 m high and 3 m wide at the base. The pile is then covered with faecal sludge, cow dung and cattle urine to add more organic matter, and as the starter of biological action. Due to aerobic biodegradation, heat starts to develop and the temperature in the pile rises to about 75°C. Hence, the microbial reaction shifts from mesophilic to thermophilic. After this, the pile is turned for cooling and aeration to avoid anaerobic reactions. The process of turning, cooling and aerating is repeated. The turning frequency varies with the moisture content and other factors. The complete process may take about 4–6 weeks, after which the compost is ready for use as manure when the temperature falls considerably.

Mechanical Composting

The open windrow method of composting is a very laborious and time-consuming process, and it requires a large area of land, which may not be available in big cities. These difficulties are overcome by adopting mechanical compositing in which the process of stabilization is expedited by mechanical devices that turn the compost. The mechanical method stabilizes the refuse compost within just three to six days. The operations involved in a large-scale composting plant are the reception of refuse, segregation, shredding or pulverizing, stabilization, and marketing the humus. The refuse is received at the plant site in quantities of 2 to 6 tons per vehicle. Hence, the plant site must have a storage capacity of about 25 to 50% of the total daily arrival before it can be segregated and shredded. Segregation is done by hand-picking in smaller plants and mechanical devices in large plants, and removes paper, rags, non-ferrous metals and large objects. Ferrous metals are removed by magnetic separators. Finer materials such as ash and particles of garbage are removed by passing the refuse over shaker screens. The remaining refuse is then shredded and pulverized mechanically. The processing of refuse by mechanical composting is shown in **Figure 35.3**. The prepared refuse is then decomposed or stabilized under controlled conditions of temperature and moisture content.

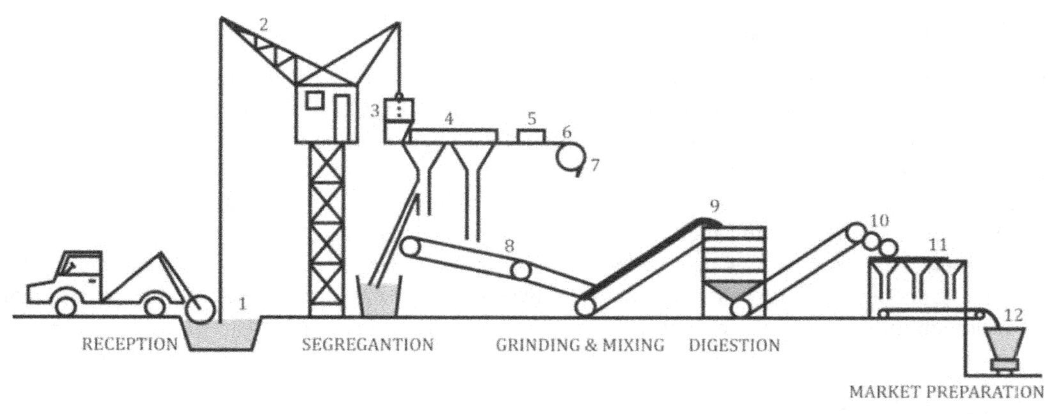

1. RECEIVING PITS	2. CRANE	3. HOPPER
4. ROTARY/VIBRATING SCREENS	5. MAGNETIC SEPARATOR	6. SORTING BELT/TABLE
8. MIXING OF SLUDGE	9. WINDOW PITS	10. ROLLERS
11. SCREENS	12. MARKET PREPARATION OF FINE, INTERMEDIATE AND COARSE HUMUS	

FIGURE 35.3 Mechanical Composting

Mechanical digestors of various types include pits or cells, windows or stacks, and vertical cylinder, horizontal cylinder or silo-type closed digestors. Closed digestors are the most hygienic of these and occupy less space. In digestors, the refuse is digested and converted into humus and stable mineral compounds. The digestion period varies between two to five days for refuse containing low cellulose or low carbon-nitrogen (C/N) ratio and seven to nine days for refuse having a higher quantity of cellulose or a high C/N ratio. The stabilized brown mass (humus) is collected, sieved and sold in packets. Sometimes the stabilized mass is enriched by adding chemical nutrients such as phosphorus and nitrogen.

35.3.8 Composting in India

In India, there are two common methods of mechanical composting: the Indore method and the Bangalore method.

Indore Method

In the Indore method, refuse, faecal sludge and animal dung are placed in small brick-lined pits, 3 m × 3 m × 1 m deep, in alternate layers of 7.5 to 10 cm heights, to make a total height of 1.5 m. Chemicals are added to prevent fly breeding. The material is turned regularly for a period of about 8 to 12 weeks and then stored on the ground for four to six weeks. In about six to eight turnings, and after about four months, the compost becomes ready for use as manure.

Bangalore Method

In the Bangalore method, refuse is stabilized anaerobically. Earthen trenches of size 10 m × 1.5 m × 1.5 m deep are filled in alternate layers of refuse and cow dung and other organic waste. The material is covered with a 15 cm layer of good earth and left to decompose. In about four to five months, the compost becomes ready for use. Normally, a city produces 200 to 250 kg/person·annum of refuge and 8 to 10 kg/person·annum of faecal sludge. The amount of compost produced is about half the amount of refuse plus faecal sludge and cow dung added in the beginning.

35.4 ENGINEERED LANDFILLING

As mentioned earlier, modern landfills are not simply dumps but properly planned, designed and operated sites. An engineered landfill requires careful site selection and management of landfill gases and leachate. Site selection should include factors such as land area availability, haul distance, soil conditions, surface and subsurface hydrology, climatic conditions and local environmental conditions. The principal methods used for landfilling may be classified as area method, trench method and depression method. The generation of landfill gases and leachate depends on the type and characteristics of the waste. The gas can be managed with vents and barriers or by recovery. Since the leachate is a potential groundwater contaminant, provisions for the containment and treatment of leachate are an important component of engineered landfills.

Example Problem 35.3

Consider a town where 10 000 households each fill one 300 L container of refuse per week. What volume does it occupy in a landfill? Assume that 10% of the landfill volume is occupied by the cover soil and the density of the refuse when collected is 100 kg/m³ and compacted in landfill to 700 kg/m³.

Given: Sub 1 = collected
Sub 2 = compacted landfilled
V_1 =10000 Houses at 300 L/unit
ρ_1 = 100 kg/m³
ρ_2 = 700 kg/m³

SOLUTION:

$$V_1 = \frac{300\ L}{unit.week} \times 10000\ units \times \frac{m^3}{1000\ L} = 3000\ m^3\ /\ week$$

$$V_2 = V_1 \times \frac{\rho_1}{\rho_2} = \frac{3000\ m^3}{week} \times \frac{100\ kg\ /\ m^3}{700\ kg\ /\ m^3} = 428.6\ m^3\ /\ week$$

$$V_{soil} = \frac{10\%}{100\%} \times \frac{428.6\,m^3}{week} = 4.28\,m^3\ /\ \text{week}$$

$$V_{fill} = 1.1 \times \frac{428.6\ m^3}{week} = 471.4 = 470\ m^3\ /\ week$$

Example Problem 35.4

A town with a population of 45 000 generates refuse at a rate of 300 g/c.d. A 12-ha landfill site is available with an average depth of compacted refuse limited to 3.5 m by local topography. It is assumed that the compacted refuse will have a density of 560 kg/m³ and that 15% of the volume will be taken up by the cover material. What is the anticipated useful life of the landfill?

Given: Population = 45 000 people at 300 g/person.d
Site: 12 ha × 3.5 m
ρ = 560 kg/m³
Soil cover = 15%

SOLUTION:

$$M_{waste} = \frac{300\ g}{person.d} \times 45000\ persons \times \frac{kg}{1000\ g} = 13500\ kg\ /\ \mathrm{d}$$

$$V_{compact} = \frac{M}{\rho} = \frac{13500\ kg}{d} \times \frac{m^3}{560\ kg} = 24.1\ m^3\ /\ d$$

$$V_{fill} = \frac{115\%}{100\%} \times \frac{24.1 m^3}{d} = 27.7 m^3\ /\ d$$

$$Life = 12\ ha \times 3.5\ m \times \frac{d}{27.7\ m^3} \times \frac{year}{365\ d} \times \frac{10000\ m^2}{ha} = 41.5 = \underline{42}\ years$$

Example Problem 35.5

How many hectares of land is required to construct a sanitary landfill with a design period of 30 years? Assume a municipal solid waste generation rate of 2.0 kg/person.d and the average fill depth on completion is 10 m. This landfill is to serve a population of 50 000 in a developed country. After compacting the waste to a density of 500 kg/m^3, it is covered with soil such that 20% of the volume is taken by the soil cover.

Given: Population = 50 000 people at 2.0 kg/person.d
d =10 m
ρ = 500 kg/m^3 = 0.5 t/ m^3
Soil cover = 20%
Design life = 30 years

SOLUTION:

$$M_{waste} = \frac{2.0\ kg}{person.d} \times 50000\ persons \times \frac{t}{1000\ kg} = 100\ t\ /\ d$$

$$V_{compact} = \frac{M}{\rho} = \frac{100\ t}{d} \times \frac{m^3}{0.5\ t} = 200\ m^3\ /\ d$$

$$V_{fill} = \frac{120\%}{100\%} \times \frac{200 m^3}{d} = 240 m^3\ /\ d$$

$$A_{fill} = \frac{V}{d} = \frac{240\ m^3}{d} \times 30\ years \times \frac{365\ d}{year} \times \frac{1}{10\ m} \times \frac{ha}{10000\ m^2} = 26.2 = \underline{26}\ ha$$

35.4.1 Landfill Containment and Monitoring System

In engineered landfills, there is provision for the collection of leachates and the motoring of groundwater to check for contamination. The volume of leachate produced is a function of the amount of water percolating through the refuse. For the same reasons, humid areas will generally produce more leachates, whereas arid regions may have very little or no leachates at all. In most cases, landfills are constructed above the water table, as shown in **Figure 35.4**. This allows considerable attenuation of leachates into the vadose zone.

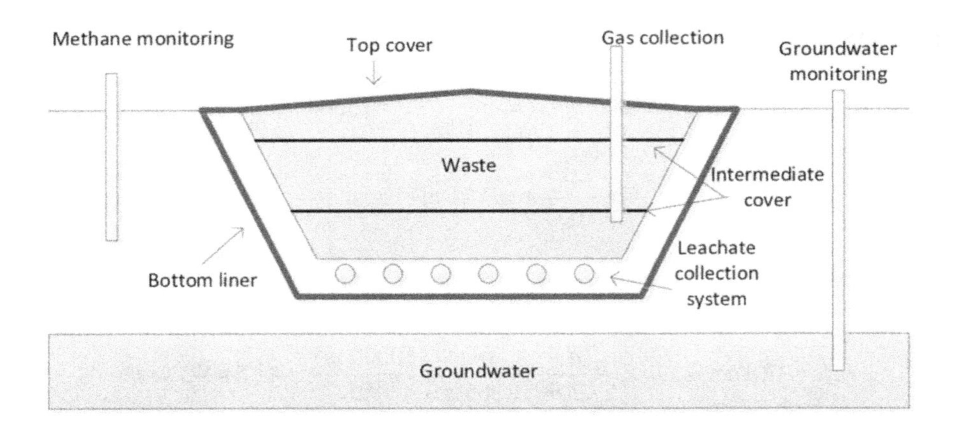

FIGURE 35.4 Components of Sanitary Landfills

Landfills are designed to:

- Minimize the formation of leachates.
- Minimize the amount of leachate leaving the landfill.
- Collect and treat leachates.

The three main components of designed landfills are:

1. Liners that are typically clay layers of very low permeability compacted clays. The thickness of these liners is in the range of 1–3 m of compacted clays with conductivity, $k < 10^{-6}$ mm/s.
2. A collection system usually comprises drainage tiles covered with a blanket of sand and gravel above the clay layer. The collected leachates are treated on-site or sent to a sewage treatment plant after pre-treatment.
3. An underdrainage system that collects leachates leaking through the clay layer. This also controls the zone of saturation.

35.4.2 Seepage Rate

Seepage rate is calculated using Darcy's law. Under saturated conditions, flow through porous media is governed by the coefficient of permeability (hydraulic conductivity) and the hydraulic gradient. For a given hydraulic head, the seepage rate depends on the hydraulic conductivity and thickness of the clay liner.

$$\boxed{Flow\ velocity, v = \frac{Q}{A} = k \times I}$$

v = flow velocity (Darcy's velocity) = specific discharge (Q/A)
k = hydraulic conductivity = constant of proportionality
I = hydraulic gradient
A = flow area

Hydraulic gradient is the difference in hydraulic head per unit length/thickness of a porous medium through which the flow is taking place.

Flow velocity is the flow rate per unit area and is smaller than actual velocity or pore velocity since flow takes place through the pores. Actual flow area is smaller than the total cross-sectional area as some area is occupied by the solid particles. Therefore, pore velocity will be greater than flow velocity depending on the porosity, n, of the medium.

$$v_n = \frac{Q}{A \times n} = \frac{k \times I}{n}$$

Pore velocity, v_n represents the velocity with which leachate would move through the saturated layer. This is further illustrated by the following example problems.

Example Problem 35.6

In a sanitary landfill, a 50 cm thick clay layer with a coefficient of permeability of 10^{-7} m/s is provided. Determine the seepage rate of leachates from the bottom of the landfill. Assume the underdrain system is laid at a depth of 80 cm below the leachates collection system at the bottom of the landfill.

Given: $\Delta h = h_L = 0.8$ m
$v = ?$
$k = 10^{-7}$ m/s
$L_c = 0.5$ m

SOLUTION:

1) The material between the clay layer and the underdrain system is highly permeable sand; hence it is safe to assume that negligible head loss occurs through this portion of the soil.
2) Flow takes place vertically from the leachates collection system to the underdrain system, and a head loss of 0.8 m mainly occurs across the clay layer.

$$I = \frac{h_L}{L_C} = \frac{0.8\ m}{0.5\ m} = 1.6\ m/m$$

$$v_{seepage} = k \times I = \frac{10^{-7} m}{s} \times 1.6 \times \frac{1000L}{m^3} \times \frac{3600\ s}{h} \times \frac{24\ h}{d} = 13.8 = \underline{14}\ L/s \cdot m^2$$

The seepage rate is inversely proportional to the thickness of the clay layer and is directly proportional to k.

Example Problem 35.7

Determine the thickness of a clay layer that must be placed at the bottom of a landfill if the seepage flow rate is to be limited to about 2.0 L/m²·d. Assume that the water table is located at the bottom of the landfill. The leachate level above the clay layer is to be maintained at 60 cm by pumping. Assume the k of the compacted clay is 0.80 L/m²·d.

Given: Difference in hydraulic head, $\Delta h = 0.6$ m + L_c(thickness)
$v = Q/A = 2.0$ L/m²·d
$k = 0.80$ L/m²·d

SOLUTION:

$$I = \frac{\Delta h}{L_C} = \frac{0.60\ m + L_C}{L_C}\ or\ v = k\left(\frac{0.60\ m + L_C}{L_C}\right) or\ L_C = \frac{k(0.6\ m + L_C)}{v}$$

$$L_C = \frac{0.6\ \text{m} \times k}{(v-k)} = \frac{0.6\ m \times 0.8}{(2-0.8)} = 0.40\ m = \underline{40\ cm}$$

35.5 RESOURCE AND ENERGY RECOVERY

An integrated waste management system depends on the materials and energy recovery systems used in the system. Recoverable materials such as paper, cardboard, plastics, glass, metals, etc., contained in waste can be recovered. However, the decision to recover these materials is usually based on economic evaluation and local considerations. The fraction of light combustible materials present can be converted into refuse-derived fuel (RDF).

Thermal conversion processes such as incineration, gasification, pyrolysis or anaerobic methane production, as discussed earlier, can be considered as potential options for energy recovery. On the derivation of the conversion products (steam, heat, gases and oils) by one of the thermal or biological processes listed, energy is harnessed using suitable turbines or generators. Waste tyres must be dropped off at a solid waste facility permitted to accept waste tyres. There are many permitted facilities around the state where waste tyres are accepted for temporary storage and then transported to processing or disposal facilities.

35.6 CONSTRUCTION MATERIALS WASTE

Building demolitions or renovation jobs at any institutional, commercial, public or industrial buildings are subject to notification requirements and emission control requirements. In addition, inspections may be conducted. Residential buildings with four or fewer dwelling units, including houses, duplexes, barns and other outbuildings, do not need to meet these requirements.

Steps to practise waste management at home

- Keep separate containers for dry and wet waste in the kitchen.
- Keep two bags for dry waste collection – paper and plastic – for the rest of the household waste.
- Keep plastic from the kitchen clean and dry and drop it into the dry waste bin. Keep glass and plastic containers rinsed of food matter.
- Keep a paper bag for disposing of sanitary waste.

35.7 SOLID WASTE: INDIAN PERSPECTIVE

Solid waste in India, as a developing country, is very different from developed countries in Europe and North America. From an economics point of view, waste is something of zero or negative monetary value. For example, many plastic containers that might be waste in developed countries are reused in India. At the same time, common people's attitude towards waste is also very different. Even a villager in Europe knows how to segregate waste. In urban centres, a housemaid usually handles waste and dumps everything into one container. Things like paper and cardboard are generally kept separate and are sold and hence recycled. Another important difference is that people throw waste anywhere as long as it is not in their house. It is a very common practice to clean the yard and floor and throw the waste onto the street. In India, behavioural changes regarding waste are badly needed. India can learn recycling and waste management techniques from the West, but this does not necessarily mean learning Western consumption habits.

Discussion Questions

1) Explain in detail any five types of solid waste disposal.
2) What is the overall objective of solid waste management? List the parameters that require careful consideration in solid waste management.
3) What are the advantages of composting? Briefly describe the common methods adopted for composting.
4) Briefly describe the main components of a sanitary landfill with the help of a sketch.
5) Discuss the interrelationship of the functional elements comprising a solid waste management system in your own words.
6) What are the operations involved in a large-scale composting plant?
7) List some important points that need to be taken into consideration while performing incineration.
8) Briefly describe the common methods of solid waste disposal.
9) Briefly discuss the technical and environmental factors involved in choosing a new sanitary landfill site. Why is the study of the hydrogeology of the area important in site selection?
10) How does solid waste management in India compare with that of Western countries?
11) What recommendations would you make to improve solid waste management in India?
12) What are the three key characteristics of a sanitary landfill that distinguish it from an open dump?
13) What materials in the municipal solid waste stream are recyclable? How do recycling adaption and practices in India compare with those in the US and Canada?
14) Describe the quantity and quality of waste generated at your home. Describe the method of its collection, segregation and recycling. What ways do you suggest homeowners take to improve upon current practices?
15) What are the main methods of composting in India? Make a comparison.

Practice Problems

1. How thick a clay layer is required to control the leachates rate from the bottom of a landfill to a value of 2.0×10^{-4} L/s.m^2? Assume the hydraulic conductivity of the clay layer is 1.0×10^{-7} m/s and the underdrain system is laid at a depth of 1.0 m below the leachates collection system at the bottom of the landfill. (0.50 m)
2. Consider a town with 12 000 households, each producing 350 L of refuse per week. What volume does it occupy in a landfill? Assume that 15% of the landfill volume is occupied by the cover soil and the density of the refuse when collected is 100 kg/m^3 and compacted in landfill to 700 kg/m^3. (690 m^3/wk)
3. A town with a population of 45 000 generates refuse at a rate of 350 g/person.d. A 10-ha landfill site is available with an average depth of compacted refuse limited to 3.2 m by local topography. It is assumed that the compacted refuse will have a density of 650 kg/m^3 and that 12% of the volume will be taken by the cover material. What is the anticipated useful life of the landfill? (32 years)
4. How much land area is required to construct a sanitary landfill in a developed country with a design period of 30 years? Assume a municipal solid waste generation (MSW)

rate of 2.5 kg/person.d and the average fill depth on completion is 7.5 m. This landfill is to serve a population of 40 000. After compacting the waste to a density of 500 kg/m^3, it is covered with soil such that 20% of the volume is taken by the soil cover. (35 ha)

5. In the bottom of a sanitary landfill, a 50 cm thick clay layer with a coefficient of permeability of 10^{-5} mm/s is provided. Find the seepage rate of leachates knowing that the underdrain system is laid at a depth of 1.0 m below the leachates collection system at the bottom of the landfill. (2.0×10^{-5} L/s.m^2)
6. Determine the thickness of a clay layer that must be placed in the bottom of a landfill if the seepage flow rate is to be limited to about 2.0 L/m^2·d. Assume that the water table is located at the bottom of the landfill. The leachate level above the clay layer is to be maintained at 70 cm by pumping. Assume the k of the compacted clay to be 0.50 L/m^2·d. (0.23 m)
7. How many hectares of land are required to construct a sanitary landfill with a design period of 35 years? Assume an MSW generation rate of 400 g/person.d and the average fill depth on completion is 5.5 m. This landfill is to serve a population of 50 000. After compacting the waste to a density of 600 kg/m^3 it is covered with soil such that 20% of the volume is taken by the soil cover. (9.3 ha)
8. In a mid-size town, household waste is collected once every week. On average, each household produces 10 kg of waste per week and the packer vehicle can compact the waste to a density of 350 kg/m^3. Estimate the number of households that have their waste collected prior to making a trip to the landfill. Assume the truck capacity is 12 m^3. (420 houses)
9. A garbage collection truck and its crew service customers at a rate of 12 customers/10 minutes. Out of the total 8.0-hour shift, the actual time spent collecting household waste is 6.0 hours. The rest of the time is used to make trips to the landfill site and for breaks. Estimate the number of customers that can be served per shift. (430 households)
10. A community of 45 000 people generated MSW at a rate of 2.0 kg/person.d. It is compacted to a density of 540 kg/m^3 in a sanitary landfill. After one year of operation, to what depth will a 12-ha landfill be covered? Assume the soil cover is 20% of the volume. (0.61 m)

36 Hazardous Waste

Hazardous waste may be defined as any waste or combination of wastes that pose substantial danger to human beings, plants, and animals.

The US Environmental Protection Agency (EPA) defines hazardous waste as any substance that "because of its quantity, concentration, or physical, chemical, or infectious characteristic may cause or significantly contribute to an increase in mortality, serious irreversible illness, or incapacitating reversible illness; or pose a substantial present or potential hazard to human health or the environment when improperly treated, stored, transported or disposed of, or otherwise managed."

Hazardous waste is a name given to material that, when intended for disposal, meets one of the following two criteria.

It contains one or more of the criteria pollutants or those chemicals that have been explicitly identified as hazardous. There are presently over 50,000 chemicals thus identified. If a material is not on the list, it does not imply that it is non-hazardous. It still may be defined as hazardous if it exhibits any of the measurable characteristics of a hazardous waste.

36.1 CHARACTERISTICS OF HAZARDOUS WASTE

The four primary characteristics are based on the physical or chemical properties of toxicity, reactivity, ignitability and corrosivity. Two additional types of hazardous materials include waste products that are either infectious or radioactive.

Toxic wastes are poisonous, even in trace amounts. Some may have an immediate effect on humans or animals, causing death or serious illness, while others may have long-term effects. The toxicity of any particular waste is determined by an EPA-specified test called the toxicity characteristic leaching procedure (TCLP). The TCLP is used to determine the mobility of organic and inorganic substances in the waste.

Flammable materials are those liquids that burn at relatively low temperatures (below 60°C) and are capable of spontaneous combustion during storage or transportation. Many waste oils and solvents are ignitable.

Reactive wastes are unstable and tend to react vigorously with air, water or other substances. The reactions cause explosions or form toxic vapours and fumes.

Corrosive wastes, including strong alkaline or acidic materials, destroy materials and living tissues by chemical reaction. Typically, in an aqueous solution, these wastes have pH values outside the range of 2.0 to 12.5, or any liquid that exhibits corrosivity to steel at a rate greater than 6.35 mm per year.

Radioactive waste, particularly high-level wastes from nuclear plants, is also of special concern as a hazardous waste. Radioactivity may be defined as the spontaneous breakup of the nucleus of an atom. Radioactive wastes are handled separately and are governed by separate rules and regulations.

36.2 HAZARDOUS WASTE MANAGEMENT

Several options are available for the management and disposal of hazardous waste. In order of preference, these can be summarized as follows:

DOI: 10.1201/9781003231264-36

1. Reduce waste at the source by modifying industrial processes and other techniques.
2. Reclaim and recycle the waste, using it as a resource for other industrial or manufacturing processes.
3. Stabilize the waste, rendering it non-hazardous, using appropriate chemical, biological or physical processes.
4. Incinerate the waste at temperatures high enough to break it down into non-hazardous substances.
5. Apply modern land disposal methods, preferably after some form of treatment and containment.

36.2.1 Treatment of Hazardous Wastes

The treatment of materials deemed hazardous is specific to the material and the situation. Therefore, there are several alternatives that engineers may consider in such treatment operations.

Some types of hazardous wastes can be detoxified by chemical, biological or physical treatment. Treatment of hazardous waste is costly, but it can prepare the material for recycling or safer disposal.

Chemical treatments include incineration, ion exchange, neutralization, precipitation and oxidation-reduction. Incineration is a thermal-chemical process that can detoxify certain organic wastes or essentially destroy them. The burning of organic wastes at high temperatures converts them to ash residue and gaseous emissions. Not all wastes can be incinerated. Heavy metals, for example, are not destroyed but enter the atmosphere in vapour form.

Plasma gasification is an innovative waste-to-energy technology now being applied and tested for hazardous waste management. Operating temperatures exceed 3000°C, and a super-heated column of plasma completely destroys the waste material while producing synthesis gas (syngas) and a non-hazardous residue (slag). The syngas primarily consists of carbon monoxide and hydrogen and can be used for chemical manufacturing processes or converted into hydrocarbon fuel for energy production.

In some cases, a simple neutralization of the hazardous material will render the chemical harmless. In other cases, oxidation is used, such as for the destruction of cyanide. Ozone is often used as the oxidizing agent. In a case where heavy metals must be removed, precipitation is the method of choice. Most metals become extremely insoluble at high pH ranges, so the treatment consists of the addition of a base, such as lime or caustic, and the settling of the precipitate. Other physiochemical methods employed in the industry include reverse osmosis, electrodialysis and solvent extraction.

Biological treatment processes involve the breakdown of waste by microorganisms. If the hazardous material is organic and readily biodegradable, often the least expensive and most dependable treatment is biological. It can be accomplished by adopting engineered natural processes, as explained earlier. Alternatively, it is now increasingly likely that specific microorganisms, or designer bugs, can be created by gene manipulation to attack certain especially difficult-to-treat organic wastes. Depending on the type of wastes, any one or combination of these methods may be used.

Organic waste from the petroleum industry can be treated biologically. In addition to conventional treatment such as activated sludge and trickling filtration, land treatment can be used to treat organic hazardous waste. Hazardous waste is mixed with surface soil and nutrients, and microorganisms may be added to the mixture as needed. Organics are biologically degraded, and inorganics are adsorbed and retained in the soil.

Landfarming is a relatively inexpensive method for treatment and a way to ultimately dispose of certain types of hazardous wastes. However, food or forage crops cannot be raised on this land because of the possibility of the uptake of toxic compounds. One major drawback of this method is that a large area of land is required, and surface topography and geologic conditions should be such that any possibility of surface or subsurface contamination is minimized.

36.2.2 Clean-Up of Old Sites

The type of work conducted at these sites depends on the severity and extent of the problem. In some cases where there is an imminent threat to human health, the hazardous material must be removed for safe disposal or treatment. This is called off-site remediation. Where off-site remediation is most desirable for people living in the area, it is expensive and still involves some risk in moving the hazardous material from one location to another.

On-site remediation does not involve the removal of waste from the site. In less acute instances, the site is stabilized so that it is less likely to cause health problems. On-site remediation largely focuses on minimizing the production of leachate and eliminating groundwater pollution. Another goal is to mitigate further the propagation of groundwater pollution that might already have occurred.

When leachates from a landfill mix with the groundwater, they form a plume that spreads in the direction of the flowing groundwater. As one goes away from the source, the concentration decreases owing to hydrodynamic dispersion and retardation.

Remedial action, therefore, implies that the site has been identified and action has been taken to minimize the risk of having the hazardous material cause human health problems. Depending on the seriousness of the situation, several remedial action options are available. If there is no threat to life and if it can be expected that the chemical will eventually metabolize into harmless end-products, then one solution is to do nothing except continuous monitoring. This method is known as natural attenuation. In most cases, this is not so, and direct intervention is necessary.

Containment is used where there is no need to remove the offending material or if the cost of removal is prohibitive, as was the case at Love Canal in New York state. Containment usually involves the installation of slurry walls, which are deep trenches filled with bentonite clay or some other highly non-permeable material, and continuous monitoring for leakage out of the containment. In time the offending materials might slowly biodegrade or chemically change to a non-toxic form, or new treatment methods may become available for detoxifying this waste.

Extraction and treatment involves pumping contaminated groundwater to the surface for either disposal or treatment or the excavation of contaminated soil for treatment. Sometimes air is blown into the ground and the contaminated air is collected. Some soils may be so badly contaminated that the only option is to excavate the site and treat the soil off site. This is usually the case with polychlorinated biphenyl (PCB) contamination because no other method seems to work well. The soil is dug out and usually incinerated to remove the PCBs and then returned to the site or landfilled. Depending on the contaminant, biodegradation in reactors or piles may be used.

In-situ treatment of contaminated soil involves the injection of bacteria or chemicals that will destroy the offending material. If heavy metals are of concern, these can be tied up chemically (fixed) so that they will not leach into the groundwater. Organic solvents and other chemicals can be degraded by injecting freeze-dried bacteria or by making conditions suitable for indigenous bacteria to degrade the waste (e.g., by injecting air and nutrients). The past few years have seen an amazing discovery of microorganisms able to decompose materials that were previously thought to be refractory or even toxic.

36.3 STORAGE TANKS AND IMPOUNDMENTS

Proper storage of hazardous waste is imperative because of the potential for it to cause harm to public health and damage the environment. Many hazardous waste generators store the waste on-site for varying lengths of time. Relatively large quantities may be stored in above-ground basins or lagoons. Above-ground basins may be steel or concrete but are not suitable for storing reactive and ignitable wastes.

Relatively small amounts of wastes generated on an interim basis can be placed in 200 L fibre glass, plastic or steel drums for ease of handling, temporary storage and transportation.

36.3.1 Underground Storage Tanks

There are numerous examples of surface and groundwater contamination caused by the leakage of containers storing hazardous wastes. Underground tanks pose a greater risk than above-ground tanks. Due to environmental concerns, underground tanks, piping and pumping systems are now regulated. Such regulations intend to protect leaks and spills, as well as to detect them if they do occur.

36.3.2 Surface Impoundments

Before any regulations were implemented, large volumes of liquid hazardous waste were deposited in surface excavations such as pits, ponds and lagoons (PPLs). Most PPLs were unlined and hence provided no protection against leakage. Except for sedimentation, evaporation of volatile organics and possibly some aeration, they provided no treatment to waste.

The surface impoundment of liquid hazardous waste is allowed for temporary storage and treatment if it meets stringent criteria. All such impoundments must have at least one liner and be located over an impermeable layer of clay or bedrock. New impoundments must have two liners and a leachate collection system. In addition, monitoring wells and dikes or berms are required. Accumulated sludge must be periodically removed and provided further handling as a hazardous waste.

36.3.3 Waste Piles

Generators of certain wastes are allowed to use a waste pipe for the temporary accumulation of hazardous wastes. Only containerized, solid and non-flowing material can be stored in a waste pile. Material must be landfilled when the pile becomes unmanageable. Waste piles must be constructed over an impervious base and comply with the requirements for landfills.

36.4 LAND DISPOSAL

Land disposal of hazardous waste is not an attractive option because of environmental concerns and future liability. As mentioned earlier, hazardous waste sites did not provide adequate environmental protection, and the clean-up of old sites is a very tedious and expensive process. However, with proper site selection, engineering design and operational safeguards, land disposal is the least expensive alternative for many types of hazardous waste. Certain hazardous wastes, such as dioxins, PCBs, cyanides, halogenated organic compounds and strong acids, are banned from landfills. These wastes must be treated and stabilized before land disposal.

36.4.1 Secure landfill

A secure landfill must have a minimum of 3 m of height separating the base of the landfill from the underlying bedrock or groundwater aquifer. A secure landfill must have:

1. A double liner.
2. A leachate collection system.
3. Monitoring wells.
4. An impermeable cap or cover after completion.

A cross-sectional view of a typical secure landfill is shown in **Figure 36.1**.

An impermeable layer on the bottom and sides of a secure waste landfill serves as a barrier, preventing groundwater from entering or any leachate from leaving the landfill. However, no liner is perfect, and some leachate is expected over the life of the landfill. The second liner and leachate

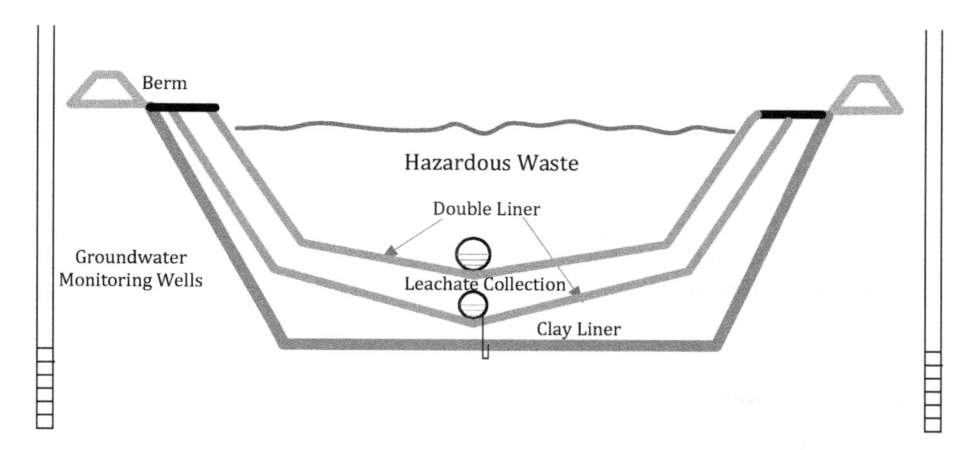

FIGURE 36.1 Sectional View of a Secure Landfill

collection system provide an additional safeguard for protecting environmental quality and public health.

A synthetic flexible fibre called a geotextile may be used to separate soil and waste material from the primary leachate collection system. Under the geotextile is the primary liner, which must be a synthetic material with hydraulic conductivity equal to or less than 10^{-6} mm/s. This is called a geomembrane or a flexible membrane liner (FML). The lower or secondary liner is a composite layer of compacted clay and an FML. The secondary leachate collection system serves as a leak detection and backup system. Collected leachate is pumped to a treatment facility for further processing.

A secure landfill is considered to have four phases in its total operation. During the active phase, hazardous waste is deposited, making sure incompatible wastes are placed at separate locations to avoid explosions. During the second or closure phase, an impermeable cover is constructed over the landfill site. A third post-closure phase, defined as the 30-year period after the closure of the site, involves the continuous operation of the monitoring well. This is a second line of defence. A fourth and last phase, called the eternity phase, is expected to involve some leakage of waste material into the environment. When leakage is detected, it is pumped from monitoring wells and treated. This last line of defence is intended to prevent the contamination of groundwater.

36.4.2 Underground Injection

A deep well injection involves the pumping of hazardous liquid into the porous layer of deep underground rock. This layer must be sandwiched between two impermeable layers. The injection well must be at least 400 m from an underground source of drinking water, and waste must be injected into a separate geological formation free of faults or fractures.

The deep well injection method is more common with chemical and petroleum refining industries. Another criterion for these kinds of wells is that there must be a water-bearing stratum of non-beneficial use that has sufficient volume, porosity and permeability to accept the injected waste. Though this method requires very little land and no pre-treatment of waste is needed, it is pertinent to have a thorough engineering study and design to make sure the drinking water supply is not affected.

36.4.3 Radioactive Waste Disposal

Ocean disposal and deep well injections are not permitted for the disposal of radioactive waste. The most suitable method for the disposal of low radioactive waste is called near-surface land disposal. In this method, waste can be stored above or below grade. However, waste must be isolated to avoid contact with water and the release of radiation into the air. Multiple engineered barriers must be incorporated into the facility for obvious reasons.

Vaults made of reinforced concrete, metal or masonry can be constructed above or below grade. Such vaults must be equipped with sump pumps and monitoring equipment. Another disposal method, improved shallow land burial, involves the placement of low-level waste in trenches typically 10 m deep and 30 m wide. The bottom of the trenches must be above the water table, and proper drainage is provided to keep the foundation dry. Spaces between waste containers are filled with sand and the filled trench is covered with compacted clay. Trenches must be equipped with a monitoring system to detect any leakage of radioactive material.

Discussion Questions

1. List five options for the management and control of hazardous waste. Which is the least and which is the most desirable? Explain why.
2. Give a workable definition of hazardous waste, including the basic properties that are characteristic of such waste. Briefly describe these properties.
3. What is the difference between hazardous waste and hazardous materials? Should the term toxic waste be used interchangeably with the term hazardous waste?
4. Briefly describe the restrictions on the land disposal of hazardous waste.
5. In what sense is a sanitary landfill different from a secure landfill? Sketch the cross-section of a secure landfill and describe the bottom liner and leachate collection system.
6. Briefly describe the biological methods of treating hazardous waste. What is the difference between landfilling and land farming?
7. What is the role of incineration in the treatment of hazardous waste? What other chemical treatment processes can be applied to hazardous waste?
8. What is meant by site remediation? Briefly describe the differences between off-site and on-site remediation of hazardous waste sites.
9. What are the necessary conditions to apply deep well injection as a land disposal option for hazardous waste? Are there any other options that may be preferable? Explain.
10. There have been many episodes of damage to public health and the environment due to the mismanagement of hazardous waste, including Love Canal in the US, and the gas leak in Bhopal, India. Search the internet to find one such episode and write down your findings.

37 Noise Pollution

Sound is the sensation of acoustic waves that develop pressure fluctuations in a medium. Pure sound is described by pressure waves travelling through a medium. These pressure waves are described by their amplitude and their frequency. With reference to **Figure 37.1**, a pure sound wave can be described as a sinusoidal curve, having positive and negative pressures within one cycle. The number of these cycles per unit time is called the sound frequency, often expressed as cycles per second, or hertz (Hz). A sound wave with a frequency of 1000 Hz, for example, is one in which the pressure wave will pass a given point 1000 times, or 1000 c/s. Typical sounds that healthy human ears hear range from about 15 Hz to about 20000 Hz. Low-frequency sound is deep (low-pitched), while high-frequency sound is high-pitched. For example, middle A on a piano is at a frequency of 440 Hz. Speech is usually in the range of 1000 to 4000 Hz.

37.1 NOISE

Noise is defined as unpleasant, unwanted and disturbing sound. Unwanted sound produces certain undesirable effects and hence is called a pollutant. Noise slows down productivity and can annoy or hurt people psychologically and physiologically. The loudness of noise is expressed as its amplitude (denoted as A in **Figure 37.1**). Amplitude is the heights of the peaks related to the pressure intensity and the volume or loudness of the perceived sound. The wide frequency spectrum of audible sound is significantly reduced by age and environmental exposure to loud noise. The most significant sources of such damaging noise come from occupational sources and loud music, particularly rock concerts and earphones turned very high. Young, healthy people (particularly young women) can hear very high frequencies, often including such signals as automatic door openers. With age, the ability to detect a wide frequency range drops. Older people especially tend to lose the high end of the hearing spectrum.

37.1.1 Speed of Sound

Frequency should not be confused with speed of sound. Speed of sound is constant for a given medium. In denser mediums (where molecules are close together), sound travels faster. Thunder is heard after lightning is seen since light travels much faster than sound. In air, at standard pressure and temperature, the speed of sound is constant at 340 m/s. There is a mathematical relationship between frequency, f, wavelength, λ, and speed of sound, $v = f \times \lambda$

Since the speed of sound in each medium is constant, there is an inverse relationship between frequency and wavelength. In other words, short wavelength means higher frequency and vice versa.

Example Problem 37.1

Sound in air at 22°C travels at a speed of 343 m/s. How long will it take to hear thunder from lightning that occurred 4.5 km away in the sky?

DOI: 10.1201/9781003231264-37

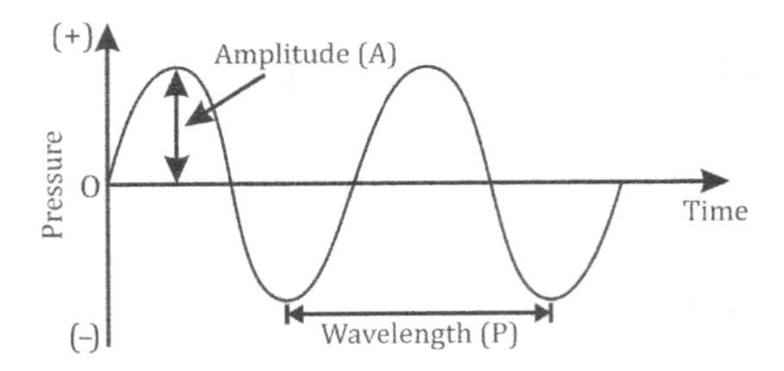

FIGURE 37.1 Sinusoidal Sound Wave

SOLUTION:

$$t = \frac{s}{v} = 4.5\ km \times \frac{1000\ m}{km} \times \frac{s}{343\ m} = 13.12 = \underline{13\ s}$$

Example Problem 37.2

The frequency of sound caused by a moving train is 450 Hz. Assuming that the speed of sound in steel rails is 5.0 km/s, what is the wavelength of the travelling sound?

SOLUTION:

$$\lambda = \frac{v}{f} = \frac{5.0\ km}{s} \times \frac{1000\ m}{km} \times \frac{1}{450\ Hz} \times \frac{Hz.s}{1} = 11.1 = \underline{11\ m}$$

37.1.2 Loudness and Pitch

Two common terms used to describe the human perception of sound are loudness and pitch. Loudness is related to the amplitude of the wave, as described earlier. The pitch of a sound is related to the frequency of the wave that sound produces. A high-pitched sound, like a whistle, has a relatively high frequency. A foghorn, on the other hand, is low-pitched. In other words, low-pitched sound waves have a long wavelength and high-pitched sounds have a short wavelength. The human ear can detect sounds in the frequency range of 20 to 20000 Hz.

37.2 SOUND PRESSURE

With reference again to **Figure 37.1**, the energy in a pressure wave is the total area under the curve. Because the first half is a positive pressure and the second is a negative pressure, adding these would produce zero net pressure. The trick is to find the root mean square analysis of a pressure wave, multiplying the pressure by itself and then taking its square root. Since the product of two negative numbers is a positive number, the result is a positive pressure number. In the case of sound waves, the pressure is expressed as newtons per square metre (N/m^2 = Pa), although sound pressure is sometimes expressed in bars or atmospheres. Sound pressure is so small that it is more conveniently expressed in μPa.

The human ear is a sensitive organ. The average person can detect a sound with a pressure as low as 20 μPa. The highest sound pressure a human ear can tolerate without causing pain is 100 Pa. This means the range of sound pressure has a wide spectrum. Thus, measuring sound with pressure units that can vary over such a wide range is impractical and inconvenient. Another disadvantage is that the ear responds non-linearly to pressure. Doubling the sound pressure does not cause a doubling in loudness.

37.2.1 Standard Unit of Noise

With respect to noise, the energy level of the sound is considered rather than the pressure. The energy level is expressed as energy per unit area, which also indicates the sound intensity. The sound pressure intensity is proportional to the sound pressure squared; hence this ratio can also be expressed as:

$$Ratio = \log\left(\frac{I}{I_0}\right) = \log\left(\frac{p^2}{p_0^2}\right) = \log\left(\frac{p}{p_0}\right)^2 = 2 \times log\left(\frac{p}{p_0}\right)$$

Where I = sound intensity, W/m^2, p = sound pressure in N/m^2 and P_0 = reference sound pressure in N/m^2.

Reference pressure is taken as the minimum pressure a human ear can detect, which is 2×10^{-5} N/m^2 (20 μPa). This ratio is used to measure sound pressure level and its unit is called bel (B).

Sound Pressure Level

To avoid the disadvantage of using pressure for sound or noise measurement, a logarithmic relationship called the decibel scale is used to measure the sound pressure level (SPL). Since this is a ratio, it does not represent a physical quantity and requires logarithms to convert it to more manageable numbers or a scale. Dividing this unit into 10 makes it easier to use, avoiding fractions, and is known as the decibel (dB).

$$SPL(dB) = 10 \times \log\left(\frac{I}{I_0}\right) = 10 \times \log\left(\frac{p}{p_0}\right)^2 = 20 \times log\left(\frac{p}{p_0}\right)$$

Factor 10 in the above equation is to convert to decibels. The SPL of threshold hearing ($p = p_0$) is 0 dB since log(1) = 0. Hence an SPL of 0 dB does not represent the complete absence of sound and represents the faintest sound a human ear can perceive. Since it is based on logarithms, each tenfold increase in sound intensity is represented by 10 dB. A sound level of 20 dB is 100 times and of 30 dB is 1000 times that of the reference sound level. If sound intensity doubles, the SPL increases by three decibels, as shown below.

$$SPL = 10 \times \log\left(\frac{I}{I_0}\right) = 10 \times \log(2) = 3.01 = 3.0\,dB$$

Example Problem 37.3

A fire truck causes a sound pressure of 22 Pa. What is the SPL of this sound?

SOLUTION:

$$SPL = 20 \times \log\left(\frac{p}{p_0}\right) = 20 \times \log\left(\frac{22\,Pa}{20\,\mu Pa} \times \frac{\mu Pa}{10^{-6}\,Pa}\right) = 121 = \underline{120\,dB}$$

TABLE 37.1
Typical Sound Pressure Levels

Sound	SPL (dB)
Threshold of hearing	0
Inside of hearing test chamber	10
Remote area of Yellowstone	20
Library	40
Suburban subdivision	45
Typical classroom	50
Normal speech	60
Busy office	65
Ringing alarm clock next to head	80
Average street traffic	85
Lawnmower	90
Individual truck passes by	90
Rock concert	110
Fighter jet at take-off	120
Saturn rocket on take-off	134
Threshold of pain	140
Maximum SPL in air	194

Typical sound pressure levels are shown in **Table 37.1**. The highest possible SPL, at which point the air molecules can no longer carry pressure waves, is 194 dB, while 0 dB is the threshold of hearing. The loudest sound recorded seems to be the Saturn rocket at 134 dB, not far from the threshold of pain at 140 dB.

37.2.2 Cumulative Sound Pressure Levels

When more than one source producing sound exists simultaneously in a space, the overall sound pressure level needs to be evaluated. Because sound pressure levels are logarithmic ratios, they cannot be added directly. If two sources of sound are combined to find the cumulative SPL, the ratio p/p_0 must be calculated from the SPL equation first. These ratios are then added and a new SPL is calculated as:

$$Cumulative, SPL_c = 10\log\left(\Sigma\left(10^{\frac{SPL}{10}}\right)\right)$$

This formula can be further modified for n similar machines operated in a common area.

$$(Similar\ sounds)\, SPL_c = 10\log\left(n \times 10^{\frac{SPL}{10}}\right)$$

Example Problem 37.4

A machine shop has five similar machines, each producing a sound pressure level of 75 dB. What is the new SPL in the room?

SOLUTION:

Five similar machines' combined SPL:

$$SPL_c = 10\log\left(n10^{\frac{SPL}{10}}\right) = 10\log\left(5\times10^{\frac{75}{10}}\right) = 81.99 = \underline{82}\,dB$$

Example Problem 37.5

A machine shop has two machines, one producing an SPL of 70 dB and one producing an SPL of 58 dB. A new machine producing 70 dB is introduced into the room. What is the new SPL in the room?

SOLUTION:

When two machines with 70 dB and 58 dB produce sound simultaneously:

$$SPL_c(\# = 2) = 10\log\Sigma\left(10^{\frac{SPL}{10}}\right) = 10\log\left(10^{\frac{70}{10}} + 10^{\frac{58}{10}}\right) = 70.2 = 70\,dB$$

$$SPL_c(\# = 3) = 10\log\left(10^{\frac{70.2}{10}} + 10^{\frac{70}{10}}\right) = 73.1 = \underline{73}\,dB$$

i. Note that the cumulative level of the two machines increased by only 0.2 dB over the one with the highest SPL, that is, 70 dB.
ii. The cumulative SPL of all three machines resulted in an increase of 3 dB.

Rules of Thumb

In the previous two example problems, it was demonstrated that cumulative SPL is much less than individual SPL. Based on this, the following conclusions can be made.

If the difference between two sounds is greater than 10 dB, the lesser of the two does not contribute to the overall level of sound. Using the 10 dB difference rule of thumb, 70 - 58 = 12, which is greater than 10, the effect of the 58 dB sound is negligible and will result in an overall sound pressure level of 70 dB.

Second, if two equal sounds are added, they result in a 3 dB increase in overall sound level (which is just barely noticeable). Earlier, it was shown that if sound intensity doubles, SPL increases by 3 dB.

When another 70 dB machine is brought in, the two sounds are equal, producing an increase of 3 dB. Thus, the SPL in the room would be 73 dB. Again, using the equation above, the answer is 73.1 dB for the three machines, which rounds to 73 dB.

37.2.3 Inverse Square Law

Sound in the atmosphere travels uniformly in all directions, radiating out from its source. The sound intensity is reduced as the square of the distance away from the source of the sound, according to the inverse square law. As a sound wave travels from a point source, it spreads around in concentric spheres. Since the surface area of a sphere is proportional to its radius squared, sound intensity decreases proportionally to the distance squared. In other words, doubling the distance will cause the sound intensity to decrease four times. That is, the SPL is proportional to $1/r^2$, where r is the radial distance from the point source. An approximate relationship can be developed if the sound power is expressed as a logarithmic ratio based on some standard reference power, such that:

$$Point\,source, SPL_r = SPL_0 - 20\log r$$

Line source of sound

Most noise sources are not point sources, and reflections of sound from walls and ceilings prevent the sound from travelling in spherical waves. A highway, for example, is considered to be a line source from which sound is propagated in the form of half a cylinder. Since the lateral surface area of a cylinder is proportional to its radius, sound intensity decreases in direct proportion to its distance from the line source.

$$Line\,source, SPL_r = SPL_0 - 10\log r \qquad SPL_{r2} = SPL_{r1} - 10\log\left(\frac{r_2}{r_1}\right)$$

Example Problem 37.6

A point source of sound generates 70 dB. What would the SPL be at 50 m from the source?

SOLUTION:

$$SPL_r = SPL_0 - 20\log r = 70 - 20 \times \log(50) = 36.02 = 36\,dB$$

Example Problem 37.7

The sound level measured 4.5 m from the centre line of a busy highway is 90 dBA. What will the sound level be at 10 m from the centre line? Find the distance from the centre line at which the sound level is reduced to 80 dB.

SOLUTION:

$$SPL_{r2} = SPL_{r1} - 10\log\left(\frac{r_2}{r_1}\right)$$

$$SPL_{10} = SPL_{4.5} - 10\log\left(\frac{10m}{4.5m}\right) = 90\,dBA - 10\log 2.2 = 86.5 = \underline{87dBA}$$

$$10log\left(\frac{r_2}{r_1}\right) = SPL_{r1} - SPL_{r2} = 90\,dBA - 80\,dBA = 10\,dBA$$

$$r_2 = r_1 \times 10^{\frac{10}{10}} = 4.5m \times 10 = 45m$$

37.3 SOUND METERS

Sound is measured with an instrument called a sound level meter that converts the energy in the pressure waves to an electrical signal. The basic parts of a sound level meter include a microphone, amplifiers, weighing networks and a display reading in dB. **Figure 37.2** shows the schematic diagram of a sound level meter. The microphone picks up the pressure waves coming from the source and a meter reads the SPL, directly calibrated into decibels. The data thus obtained with a sound pressure level meter represent a measurement of the energy level in the air.

The sound pressure that the meter receives is not the same as perceived by the human ear. While we can detect frequencies over a wide range, this detection is not equally effective at all frequencies. If the meter is to simulate the efficiency of the human ear in detecting sound, the signal must

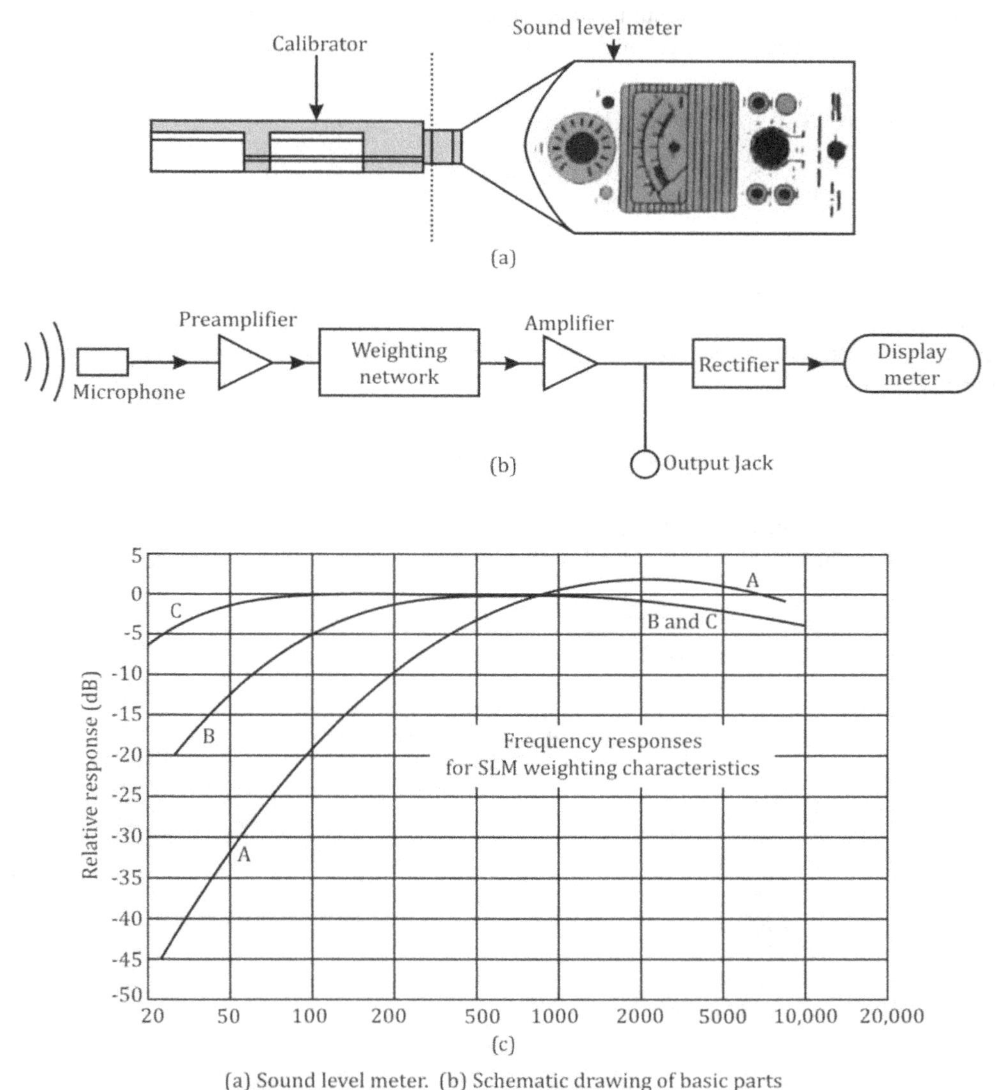

(a) Sound level meter. (b) Schematic drawing of basic parts of sound level meter. (c) Plot of weighting networks

FIGURE 37.2 Schematic Diagram of a Sound Level Meter

be filtered. For this reason, an electronic circuitry called a weighting network is built into the meter to produce a read-out that closely resembles a human response. A series of three internationally accepted weighting scales has been adopted.

There are the A, B and C weighting networks. A-weighted networks filter out low-frequency and very high-frequency sounds, where the human ear is less efficient. This helps to match the meter readings with the sensitivity of the ear and with the average person's judgement of the relative loudness of various sounds. Two sounds of equal dB, but different frequencies, have different dBA levels; the lower frequency sound has the lower dBA.

Depending on whether the network used is A, B or C, read-outs are in dBA, dBB or dBC, respectively. Since the readings do not represent true sound pressure levels, read-outs from A, B and C networks are called simply sound levels. For this reason, meters are called sound level meters and not sound pressure level meters. The A network is more commonly used and is taken as the best measure of environmental or community noise. The D-weighted network is recommended as closely resembling the human response to noise at airports.

37.4 FREQUENCY BAND ANALYSIS

Frequency in cycles per second (Hz) and amplitude in decibels describe a pure sound at a specific frequency. However, environmental sounds are quite "dirty", with many frequencies. Often a frequency analysis is useful in noise control because the frequency of the highest SPL can be identified and corrective measures can be taken. By convention, frequencies are designated as octave bands that represent a narrow range of frequencies, as shown in **Table 37.2**.

The average sound pressure can be calculated from a frequency diagram by adding SPLs at individual frequencies. Again, because SPL is not an arithmetic quantity but rather a logarithmic ratio, such addition must be done by the following equation:

$$SPL_{Avg} = 10\log(\Sigma(10^{SPL/10})/N)$$

Where, SPL_{avg} = average SPLin dB and N = number of measurements.
SPL_i = the ith SPL in dB and I = 1, 2, 3, N

TABLE 37.2
Octave Bands

Frequency Range (Hz)	Geometric Mean (Hz)
22–44	31.5
44–88	63
88–175	125
175–350	250
350–700	500
700–1400	1000
1400–2800	2000
2800–5600	4000
5600–11200	8000
11200–22400	16000
22400–44800	31500

Example Problem 37.8

The SPL measurement of an industrial unit indicates the following pattern of frequency. Find the average SPL.

Octave Band (Hz)	SPL (dB)
31.5	8
63	10
125	15
250	16
500	21
1,000	53
2,000	31
4,000	42
8,000	26
16,000	24
31,500	36

SOLUTION:

Band (Hz)	SPL (dB)	$10^{(SPL/10)}$
31.5	08	6.3
63	10	10
125	15	31.6
250	16	39.8
500	21	125.9
1,000	53	199526.2
2,000	31	1258.9
4,000	42	15848.9
8,000	26	398.1
16,000	24	251.2
31,500	36	3981.1
		221478

Average sound pressure level

$$SPL_{Avg} = 10\log(\Sigma(10^{SPL/10})/N) = 10log\left(\frac{221478}{11}\right) = 43.03 = 43$$

Discussion Questions

1. Define and contrast sound and noise.
2. Draw a typical sinusoidal sound wave and define amplitude and frequency.
3. What is the difference between sound pressure level and sound level?
4. State and explain the importance of the inverse square law.
5. Sketch and explain the working of a sound level meter.
6. Explain the importance of weighted filtering in sound level measurement.
7. What is meant by octave bands?

Practice Problems

1. What is the overall sound pressure level (SPL) of two sources producing sounds of 68 dB and 70 dB? (72 dB)
2. A point sound source generates a noise level of 80 dB. What would the SPL be 1 km from the source? (20 dB)
3. Repeat Practice Problem 2, assuming the source of sound is a line source. (50 dB)
4. A sludge pump emits the following sound as measured by a sound level meter at different octave bands:

Octave Band (Hz)	SPL (dB)
31.5	39
63	42
125	40
250	48
500	22
1,000	20
2,000	20
4,000	21
8,000	10

   What is the average sound level in dB? (40 dB)
5. Sound exceeding 80 dB is considered harmful. How many machines each producing 60 dB can be put in a room without exceeding the limit? (100)

38 Noise Pollution and Control

Noise has been identified as a pollutant for its adverse effect on human health. Noise, an unwanted or annoying sound, can originate from any sound-producing activity, natural or man-made. Natural sources include thunder, cyclones, storms and wind, volcanoes, heavy water fall and earthquakes. The major man-made noise sources are transport, industrial and neighbourhood noise.

38.1 MAN-MADE SOURCES OF NOISE

38.1.1 Transport Noise

Transport noise includes air traffic, road traffic, and seashore and inland water traffic. The amount and type of noise produced by the means of transport largely depend upon the type of traffic. Some typical noise levels of traffic are listed in **Table 38.1**. Accordingly, several agencies have stipulated limiting values for noise from different types of vehicles.

38.1.2 Industrial Noise

Noise is the essential by-product of industry, its intensity and nature being dependent upon the industry type. Industrial noises are usually produced by rotating, reciprocating or other types of machinery, or by high-pressure, high-velocity gases, liquids or vapour involved in the industrial processes. The usual noise level of these industries is in the order of 60 to 95 decibels.

38.1.3 Noise Produced by Other Sources

Several other human activities, such as the use of loudspeakers and sirens, market hawkers shouting, children playing, general life and activity, ringing of temple and church bells, etc., produce noises of different levels, tones and spectra.

Based on its location, noise pollution is classified as:

- Indoor: babies crying, radios, TVs, machines, doors banging, etc., in an indoor environment.
- Outdoor: loudspeakers, transport-related noise, etc., in an outdoor environment.

38.2 EFFECTS OF NOISE ON HUMAN HEALTH

The exposure of human beings to noise of different amplitudes is increasing day by day. Since noise is a subjective feeling, the specific effects of noise on human health are yet to be precisely correlated. However, much thought has been put into their ill effects on human health. To correctly appreciate the effects of noise pollution on human health, we need to understand the mechanism of human hearing.

As shown in **Figure 38.1**, the ear is composed of the outer, middle and inner ears. The pinna, auditory canal and eardrum (tympanic membrane) are part of the outer ear. Air pressure waves first hit the tympanic membrane, causing it to vibrate. The cavity leading to the tympanic membrane and the membrane itself are often called the outer ear. The tympanic membrane is attached to three small bones in the middle ear that move when the membrane vibrates. The purpose of these bones is

DOI: 10.1201/9781003231264-38

TABLE 38.1
Noise Levels of Different Sources of Traffic

Source of Noise	dB
Air traffic	
(i) Jet aircraft at take-off stage at about 300 m	100–110
(ii) Propeller aircraft at take-off stage at about 300 m	90–100
Rail traffic (at about 30 m)	90–110
Heavy road traffic (highway)	80–90
Medium road traffic (main streets)	70–80
Light road traffic (side streets)	60–70

to amplify the physical signal. This air-filled cavity is called the middle ear. The amplified signal is then sent to the inner ear by first vibrating another membrane called the round window membrane, which is attached to a snail-shaped cavity called the cochlea.

Within this fluid-filled cochlea, a cross-section of which is shown in **Figure 38.1**, is the basilar membrane, which is attached to the round window membrane. Attached to the basilar membrane are two sets of tiny hair cells pointing in opposite directions. As the round window membrane vibrates, the fluid in the inner ear is set in motion, and the thousands of hair cells in the cochlea shear past each other, setting off electrical impulses that are sent to the brain through the auditory nerves. The frequency of the sound determines which of the hair cells will move. The hair cells close to the round window membrane are sensitive to high frequencies, and those in the far end of the cochlea respond to low frequencies.

Damage to the human ear can occur in several ways. First, very loud impulse noises can burst the eardrum, causing mostly temporary loss of hearing, although frequently torn eardrums heal poorly, resulting in permanent damage. The bones in the middle ear are not usually damaged by loud sounds, although they can be hurt by infections. Because our sense of balance depends very much on the middle ear, an infection in that area can be debilitating. Finally, the most significant and permanent damage can occur to the hair cells in the inner ear. Very loud sounds will stun these hair cells and cause them to cease functioning. Most of the time this is a temporary condition and time will heal the damage. Unfortunately, if the injury to the inner ear is prolonged, damage can be permanent. This damage cannot be repaired by an operation or corrected by hearing aids. It is this permanent damage to young people, inflicted by loud music, that is the most frequent and insidious. However, loud noise does more than cause permanent hearing damage.

In a Darwinian sense, noise is synonymous with danger. Thus, the human body reacts to loud noises to protect itself from imminent danger. The eyes dilate, adrenaline flows, the blood vessels dilate, the senses are alerted, the heartbeat is altered, blood thickens, all to make the person more alert. Such an alert state, if prolonged, could be quite unhealthy. People who live and work in noisy environments have measurably greater general health problems, are grouchy and ill-tempered, and have trouble concentrating. Noise that deprives a person of sleep carries with it an additional array of health problems. The notion that humans adapt to loud noise is largely a myth because many people might have gone deaf to the prevailing frequency of sound in the working environment. The major health and environmental effects of noise pollution include direct and indirect human health effects.

38.2.1 Auditory Effects

The most direct effect of excessive noise is physical damage to the ear and the temporary or permanent hearing loss that results from such damage. Temporary hearing loss, often called a noise-induced temporary threshold shift (NITTS), refers to a reduced ability to hear weak sounds, such as

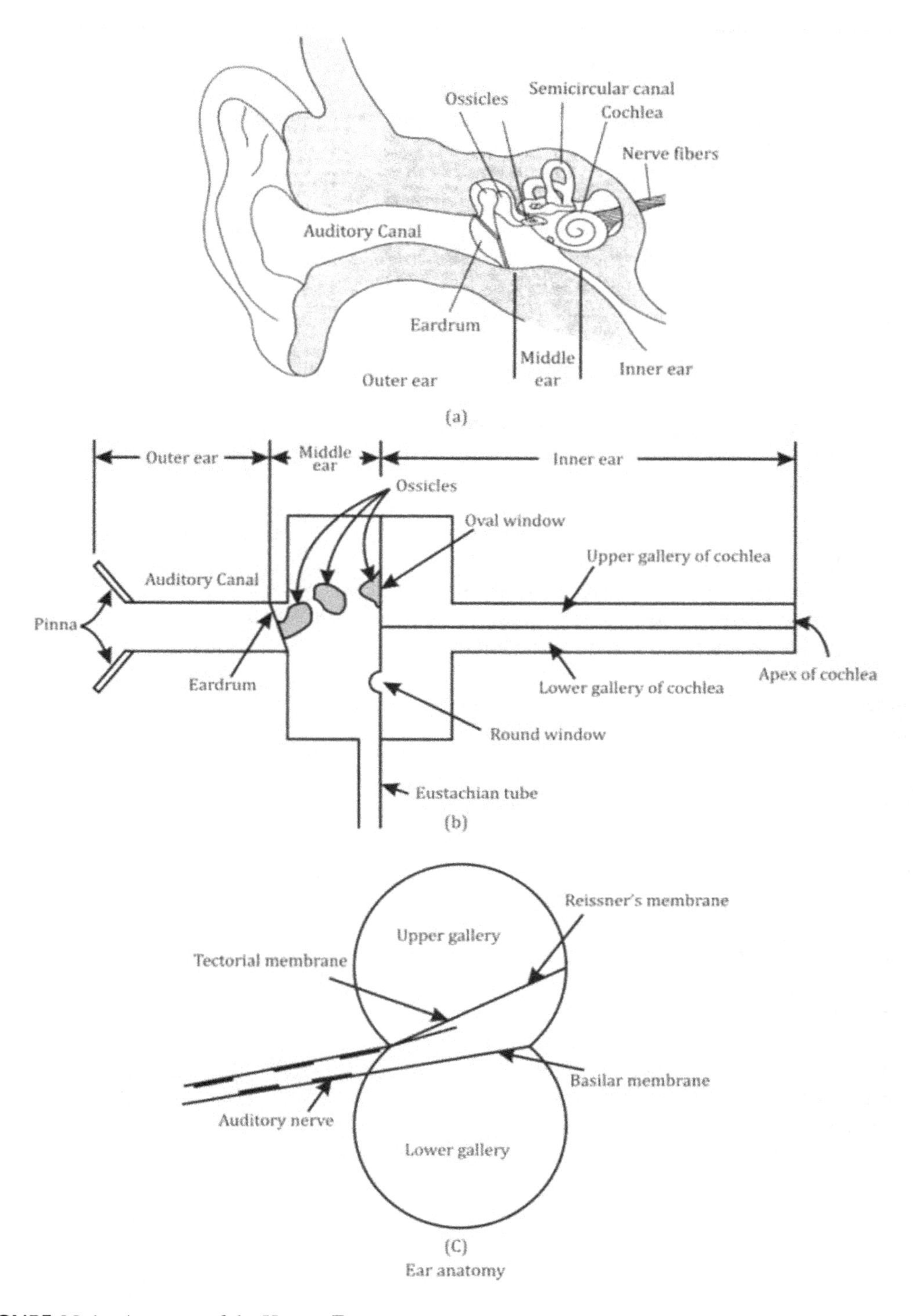

FIGURE 38.1 Anatomy of the Human Ear

hearing loss due to exposure to loud noise, like the bursting of crackers. This situation is temporary and can be reversed usually within a month after the exposure.

Permanent deafness, or a noise-induced permanent threshold shift (NIPTS), occurs because of exposure to loud noise for a long period of time. This results in permanent hearing loss from which there is no recovery.

Below a sound level of 80 dBA, hearing loss does not occur at all. However, temporary effects are noticed at a sound level of between 80 and 130 dBA. About 50% of people exposed to 9 dBA levels at work will develop NITTS, and most people exposed to more than 105 dBA will experience permanent loss to some degree.

38.2.2 Non-Auditory Effects

The major non-auditory effects reported due to excessive and continuous exposure to noise are:

- Interference with speech and communication.
- Annoyance, such as feelings of displeasure, restlessness, irritation, various types of visible, mental and physical disturbances, and anger.
- Mental disorientation, giddiness.
- Hypertension, damage to the heart, kidneys and liver, an increased heart rate, and a change in the hormone content of the blood.
- Fatigue, violent behaviour, dilation of the pupils.
- Upsets the chemical balance of the body, causes nervous breakdowns and interferes with sleep.
- Increased sweating, reduction in work efficiency, and loss of concentration.

Indirect effects include habitat loss, loss of value to property and increased health cost of people living in noise prone areas.

38.3 NOISE ABATEMENT AND CONTROL

It is clear from the preceding discussion that nuisances caused by noise pollution must be reduced and abated, and its adverse effects on human health controlled. Society must therefore be protected from the harmful effects of noise by devising and implementing ways and means for noise to be abated.

Certain noises can be kept under control by legal laws and ordinances, and others have to be damped and attenuated using technology and other engineering interventions such as town planning. Noise produced by motor vehicles can be controlled to some extent by the proper maintenance of vehicles, which can be ensured by prescribing maximum permissible noise levels for different types of automobiles through a Motor Vehicles Act. Similarly, industrial noises can be brought under control by a Factories Act that specifies maximum permissible noise levels and other checks. The Occupational Safety and Health Administration (OSHA) standards are shown in **Table 38.2**.

Similarly, publicly blaring loudspeakers and radio sets at loud levels can be prevented by general legal laws of public nuisance or laws specifically made for noise pollution. However, there exist several noises that have become a part of modern life. All such noises are to be reduced by better-designed technology used in modern-day gadgets, such as fans, air-conditioners, washing machines, refrigerators, mixers, grinders, etc.

Another important method for abating noise effects on mankind is to use proper town-planning techniques and thus to ensure the construction of houses and offices away from the major sources of noise. The proper segregation, zoning and separation of residential complexes from commercial and industrial ones with physical barriers, roads, railway lines, parks or green belts constitute an important aspect of good town planning.

TABLE 38.2
OSHA Industrial Noise Limits

Duration (hours)	Sound Levels (dBA)
8	90
6	92
4	95
3	97
2	100
1	105
0.5	110
0.25	115

Noises produced by automobiles and trains, being the biggest noise nuisance in modern city life, can be abated by constructing walls on both sides of roads and railway lines. Raising such obstructions and barriers between the noise sources and residences may considerably reduce the noise levels reaching those residences. Attenuation of up to 15 decibels is possible in this manner. Raising thick, high vegetation and trees along the sides of roads and railway lines offers cheaper barriers to cause such noise reductions.

Locating the noisy sources downwind of residences may be another important consideration in good town planning because the noise will travel further in the downwind direction, away from the residences.

Noise levels in residential buildings can be reduced to some extent by offsetting the building from the main or street roads by a suitable distance. The further the distance, the better the attenuation, because the intensity of noise gets reduced with an increase in distance.

38.4 NOISE CONTROL STRATEGIES

Noise levels can be reduced using one of three strategies: protect the recipient, reduce the noise at the source or control the path of sound.

38.4.1 Protect the Recipient

Protecting the recipient usually involves the use of earplugs or other ear protectors. Small earplugs, although easy and cheap, are not very effective for many frequencies of noise. The ear can detect sound not only coming through the ear canal but also from the vibration of the bones surrounding the ear. Thus, small earplugs are only partially effective. Earmuffs that cover the entire ear are better and protect the wearer from most of the surrounding noise. However, people often shun what they consider clumsy and uncomfortable ear protectors and decide to take their chances, thus negating the effectiveness of the protection. Workers tend not to wear them regularly, despite company requirements for their use.

38.4.2 Reduce Source Noise

Reducing the source of noise is often the most effective means of noise control. The redesign of commercial airplanes is an example of the effectiveness of this control strategy. Changing the types of motors used in and around the home can also effectively reduce noise. For example, changing from a two-stroke gasoline lawnmower to an electric lawnmower effectively eliminates a common and insidious source of neighbourhood noise.

Traffic noise can be reduced through redesigning vehicles and pavements. It can also be reduced to some extent by increasing the use of alternative modes of transportation, including mass transit, walking and cycling. For residential areas and other areas where higher traffic noise levels are unacceptable, traffic noise can be lessened by reducing vehicle speeds and encouraging the use of alternate routes, either through speed control devices or road designs.

Traffic volume and speed has had significant effects on overall sound levels. Doubling speed increases the sound level by about 9 dBA, doubling the traffic volume sound level by about 3 dBA. A smooth flow of traffic causes less noise than a stop-and-go traffic pattern.

The construction industry has long been a focus of complaints related to excessive noise. Noise levels at construction sites can be controlled using proper construction planning and scheduling techniques. For example, locating noisy air compressors and other equipment away from the site boundary may help mitigate noise levels outside the site. Construction managers must be aware of noise ordinances that may restrict the hours of construction activity and of the permissible noise levels emitted from various types of equipment.

38.4.3 Control Path of Noise

Changing the path of the noise is a third alternative. As described earlier, noise levels drop significantly as the distance from the noise source is increased. Increasing the path length between the source and the recipient offers a passive means of control and requires no effort on the part of the recipient. For a point source, sound intensity decreases with the square of the distance. In other words, by doubling the distance, the sound intensity will decrease by a factor of four. For a line source, noise level decreases in direct proportion to the distance as described by the following relationships.

$$\textit{Point source}, SPL_r = SPL_0 - 20\log r$$

$$\textit{Line source}, SPL_r = SPL_0 - 10\log r$$

Another approach is to place a barrier in the path of the sound wave. The most visible evidence of this tactic is the growth of noise walls, or barriers, along our highways. A variety of materials and designs are used for noise barriers. While earth berms are the cheapest noise barrier to construct, they require large areas of land, something that is typically lacking in urban areas. Currently, most walls are built with concrete or concrete block.

FIGURE 38.2 Noise Barrier for Highways

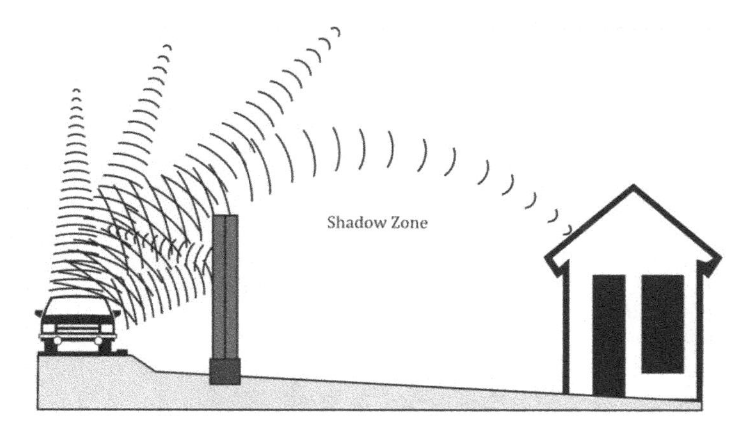

FIGURE 38.3 Noise Barrier Shadow Zone

Noise walls provide the most noise reduction to only the first row of receivers (e.g., houses). The reduction is half as much for the second row and negligible for others. If a barrier is installed on only one side of a road, the receivers on the opposite side of the road actually receive more noise. These effects are due to the properties of sound and noise walls. Noise walls create a shadow zone (**Figure 38.2**) in which the maximum noise reduction occurs. Unfortunately, noise is not like light. A noise shadow is not perfect, and noise can bend, bounce off the barrier and even off the air, depending on atmospheric conditions. Sometimes noises miles away can be heard as the pressure waves bounce off inversion layers. Highway noise barriers (**Figure 38.3**) are considered for new highway constructions when noise impacts are found.

Noise impacts are considered to occur when future noise levels are predicted (through computer modelling) to significantly exceed existing levels (defined by state departments of transportation) or when future levels are predicted to equal or exceed 66 dBA. However, noise barriers are constructed only when it is reasonable and feasible.

Highway sound can also be abated by heavy natural growth (arboriculture). Trees by themselves are not very effective, but a dense growth will reduce the sound pressure level by several decibels. Cutting down natural growth to widen a highway invariably will cause increased noise problems. In most urban settings, of course, there is no room for this type of dense growth, so plantings are more for aesthetics and psychological benefit than actual noise reduction.

Discussion Questions

1. Draw and explain the hearing mechanism in human beings.
2. Establish how noise can be a pollutant affecting human health.
3. List and differentiate between direct and indirect impacts of noise pollution.
4. Why is it necessary to abate noise pollution?
5. What are the major strategies for noise pollution control?
6. Search and list the OSHA standards for noise exposure.
7. List the noise pollution standards in your region.
8. What are the methods for the prevention of noise at the source?
9. Sketch and explain how noise barriers help in abating noise pollution.
10. How does arboriculture help noise pollution control?
11. How can we detect noise pollution?
12. What are the causes and effects of noise pollution on animals?

39 Environmental Impact Assessment (EIA)

Environmental impact assessments (EIAs) started in the US in the second half of the 20th century. It was not until 1970, when the National Environmental Policy Act (NEPA) took effect, that environmental impacts were given due consideration in project planning processes. The NEPA regulations focused on major federal projects that could damage the environment. Although it was initiated primarily for industrial projects, these days it extends to all major developmental activities. Originally it was developed as a tool for evaluating the environmental impacts of a proposed project, but it is equally important for ongoing or existing projects and even events such as earthquakes, landslides, floods and wars. It is a decision-making process that systematically evaluates the possible significant (negative or positive) effects that a proposed project action may exert on the natural, social and human environment of a particular geographic area.

An EIA is intended as an instrument of preventive environmental management. It provides a framework and information for decision-making on activities affecting the environment. An EIA also compares various alternatives for a project and seeks to identify the ones that represent the best combination of economic and environmental costs and benefits.

The environment is composed of biotic and abiotic components between which there exists an adynamic equilibrium. Whenever a project is undertaken, it tends to disturb these components. To maintain the natural environment, perspectives are studied about possible effects and remedial measures, e.g., a forest ecosystem disturbed due to the construction of roads or dams. The impact of an activity is a deviation from the baseline situation that is caused by that activity.

An EIA applies to the assessment of the environmental effects of those public and private projects that are likely to have significant effects on the environment. An EIA project is the execution of construction works or of other installations or schemes in the natural surroundings and landscape, including those involving the extraction of minerals.

Many existing industrial facilities are audited for their compliance with environmental regulations and other purposes. It is important that professionals involved in the design and operation of these facilities understand what environmental studies and audits are and how they are used.

39.1 PURPOSE OF EIA

The various purposes of an EIA are listed and briefed in the following sections.

39.1.1 Decision-Making Tool

An EIA helps to conduct an orderly and holistic examination of the environmental impacts of the various proposed alternatives of a project. This further leads to the development of an Environmental impact statement (EIS) and aids decision-makers in making a judgement.

39.1.2 Aid in Establishment of Development

Though many developers initially viewed EIAs as a hurdle, they have proved to be a great help when selecting a location and considering the design and environmental issues simultaneously. They aid

DOI: 10.1201/9781003231264-39

in implementing environmental considerations in the planning stage and thus lead to eco-sensitive development. They help build a better relationship between the developer, local authority and concerned community.

39.1.3 Device for Sustainable Development

Sustainable development means considering how to maintain a good quality of life, providing constant access to natural resources and avoiding irreversible environmental impacts.

Governing and decision-making bodies recognize the collaboration between economic and social development and the ecosystem, and between human action and the bio-geological world. While attempting to manage this interaction effectively, investigations reveal disturbing trends that can have troubling consequences for the ambient environment.

39.2 TYPES OF ENVIRONMENTAL IMPACTS

Environmental impacts of projects may be primary, having a direct effect, or secondary, having an indirect effect on human beings, fauna and flora, material assets or cultural heritage. Primary impacts include first-round impacts associated with projects such as air and water pollution, employment generation and displacement of people. Secondary impacts are those that are induced by the primary or direct impacts. For example, when people are displaced to a new location, subsequent changes in their socio-economic conditions can be treated as indirect impacts. Enhanced economic activity arising from the construction of a highway can also be viewed as a secondary impact. In short, the inputs to a project cause primary impacts and the outputs lead to secondary impacts. Environmental impacts may also be classified as short-term and long-term. The impacts exhibited over a short period of time, such as during the construction period or in the early phase of operation, are classified under short-term impacts, and the impacts felt over a long period of time are considered long-term effects. For example, clearing a forest as part of a project has a short-term effect on canopy loss, and changes in the hydrological cycle or microclimate have a long-term effect. However, a precise cut-off period separating short-term and long-term impacts cannot be defined. Considering the nature of short-term and long-term environmental impacts, one may find a trade-off between the two.

39.3 EIA MODELS

There are two types of EIA models: the statutory model, which makes an assessment of impacts and is compulsory under an enacted law, and the administrative model, under which an administration exercises its discretion to determine whether an impact study is necessary.

39.4 INFORMATION REQUIRED FOR EIA

The information required for an EIA includes:

- A description of the project: its physical characteristics, land-use requirements during construction and production processes, materials used, operation estimate of expected residues and emissions (water, air, soil pollution, noise, light, heat, radiation, etc.).
- An outline of the main alternatives and the main reasons for a particular choice, including environmental effects.

During this process, the impact of the project on population, fauna, flora, soils, water, air, climatic factors and material assets, including architectural and archaeological heritage, needs to be investigated.

39.5 SIGNIFICANT ENVIRONMENTAL IMPACTS

The likely significant effects of the existence of a project are in its use of natural resources and emission of pollutants. A description of the methods used to assess the effects to prevent, reduce and, where possible, offset any significant adverse effects on the environment needs to be demonstrated. The procedural steps to be followed in an EIA are: a description of the project, a description of the environment, identification of the environmental impacts, evaluation of environmental impact management and control of impact, a presentation of the study, public participation, and judgment by authorities.

39.5.1 Objectives of EIA

The primary objectives of an EIA are:

- To ensure that eventual effects on the environment are considered before approval of a project.
- To encourage the implementation of relevant procedures before finalizing a project.
- To encourage the creation of procedures for information exchange, reporting and consultation amongst countries.

39.5.2 Core Values of EIA

The three core values of an EIA are integrity, utility and sustainability. The EIA process should be fair, objective, unbiased and balanced. Secondly, the EIA process should provide balanced, credible information for decision-making. Finally, the EIA process should result in environmental safeguards.

39.6 EIA PROCEDURE

Steps involved in the preparation of an EIA are displayed as a flow chart in **Figure 39.1**. A description of each step is outlined in the following paragraphs.

39.6.1 Screening

The screening process determines whether a particular project requires the preparation of an EIA or not. Only projects with significant environmental impacts are required to go through the entire process mentioned above; thus, project screening narrows down the applications and makes the process more efficient. The screening criteria are partially decided by the EIA authority functioning in the country.

39.6.2 Scoping

Scoping identifies the key environmental issues that should be addressed in an EIA and usually involves the public and other interested parties. This stage identifies, at an early stage, the significant and critical impacts of the project and its possible alternatives.

39.6.3 Consideration of Alternatives

Here, the methodology for selecting the proposed project is determined and analyzed. All the data of the various alternative and feasible project locations, sources of materials, operations, layout, size and the "no action" scenario are compared to ensure the proposed action has a minimal environmental impact.

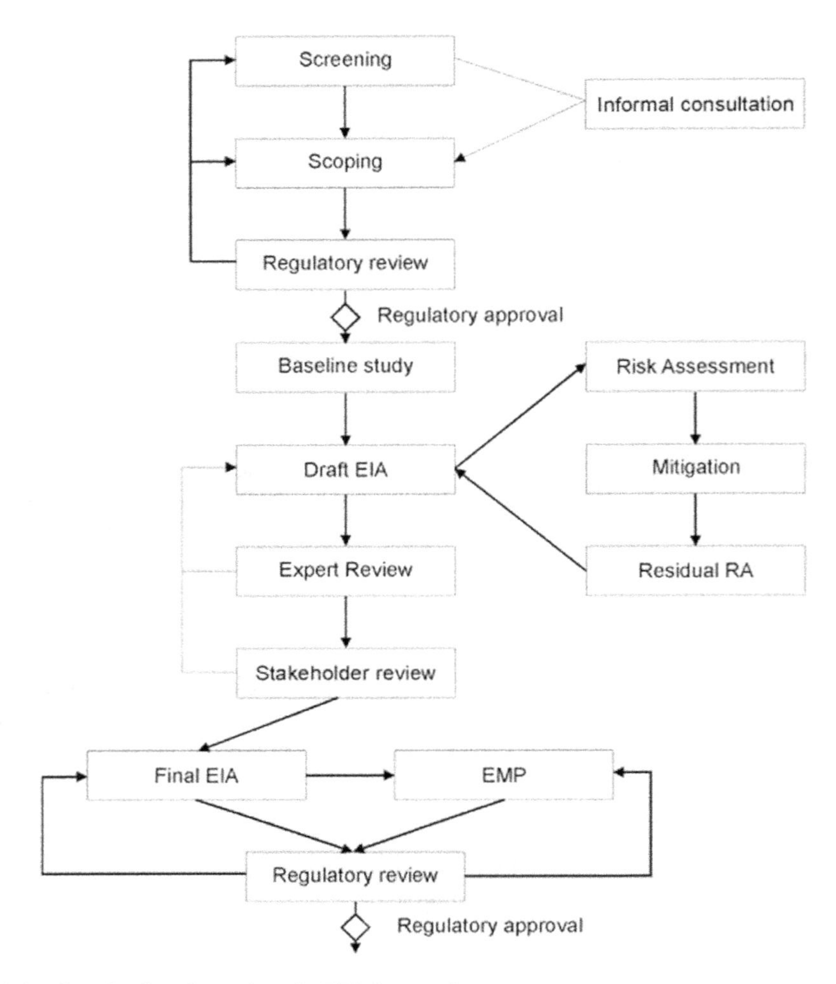

FIGURE 39.1 Step-by-Step Procedure for EIA Preparation

39.6.4 Description of the Project

This stage includes clarification of the aims and objectives of the project. The characteristics of the project, including its various phases of development, are explained in detail.

39.6.5 Baseline Study

The environmental baseline study includes a depiction of the present and future state of the environment, in the absence of the project, based on the data collected and taking account of anthropogenic activities and natural processes.

39.6.6 Impact Analysis

The type, extent, duration, timing, severity and significance of the impact is analyzed. In this stage, the data collected from the previous stages is used and analyzed to pinpoint the most significant environmental impacts and their incorporation into the process.

Based on the baseline data collected, the environmental impacts are predicted in comparison with the possible outcome of the project's implementation.

Residual impacts are checked to make sure they are within acceptable levels and enhance environmental and social benefits. Measures are implemented that are designed to reduce the undesirable effects of a proposed action on the environment.

39.6.7 Evaluation and Assessment of Significance

This stage includes a comparative evaluation of the significance of various predicted impacts. Here, significance is defined as the product of the consequence of the impact and its likelihood.

$$\text{Significance} = \text{Consequence} \times \text{Likelihood}$$

39.6.8 Mitigation

This stage encompasses the implementation of mitigation measures to circumvent, diminish, modify or compensate for the critically adverse impacts of the project or development action.

39.6.9 Public Consultation and Participation

An EIA is a community-focused process, and public consultation and participation is important for ensuring quality and usefulness. This step makes sure that communities and stakeholders are involved in the process of the EIA and that the public's opinions are satisfactorily considered during decision-making.

39.6.10 Environmental Impact Statement

An environmental impact statement (EIS) is an effective and concise presentation of an EIA study and is one of the vital phases. An ill-formed EIS can reduce the efficiency of the decision-making process.

39.6.11 Reporting

An EIA report is prepared and submitted to the concerned authority and includes the following information:

1. A description of the project.
2. An outline of the main alternatives studied by the developer and an indication of the main reasons for the choices made.
3. A description of the aspects of the environment likely to be significantly affected by the proposed project.
4. Measures to prevent, reduce and possibly offset adverse environmental effects.
5. A non-technical summary.
6. An indication of any difficulties (technical deficiencies or lack of know-how) encountered while compiling the required information.

39.6.12 Review

In this step, the report is studied thoroughly for any deficiencies and corrections are made. Public participation is an important part of this step.

- Terms and conditions: Deciding whether the project is acceptable or not. The terms and conditions are finalized in this step.

- Monitoring: Ensuring the implementation of the conditions attached to a decision, verifying the impacts are as predicted or permitted, confirming the mitigation measures are working as expected and taking action to manage any unforeseen changes.
- Decision-making: The relevant authority, during this stage, decides to accept, reject or ask for modification of the proposed project based on the presentation of the EIS and other relevant documentation.
- Post-decision monitoring: After the decision-making process, the recordings of the outcomes related to the development actions are monitored. This step helps in efficient project management and quality assurance.
- Auditing: After collecting the required data (during the operation phase) in the monitoring process, auditing involves comparing the actual consequences with the foretold consequences. The deviation between the two is analyzed to assess the quality of the prediction and the mitigation efficiency.

39.7 PARTICIPANTS OF EIA

EIAs apply to public and private sections. The main players are:

1. Those who propose the project.
2. The environmental consultant who prepares the EIA on behalf of the project proponent.
3. Pollution control statutory agencies.
4. The public, who have the right to express their opinion.
5. The Impact Assessment Agency.

39.8 EXPERT COMMITTEES FOR EIA

The expert committees for an EIA consist of experts in disciplines including water resources, air and water pollution, social sciences, subject specialists, environmentalists and land-use planners. The chairman will be an outstanding and experienced ecologist, environmentalist or technical professional with wide managerial experience in the relevant development. The representative of the Impact Assessment Agency will act as member secretary. The chairman and members will serve in their individual capacities except for those specifically nominated as representatives. The membership of a committee shall not exceed 15 members.

Discussion Questions

1. Define an environmental impact assessment (EIA). What are its main purposes?
2. Briefly describe the steps to conduct an environmental EIA.
3. What are the areas from which members of the EIA committee are drawn? What is the typical formation of the committee?
4. Do all development projects require an EIA? If the answer is no, which type of project usually requires an EIA?
5. “An EIA is usually required for a development project.” Justify this statement.
6. What is an environmental impact statement (EIS)? What must be included in an EIA report?
7. What are the key objectives of an EIA review?
8. What are the strengths and limitations of an EIA?
9. How can the impact of a project’s activity on the biological environment be predicted?
10. When conducting an EIA study, what kinds of impacts are considered?

40 Rural Sanitation

Rural sanitation is an issue of specific importance to village areas or small towns having no amenities related to a water carriage system of sewerage. The major objective of such systems is to remove waste material, including human excreta, in an efficient and hygienic manner. As health is the main criterion while designing such systems, the following specific points must be carefully considered:

- All waste and human excreta should be removed to an isolated area.
- The waste materials should not be accessible to flies, insects or other animals.
- The waste should not contaminate any surface or groundwater supply.
- The method of disposal should be simple and economical.

Methods of rural sanitation cannot be identical to urban sanitation. They must be decided based on the socio-economic set-up of the area and sustainability. They include proper arrangements for a protected water supply, the efficient drainage of rain and used water, and the expeditious disposal of house refuse, animal waste and human excreta. For areas where disposal by the water carriage of excreta is not available, some of the commonly used methods used are discussed below.

40.1 PIT PRIVY

A pit privy consists of a hand-dug hole in the ground covered with either a squatting plate or a slab provided with a riser and a seat. A superstructure or house is then built around it, as shown in **Figure 40.1**.

40.1.1 The Pit

The function of a pit is to isolate and store human excreta in such a way as to avoid unsightly appearance and the transmission of harmful pathogens from there to a new host. The pit is usually round or square forban individual family installation and rectangular for a public latrine.

Dimensions vary from 90 cm to 120 cm in diameter or square. A common figure for a family latrine is 90 cm in diameter or 1.06 m square. For public installation, the pit will be 90 cm to 100 cm wide and its length will depend upon the number of holes provided. The depth is usually about 2.50 m but may vary from 1.80 m to 5 m.

A capacity of 40–60 L/person/year is provided. The storage period is usually 4–6 years. A pit capacity of 1500 L for an average family can last several years.

40.1.2 Life of a Pit

One of the most important aspects of a pit is its useful life. The longer a pit privy will serve a family without being moved or rebuilt, the more certain is the health protection it can give and, therefore, the more value it has to the family and community. It is important to extend the useful life of a privy pit, and thereby reduce the annual cost per person of its installation, by increasing its capacity and efficiency. The life of a privy depends on the care with which it is built, the materials used in its construction and the time required for the pit to fill.

DOI: 10.1201/9781003231264-40

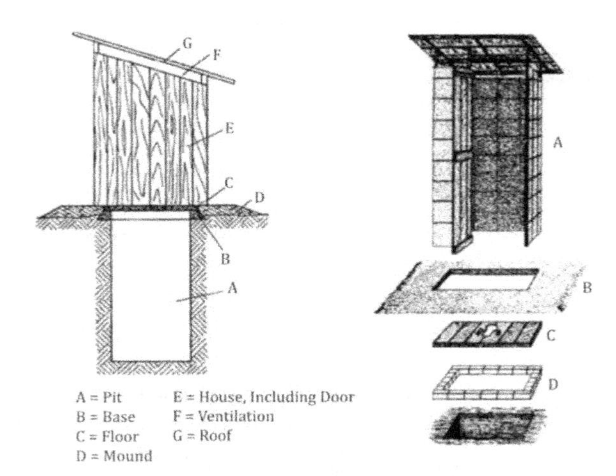

FIGURE 40.1 Parts of a Sanitary Privy

40.1.3 Location of Pit Privy

The site should be dry, well-drained and above flood level. In homogeneous soils, the chance of groundwater pollution is virtually nil if the bottom of a latrine is more than 1.5 m above the groundwater table. The immediate surroundings of the latrine, i.e., an area 2 m wide around the structure, should be cleared of all vegetation, wastes and other debris. It is of great importance to locate the privy or cesspool downhill, or at least on some level piece of land, and to avoid, if possible, placing it directly uphill from a well. Where uphill locations cannot be avoided, a distance of 15 m will prevent bacterial pollution of the well.

40.1.4 Advantages

A pit privy is used almost exclusively throughout the Western Hemisphere and Europe and is common in parts of Africa and the Middle East. With minimal attention to location and construction, there will be no soil pollution or surface or groundwater contamination. A pit privy is simple in design and easy to use, and does not require operation. Its lifespan will vary from 5 to 15 years, depending upon the capacity of the pit and the use and abuse to which it is put. Its chief advantage is that it can be built cheaply, in any part of the world, by a family with little or no outside help using locally available materials.

40.2 BOREHOLE LATRINE

The borehole latrine is a variation of the pit privy that differs by having a much smaller cross-sectional pit area. The latrine floor or slab and the superstructure are the same for both types of installation. In a borehole latrine, excreta goes directly into the pit and get digested in due course, while the liquid portion gets absorbed.

40.2.1 Borehole

This consists of a circular hole, usually 40 cm in diameter, bored vertically into the ground by an earth auger or borer to a depth of 4–8 m, most commonly 6 m. Holes of 30 cm and 35 cm have also been used extensively and are easier to bore than the larger 40 cm size, but experience shows that their capacities are much too small. In fact, the volume of the 40 cm diameter hole is considerably

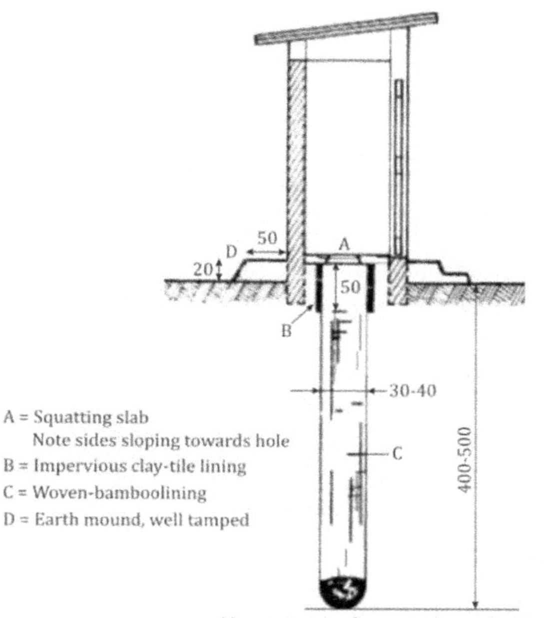

FIGURE 40.2 Typical Borehole Latrine

smaller than that of a pit privy of the same depth, the ratio being 1 to 6.5 in favour of a pit 90 cm square. Because of its small capacity, a borehole latrine dug into dry ground and used only by a family of five or six persons does not last more than 1.5–2 years in most instances, and less where bulky cleansing materials are used.

40.2.2 Location of Borehole Latrine

With borehole latrines, the danger of polluting the groundwater is obvious since it is generally desirable that the borehole penetrates it deeply for more efficient and durable operation. However, the rules governing the location of borehole latrines are the same as those for pit privies.

40.2.3 Advantages

Given proper construction and location, a borehole latrine satisfies most requirements. In particular, they are cheap and easy to construct in ordinary soils.

40.2.4 Disadvantages

A borehole latrine requires special equipment for its construction. If it does not penetrate groundwater over approximately one-third of its depth, its life span is extremely short. In many countries, it is difficult to secure cheap but strong and durable materials to support the walls of a borehole against caving.

40.3 CONCRETE VAULT PRIVY

A concrete vault privy is an improved form of borehole privy suitable for areas with a high water table and highly pervious sandy soils. As seen in **Figure 40.3**, it essentially consists of a watertight

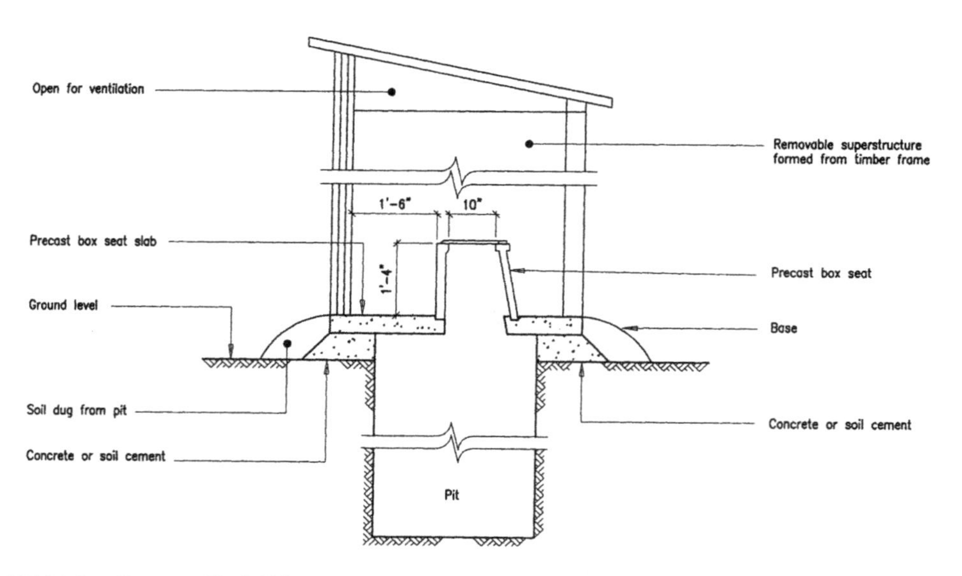

FIGURE 40.3 Concrete Vault Privy

concrete vault or chamber constructed underground. A squatting pan with a compartment is placed over the vault. To prevent odours, after each use, soil is thrown over the excreta. Soil mixed with excreta helps in the speedy breakdown of the organic matter. Though not shown in the figure, the chamber may be vented to control odours. When the chamber is full, it is emptied and cleaned.

40.4 SULABH SHAUCHALAYA

A Sulabh privy is an improvement over the other privies discussed so far. Odours are controlled by providing a water seal, and it requires significantly less water than the septic units discussed in the previous chapters.

Sulabh latrines can be constructed for individual houses or a group of houses, as for a colony. They consist of a specially designed 500 mm long WC pan made of moulded mosaic or china clay mounted on a 200 mm water seal with a groove. A Sulabh pan with a Sulabh seal is fixed in a raised pedestal to provide a WC seat. The WC is connected to the two underground tanks by covered drains.

There are two tanks of varying size and capacity, depending on the number of users. The two underground tanks are either circular (1 m diameter × 1 m deep) or rectangular (1 m × 1 m × 1.2 m deep) and are used in turn. The rectangular tanks require less space as they can be constructed side by side.

Both tanks are connected by a 12 mm diameter cast iron pipe so that gases can pass from one tank to another. The roof level of the tanks is kept above the ground surface to prevent rainwater from entering. Three of the four walls of each tank are made of honeycombed brickwork in the middle. This arrangement allows sewage water to leach laterally into the adjoining soil. The common wall has solid brickwork to avoid leakage from one tank to the other.

The bottom base of both the tanks is kept earthen to allow sewage water to infiltrate. This helps to accelerate biological decomposition by soil bacteria and transform the excreta into manure.

Each tank will serve a family of five members for a period of five years, after which time the first-used tank is be closed by blocking the inlet drain and opening the inlet of the other tank. Both tanks are used alternately. When one pit is full, the incoming excreta is diverted into the second pit. Within a couple of years, the sludge is digested and is almost dry and pathogen-free, and thus safe for handling as manure. Digested sludge is odourless and is a good manure and soil conditioner.

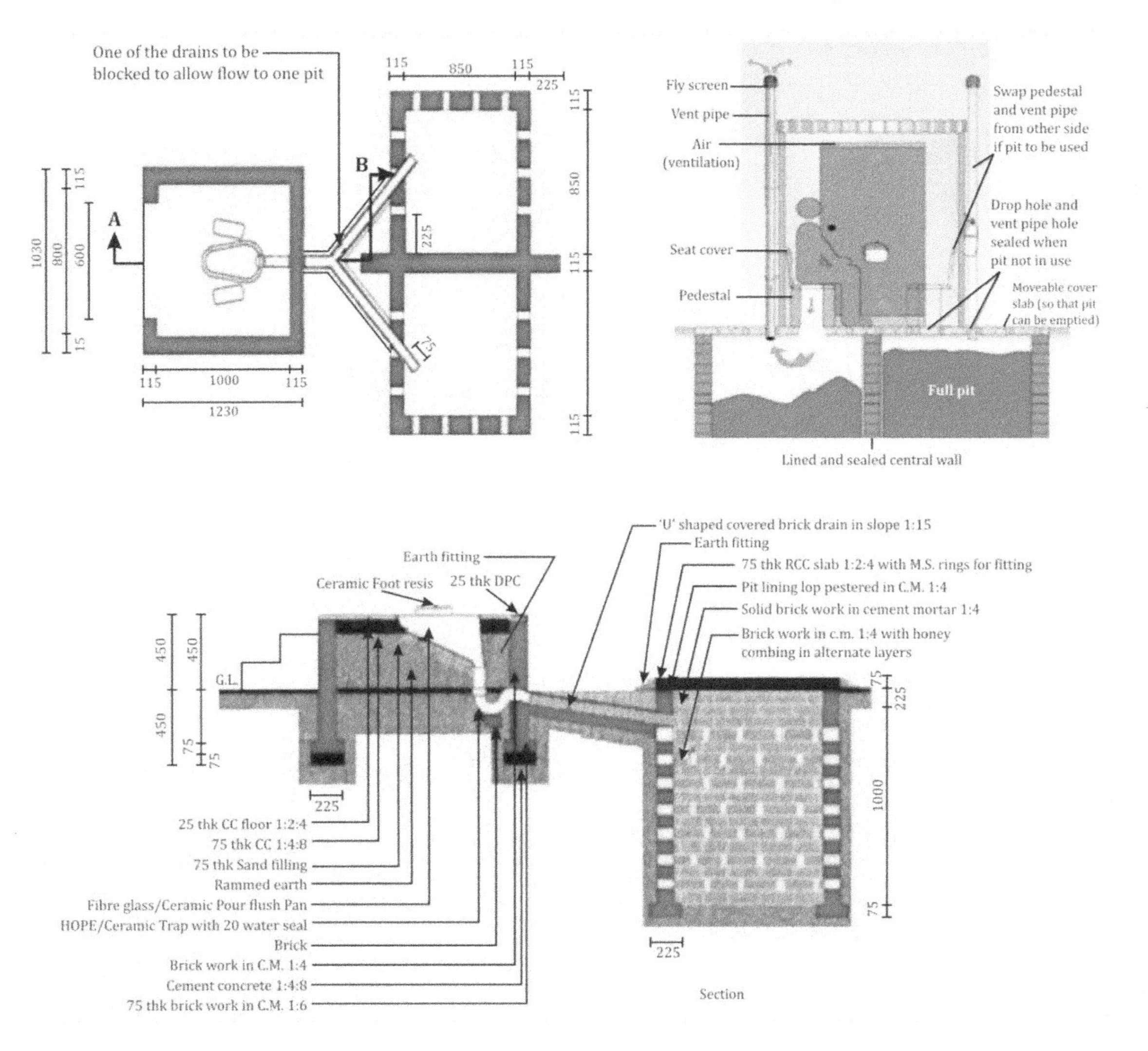

FIGURE 40.4 Sulabh Shauchalaya

The manure can be dug out easily and used for agricultural purposes. The cost of emptying the pit can be met partially from the sale of manure made available.

40.4.1 Design

In the design of a Sulabh privy, two leach pits are connected to a single pour-flush toilet. When one pit is emptied, the second pit can be used, making this a highly convenient design.

The storage volume required to accommodate the sludge and scum that accumulates in the pit during its operational life can be calculated from:

$$V_{pit} = N \times P \times R$$

V_{pit} = the effective volume of the pit.
N = the effective life of the pit (years).
P = the average number of people who use the pit each day.
R = the estimated sludge accumulation rate for a single person.

40.4.2 Advantages and Disadvantages

- A twin-pit set-up improves conditions for pit emptying and has greater potential for reuse than a single pit.
- Pits are alternated every six months or, better, every 12 months.
- However, a lack of faecal sludge management and the potential for groundwater pollution are still problems.
- Although a slightly larger space may be needed to construct this toilet, the cost is still low.
- Having two pits is an advantage as when the first pit is full, the flow of excreta is diverted into the second pit.
- It is hygienically and technically appropriate, and socioculturally acceptable.
- It is affordable and easy to construct with locally available materials.
- Its design and specifications can be modified to suit a household's needs and affordability.
- It eliminates mosquito, insect and fly breeding.
- It can be constructed in different physical, geological and hydrogeological conditions.
- They are free from health hazards and will not pollute surface or groundwater if the proper precautions and safeguards are taken during construction.

A typical design of a pit privy is explained in Example Problem 40.1.

Example Problem 40.1

A family of six want to dig a pit latrine with an operational period of 15 years. Water is used for anal cleansing and the ground is mainly composed of fine sand. The sludge accumulation rate is 60 L/person/year and the infiltration capacity of the soil is 33 L/m^2/d. What should the depth of the pit be? (Pit depth = sludge depth + infiltration depth + soil seal depth (assume 0.5 m))

SOLUTION:

$$V_{pit} = N \times P \times R = 15\ y \times 6\ persons \times \frac{60\ L}{person.y} \times \frac{m^3}{1000\ L} = 5.4\ m^3$$

Assume the pit is rectangular with internal dimensions of 1.0 m by 2.0 m.

$$d_{storage} = \frac{V_{pit}}{A} = \frac{5.4\ m^3}{1.0\ m \times 2.0\ m} = 2.7\ m$$

The soil is composed of sandy soil having an infiltration capacity of 33 L/m²·d. Assume the volume of water entering the pit is 30 L/person·d.

$$A_{lateral} = \frac{Q}{v} = \frac{30\ L}{person \cdot d} \times 6\ persons \times \frac{m^2 \cdot d}{33\ L} = 5.454 = 5.4\ m^2$$

$$d_{infil} = \frac{A_{lateral}}{Perimeter} = \frac{5.45\ m^3}{2 \times (1.0\ m + 2.0\ m)} = 0.09 = 0.91\ m$$

Provide a free board of 0.5 m and additional soil seal depth of 0.5 m.

$$d_{tot} = d_{storage} + d_{infil} + d_{seal} + f.b. = 2.7\ m + 0.91\ m + 0.5\ m + 0.5\ m = \underline{4.61\ m}$$

Final dimensions 2.0 × 1.0 × 4.6 m.

40.5 AQUA PRIVY

The aqua privy is also known as a wet latrine. As shown in **Figure 40.5**, an aqua privy consists of a tank filled with water into which plunges a chute or drop-pipe hanging from the latrine floor. The faeces and urine fall through the drop-pipe into the tank where they undergo anaerobic decomposition, as in a septic tank. The digested sludge, which is reduced to about a quarter of the volume of the excreta deposited, accumulates in the tank and must be removed at intervals.

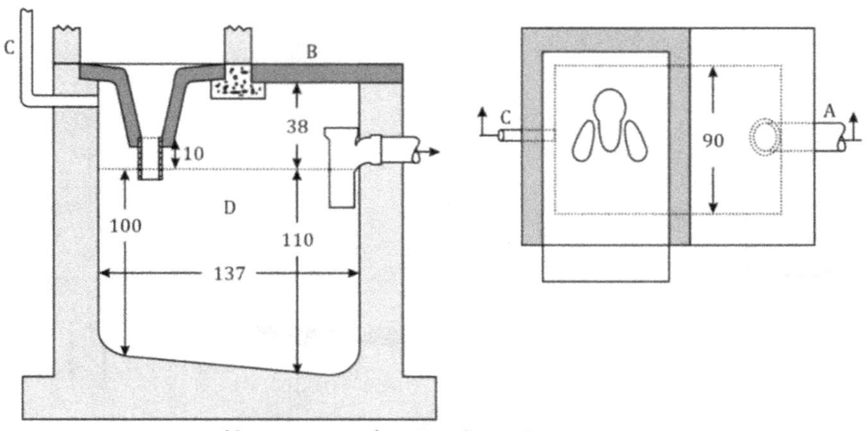

FIGURE 40.5 Single-Tank Aqua Privy

40.5.1 Single Tank

The function of a tank is to receive, store and digest excreta, and keep it away from flies and other vermin. The shape of a tank depends on the local construction facilities and materials and may be round, square or rectangular.

Concrete tanks are built in place and usually square or rectangular since those shapes are easier to construct. Round tanks may be made of plain concrete sewer pipes 90 cm or 120 cm in diameter placed vertically in an earth pit and sealed at the bottom with concrete. The capacity of a family-size aqua privy should preferably be not less than one cubic metre, allowing for six years or more between cleaning operations. For public latrines of this type, experience dictates a design figure of 115 L per person for the maximum number of persons to be served. The usual practice is to provide a water depth of 1.0–1.5 m, with 1 m being considered the minimum. Materials commonly used for the construction of the tank include plain or reinforced concrete, brick, or stone masonry with a plaster cover.

For each litre of water added to the watertight tank of an aqua privy, a corresponding amount of sewage must be evacuated and disposed of as effluent. The latter is septic and loaded with finely divided faecal matter in suspension that is in the process of decomposition. The average amount of water to be evacuated from an aqua privy has been estimated at about 4.5 L per person per day. However, a capacity of 9.0 L is recommended for the design of the disposal system.

40.5.2 Two Chambers

As seen in **Figure 40.6**, in addition to a digestion tank, a smaller tank is provided that is hydraulically connected to the larger tank. The purpose of this tank is to provide aerobic decomposition and clarification. In a way, this tank provides secondary treatment. Note that biological action is anaerobic in a digestion tank and aerobic in a clarifying tank. The gases from the decomposition chamber are vented. The clarifier chamber is also vented to allow the mixing of air to keep conditions aerobic.

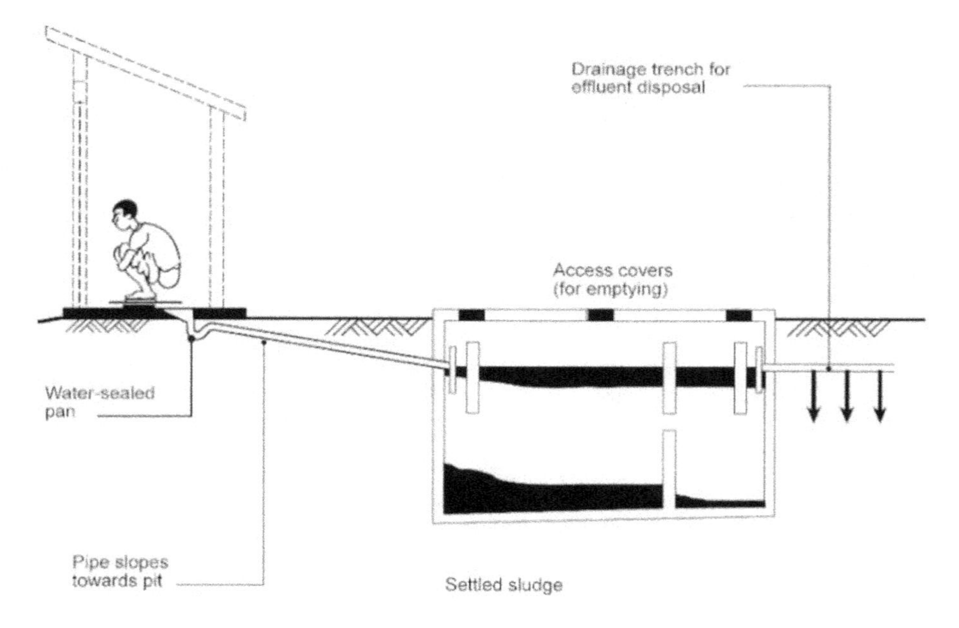

FIGURE 40.6 Aqua Privy (Two Tanks)

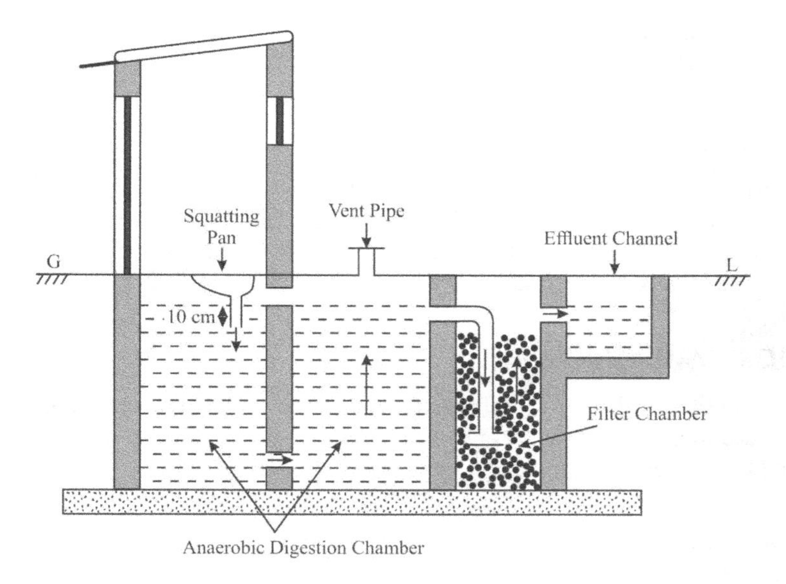

FIGURE 40.7 Aqua Privy (Three Chambers)

As water goes through aerobic decomposition, the effluent quality improves and becomes free of bad odours. The effluent is discharged into a drainage trench. As seen in **Figure 40.6**, sludge is deposited in the first and second chambers. The volume of sludge is significantly reduced on decomposition. Sludge is removed every couple of years.

40.5.3 Three Chambers

A three-chamber wet latrine or aqua privy is very similar to a septic tank system. As shown in **Figure 40.7**, effluent from the second chamber is discharged into the third chamber, called the filter chamber. As the water rises, a further reduction in biochemical oxygen demand (BOD) takes place, and the quality of effluent is further improved. This effluent can be used for kitchen gardening or farming.

An aqua privy is very suitable in places that are connected to a sewer system and is quite economical and hygienic.

40.5.4 Advantages

- If properly used and maintained, an aqua privy is better for health and aesthetic considerations.
- It is a permanent type of installation that is relatively simple and inexpensive.
- It can be placed near a dwelling.
- It will withstand abuse.

40.5.5 Disadvantages

- Its rather high initial cost may prevent its extensive use in rural areas in certain parts of the world.
- It may not be successful in rural areas with no organized sanitation and health education services.
- It requires water (although a small volume only will suffice) for its operation.

- It requires operation and maintenance on a daily basis.
- It cannot be used in cold climates.

40.5.6 Construction

As with a pit privy, locally available materials should be used to the greatest possible extent in the construction of the aqua privy. The vital part of an aqua privy is its tank, which must be watertight. Concrete is to be preferred for the construction of its floor, although wood may also be used. For the housing, any locally available material may be used.

40.6 PROGRAMME SUCCESS

To make rural sanitation projects successful and useful to society, modifications are necessary as per the skill and educational level of the people using them. Locally available economics and building materials should be preferred.

40.6.1 Appraisal of Option

It is important to identify situations where improvement is needed. Therefore, a careful assessment of problems be made along with an appraisal of all the options for improved sanitation. Some aspects of the appraisal should take place at the same time as the local needs are being assessed and should involve the investigation of local craft skills that could be used in the project.

40.6.2 Appropriate Technology

The equipment and techniques used should favour local resources as well as local needs. Careful appraisal of local resources and the options they present should be undertaken.

40.6.3 Socially Acceptable

Everyone has their own opinion. Planners of sanitation programs have expert knowledge of medicine or engineering, but local people see problems differently, and new technology is not always accepted socially. Experts can fail to recognize and reconcile these different types of views. One should not pretend that one always knows what is best for other people, as all of their problems may not be all apparent.

Discussion Questions

1. What are the different types of human excreta disposal systems without a water carriage?
2. What considerations should be made when choosing and installing on-site excreta disposal systems?
3. Sketch and differentiate between the features of a single- and a two-pit privy?
4. Discuss the advantages and disadvantages of Sulabh Shauchalaya.
5. What is meant by an ecological sanitation system?
6. Search the internet and make notes on the zero discharge toilet system (ZDTS) and the status of its application by Indian Railways?
7. What are the social issues associated with the implementation of on-site sanitation strategies in India?

8. Sketch and explain the workings of an aqua privy.
9. Differentiate between a two-chamber and a three-chamber aqua privy.
10. Discuss the steps to be adopted for making sanitation projects socially acceptable.

Practice Problems

1. A family of five wants a single-pit privy 1.2 m in diameter for an operational period of 10 years. Water is used for anal cleansing and the ground is mainly composed of fine sand. Assuming the sludge accumulation rate is 50 L/person/year, daily water use is 30 L/person.d and the infiltration capacity of the soil is 35 $L/m^2/d$, what is the desired depth of the pit? Assume a soil seal depth of 0.5 m. (4.2 m)
2. A single-chamber aqua privy is planned for a family of eight in a rural set-up. Based on the ministry recommendations, the design capacity of the digestion chamber is to be 120 L/person. To maintain a water depth of 1.2 m and space for the storage of sludge, the effective depth of the chamber should be no less than 1.2 m. If a rectangular tank with a length twice its width is planned, what should the dimensions of the digestion chamber be? (1.3 m × 0.65 m, 1.7 m deep)

Appendices

A. PROPERTIES OF WATER

Temp.	Specific gravity	Mass Density	Weight Density	Surface Tension	Dynamic Viscosity	Kinematic Viscosity	Saturated DO	Vapour Pressure	
°C		kg/m^3	kN/m^3	N/m	Pa.s	m^2/s	mg/L	kPa	bar
0	1.000	999.9	9.809	7.56E-02	1.79E-03	1.79E-06	14.60	1.946E+00	6.23E+00
5	1.000	1000.0	9.810	7.49E-02	1.52E-03	1.52E-06	12.80	1.706E+00	8.89E+00
10	1.000	999.7	9.807	7.42E-02	1.31E-03	1.31E-06	11.30	1.506E+00	1.25E+01
15	0.999	999.1	9.801	7.35E-02	1.14E-03	1.14E-06	10.10	1.346E+00	1.74E+01
20	0.998	998.2	9.793	7.28E-02	1.00E-03	1.00E-06	9.10	1.213E+00	2.38E+01
25	0.997	997.1	9.781	7.20E-02	8.90E-04	8.93E-07	8.30	1.106E+00	3.23E+01
30	0.996	995.7	9.768	7.12E-02	7.98E-04	8.01E-07	7.54	1.005E+00	4.33E+01
35	0.994	994.1	9.752	7.04E-02	7.19E-04	7.23E-07	6.93	9.237E-01	5.73E+01
40	0.992	992.3	9.734	6.96E-02	6.53E-04	6.58E-07	6.41	8.544E-01	7.52E+01
50	0.988	988.1	9.693	6.79E-02	5.47E-04	5.54E-07	5.90	7.864E-01	1.26E+02
60	0.993	993.2	9.744	6.62E-02	4.66E-04	4.69E-07		0.000E+00	2.03E+02
70	0.978	977.8	9.592	6.44E-02	4.04E-04	4.13E-07		0.000E+00	3.18E+02
80	0.972	971.8	9.534	6.26E-02	3.55E-04	3.65E-07		0.000E+00	4.83E+02
90	0.965	965.3	9.470	6.08E-02	3.15E-04	3.26E-07		0.000E+00	7.15E+02
100	0.958	958.4	9.402	5.89E-02	2.82E-04	2.94E-07	0.00	0.000E+00	1.03E+03

B. CONVERSION FACTORS (SI TO USC)

Symbol	When you know	Multiply by	To find	Symbol
mm	millimeter	0.039	inch	in
m	meter	3.28	feet	ft
m	meter	1.09	yard	yd
km	kilometer	0.621	mile	mi
mm^2	square millimeter	0.0016	square inch	in^2
m^2	square meter	10.764	square feet	ft^2
m^2	square meter	1.196	square yard	yd^2
ha	hectare	2.47	acre	ac
km^2	square kilometer	0.386	square mile	mi^2
mL	millilitre	0.034	fluid ounce	fl oz
L	litre	0.264	Gallon	gal
m^3	cubic meter	35.314	cubic feet	ft^3
m^3	cubic meter	1.307	cubic yards	yd^3
g	gram	0.035	ounces	oz
kg	kilogram	2.202	pound	lb
Mg	megagram	1.103	short tons	T
t	metric ton		2000 pound	
°C	Celsius	1.8C +32	Fahrenheit	°F
N	newton	0.225	pound force	lbf
kPa	kilopascal	0.145	pound per square inch	lbf/in^2

C. CONVERSION FACTORS (USC TO SI)

Symbol	When you know	Multiply by	To find	Symbol
In	Inch	25.4	millimeter	mm
ft	Feet	0.305	meter	m
yd	Yard	0.914	meter	m
mi	Mile	1.61	kilometer	km
in^2	square inch	645.2	square millimeter	mm^2
ft^2	square feet	0.093	square meter	m^2
yd^2	square yard	0.836	square meter	m^2
ac	acre	0.405	hectare	ha
mi^2	square mile	2.59	square kilometer	km^2
fl oz	fluid ounce	29.57	millilitre	mL
gal	gallon (US)	3.785	litre	L
ft^3	cubic feet	0.028	cubic meter	m^3
yd^3	cubic yard	0.765	cubic meter	m^3
oz	ounce	28.35	grams	g
lb	pound	0.454	kilogram	kg
T	short ton	0.907	megagram	Mg
	(2000 lb)		metric ton	t
°F	Fahrenheit	5(F-32)/9	Celsius	°C
lbf	pound force	4.45	newton	N
lb/in^2	pound per square inch	6.89	kilopascal	kPa

D. CT VALUES

D.CT Values in (mg/L).min for inactivation of Giardia lambia cysts

Free Chlorine		Water Temperature in °C			
Res., mg/L	pH	0.5	5	10	20
<0.4	6.5	163	117	88	44
	7.0	195	139	104	52
	7.5	237	166	125	62
	8.0	277	198	149	74
1.0	6.5	176	125	94	47
	7.0	210	149	112	56
	7.5	253	179	134	67
	8.0	304	216	162	81
2.0	6.5	197	138	104	52
	7.0	236	165	124	62
	7.5	286	200	150	75
	8.0	346	243	182	91
3.0	6.5	217	151	113	57
	7.0	261	182	137	68
	7.5	316	221	166	83
	8.0	382	268	201	101

CT Values in (mg/L).min for log-0.5 and log-1 inactivation of Giardia lambia cysts

Disinfectant	Log inactivation	pH	Water Temperature in °C				
			0.5	5	10	15	20
Free Chlorine	0.5	6.0	25	18	13	9	7
	1.0	6.0	49	35	26	18	13
	0.5	7.0	35	25	19	13	9
	1.0	7.0	70	50	37	25	18
	0.5	8.0	51	36	27	18	14
	1.0	8.0	101	72	54	36	27
Combined	0.5	6-9	640	370	310	250	190
	1.0	6-9	1300	740	620	500	370
Chlorine	0.5	6-9	10	4.3	4.0	3.2	2.5
Dioxide	1.0	6-9	21	8.7	7.7	6.3	5.0
Ozone	0.5	6-9	0.48	0.32	0.23	0.16	0.12
	1.0	6-9	0.97	0.63	0.48	0.32	0.24

CT Values in (mg/L).min for inactivation of Viruses at pH 6-9

Disinfectant	Log inactivation	Water Temperature in °C				
		0.5	**5**	**10**	**15**	**20**
Free Chlorine	2.0	6	4	3	2	1
	3.0	9	6	4	3	2
	4.0	12	8	6	4	3
Combined	2.0	1200	860	640	430	320
	3.0	2100	1400	1100	710	530
Chlorine Dioxide	2.0	8.4	5.6	4.2	2.8	2.1
	3.0	25.6	17.1	12.8	8.6	6.4
Ozone	2.0	0.9	0.6	0.5	0.3	0.2
	3.0	1.4	0.9	0.8	0.5	0.4

Source: EPA-USA

Index

A

B

E

F

S

W

Y

Z

CPSIA information can be obtained
at www.ICGtesting.com
Printed in the USA
BVHW010335190422
634676BV00003B/103

9 780367 750503